Divided Spheres

Divided Spheres

Geodesics and the Orderly Subdivision of the Sphere

Edward S. Popko

CRC Press
Taylor & Francis Group
Boca Raton London New York

CRC Press is an imprint of the
Taylor & Francis Group, an **informa** business
AN A K PETERS BOOK

CRC Press
Taylor & Francis Group
6000 Broken Sound Parkway NW, Suite 300
Boca Raton, FL 33487-2742

© 2012 by Taylor & Francis Group, LLC
CRC Press is an imprint of Taylor & Francis Group, an Informa business

No claim to original U.S. Government works

Printed and bound in India by Replika Press Pvt. Ltd.
Version Date: 20120409

International Standard Book Number: 978-1-4665-0429-5 (Hardback)

Library of Congress Cataloging-in-Publication Data

Popko, Edward.
    Divide spheres : geodesics and the orderly subdivision of the sphere / Edward S. Popko.
      p. cm.
    Includes bibliographical references and index.
    ISBN 978-1-4665-0429-5 (hardback)
      1. Polyhedra. 2. Geodesic domes. 3. Spherical projection. 4. Fuller, R. Buckminster (Richard Buckminster), 1895-1983. I. Title.

    TA660.P73P67 2012
    516'.156--dc23
                                                                                        2011048704

Visit the Taylor & Francis Web site at
http://www.taylorandfrancis.com

and the CRC Press Web site at
http://www.crcpress.com

To my mother Frances and father Edward, who gave me the freedom to expore.
Their childhood gift of *Fun with Figures* probably started it all.[1]

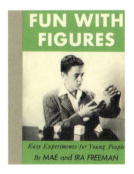

[1] Book cover by Mae and Ira Freeman, © 1946, renewed in 1974 by Random House, Inc. Used by permission of Random House, Inc.

Table of Contents

3. Putting Spheres to Work   49

4. Circular Reasoning   79

10. Computer-Aided Design   275

11. Advanced CAD Techniques   331

A. Spherical Trigonometry   363

Preface

This book summarizes the key spherical subdivision techniques that have evolved over 50 years to help today's designers, engineers, and scientists use them to solve new problems.

I became enthusiastic about geodesic domes in the mid-1960s, through my association with Buckminster Fuller and his colleagues. I was an architectural intern at Geometrics, Inc., a Cambridge, Massachusetts, firm whose principals, Bill Ahern and Bill Wainwright, had pioneered geodesic radome designs. During my internship, Ahern and Wainwright were collaborating with Fuller and Shoji Sadao on the US Pavilion dome at Expo'67 in Montreal. I was immediately drawn to the beauty and efficiency of designs based on geodesic principles.

I first met Fuller at Geometrics. He surprised me with his stature and energy, his easy rapport with audiences, and his discourse on design. It was, however, a struggle to understand him. He had his own language, which combined geometry with physics and design into a personal philosophy he called *synergetics*. Later, I visited the original geodesics "skunk works"—Synergetics, Inc. in Raleigh, North Carolina. At Synergetics, Jim Fitzgibbon, T. C. Howard, and others showed me projects from their start-up years in the early 1950s, as well as their latest work. They were making history with their innovative spherical designs. Duncan Stuart, another early member of the firm, had made his own history in the early 1950s, when he invented a method (*triacon subdivision*) that made geodesic domes practical to build. The spherical grids this method produced had the fewest number of different parts of any subdivision technique. It is still one of the best gridding methods and it is detailed in this book.

More recently, I met Manuel Bromberg and Chizuko Kojima. In the late 1940s, Bromberg, along with Jim Fitzgibbon, Duncan Stuart, and other faculty members at North Carolina State University, formed the first geodesic start-up company, Carolina Skybreak. When architectural commissions materialized, the original company evolved into two others, Synergetics, Inc. and Geodesics, Inc., and Raleigh quickly became the epicenter of geodesic design. Kojima was a "computer" for Geodesics in the late-1950s. In those days, "computer" was a job title for someone who calculated. Kojima's job was to calculate hundreds of angles and grid coordinates for early dome projects—by hand.

A few years ago, I met Magnus Wenninger, a monk, teacher, mathematician, and polyhedral model builder *par excellence*. Wenninger has built thousands of models, and his classic books on polyhedral and spherical models have inspired generations of schoolchildren, artisans, and mathematicians. He showed me how to work with some lesser-known skewed spherical subdivisions; the main techniques are covered in this book.

My professional work with computer-aided design (CAD) systems for architects and engineers led me to more applications of geodesics. My hobbies in sailing and celestial navigation with a sextant sharpened my understanding of spherical trigonometry and improved my skills in geodesics. I had always been interested in world maps and their varied graphic projections as well as the delicate spherical structures found in microorganism exoskeletons. But what surprised me most were spherical designs in things like ocean-bobbing fish pens, panoramic photography, underground neutrino observatories, and virtual-reality simulators for the military—all of which used geodesic geometry. Geodesics appeared again in virus research, astronomy catalogs, weather forecasting, and kids' toys. But the most unexpected geodesic application I found was in the innovative design and layout of golf ball dimples. Dozens of different dimple patterns resembled small geodesic domes. These applications and many more are detailed in a later chapter.

Divided Spheres

For designers, the principles of spherical design are, at first, counterintuitive and somewhat obscured by a unique vocabulary. Chapter 1, "Divided Spheres," highlights the major challenges and approaches to spherical subdivision. The chapter states the design objectives that many designers use. We will meet them in later chapters. Key concepts and terms are introduced.

Buckminster Fuller's pioneering work in the late 1940s and the research and development of his colleagues in the 1950s led to many of the techniques we use today. Chapter 2, "Bucky's Dome," examines how Fuller's design cosmology, synergetic geometry, was first applied to cartography and then to geodesic domes. The interplay between Fuller and key associates who worked out practical solutions to geometry and construction problems is particularly important. Many of today's subdivision techniques were developed at this time. This chapter contains the first commentary on how geodesic domes were originally calculated.

Chapter 3, "Putting Spheres to Work," provides a brief glimpse of the wide diversity of today's spherical applications in fields like biology, astronomy, virtual-reality gaming, climate modeling, aquaculture, supercomputers, photography, children's games, and sports balls. If you are a golfer, you will enjoy seeing how manufacturers use spherical design and unique dimple patterns to maximize player performance.

Spherical geometry is quite different from Euclidean geometry, though they share common principles. Chapter 4, "Circular Reasoning," develops circular reasoning with points, circles, spherical arcs, and polygons. *Spherical triangles* are the most common *polygon* created when subdividing spheres. We look closely at their properties to establish an understanding of their areas, centers, type, and orientation.

It is easy to evenly divide the circumference of a circle on a computer to any practical level of precision. It's not so easy to evenly subdivide spheres, computer or not. Chapter 5, "Distributing Points," focuses on the challenge of evenly distributing points on a sphere and, in so doing, defines specific design optimizations that this book will develop.

Spherical polyhedra offer a convenient starting point for subdivision. Chapter 6, "Polyhedral Frameworks," describes useful Platonic and Archimedean solids. We are particularly interested in their *symmetry* properties and their spherical versions.

Many readers will be surprised by the rich and varied spherical subdivisions that golf ball dimple patterns demonstrate. Chapter 7, "Golf Balls," references some of the famous US government patents, showing the golf ball industry's amazing diversity of designs and illustrating why a seemingly small thing such as dimple patterns have become a key part of a multimillion-dollar sports ball industry.

Divided Spheres presents six classic subdivision techniques grouped into three classes. In Chapter 8, "Subdivision Schemas," each technique is presented in the same step-by-step format. These techniques are flexible and apply to many different spherical polyhedra. Depending on the designer's requirements, certain combinations of spherical polyhedra and subdivision techniques may be more appropriate to use than others. Thus, one objective of this chapter is to show how various techniques affect the final subdivision.

With so many design choices, it is natural to ask which combination is best. Of course, the answer depends on the application. Chapter 9, "Comparing Results," shows how to cut through all the design variables and select a combination that best fits your design requirements. This chapter relies extensively on graphics rather than taking a statistical approach. This book is also the first to use graphical analysis, such as Euler lines and stereographics, to highlight the subtle differences between subdivision techniques.

For many years, computing spherical geometry was a major challenge. When I first worked with geodesics in the mid-1960s, programming mainframe computers was time-consuming and expensive. Today, every engineer, architect, product designer, cartographer, biologist, and scientist relies on computers, and most use CAD or geometric modelers and 3D visualization systems for their work. Chapter 10, "Computer-Aided Design," describes how to use CAD for spherical subdivision. The benefits of using 3D design systems are explained, and examples show various types of visualization, geometric calculation, automated drawings, photorealism, analysis, and data for manufacturing. The text illustrates how, with just a few spherical points, a user can generate common spherical parts that can be assembled into larger groupings until they cover the entire sphere without leaving gaps or causing overlaps. The techniques and terminology used are common to almost all CAD systems.

Designers often work collaboratively with other designers, project owners/developers, or consultants, and no CAD system today is specifically designed for spherical work. Chapter 11, "Advanced CAD Techniques," explains techniques that make it easier for designers to share their work, manage complex projects, and retain 3D designs for future reuse. This chapter describes how to customize general CAD systems to be more spherical-friendly and to work the way a spherical designer expects them to work. Customization makes CAD easier to use and work collaboratively with others to achieve consistent results and save time. This chapter also describes some modestly priced and free 3D modeling software and visualization programs that a user without access to a CAD system might want to try.

Three primers are included in the appendices: "Spherical Trigonometry," "Stereographic Projection," and "Coordinate Rotations." Spherical trigonometry is not emphasized in school anymore, but it is essential for anyone working with geodesics. This appendix covers everything you need to know to subdivide spheres. Stereographic projection is a graphical technique for making 2D drawings on the surface of the sphere. Stereograms appear in several sections of the book; we use them to compare the results of each subdivision method. An appendix explains the theory behind this classic graphical technique. Each of the six subdivision techniques explained in this book results in a grid that covers only a small part of the sphere. This geometry must be replicated to cover the entire sphere, without causing overlaps or gaps. Appendix E explains how this is done.

This book is intended to be a complete resource for dividing spheres. Important terms are *highlighted* the first time they are used and are collected in an illustrated glossary.

You will not find a few things in this book. Complete computer programs are omitted. Instead, an appendix called "Geodesic Math" collects the basic algorithms. These short computer program fragments in Appendix C show how to perform each subdivision method described in this book. These are all that is needed for those interested in writing their own geodesic computer programs. The resulting geodesic grids for every technique explained are also listed in Appendix D You can input this data into display programs or compare it with the results of programs you might write.

Graphic Conventions

This book is about 3D spherical geometry. The foundations for the most uniform subdivisions are based on the Platonic and Archimedean solids—forms discovered by the Greeks and made from combinations of *regular polygons* (polygons in which all edges and angles are equal), such as equilateral triangles, squares, pentagons, or hexagons. These 3D forms have pure "theoretic" definitions, but we use a number of graphic conventions in this book to illustrate and explain their features.

Figure 1 shows six different graphic conventions for the common *dodecahedron* (12 faces, 30 edges, and 20 vertices). Each graphic is the same scale and is shown from the same viewpoint. If placed on top of one another, the vertices of all six figures would be coincident. Each convention emphasizes a different aspect of the solid.

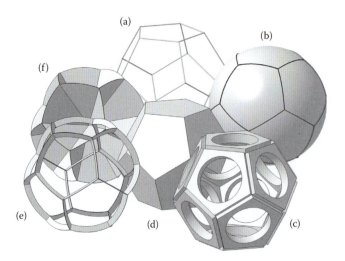

Figure 1. Polyhedra graphic conventions used in this book.

The wireframe (a) and planar versions (d) are the most familiar and come closest to the pure definition of a planar dodecahedron. Its spherical cousin (b) makes the differences in volumes and face angles apparent. The holes in version (c) reveal the relationships of interior faces, show what lies behind the form, and draw attention to the centers of the faces (there is more than one type of center). Thick faces emphasize the angles (dihedral) between adjacent faces. Graphic convention (e) emphasizes the spherical arcs and planes of the dodecahedron's edges, while the graphic convention in (f) is useful when examining the surface angles where arcs meet. The planes and arcs are part of the volume encompassed by the form. These six graphic conventions appear throughout this book.

I have covered the basics of what I have learned over the years in my own spherical work. I hope that *Divided Spheres* makes it easier for you to understand the principles of spherical subdivision and to develop your own spherical designs. Spherical applications are as limitless as they are beautiful.

—Edward S. Popko
Woodstock, New York
2012

Acknowledgments

T his book is a tribute to the work of many people. In a way, it does not seem fair that my name is the only one on the cover. It is a pleasure to recognize those who contributed so much and supported me along the way.

No one has been more patient or taught me more about putting a book together than Jim Morrison. I cannot say enough about his enthusiastic support. Jim was my sounding board for new ideas, an instructor *par excellence* on PostScript graphics programming, and my mentor on manuscript preparation. His recent book, *The Astrolabe*, paved the way for mine, and I am grateful for all the advice and experience he shared with me.

I have been fortunate to have known and worked with experts in polyhedral geometry and spherical design. Father Magnus Wenninger, OSB, stands out. He has built thousands of models over the years and his books, including *Polyhedron Models*, *Spherical Models*, and *Dual Models,* have made these subjects accessible. I am grateful for his friendship and for the advice he offered throughout this project. The polyhedral chapter and skewed grid sections owe a great deal to him.

Buckminster Fuller's invention of the geodesic dome was the biggest stimulus for spherical subdivision research and development. This book samples some of his amazing work and the work he inspired in others. I was fortunate to have met him on several occasions and to have worked in one of his affiliate offices in the 1960s. Those experiences stayed with me and greatly enriched my life. I also came in contact with many of Fuller's associates and early geodesic pioneers. I am particularly indebted to Manuel Bromberg, the late Jim Fitzgibbon, T. C. Howard, Bruno Leon, Chizko Kojima, Shoji Sadao, Bill Ahern, Bill Wainwright, and the late Duncan Stuart. All of them were part of early Fuller start-ups like Skybreak Carolina, Synergetics, Geodesics, and Geometrics. Each taught me different aspects of geodesics and the early history of the dome development. Manuel Bromberg and Bruno Leon shared their personal experiences with Fuller in the late 1940s early 1950s. These were the early days of research and development and Fuller's first major dome customer, the Marine Corps. I am fortunate to be able to include some of their accounts in this book. Duncan Stuart revolutionized spherical subdivision with his triacon method. He taught me how the method worked and encouraged me to write one of the first computer graphics programs based on it in the 1960s. A section of this book outlines Stuart's method and shows how it made domes practical to build. Stuart's method is used today in many applications unrelated to geodesic domes. Chizko Kojima was a "computer" (a job title) at Geodesics in the 1950s. She gave me a sense of how exacting and tedious it was to calculate the geometry of Fuller's early domes *by hand*. Bill Ahern and Bill Wainwright pioneered radomes. They taught me the principles of *random subdivision*. Both have unique patented subdivision systems. I am grateful for every lesson.

Important records of Fuller's life and geodesic dome projects are cataloged in the R. Buckminster Fuller Papers collection at Stanford University's Department of Special Collections and University Archives. Staff librarians were particularly helpful in guiding me through this enormous collection and in securing rights to use and reference material. I am grateful for their help and welcoming spirit each time I visited.

I have never met Joseph Clinton, Hugh Kenner, or Amy Edmondson, but their influence is evident in this book. Joseph Clinton wrote many of the early primers on geodesic geometry and he developed several unique subdivision methods. He also created the class system we use today to characterize different spherical grid systems. I make frequent use of the class system in this book. Hugh Kenner has written extensively on Fuller and geodesics and his wonderful book, *Geodesic Math and How to Use It*, shows that you don't need much more than a conceptual framework, basic trigonometry, and a good calculator to subdivide a sphere. Amy Edmondson's book, *A Fuller Explanation: The Synergetic Geometry of R. Buckminster Fuller*, is one of the clearest presentations of the subject.

Mathematics is precise and clarity is important. I am grateful to Professors David W. Henderson, Cornell University Mathematics Department; Gregory Hartman, Virginia Military Institute Mathematics and Computer Science Department; and to A K Peters reviewers for their advice and suggestions in clarifying terms and improving the text.

Many of the CAD images in this book were created with CATIA, a powerful 3D modeler from the French multinational Dassault Systèmes. Ricardo Gerardi taught me how to write CATIA macros that automatically create 3D geometry. Without his instruction, I would never have been able to create the number of detailed illustrations I have created for this book. Dassault Systèmes and Alain Houard facilitated my access to CATIA, without which I could not have included the chapters on CAD.

Good editing clarifies a subject and makes it easier for readers to understand. This is particularly important in a technical primer like this one. Bernadette Hearne did an amazing job of editing. She tolerated many rough drafts and is an artist with scissors and paste pot. Valerie Havas and Lynne Morrison picked up where my high school English teacher left off and showed me the value of consistent style. Michael Christian provided invaluable advice about publishing. I greatly appreciate his positive support along the way.

Book design is an artform that makes reading and learning a pleasure. I can't say enough for A K Peters production editor Kara Ebrahim's wonderful layouts and compositions. It was a pleasure working with her and seeing a plain manuscript transform into a beautiful book.

Like most writers, I greatly underestimated how much time and effort it would take to write this book. At times, domestic projects and family were ignored. My family was patient and tactful; never once did they ask if I was done yet. My wife, Geraldine, and sons Ed ("Turtle") and Gerald acted as a fan club that kept me going. Clearly, this book would not have happened without their support. Love to them all.

① Divided Spheres

W e owe the word *sphere* to the Greeks; *sphaira* means ball or globe. The Greeks saw the sphere as the purest expression of form, equal in all ways, and placed it at the pinnacle of their mathematics. Today, the sphere remains a focus of astronomers, mathematicians, artisans, and engineers because it is one of nature's most recurring forms, and it's one of man's most useful.

The sphere, so simple and yet so complex, is a paradox. A featureless, spinning sphere does not even appear to be rotating; all views remain the same. There is no top, front, or side view. All views look the same. The Dutch artist M. C. Escher (1898–1972) ably illustrated this truth in his 1935 lithograph where the artist holds a reflective sphere in his hand as he sits in his Rome studio.[1] It's easy to see that even if Escher were to rotate the sphere, the reflected image would not change. His centered eyes could never look elsewhere; the view of the artist and his studio would be unchanging. His sphere has no orientation, only a position in space.

---

[1] 1935 Lithograph, M. C. Escher's "Hand with Reflecting Sphere" © 2009 The M.C. Escher Company Holland. All rights reserved.

The sphere is a surface and easy to define with the simplest of equations, yet this surface is difficult to manage.[2] A sphere is a closed surface with every location on its surface equidistant from an infinitely small center point. A mathematician might go one step further and say the sphere is an unbounded surface with no singularities, which means there are no places where it cannot be defined. There are no exceptions.

The sphere is unusual, it has no edges, and it is *undevelopable*. By undevelopable, we mean that you cannot flatten it out onto a 2D plane without stretching, tearing, squeezing, or otherwise distorting it.[3] You can test this yourself by trying to flatten an orange without distorting it. You can't without stretching or tearing it no matter how small or large the orange is.

Any plane through the center of the sphere intersects the sphere in a *great circle*. For any two points, not opposite each other on the sphere (opposite points are called *antipodal points*), the shortest path joining the points is the shorter arc of the great circle through the points, and this arc is called a *geodesic*. Geodesics are the straightest lines joining the points; they are the best we can do since we can't use straight lines or *chords* in three dimensions when we measure distance along the sphere. The length of a geodesic arc is defined as the distance between the two points on the sphere.[4]

Like the *equator* on a globe, any great circle separates the sphere into two *hemispheres*. A plane not passing through the origin that intersects the sphere either meets the sphere in a single point or it intersects the sphere in a *small circle*, also called a *lesser circle*. Small circles separate the sphere into two *caps*, one of which is smaller than a hemisphere and resembles a contact lens. We use spherical caps later in the book to compare spherical subdivisions.

Great circles play a dual role when subdividing spheres. Points define great circles and great circle intersections create more points. When we start to subdivide a sphere, we typically start with just a few points that define a relatively small number of great circles. We define more points by intersecting various combinations of the great circles. The new points derived from intersections can now be used to define more great circles, and the cycle repeats. You can see already that we are going to make great use of the dual role of great circles when we describe the various techniques and their resulting grids in Chapter 8. The difference in techniques is primarily how we define the initial set of points and great circles, and what combinations of great circles we select to intersect to define additional points.

Spherical polygons are polygons created on the surface of a sphere by segments of intersecting great circles. Spherical polygons demonstrate other differences between spherical and plane Euclidean geometry. The sides of spherical polygons are always *great circle arcs*. As a result, two-sided polygons are possible. Just look at a beach ball or slices of an orange or apple for examples. These two-sided polygons are called *lunes* or *bigons* (*bi* in-

---

[2] The equation of a sphere is very simple. For a sphere whose 3D Cartesian origin is (0, 0, 0), a point on the sphere must satisfy the equation $r = \sqrt{x^2 + y^2 + z^2}$.

[3] The Swiss mathematician and physicist Leonhard Euler (1707–1783) did intense research on mathematical cartography (mapmaking). In a technical paper, "On the geographic projection of the surface of a sphere," published by the St. Petersburg Academy in 1777, he proved that it was not possible to represent a spherical surface exactly (preserving all distances and angles) on a plane. The surface of a sphere is undevelopable.

[4] The word *geodesic* comes from geodesy, the science or measuring the size and shape of the earth. A geodesic was the shortest distance between two points on Earth's surface, but today, it is used in other contexts such as geodesic domes.

stead of *poly*gons). Spherical triangles are also different. They can have one, two, or three right angles. And one of the oddest differences between spherical and Euclidean geometry is that there are no *similar triangles* on a sphere! They are either *congruent* or they are different. In plane Euclidean geometry, three angles define an infinite number of triangles differing only by the proportional length of their sides. But on a sphere, triangles cannot be similar unless they are actually congruent.

The sphere is a challenging but fascinating place to work. With all of these differences, how do we work with spheres and what Euclidean principles, if any, can we use?

1.1   Working with Spheres

A sphere can be any size at all. Its radius, *r*, could be any distance and range from sub-atomic dimensions, to the size of a playground dome, to light-years across the observable universe. To make spheres easier to work with when their radius could be anything, we treat *r* as a positive real number and make it equal to one unit. A sphere with a *unit radius* is called a *unit sphere*. One what? Are we talking about one mile, one foot, one inch, or one anything? Yes to all these questions. Unit spheres, ones where $r = 1$, are easy to cal-culate, and any spherical result is easily converted to an actual dimension such as miles, feet, inches, or whatever. Angles do not have to be converted; they are used as is no matter the size of the sphere. However, for distances, lengths, and areas, we need to convert unit sphere dimension into the true radius of the sphere our application needs.

1.2   Making a Point

Although the sphere is an infinite set of points all equally distant from its center, practi-cal design applications require us to locate specific points on the surface that relate to the design we have in mind. Locating points requires us to define a reference system and orientation for our work. In the simplest case, placing the sphere's center at the center of the Cartesian axis system, the familiar *xyz*-coordinate system we use most often, we have defined at least six special reference points on the sphere's surface, one for the positive and negative points where each coordinate axis intersects the sphere's surface. In so doing, we have also adopted standard design conventions where we can refer to a top, bottom, side, or front, if we need to. We are off to a good start. Out next challenge is to define points on the surface that help us with our design. So how do we do this?

It's natural to think of points on a sphere like points around a circle. While it is easy to *evenly* distribute any number of points around the circumference of a circle, doing so on the surface of a sphere is actually quite difficult. Figure 1.1 gives us a sense of the problem when we try to distribute points. In Figure 1.1(a), an equal number of points are arranged around rings, or lesser circles, similar to the lines of latitude on Earth. Small circles sur-round each point to give a visual indicator of their spacing. We see that as the lesser circles get closer and closer to the sphere's two poles, the points around them are getting closer and closer together; the circles surrounding them are overlapping. At the two poles, dozens of points are nearly superimposed. In this subdivision, the points are not uniformly distrib-uted at all.

In Figure 1.1(b), we see a dramatically different distribution for the same number of points as in (a). Clearly, they are more uniformly spaced with a consistent symmetry and appearance. Notice that while some circles touch, none overlap. A grid connecting each

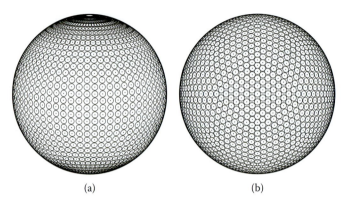

(a)   (b)

Figure 1.1. Distributing points on a sphere.

point with its nearest neighbors would produce consistently shaped triangles, whereas in the first layout they won't. These two layouts illustrate our geometric challenge when we subdivide a sphere—how do we define arrangements like the one in (b) that give us freedom to have more or fewer points that allow us to make different shapes on the sphere's surface that meet our design requirements?

We know that on a circle, we can evenly space an arbitrary number of points—say, 256, or 2,011, 9, or 37 points. In a sense, this logic is what was used in Figure 1.1(a). So why can't we evenly space an arbitrary number of points and get the result shown in (b)? The answer lies in the number of points we try to distribute and how symmetrical we want their arrangement to be. Let's look at each of these related design issues and how we intend to address them.

1.3   An Arbitrary Number

Physicists and mathematicians have theorized how to distribute an arbitrary number of points on a sphere for some time. And today, with the help of computers, there are some partial solutions. One approach, familiar to physicists, sees the point distribution problem much the way they see the way particles interact. One technique uses particle repulsion to distribute points. It beautifully illustrates the problem of trying to evenly space an arbitrary number of points on a sphere.

Particles with the same charge (positive or negative) repel each other with strength inversely proportional to the square of their distance apart. The closer they are together, the stronger they repulse each other. This natural law, called the *inverse-square law*, can be exploited to subdivide a sphere when particles represent points that can be connected to form spherical grids.[5]

If you cannot relate to charged particles, think of hermits instead of particles. Each hermit is standing on the earth and each wants to get as far away from everyone else as they can.

Imagine a given number of particles randomly distributed over the surface of the sphere, say 252 of them. Figure 1.2(a) shows them surrounded by the same diameter sphere (white),

---

[5] Inverse-square law—a physical quantity like the intensity of light on a surface or the repulsion strength of two like-charged particles is inversely proportional to the square of the distance from the source of that quantity. For example, if the distance to a source of light is doubled, only one quarter of the light now reaches the subject.

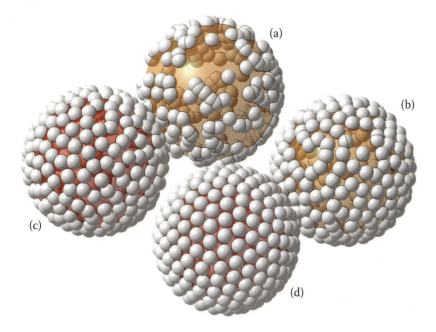

Figure 1.2. Distributing an arbitrary number of points on a sphere.

each centered on the surface of an inner sphere (transparent gold). One particle's position is fixed, but all others can move on the surface relative to the fixed one. A computer program simulates the effect of the particles repulsing each other. For each particle, the strength and direction of repulsion of every other particle acting on it is found using the inverse-square law. After every particle has been evaluated, they are allowed to move a little in the direction found so that the repulsion forces acting on them are lessened. Although the particles move a little at this stage, they always remain on the surface of the sphere. The computer once again simulates the effect of the particles repulsing each other and again, they are allowed to reposition a little to decrease the forces acting on them. In each cycle, all the particles are closer and closer to maximizing their distance to their neighbors; that is, each one is trying to get away from its neighbors and yet remain on the sphere's surface.

Figure 1.2(b) and (c) show their progress. At each stage, the particles are progressively more evenly distributed; the spaces between them are more uniform and fewer small spheres are intersected. After a number of iterations, a program parameter, the particles reach an equilibrium state. Figure 1.2(d) shows that within tolerances, another program parameter, they have established their final position on the sphere and the repositioning cycles end. None of the spheres surrounding the particles interfere now, though some touch each other. We have a visual check that the particles are as equally spaced from one another as they can be. A grid connecting neighboring particles (points) defines a triangular subdivision grid.[6]

The particle repulsion technique looks promising. The result in Figure 1.2(d) looks quite good and we have the advantage of distributing any number of points our design requires. So why not go with this approach? No need for great circles here!

---

[6] Particle repulsion simulation program Diffuse, courtesy of Jonathan D. Lettvin.

As attractive as this approach is, there are some serious drawbacks that eliminate this approach for all but the most specialized applications. First, there are an infinite number of final arrangements for the same number of particles. Resprinkle the same number of them again, let them find their equilibrium positions again, and for sure, they will settle down in a different arrangement. This means you cannot repeat the process for the same number of points and get the same design outcome. Randomization and the sequence of rebalancing in each cycle guarantees that no two simulations for the same number of particles will produce the same end result. Second, for certain numbers of particles and initial placements, it is possible they will jostle around forever and never find their equilibrium state. Forever is not a good timeframe to wait if you are trying to design something. This is one reason we fixed one point's position before we started letting them rearrange themselves. Without one fixed point, each cycle would simply continue to move particles around and they would never balance. Third, and this is a very big drawback, there is no symmetry in the final arrangement of points. Except for an occasional lucky number of points, there may be no stable pattern. The final arrangement of points depends on the initial random distribution, how forces are simulated and the iterative effects of relaxation and re-positioning. There is no way to anticipate if multiple points will lie on the same great circle, form antipodal points (two points on opposites sides of the sphere), or create any symmetric arrangements at all. This also means there is no way to prove whether a final solution is unique or not.

None of the above outcomes are good, especially if the points are to be used in designs we intend to manufacture. What we are really looking for is a way to subdivide a sphere by

- distributing points evenly and define grids as course or fine as needed;

- minimizing the variation within the grid (chords and areas);

- creating grids where some members form continuous great circles for applications that require them;

- maximizing symmetry and reuse of local grids;

- defining coordinates that uniquely define any point on the grid;

- developing simple ways to convert from one coordinate system to another; and

- defining metrics for comparing one subdivision method with another.

So how can we do this? The Greeks will show us.

1.4 Symmetry and Polyhedral Designs

Like many things in geometry, the Greeks seem to have gotten there first. Although the five *Platonic solids* have been known since prehistoric times, the Greeks were the first to recognize their properties and relationships to one another.[7] Figure 1.3 shows them with holes

---

[7] Stone figures resembling polyhedra, found on the islands of northeastern Scotland, have been dated to Neolithic times, between 2000 and 3000 BC. These stone figures are about two inches in diameter and many are carved into rounded forms resembling regular polyhedra such as the cube, tetrahedron, octahedron, and dodecahedron. By 400 BC, the time of Plato, all five regular polyhedra were known.

Figure 1.3. Platonic solids.

in their faces to make it easier to see how their parts relate. Starting in the back and going clockwise, they are the tetrahedron, cube, dodecahedron, icosahedron, and octahedron. The Platonics are one family of polyhedra (there are dozens). *Polyhedra* are 3D forms with flat faces and straight edges and they take on an amazing variety of forms, many of which are very beautiful in their symmetry and spatial design. They are one of the most intensely studied forms in mathematics.

The Platonics are unique among polyhedra. Every Platonic solid has regular faces (equilateral triangles, squares, or pentagons) and every *edge* on a Platonic face has the same length. All *face angles* are the same and the angle between faces that meet at an edge are the same.

Every Platonic solid is highly symmetrical. This means that they can be orientated in many different ways and still retain their appearance. Certain pairs of Platonics can be placed inside one another and this characteristic demonstrates how they are interrelated. The Platonics and the properties just mentioned are so important in subdividing spheres, we devote an entire chapter to them and explain them in great detail.

Another key property of the Platonics, particularly useful in spherical subdivision, is their vertices or points (we use the two words interchangeably). They are evenly spaced and lay on the surface of sphere that surrounds them and share the same center point as the polyhedron. This circumscribing sphere is called the polyhedron's *circumsphere*. Figure 1.4 shows the spherical versions of the five Platonics. The planar versions are shown inside to make the association between planar edges and spherical great circle arcs easier to see. Notice that in each spherical version, some pairs of vertices are on opposite sides of the sphere; they are antipodal points. Notice also that every spherical *face* in each Platonic is bounded by arcs of great circles and that a single Platonic has but one spherical face type (equilateral triangle, square, or pentagon). This is important because any design we develop on one face can be replicated to cover the others, thus covering the entire sphere with a pattern that has no overlaps or gaps.

Figure 1.4. Spherical Platonics.

1.5   Spherical Workbenches

Spherical Platonics give the designer a huge head start in subdividing the sphere because the polyhedron's vertices are already evenly distributed on its circumsphere and we can use them immediately to define great circles and reference points for further subdividing. Let's take a quick look at how we are going to develop a design starting with a simple polyhedron. Later chapters in the book will explain the detail of how they are done.

Referring to Figure 1.5, we start the subdivision process by first selecting one of the Platonic solids that best meets our design requirements. In this example, we select the icosahedron (a). It is the most used in spherical work, by far, but we could have used any of the other Platonics. We define the icosahedron's spherical version (b); it's an easy step given the planar one. The spherical version is now our subdivision workbench. We pick a conveniently positioned face to work with, or some symmetrical area within it (c). In this example, a complete icosahedral face is selected; one of its vertices is at the sphere's zenith (top); this can simplify our calculations. Most of our efforts will be spent here, subdividing this single face. This face is not our only working area option, but it is an obvious choice. Within this face, we can use any one of several gridding techniques. In Chapter 8, we describe six techniques. Each technique produces a different grid and each grid's set of points offer its own benefits, depending on your application. Once the subdivision is complete, we replicate the resulting points and grid to cover the rest of the sphere as shown in (d).

We have considerable flexibility in (a) through (c). We can use any of the five Platonics as our base, we can pick different standard areas to subdivide, and we can use different techniques to subdivide this area, making the grid as course or fine as we wish. Once we have a set of points, we have more flexibility in how we join combinations of points to make triangles, hexagons, pentagons (all shown in Figure 1.5(f)), diamonds, or any other spherical shape. The design possibilities are limitless. With so many choices, one might ask, which is best? The answer, of course, depends on the application you have in mind. In Chapter 9, we will describe a series of metrics that you can use to evaluate the appropriateness of one layout or another to your needs. Various metrics show the differences

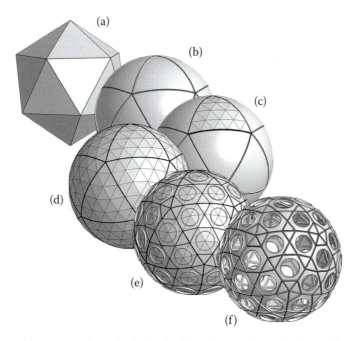

Figure 1.5. Progression of spherical subdivision to basic design application.

between subdivision methods and this helps you decide which combination best fits your requirements.

For many users, the subdivision grid in Figure 1.5(d) is just the starting point for further refinement. It must be developed into a physical design. The grid might define locations for openings, panels, struts, or even positions of dimples on a golf ball, and not every subdivision point or grid member may be part of the final design.

Figure 1.5(e) shows one of many possible designs for this particular grid. Triangles might be combined into diamonds, hexagonal, or pentagonal patterns. Such refinements call for other programs such as CAD where points, chords, and face definitions generate, manually or automatically, prismatic shapes, structural elements, surfaces, or geometric elements, as shown in a simple design application (f). Grids can be combined or layered to create more intricate patterns or truss-like structures. If the design is for a geodesic dome, part of the subdivision may be cut off at the ground level to allow for foundations or supports. The design flexibility is limitless, and CAD provides powerful visualization and analysis capabilities as well. If the design is to be manufactured, the 3D geometry can be used to automate cutting, welding, molding, or bending equipment during production. CAD is an accessible technology and a powerful tool for the *spherist*. We devote two chapters to CAD.

1.6   Detailed Designs

The grid shown in Figure 1.5(d) is quite modest. The surface of the sphere is covered with 262 points. However, from the same starting point, the initial 12 vertices of the spherical icosahedron, shown in (b), we can distribute fewer or more points to make our grid. Figure 1.6 shows how. In each figure, the *frequency* of points, a term explained later, steadily

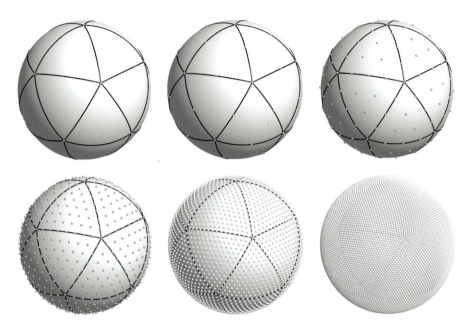

Figure 1.6. Increasing the number of points over the sphere.

increases from the spherical icosahedron's 12 vertices (top left figure) to 10,242 (bottom right figure) by progressively placing points between the points of the previous layout. We could continue the process indefinitely, but most spherical applications do not need more points. Notice also, as more points are added, we can generate more subdivision triangles by connecting points to their nearest neighbors.[8]

In this series, the number of points from left to right, top to bottom, is 12, 42, 162, 642, 2,562, and 10,242, respectively. The radius of all accent spheres is the same in all illustrations. In the last illustration, some accent spheres touch, but none interfere. In later chapters, we will use accent spheres such as these to visualize and analyze the point distributions that result from the spherical subdivision techniques we present.

1.7   Other Ways to Use Polyhedra

The polyhedral subdivisions we have been discussing are based on points defined by intersecting great circles. Great circle techniques are the ones we emphasize in this book, but it is possible to use polyhedra to define points on a sphere by intersecting lesser circles instead. Here's how.

We will use the icosahedron again but instead of focusing on its spherical faces, we will use its vertex axes to develop our grid. Every one of the icosahedron's six vertex axes passes through its center and they are perpendicular to the polyhedron's circumsphere. A cylinder is positioned around each axis. Figure 1.7(a) represents the icosahedron's axes with arrows; a

---

[8] The total number of points in any one of these spherical triangles is $n(n + 1)/2$. A single face in each of the images in Figure 1.6 contains 3, 6, 15, 45, and 153 vertices, respectively. In a later chapter, we will discuss a very useful formula called the *triangulation number* formula, which tells how many points result from any spherical subdivision.

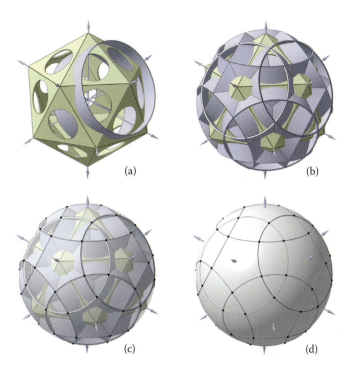

(a)   (b)

(c)   (d)

Figure 1.7. Lesser circle spherical subdivisions.

single cylinder is positioned around one of them. If the cylinder radius is large enough, every cylinder will intersect its neighbor's cylinder, as shown in (b). Each cylinder will also intersect the polyhedron's circumsphere and define a lesser circle, as shown in (c). A point can be defined at each lesser circle intersection, as shown in both (c) and (d).

The lesser circle arcs between the points in (d) are not geodesics. Only an arc of a great circle can be a geodesic. The lesser circle arcs, technically speaking, do not define spherical polygons either. Again, only great circle arcs define the sides of spherical polygons. It is possible, however, to define great circles between these points by passing a plane through pairs of them (and the origin).

There are many variations on this technique. For example, one can substitute cones for cylinders. The cone's apex is at the center of the sphere and the cone's surface intersects the sphere defining a lesser circle. Equations can systematically generate lesser circles on the sphere as well, and where circles intersect, subdivision grid points can be defined. Gary Doskas has done extensive research in this area; the patterns he explores and their symmetry are beautiful and many are utilitarian.[9] We mention lesser circle subdivision of spheres for completeness; they are not developed further in this book.

1.8   Summary

In this chapter, we looked at the challenges of subdividing spheres and the fact that spheres have unique geometric properties. We have seen that it is impossible to evenly distribute

[9] See (Doskas 2011) for more information.

an arbitrary number of points on a sphere and achieve symmetric and predictable results. Spherical polyhedra, particularly the Platonics, offer a symmetrical framework and jump start the subdivision process. The vertices of spherical Platonics are evenly spaced reference points on a sphere that surrounds the polyhedron. A small work area can be defined and grids developed there, as course or fine as a design requires. The grid can then be replicated to cover the rest of the sphere without overlaps or gaps.

Designers can use any of the spherical Platonics as a design framework, define working areas on its faces, and use any one of a number of different subdivision techniques to develop a grid over that face. The variety is limitless.

In the next chapter, we will look at the fascinating history of spherical subdivision. Buckminster Fuller, the inventor of the geodesic dome, was the first person to recognize the value of spherical polyhedra and subdivision grids to general architectural construction. Most of the subdivision techniques we use today were developed in the late 1940s and 1950s by Fuller and his associates to build geodesic domes. In their day, some domes were the largest free-span structures on Earth. Fuller's work is particularly important because he achieved highly creative results while leveraging manufacturing techniques. The spherical subdivision techniques that evolved from the 1950s are as relevant today as they were then and are applied in a wide range of science and industrial applications that have nothing to do with geodesic domes. We survey some of these applications in Chapter 3.

Additional Resources

Doskas, Gary. *Spherical Harmony—A Journey of Geometric Discovery.* LuLu Marketplace: Hedron Designs, 2011.

Messer, Peter W. "Polyhedra in Building." *Beyond the Cube: The Architecture of Space Frames and Polyhedra*, ed. J. Francois Gabriel, New York: John Wiley & Sons, 1997.

Stuart, Duncan R. "The Orderly Subdivision of Spheres." *The Student Publications of the School of Design,* North Carolina State University, monograph, 1963.

 Bucky's Dome

Buckminster Fuller (1895–1983), seen in the painting above,[1] was a true American polymath. Bucky, as he was called, was a philosopher, designer, engineer, architect, author, futurist, and prolific inventor. His earliest inventions include building blocks, temporary shelters, and an environmentally friendly bathroom where you could take a "fog shower" utilizing pressurized mist and less than a gallon of water. He even invented a bullet-shaped, three-wheeled car that could turn complete circles in its own length. However, he is best remembered for the invention of the geodesic dome and designs that do more with less.[2]

   Millions of visitors have seen the US Pavilion at Expo'67 in Montreal, Epcot Center at Disney World in Orlando, or the La Géode Theater in Paris. Countless others live, shop, or

---

[1] Painting of inventor and philosopher, R. Buckminster Fuller, Jr., by Boris Artzybasheff (1899–1965). The media is tempera on board, 21.5 × 17 in., circa 1964. *Time Magazine* cover, January 10, 1964, and gift of Time, Inc. to the National Portrait Gallery, Smithsonian Institution, Washington, DC, USA Used with permission of the National Portrait Gallery and the Estate of R. Buckminster Fuller.

[2] Buckminster Fuller patents include Prefabricated Bathroom (Patent 2,220,482, 5 Nov. 1940) and Motor Vehicle (Patent 2,101,057, 7 Dec. 1937).

worship in geodesic domes. Geodesic domes enclose radar equipment at airports and air defense stations in remote polar regions. They corral fish in ocean-bobbing pens and record neutrinos from outer space, as they zip through underground spherical observatories. Geodesic tents shelter Boy Scouts and Mount Everest climbers alike. A nearby playground might even have a geodesic jungle gym for kids to climb on. No single construction system has been built in so many sizes and of such diverse materials—wood, pipes, sheets of plastic and metal, foam panels, cardboard, plywood, bamboo, fiberglass, concrete, and even bicycle wheels and the tops of junked cars.[3]

These applications and many others use spherical techniques that Buckminster Fuller, or one of his associates, developed in the late 1940s and 1950s. They are as useful today, as they were then, because they continue to solve new problems in fields that have nothing to do with domes, such as astronomy, weather prediction, materials science, virology, product design, and PC game development. Even the dimple patterns on golf balls owe a debt to Buckminster Fuller.

The concept of geodesics is not entirely new. For mathematicians, they are the shortest path between two points on a curved surface, *any* curved surface, not just a sphere's. Spherical triangles and grids are hardly new, and neither is geodesic construction. The earliest example dates to 1922 when Walther Bauersfeld, an engineer for Carl Zeiss optical company, developed the world's first reinforced concrete dome in Jena, Germany,[4] for Zeiss' planetarium. The dome's steel reinforcing grid resembles the lattice we associate with today's geodesic dome.

Bauersfeld's structure was highly innovative at the time. However, unlike Fuller's domes, Bauersfeld's dome was never developed into a generalized construction system or used elsewhere. Fuller was the first to establish geodesics in a framework he called *energetic synergetic geometry*, or synergetics for short. For sure, Fuller's relentless promotion of geodesics had a lot to do with the success of the geodesic dome, but this success was also the result of synergetics. Synergetics acted as a vehicle for moving concepts in physics (spin, charge, attraction), mathematics (plane and spherical geometry), materials science (tension and compression), natural building processes (triangulate and space-filling forms), and design intent (conservation of energy, high strength-to-weight structures, and industrial processes) to different settings, such as mapping, long-span truss systems, and, most important, geodesic domes. The fact that these innovations were based on a broader framework for design increased their appeal and encouraged their application in new fields.

Fuller did not invent the geodesic dome in isolation. In the late 1940s, he taught at several colleges where he came into contact with highly creative and intelligent people. His style, somewhat non-academic for the times, mixed seminars, workshops, and hands-on projects, in which his ideas were freely associated with those of students and colleagues. He often made their ideas his own. This style, borrowing ideas, characterized his business life as well. There, designers, architects, engineers, and artisans, such as Kenneth Snelson, Duncan Stuart, Donald Richter, T. C. Howard, Bill Wainwright, Jim Fitzgibbon, Shoji Sadao, and Bill Ahern, also had creative skills and energy. Some were inventors and would be awarded their own geodesic patents. They were often the designers and architects on projects that only Fuller received credit for. There is no doubt that Fuller's free-association inspired others, increased the flow of ideas, and hastened the development of geodesics.

---

[3] Drop City, a hippy commune in the 1960s, was an assemblage of geodesic panel domes made from the sheet metal of automobile roofs and other inexpensive materials.

[4] "The Wonder of Jena" in *Refried Domes* (Kahn 1989, 12)

And while most relationships were symbiotic, some were not. This book recognizes the contributions of others, as often as possible.

Buckminster Fuller is a fascinating study of innovation, entrepreneurship, self-belief, and opportunism. He was a prolific inventor and writer. While some writings are razor-sharp, others read like stream of consciousness. He was always concerned about being misunderstood and, given his elliptical style and self-made vocabulary, it was a reasonable worry. Many authors and historians have tried to place this complex personality and his inventions within the broader context of twentieth-century events. Anyone interested in the genesis of geodesic domes should attempt to understand the man, Buckminster Fuller. Those wanting to know more about him can consult the additional resources at the end of this chapter.

2.1   Synergetic Geometry

Volumes have been written by and about Fuller's synergetic geometry. This book examines only the very small part of it that built the foundation for dividing spheres. By the late-1940s, Fuller was shifting his attention from developing affordable housing to exploring new ways of thinking about design. He kept a sketchbook called *Noah's Ark II*, where he diagrammed geometric relationships and spherical grids.[5] Fuller had always been interested in natural forms: rocks, crystals, shells, and so forth. He sensed that nature always found the most efficient solution to problems and he was taken by the idea that nature is in constant motion and that motion itself is relative. He looked for a way to unify fundamental laws of physics (atoms, orbits, spin, energy, charges, and bonding) with geometry (polyhedra, space-filling lattices, great circles, symmetry, and maximum packing) into a design cosmology he would call synergetic geometry. Synergetics was Fuller's summary of natural phenomena and his framework for design. He drew his principles from natural systems and used mathematics, particularly solid geometry, to show relationships—*synergy*—between systems. "The whole is always more than the sum of its parts," he would say. In his cosmology, mathematics, observation, and the analysis of natural systems were all telling us how to make the most efficient design. Fuller personalizes synergetics with his own language, which is often difficult to understand. But his basic dictum was clear: man is part of a natural system; he must learn from it, respect it, and use its principles in his work.[6]

He added, "I would not suggest that it is the role of the individual to add something to the universe. Individuals can only discover the principles and then employ them to move forward to greater understanding."

Fuller had a keen sense of how geometry increased strength and stability, and how some polyhedra could be packed together to fill space without leaving voids. He recognized the inherent stability and rigidity of the tetrahedron, octahedron, and icosahedron, three solids with only equilateral triangular faces. The cuboctahedron has both square and triangular faces, but can be constructed from just eight tetrahedra. As a result, it is highly stable and rigid. These solids were particularly important in synergetics and would appear and reappear in different combinations in his future inventions.

The tetrahedron (4 equilateral triangular faces, 6 edges, and 4 vertices) is the only polyhedron where every vertex is equidistant from every other one. Fuller visualized its geometry

---

[5] (Fuller 1950)
[6] (Edmondson 1987, 2007)

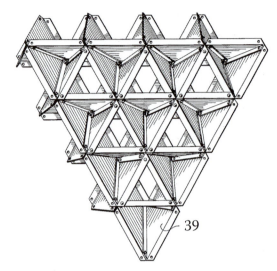

Figure 2.1. Synergetic building construction.

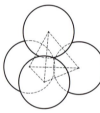

by clustering three equal-diameter spheres into a triangle and nestling a forth on top, as seen in the figure to the left. All four spheres are packed in their closest 3D configuration; their centers define the vertices of a tetrahedron. The tetrahedron is the most basic form in Fuller's synergetic geometry. It appears in numerous configurations in his later work.

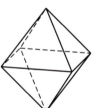

The octahedron (8 equilateral triangular faces, 12 edges, and 6 vertices) can be visualized as two back-to-back square pyramids. Octahedra and tetrahedra together can be placed side by side and perfectly fill 3D space. Figure 2.1 shows the octahedral-tetrahedral truss system Fuller later called the *octet truss*. You can see how these two solids pack together to fill all the space.[7]

Fuller describes the relationship of the tetrahedron-octahedron: "Nature's simplest structural system in the universe is the tetrahedron. The regular tetrahedron does not fill all space by itself. The octahedron and tetrahedron complement one another to fill all space. Together they produce the simplest, most powerful structural system in the universe."[8]

The icosahedron (20 equilateral triangular faces, 30 edges, and 12 vertices) has the highest number of identical regular faces of any regular polyhedron. Its faces can be subdivided into six right triangles; thus, the overall solid could be made from 120 right triangles. The subdivision of any one of them can be replicated over its surface without overlaps or gaps. It is the most used polyhedron for spherical work by far. We discuss the icosahedron in great detail later in the chapter.

The cuboctahedron (14 faces, 24 edges, and 12 vertices) is the only polyhedron out of hundreds where every edge is equal in length *and* this length is the distance between every vertex and the center of the polyhedron. You can visualize the cuboctahedron as a polyhedron made of eight tetrahedra, each sharing one vertex at the center of the polyhedron and their other vertices tangent to one another. Like the tetrahedron, the cuboctahedron can be created by close-packing spheres. Twelve spheres nested

[7] (Fuller 1961, sheet 7/7, fig. 14)

[8] (Fuller 1983, 168)

around a thirteenth, central sphere, as shown in the figure to the right. Fuller thought this configuration and angular relationship was so special he called it the *vector equilibrium* (VE), claiming that it was nature's coordinate system as opposed to normal *Cartesian coordinates* man uses.[9] Cuboctahedron and vector equilibrium all refer to the same polyhedra and we use the terms interchangeably.

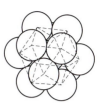

2.2   Dymaxion Projection

Perhaps due to his prior Navy service and the importance of charts and navigation, Fuller began his spherical work with cartography. Map projection, one of the longest standing geometric challenges, involves figuring out how to accurately represent features of a round earth on a flat plane such as a piece of paper. The surface of a sphere is undevelopable, which means it cannot be rolled out or flattened without distortion. This centuries-old problem has attracted scores of geographers and mathematicians and led to numerous projection schemes. Some projections and maps preserve the shape of spherical shapes and angles, while others preserve great circle distances or compass headings between points on the sphere; others represent the area of shapes accurately. No 2D projection preserves all these characteristics. In the end, all 2D maps of a sphere have some form of distortion.

At the time, and it's still true today, the most common map projection was the Mercator projection. In this projection, the earth is placed inside a cylinder (a curled-up map that wraps around the earth), and rays from the center of the earth paint the outline or land masses onto the cylinder. Only the equator, which is tangent to the cylinder, is accurately projected. The landmasses distort more and more as the projection nears the poles. Greenland and the arctic regions appear huge in comparison to their true size. The poles cannot be projected at all, since rays from the earth's center to the poles would be parallel to the surface of the cylinder and never project at all.

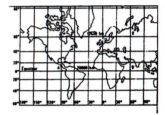

In 1944, Fuller took a totally different approach and based his map projection on the spherical subdivision of the cuboctahedron, shown in Figure 2.2(a). Essentially, Fuller developed a grid over the faces of the cuboctahedron and then projected the grid onto a sphere that surrounded the solid. The grid he developed essentially runs parallel to the edges of the cuboctahedron's regular square and triangular faces. The resulting spherical projection undeniably resembles a geodesic dome, though Fuller would not make this connection for a few more years.

In Figure 2.2(b), the projection is unfolded into a 2D view, sometimes called a *net*, and various combinations of land and ocean areas are displayed on a series of contiguous squares and triangles.[10] Some displays show a continuous ocean, others continuous land. Each serves its own purpose. Few thought it possible to invent anything new in a field so exhaustively studied and thoroughly developed as cartography but Fuller did just that. In his patent, he briefly describes the problem he solved: "The earth is a spherical body, so the only true cartographic representation of its surface must be spherical. All flat surface maps are compromises with truth."[11]

---

[9] The Cartesian coordinate system uses three mutually orthogonal axes; Fuller believed that nature's coordinate system is based on a 60-degree coordinate system. See also (Williams 1979, 164).

[10] Map image courtesy of The Estate of R. Buckminster Fuller.

[11] (Fuller 1946, p. 1, col. 1, para. 1)

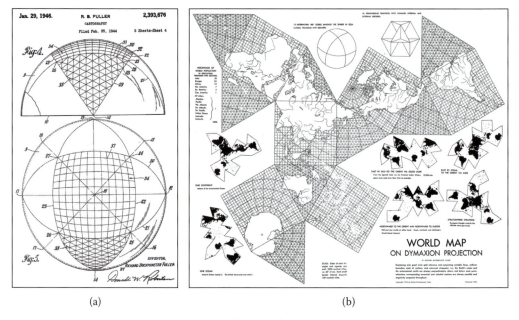

Figure 2.2. World maps with Dymaxion projection.

Fuller called this configuration a *Dymaxion projection*.[12] The projection is based on a subdivision grid that runs either parallel or perpendicular to the edges of the square and triangular faces.

After pointing out the disadvantages of current projections, including the popular Mercator projection, Fuller goes on to say: "Another expedient has been to resolve the earth's surface into a polyhedron, projecting gnomonically to the facets of the polyhedron, the idea being that the sections of the polyhedron can be assembled on a flat surface to give a truer picture of the earth's surface and of directions and distance."[13]

In essence, Fuller claims that a gnomonically projected[14] map placed onto the faces of a polyhedron, the cuboctahedron in this case, provides a truer representation of areas, boundaries, directions, and distances than any plane surface map heretofore known. By using the cuboctahedron as the base polyhedron for his projection, all vertices lie on one of four great circles.

Fuller makes three invention claims in his patent. The Dymaxion map is

- a projected map of square and triangular sections where edges are represented by projected great circles with a uniform cartographic scale;

---

[12] In the 1930s, Fuller was advised to find a better name for his 4D house invention, and with the help of an advertising wordsmith, Waldo Warren, the word "Dymaxion" was coined by selecting suitable syllables from Fuller's account of his design ideas. The roots of this name are "dynamism," "maximum," and "ions." Buckminster Fuller, *The Dymaxion World of Buckminster Fuller*, ed. Robert Marks. New York: Doubleday, 1973.

[13] (Fuller 1946, p.1, col. 1, para. 2)

[14] Gnomonic or gnomic projection is a map projection obtained by projecting points on the surface of a sphere from a sphere's center to a plane that is tangent to the sphere or a projection from the center of a sphere to the surface that surrounds an object (such as a sphere around a polyhedra). In a gnomonic projection, great circles are mapped to straight lines.

- a map with two or more sets of matching sections where equilateral triangles and square sections are constructed on a two-way and three-way great circle grid; or

- a world map comprising six equilateral square sections and eight equilateral triangular sections matching along their edges.[15]

Fuller believed that his map projection minimized distortion, provided uniform scales, and did not have a special orientation (there was no "up"). Although the context is mapping, we clearly see an initial step towards creating a spherical version of a planar polyhedron and subdividing its faces with a grid. The same method Fuller uses here for mapping would lead him to invent the geodesic dome in less than five years.

By 1954, Fuller had extended the Dymaxion projection concept to the icosahedron. The Airocean World Map, shown in Figure 2.3, is perhaps the most recognized of Fuller's maps.[16] It shows all landmasses contiguous across the faces of an unfolded version of the icosahedron. The color scheme represents the mean low annual temperature (blue regions are colder than warm red ones). While Fuller positions the icosahedron so the landmasses are continuous, he does not grid the icosahedron's equilateral faces. He retains the latitude and longitude reference grid instead. Later, we will see some recent mapping applications where latitude and longitude grids are eliminated altogether in favor of geodesic grids.

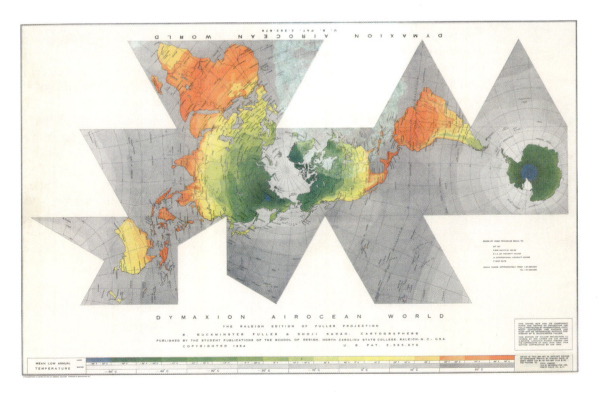

Figure 2.3. Dymaxion air ocean world projection.

---

[15] (Fuller 1946, p. 3, col. 2)
[16] Map image, courtesy of The Estate of R. Buckminster Fuller.

The world map Dymaxion and air ocean projection proved that practical grids can be made by subdividing the faces of spherical polyhedra and that these grids can be generated from great circles. Without recognizing it at the time, Fuller was already establishing the future methodology behind geodesic domes. He studied the cuboctahedron and icosahedron in detail, but at the time, he did not recognize the applicability of these grids and polyhedra frameworks to construction systems. It would be five years before they would be used in geodesic domes.

2.3   Cahill and Waterman Projections

Fuller was not the first person to use polyhedra for map projections, but he was the first to use them with grids that generalized the subdivision of spheres, that he later adapted to his famous geodesic dome construction system. Bernard Joseph Stanislaus Cahill (1866–1944), cartographer and architect, proposed using polyhedra for projections as early as 1909.[17] Cahill patented his projection in 1913, explaining his goals as follows:

> None of the methods of projection and development at present used in forming a map of the earth furnishes a representation of its land surface with any considerable degree of accuracy in regard to relative dimensions of the various parts, or is free from noticeable distortion and exaggeration, and also free from discontinuities of said land surface. The object of the present invention is to provide a map of the earth, which will avoid the above objections.[18]

Steve Waterman extended Cahill's concept to *Waterman polyhedra*, producing a low-distortion map based on a truncated octahedron that minimizes landmass separations. Applying a *gnomonic projection*, the map is neither *conformal* nor *equal-area*;[19] it does, however, have constant scale at the equator (a fold line). The four boundary meridians at 45 degrees east and west of each lobe's central meridian are standard meridian lines.

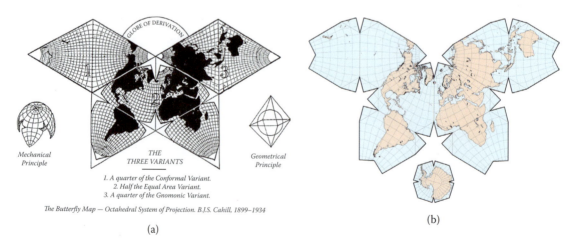

Mechanical Principle

THE THREE VARIANTS
1. A quarter of the Conformal Variant.
2. Half the Equal Area Variant.
3. A quarter of the Gnomonic Variant.

Geometrical Principle

The Butterfly Map — Octahedral System of Projection. B.J.S. Cahill, 1899–1934

(a)

(b)

Figure 2.4. Cahill octahedral world map and Waterman's butterfly world map.

[17] (Cahill 1909). See (Keyes 2009) for a comparison of Cahill's map projections to Fuller's Dymaxion map.
[18] (Cahill 25 Feb. 1913, col. 1, reference lines 9–20, sheet 1). Restored image courtesy of Gene Keyes.
[19] Conformal means the fidelity of shape and the angle between any two lines on a sphere are the same as between their projected counterparts on the map. Equal-area means that any spherical area projects approximately the same area, regardless of where the area is on the sphere.

Figure 2.4 shows Cahill's original 1909 octahedral world map projection (a)[20] alongside Carlos A. Furuti's display of Waterman's butterfly projection using a Waterman polyhedron (b).[21] Cahill's projection represents all landmasses on contiguous faces in the 2D foldout (called a net) of an octahedron. The Waterman polyhedra in (b) is named W5 and is a regular truncated octahedron in which vertex truncation results in side ratios of 1:2.[22] For the most part, Waterman's projection represents all landmasses on contiguous faces, though a part of Siberia and some Canadian islands are divided. Antarctica is treated separately, otherwise parts of it would be separated over four other map sections, as they are in Cahill's projection.

2.4   Vector Equilibrium

The Dymaxion projection is based on a subdivision of a spherical cuboctahedron. When projected on to the surface of a circumscribing sphere, the cuboctahedron's edges define a set of great circles. These great circles intersect one another, defining areas on the sphere's surface that can be subdivided into an endless variety of grids.

Figure 2.5 shows the cuboctahedron's master set of 25 great circles, the result of overlaying four sets of great circles.[23] The edges of a spherical cuboctahedron themselves define the first set of four great circles, shown in (a). Six more great circles mark the heights

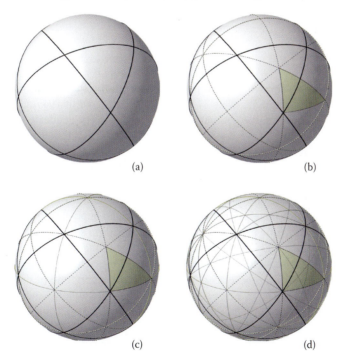

(a)   (b)

(c)   (d)

Figure 2.5. Vector equilibrium's 25 great circles.

---

[20] "Figure 11—Map with four lobes, modified from Fig. 10." (Cahill 1909, 464)

[21] Image courtesy of Carlos A. Furuti.

[22] All sides are equal on the Archimedean truncated octahedron.

[23] Robert Gray has done an extensive analysis of the primary and secondary great circles of spherical vector equilibrium and icosahedron; see www.rwgrayprojects.com.

of each equilateral and square face; they are shown dashed in (b) and define the second set. These great circles are the principal grid axes in the world map Dymaxion projection (Figure 2.2(a)).

The diagonals of the cuboctahedron's square faces define the third set of great circles. They are mutually *orthogonal* planes aligned to the Cartesian *x*-, *y*-, and *z*-planes. They are shown as light lines in (c). And finally, 12 more great circles pass through vertices of square faces and through the centers of surrounding equilateral faces that are not adjacent to these vertices. This fourth set of great circles is shown together with the other sets in (d).

In Figure 2.5(d), the whole sphere is covered by a set of spherical right triangles—24 right-handed and 24 left-handed triangles for a total of 48; one is highlighted. Fuller called this single right triangle the *least common denominator (LCD)*, because the entire surface of the sphere can be described by this single common triangle. Any pattern or subdivision developed within this small area can be replicated to cover the entire sphere without leaving gaps or causing overlaps. Mathematicians call the LCD a *Schwarz triangle* after Hermann Amandus Schwarz (1843–1921), who first discovered it (and many triangles like it) while investigating polygons that cover spheres. Schwarz triangles are very important in spherical subdivision. Chapter 4 discusses them in detail.

Vector equilibrium was a pivotal discovery for Fuller, forming the basis of several of his inventions. We will return to it in a later chapter when we build 3D models of his famous octet truss construction system.

2.5   Icosa's 31

The icosahedron is the most important polyhedron from the standpoint of spherical subdivision. Fuller's first diagrams of its 31 great circles appeared in 1947. Combinations of these circles would become the geometric references for every geodesic gridding system invented in the 1950s. James Ward, editor of a series of books on Fuller, said, "The materialization of great circles was achieved in Fuller's sketchbook *Noah's Ark II*, which alludes to the teleological role he expected his applied geometry eventually would play."[24]

The spherical icosahedron, shown in Figure 2.6(a), has 20 equilateral faces, 30 edges, and 12 vertices. Each vertex has a companion on the opposite side of the sphere. Thus, edges between vertices also have a corresponding edge on the opposite side. The pairs of opposing edges lie on the same plane and define the icosahedron's first set of 15 great circles. Figure 2.6(b) shows them as dashed lines. Note that five of these great circles always pass through a vertex and that the same great circles intersect every face edge at its center point. Just like the cuboctahedron, these 15 great circles divide every one of the icosahedron's equilateral faces into six LCDs. For a whole sphere, there are 60 left-handed LCDs and 60 right-handed ones. A typical LCD is colored gold in Figure 2.6(b), (c), and (d).

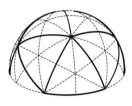

These 15 great circles have three other important relationships: they coincide with the height of a face; they define the face's three vertex axes; and they intersect every edge perpendicular at its midpoint. If any of these great circles are treated as a ground plane, the icosahedron will be oriented so that one of its edges is in the uppermost position. This so-called *edge-zenith* orientation is shown in the figure to the left. The points where great circles intersect edges define the next two sets of great circles.

[24] (Fuller 1985, 31)

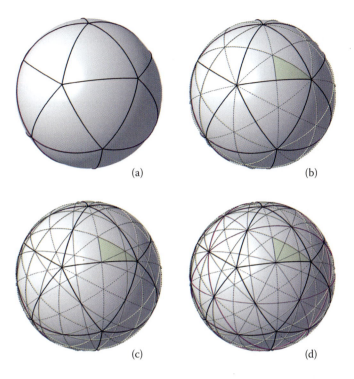

(a)   (b)

(c)   (d)

Figure 2.6. Icosahedron's 31 great circles.

Ten more great circles pass through the mid-edge points of two adjacent icosahedral faces. They are shown dotted in Figure 2.6(c). They also define 60 more points along icosahedral edges and pass perpendicular to one edge of each face. Unlike the first 15 great circles, none in this set pass through any icosahedral vertices. If any one of these great circles is treated as a ground plane, one of the icosahedron's faces is in the uppermost position. The orientation is called face-zenith and is shown in the figure to the right.

Fuller identified one more important set of great circles, a set of six that span each icosahedral face from mid-edge to mid-edge, with none of them passing through any of the icosahedron's vertices. Figure 2.6(d) shows them as solid lines along with the previous set of 25 great circles. Arcs from these great circles run parallel to the edges of the icosahedron. This grid-edge orientation will be the basis of an entire family of subdivisions called Class I. We will detail them in a later chapter. If any one of these great circles is used as a ground plane, the icosahedron will be oriented vertex-zenith, as shown in the figure to the right. Most geodesic domes are subdivided spherical icosahedra oriented vertex-zenith.

This last set of great circles illustrates an odd fact about spherical geometry. When a spherical equilateral triangle is subdivided around its mid-edges, as shown in the last example of icosahedral great circles, a curious thing happens. At first glance, it appears the grid divides the overall equilateral triangle into four smaller equilateral triangles. And in two dimensions, this is precisely what happens—but not on a sphere. On a sphere, only the center triangle remains equilateral and the other three around it are isosceles. Furthermore, the face angles for the inner and outermost equilaterals are not the same. Spherical oddities such as these are covered in Chapter 4.

Fuller's development of the 31 great circle sphere was pivotal. First, it led to the invention of the geodesic dome. Almost all domes and many other spherical applications today are based on the icosahedron. Second, two out of the three sets of 31 great circles would become reference frameworks for two different classes of subdivision grids. Each class offers distinct advantages, depending on design requirements. These developments would take place in the bucolic mountains of North Carolina.

2.6   The First Dome

Black Mountain College was an ideal place for experimenting with new ideas such as geodesic domes. The avant-garde interdisciplinary college located in Asheville, North Carolina, was founded in 1933 on the principle that the arts are central to learning. Although the school lasted only 24 years, it became the model for many future liberal arts colleges. Over the years, its faculty and lecturers would include some of the most famous artisans and thinkers of the era, including John Cage, Willem de Kooning, Ben Shahn, Robert Motherwell, Harry Callahan, Aaron Siskind, Merce Cunningham, and Buckminster Fuller.

Encouraged by others, Josef Albers invited Fuller to join the 1948 Summer Institute Program as a substitute for an architecture professor who withdrew at the last minute. It was Fuller's first teaching job and upon arrival, his seminars captivated faculty and students alike with his vision of how technology would solve world problems like hunger and the lack of adequate housing. He lectured on synergetic geometry and prepared his students to build the first geodesic dome.

Fuller, who regarded the geodesic dome as a three-way grid of great circles,[25] naturally chose to base his first dome on a hemispherical version of the *icosahedron's 31 great circles* (see Figure 2.6(d)). Bucky was not well off and was only making ends meet through his lectureships. He bought a large quantity of inexpensive aluminum slat material, a common product used in manufacturing window venetian blinds. The material seemed stiff enough to work in compression (it is slightly cupped and has some longitudinal rigidity) and is very strong in tension. The students set about cutting and marking long strips of aluminum stock using the 31 great circle model as a reference and calculations by Donald Richter. The strips were labeled and marked where they overlapped one another. Then the strips were laid out on the ground in the order they would assume in the final dome. Figure 2.7 shows Fuller in his Black Mountain studio (a) and in the field (b) supervising the dome's construction. In (b), Fuller is in the background dressed in a white shirt, the painter Elaine de Kooning is to his right, and Albert Lanier, an architecture student, is to his left.[26] The great circle reference model is clearly visible in both photographs. Notice the arrangement of strips in (b) and how they approximate their position in the final dome.

To no one's surprise, the dome collapsed as soon as it was lifted from the ground! It simply could not support its own weight. Though strong in tension, the thin-gauged slats were not very stiff in compression. They just buckled under the load. Fuller attributed the

---

[25] (Marks 1973, 60)

[26] Left photograph of Buckminster Fuller by Hazel-Frieda Larsen Archer, who documented her years at Black Mountain College in the 1940s and early 1950s. Both photographs courtesy of The Estate of R. Buckminster Fuller.

(a)   (b)

Figure 2.7. Fuller's first geodesic dome construction.

collapse to poor-quality material, but the LCD's geometry wasn't helping, either. The *arc lengths* within the LCD triangle varied in length by as much as 2:1. A different grid with shorter and more equal arc lengths as well as a smaller radius dome might have worked.

Albert Lanier said, "We worked like the devil all summer and waited for the dome to rise like the second coming of Moses, but it laid there like a bowl of wet spaghetti."[27] The class and school were deflated, but Bucky got all excited. He told the students, "Failure is a part of the process of inventing, and success is achieved when one stops failing."[28] Elaine de Kooning later dubbed it the "supine dome (lying on the back, face up)." The setback was only temporary, however; Fuller knew he was on the right track. The next year, the dome was a huge success.

That winter, Fuller continued his geodesic work with students at the Institute of Design in Chicago but returned to Black Mountain's 1949 Summer Institute Program. This time "twelve disciples" followed him back. Among them was Donald Richter, a skilled mathematician and designer who had done the calculations on the supine dome. Fuller intended to work on another concept that summer, the autonomous dwelling facility. Under Fuller's direction, another dome was built, this time with better materials. Its layout once again followed the 31 great circle model. They also experimented with insulating it with a transparent plastic covering. Other experiments that summer tried to mold a geodesic dome by making spherical sections in fiberglass, but the materials did not cure properly and did not work out. Despite mixed results, 1949 was a very successful year for Fuller. He understood the 31 great circle framework better, and he had completed enough tests to know what materials and joinery were best. Now he was ready for a full-scale project. This time, the project would be built in Montreal.

---

[27] (Thomas Zung 2011, pers. comm.)
[28] (Marks 1973, 71)

Fuller built his first large-size geodesic dome, 50 feet in diameter, in Montreal, with the help of three former Institute of Design students. Once again, Donald Richter, with the help of Jeffrey Lindsay and Ted Pope, did the calculations. This innovative dome was made of metal tubes, wire bracing, and covered in Orlon fabric.[29] It took 48 man-hours to erect and weighed approximately 1,140 pounds. A similar design was exhibited at the Museum of Modern Art in New York the following year. Richter would go on to design other domes. It was clear that he would be one of the most skilled geodesic designers ever, even in these early days. We will see his design work for Kaiser Aluminum and Chemical Company shortly.

2.7 NC State and Skybreak Carolina

In 1948, North Carolina College of Agriculture and Engineering in Raleigh, later renamed North Carolina State University, was in the process of establishing a fully accredited architectural program. Henry Kamphoefner, an architectural history professor at the University of Oklahoma, was hired to be the new dean of the program, which would later be called the School of Design (SOD). Kamphoefner, in turn, recruited two others from the University of Oklahoma, James Fitzgibbon and Duncan Stuart. At NC State, Kamphoefner was to build one of the most influential schools of design in America offering accredited programs in architecture, landscape architecture, and graphic and industrial design. Fitzgibbon and Stuart were part of that dynamic faculty and would teach classes in structural concepts and geometry for design, among others. Both would make major contributions to the development of geodesics.

Fitzgibbon was building a house for a client in Knoxville in the summer of 1949. He detoured going to Knoxville from Raleigh and went to see what Black Mountain College was all about. He met Bucky, saw the projects he was working on, and was much impressed with his vision and geodesic experiments. He told Bucky about the new School of Design in Raleigh.[30] As part of his plan to modernize NC State's curriculum, Kamphoefner planned to invite esteemed guest lecturers. Fitzgibbon, along with Duncan Stuart and Manuel Bromberg, recommended they invite Fuller. Fuller's first lecture at NC State was in March 1949.

Fuller was not well-known at the time. Some faculty members felt that he was not really an architect, and questioned whether he was qualified to lecture their students on design. The engineering faculty had their doubts about Fuller too. The prevailing philosophy of design, best characterized by the writings of renowned architectural historian, Lewis Mumford (another SOD faculty member at that time), was that architecture created spaces where people would be brought together in secure and permanent settings that were "of and fixed to the earth" as it were. Fuller believed that buildings should be lightweight, portable, and use materials in tension. They should be fabricated as pop-up tinker toys. They could be manufactured elsewhere and plunked down anywhere. These ideas were not part of the design vernacular at the time. Fuller's lectures on autonomous living packages showed that you could go anywhere you wanted *and* take your house with you; synergetic geometry and geodesic domes were literal examples of thinking outside the box.

---

[29] Orlon is a trademark for an acrylic fiber and woven cloth resistant to sunlight and atmospheric gases and commonly used for awnings and other outdoor uses.

[30] Fuller's work appealed to Fitzgibbon on many levels. In addition to Fuller's fresh approach to synergetic design and geodesic construction, Fitzgibbon had a life-long interest in "ephemeral" architecture, portable structures built by nomadic tribes or used by fairs or circuses. The portability of geodesic domes perfectly fit his interests. Fitzgibbon later wrote several books on the subject. (Brook 2005, 38)

Fuller's marathon eight-hour, run-on lecture style did not impress the entire faculty. Some characterized his lectures as semicoherent, unedited streams of thought. Fuller freely associated his ideas and moralized about the future of man and the destiny of designers. He mixed incomplete concepts in geometry with spinning particles, packed spheres, and natural laws, and delivered them in long, run-on sentences. For him, there were design imperatives such as housing mankind in the most equitable and efficient way possible. Manuel Bromberg, a faculty supporter who attended these sessions, describes him this way: "Bucky was highly intuitive and seductive with an audience. With thick glasses and fleeting stares, he seemed to quickly find the soul of people he was with. He seemed to be a thinker and con artist at the same time, creating an atmosphere of 'we're off to see the wizard.'"[31]

Although some faculty members had their doubts about Fuller, he was an instant celebrity with the architecture students. Bucky's lectures quickly took over the school. Students quit attending other classes just to hear him speak. Faculty attitudes would change as Fuller's ideas gained acceptance slowly over time. So much so that NC State University would award him his first honorary degree in design in 1954.[32]

Fitzgibbon was convinced that Fuller's basic design philosophy was correct, and became one of Fuller's business partners in several companies. In 1951, along with Stuart, Bromberg, and several other School of Design faculty members, he formed Skybreak Carolina Corp.,[33] as a way to develop new ideas about the builtenvironment and keeping contact with Fuller (who only occasionally visited Raleigh). There were no paying customers and no company mission *per se*, but Skybreak founders saw many possibilities with synergetic geometry, and the company gave them a forum for their creativity. This unusual collective of talented architects, painters, sculptors, and engineers was essentially the first geodesic start-up company. Its interdisciplinary mix would become the norm at many of Fuller's future offices.

Fitzgibbon, a highly talented architect in his own right, wanted to build a 30-foot-diameter prototype dome. He and Duncan Stuart saw the prototype as a way to test geometric layouts, understand the tolerances required when building a dome, try different ways of prefabricating, and experiment with ways of weatherizing domes. They also saw value in having a demonstration model they could show prospective clients, if they ever had a client.

With the help of students from NC State's School of Design,[34] a dome was built in Fitzgibbon's backyard. It was made of inexpensive 3/8-inch by 2-inch Carolina pine wood slats bolted together in large triangulated sections they called rafts.[35] Figure 2.8 shows the lattice dome before it was enclosed.[36] The structure was a 16 frequency subdivision of a spherical icosahedron and it was surprisingly rigid, yet lightweight. The cluster of semicomplete triangles in the center of the photograph is where the apex of five rafts meet and

---

[31] (Bromberg 2004, pers. comm.)

[32] Buckminster Fuller was awarded 47 honorary degrees in his lifetime.

[33] Skybreak Carolina Corp., Raleigh, NC, 1951, founders: James Fitzgibbon, Duncan Stuart, Manuel Bromberg, Roy Gusso, George Matsumotoa, and William Parkhurst.

[34] Two students would become notable architects. T. C. Howard became the principal architect and geodesic dome designer in Synergetics, Inc., and Bruno Leon became a practicing architect and dean of the School of Architecture, University of Detroit.

[35] A raft corresponded to a large diamond symmetrical to each edge of the icosahedron. All rafts had the same triangular geometry except for those rafts that were cut off at the ground.

[36] Image used with permission, NC State School of Design. Photo of wood lattice dome built by Fitzgibbon and students from School of Design taken by Mark Mills in 1951. Appearing from left to right: James Fitzgibbon and Duncan Stuart. See Student Publications of the School of Design, 1952.

Figure 2.8. Lattice dome.

form a pentagonal pattern; all others are hexagonal shapes around any vertex. Once the frame was completed, Fitzgibbon considered ways to weatherize it. He wanted to experiment with "cocooning," a spray process developed by the US Navy to preserve mothballed ships after World War II. It encased objects in a rubber-plastic neoprene material making them weatherproof. His idea was to apply this neoprene material over chicken wire nailed to the slats. The process worked well, and the dome, when it was completed in January 1952, weighed only 1,000 pounds, a featherweight by any construction standard.

Stuart was not satisfied with the design or prefabrication of the lattice dome. The lattice grid was based on modifications he had made to Fuller's 1948 three-way regular grid, but it was not ideal. Each raft required too many different and individually cut triangles and many arcs did not meet exactly at a common grid point. Bucky was aware of the regular grid problems and had discussed it with Stuart. Shoji Sadao, an early pioneer in geodesics and a longtime business associate of Fuller's, described the grid problem this way:

> There had always been a nagging problem with the regular grid—it had windows. That is to say, at certain vertices the three great circles did not go through a common point. Calculations were checked and rechecked, but the windows persisted to the point where Fuller thought it was a message from God that the trigonometric tables had an error in them! A copy of "Tables of Sines and Cosines to Fifteen Decimal Places at Hundredths of a Degree," published by the US Department of Commerce, National Bureau of Standards,[37] was used, to no avail.[38]

Stuart was a skilled mathematician with an extraordinary sense of 3D relationships. He could visualize and render the most complex spatial arrangements (one of Stuart's

[37] (US Department of Commerce 1949)
[38] Interview "A Brief History of Geodesic Domes" by Shoji Sadao, (Zung 2001, 25).

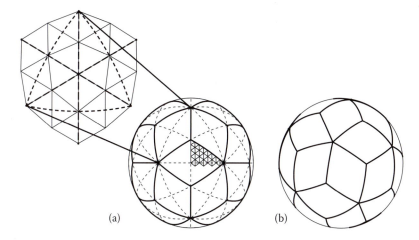

Figure 2.9. Triacon subdivision.

freehand sketches appears in Section 3.15). Stuart's dissatisfaction with the lattice dome and the regular grid made him rethink the whole approach to spherical subdivision. There was no problem with Bucky's 15 great circle icosahedron reference. The problem was how to achieve consistent gridding between these reference great circles. He was looking for a geometric layout that would minimize the number of different *chord lengths* and angles between members. Up to this time, every dome prototype started with Bucky's 15 great circle icosahedron reference but further subdivided it in slightly different ways. No two domes used the same method and even with high precision trigonometric tables, some grid members did not intersect at exactly the same point, creating small triangular windows, the ones that Sadao commented on above. Stuart noticed that if the edge of the spherical icosahedron was evenly subdivided (see Figure 2.9(a)), a set of repetitive and nested spherical right triangles could be defined and when those triangles were further subdivided, they produced a very uniform grid. Stuart worked out this new geometry for a series of increasingly dense grids and compared the results to previous methods. His analysis showed conclusively that the denser the subdivision grid, the more advantageous the new system was. Stuart wrote to Bucky in June 1952 and described his new system.[39] He also sent him a report detailing the new method and comparing its results to the previous method. His new system offered[40]

- a higher order of symmetrical relationships—rendering the system more advantageous structurally;

- a great simplification in terms of the number and kinds of parts, a greater simplicity in determining the relationships or the various parts;

- more advantageous diameter-frequency relationships; and

- more favorable erection conditions when the system is applied to structural problems.

[39] (Stuart 1952), unpublished letter to Buckminster Fuller.
[40] (Stuart 1953), unpublished technical Skybreak report.

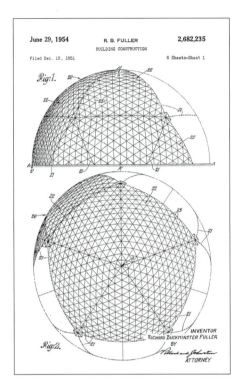

Figure 2.10. Buckminster Fuller's geodesic dome patent.

Stuart named his new breakdown *triacon* because the diamond pattern resembles the diamond faces of the rhombic triacontahedron, shown in Figure 2.9(b), another polyhedron closely related to the icosahedron.[41] Stuart's triacon was the breakthrough Fuller needed. Although the triacon was related to earlier gridding systems, it was unique in its approach and made geodesic domes practical to build. Domes using this new method would be much easier to calculate (right triangles are the easiest to solve), easier to prefabricate, and cheaper to build. Stuart saw that the repetitive diamond pattern offered additional possibilities for prefabrication. The triacon method was so key it would be employed on almost every major project in the 1950s. Today, the triacon subdivision method is one of several breakdowns where the resulting grid members essentially run perpendicular to the edges of the spherical polyhedra on which they are based. Grids such as this are categorized as Class II grids. But Stuart's triacon method is the only Class II grid that minimizes the number of different parts; it is detailed later in Chapter 8. Stuart was one of Fuller's indispensable resources and made many more contributions to dome projects in the 1950s. Although Fuller depended on his skills, he never recognized or credited Stuart for his work.[42]

Based on his geodesic experiments at Black Mountain, successful prototypes in Canada, the Pentagon Garden dome in Washington, DC, and in Lawrence, New York, Fuller

---

[41] The rhombic triacontahedron is the *dual polyhedron* of the icosidodecahedron. See Section 6.4.

[42] Stuart's contributions were not the only ones overlooked by Fuller. Press coverage often attributed projects to Fuller when, in fact, it was the indendent work done at one of Fuller's franchised offices. See (Chu et. al 2009) for an excellent analysis of the R. Buckminster Fuller archives at Stanford University.

filed for US patent protection in 1951. His building construction patent describes a general method for subdividing spheres and creating a structural truss work. In 1954, he was awarded patent 2,682,235, shown in Figure 2.10. Patent claims 3, 6, and 7 state the essence of the invention:[43]

3. A framework of generally spherical form, in which the main structural elements form a substantially uniform overall pattern of great circle arcs intersecting in a three-way grid.

6. A building framework of generally spherical form, in which the longitudinal centerlines of the main structural elements lie substantially in great circle planes whose intersections with a common sphere form grids comprising substantially equilateral spherical triangles.

7. A building framework constructed in accordance with claim 6, in which the main structural elements are interconnected to form a truss, the outermost points of which lie substantially in a common spherical surface.

It is interesting to note that in its application review, the US Patent Office found only one remotely related prior patent by another inventor.[44] This suggests just how novel and unprecedented Fuller's basic geodesic dome invention really was.

Fuller's patent announced the geodesic concept to the world and ensured Fuller the commercial rights to the idea, if it ever caught on. Stuart's triacon system gave designers at Fuller Research Foundation and Skybreak the confidence that geodesic building construction was practical. It also made Raleigh the epicenter for geodesic development.

2.8   Ford Rotunda Dome

The Ford Rotunda dome was Fuller's first real commercial project and his patron was none other than Henry Ford himself. The Rotunda was originally built as an exhibit building for the 1933 Chicago World's Fair. After the close of the fair, the building was taken apart, shipped to Dearborn, Michigan, and reassembled on Rotunda Drive, across from Ford's Central Office Building, his world headquarters at the time. After remodeling the building, it was opened to the public, and was a huge tourist attraction until the outbreak of World War II. After the war, and in preparation for Ford Motor Company's fiftieth anniversary celebration in 1953, the Rotunda again would be remodeled and reopened to the public. Plans called for covering the interior courtyard.

Various roofing schemes were considered, but all proved impractical. The Rotunda walls were not strong enough to support the weight of long span steel roofing systems estimated to weigh as much as 160 tons. Ford wasn't satisfied, and asked his engineers to contact Buckminster Fuller. Bucky considered the requirements and proposed covering the 93-foot circular opening with a lightweight aluminum truss, a geodesic dome, which he said would weigh only eight tons. The dome would incorporate a new lightweight truss system he called octet truss, which combined tetrahedral and octahedral structural units made of light extruded 5-ounce aluminum struts into a rigid frame. Ford liked the idea and signed a contract in 1952.

---

[43] (Fuller 1954, sheet 1/6, figs. 1–2)
[44] (Pantke 1930)

(a)   (b)

Figure 2.11. Ford Rotunda dome.

Fuller's associate, T. C. Howard, designed a hemispheric dome that required almost 20,000 yard-long struts. Each one was cut and marked with color codes. Workers matched the ends of struts with the same color codes and preassembled them into 160 large triangular octet truss subassemblies; each subassembly was about 12 feet on a side. In Figure 2.11(a), Fuller holds Howard's model of the dome; the subassemblies are clearly visible as darker triangles inside the model. A truss subassembly is made in the following way: three struts formed a triangle, and four triangles made an octahedral unit such as the one pictured behind Fuller in Figure 2.11(a). Ten octahedral units form a single truss subassembly. Each weighs only about four pounds and is easily handled by a single person.[45] The dome was assembled from the top down. A central mast elevated it much like an umbrella as workers added sections to its edges. They worked from a wide bridge platform above the court. In Figure 2.11(b), workers apply transparent polyester fiberglass panels to weatherize the dome.[46] The completed dome required 720 man-hours to erect and weighed approximately 17,000 pounds, or about 2.5 pounds for every square foot covered.[47]

The entire Rotunda dome project was designed, manufactured, preassembled, and installed in just four months, two days ahead of schedule. The Rotunda was open to the public again in June 1953 and quickly became one of the top ten tourist destinations in the country at the time. And Bucky's dome was in plain sight.[48]

The Ford project marked another turning point for Fuller and his associates. It was their first commercial project, and it was a resounding success. There was another first: it was the first large dome project to use a new spherical breakdown based on the icosahedron's mid-edge great circles, shown dashed in Figure 2.12. Unlike the triacon, in which grid members run perpendicular to face edges, the Rotunda's grid would run parallel to the

---

[45] (*LIFE* 1953, 67–70)

[46] Both images courtesy of The Estate of R. Buckminster Fuller.

[47] For an excellent collection of architectural drawings detailing the entire design, see (Fuller 1985, 96–151).

[48] These roofing panels were a constant source of leaks and eventually led to the demise of the dome. Workman heating a tar mixture for patching inadvertently set fire to panels, destroying the entire structure, November 1962.

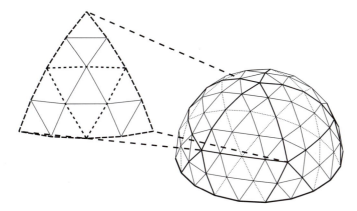

Figure 2.12. Ford or alternate breakdown.

edges. The Rotunda dome divided each face of the spherical icosahedron into 16 subfaces, and each subface was covered by a planar octet truss structure. The subdivision technique was aptly called the *Ford* or *alternate* (alternate to the triacon breakdown). Today the technique is grouped with similar ones and they are collectively known as *Class I* breakdowns. They are detailed in Chapter 8.

2.9   Marines in Raleigh

The late 1940s marked the resurgence of the Soviet Union as a post–World War II superpower and the beginning of a Cold War with the United States that would last more than 40 years. In the years to come, battles would take place on many fronts—propaganda, espionage, weapons development, industrial advances, and even in outer space, where the two superpowers aimed to prove their technical superiority in a "space race." In the 1950s, unrestrained nuclear testing and increased proliferation made an all-out nuclear war between the United States and Russia appear inevitable. Every branch of the US government was engaged to counter the threat. The US Marine Corps and Air Force were no exception—and they were taking notice of geodesic domes.

In 1953, Colonel Henry C. Lane, USMC Head of Aviation Logistics, visited NC State to interview James Fitzgibbon about the new geodesic construction system. He was looking for solutions to a wide variety of rapid deployment shelters—personnel shelters, aircraft hangars, and storage facilities, to name a few. Lane noticed a model of the lattice dome (Figure 2.8) in Fitzgibbon's office and asked about it.[49] Fitzgibbon told him how strong and light geodesic domes were, that they could span large distances without column support, and that they were very portable. He went on to explain how they can be built of many materials such as metal, wood, cardboard, or plastic, and that they can be weatherized in many ways—through cocooning, for example. They had very high strength-to-weight ratio, which meant that even a large dome was light enough that it could be built off-site and airlifted into place. They were easily prefabricated in pieces and could be shipped anywhere.

Lane was completely sold on the concept, its potential for prefabrication, simple construction, and its light weight. But he needed proof. He needed to test it and see how it met

---

[49] (Bromberg 2008, pers. comm.)

the Corps' many logistical needs. Shortly after, Lane arranged for a Marine Corps helicopter to lift the lattice dome out of Fitzgibbon's backyard and fly it across a nearby Catholic church. This made front-page headlines as far north as New York, where the *New York Times* took note. So convincing was this demonstration that Lane authorized a contract for further development.

Lane's contract with Skybreak was a turning point in geodesic dome development, but the next step wasn't so clear. Lane knew what he needed to know, but Skybreak did not have enough resources to develop all of the prototypes he wanted. Fuller recognized this too, so parallel efforts were started that involved a number of universities. In 1954, the same years Fuller's geodesic dome patent was issued, Fitzgibbon and T. C. Howard (a graduate of NC State who had worked in Fuller's New York office) teamed with Fuller to form Geodesics, Inc. A year later, the same team along with others formed a second company, Synergetics, Inc. The former developed military projects; the latter did commercial work. Fitzgibbon acted as office manager for both companies, and Howard became the principal architect and designer.

2.10 University Circuit

In the late 1940s and early 1950s, Fuller traveled extensively and lectured at many universities. His reputation was spreading, and his network of associates rapidly increased. Under pressure from the Marine Corps contract, he reached out to his university contacts and engaged a number of them in research and development. His format was similar at each institution. He would use the same seminar and hands-on approach that had proved successful at Black Mountain College. He taught the basics of synergetic geometry and then defined a dome problem the university team was to address. Projects differed in size, materials, joinery, and enclosing method. After starting up a project, he would move onto the next one, picking up ideas from one and dropping them off elsewhere. Some said he often left the impression the borrowed ideas were his own.

The cost of materials in most cases was quite low. Many prototypes were built in corrugated paperboard reinforced and water proofed with polyester resin, or used simple wood struts with wire tension members. Stuart's triacon was the favored spherical subdivision method. By February 1954, Tulane University had completed a one-sixth scale model of a six-plane hangar dome 18 feet in diameter; the full scale would have been 108 feet in diameter. In rapid succession, prototypes were completed at the University of Michigan, Virginia Polytechnic Institute, North Carolina State College, and the Marines Corps Schools in Quantico, Virginia. Domes ranged in size from a small 14-foot-diameter, six-person personal shelter made of cardboard to a large 50-foot-diameter hangar dome with a magnesium frame that was light enough to be moved 50 miles in 60 minutes without disassembly. Figure 2.13 shows its erection sequence. It could hangar any plane or helicopter the Marine Corps owned except conventional transport aircraft.[50] These pilot projects tested a wide variety of sizes, materials, and techniques. Some newspapers were carrying stories that Fuller's cardboard domes would soon replace the army's standard-issue tent.[51] Although some domes encountered problems, especially in the detailing, these demonstrations proved that the geodesic principle was essentially sound.

---

[50] (Lane 1955, 51–52, 54)
[51] (*New York Times* July 18, 1954)

Figure 2.13. Marine Corps hangar assembly.

At the same time the universities were developing prototypes, Geodesics, Inc. was designing and building three 30-foot-diameter domes made of cardboard. The Marines erected them at their Quantico post in Virginia, next to the Potomac River. Geodesics' project manager, Bruno Leon, recalls that the marines disregarded instructions to anchor the domes. The wind blew them into the Potomac River, but the enclosure concept was successfully demonstrated.[52]

By the end of the year, Lane had his proof points and enough backup data to recommend that the Marine Corps replace conventional tent structures with geodesic dome-type shelters and use the 55-foot dome for aircraft shelters wherever there was no available substitute.[53] He also recommended that the Marine Corps continue its research and development in geodesic domes.

The Marine Corps project made beachheads out of Fitzgibbon's backyard experiment. The Marine Corps went on to deploy hundreds of domes worldwide. Geodesics, Inc. would go on to design many more military projects, not all of which were domes.

2.11   Radomes

The Marines were not the only ones fighting the Cold War with geodesic domes. In 1954, President Eisenhower signed a bill authorizing the construction of the Distant Early Warning Line (DEW Line) system, an "over the pole" line of radar stations operated by the US Air Force to detect airspace intrusions. In one of the most hostile environments on Earth, with temperatures as low as −30°F and winds as high as 150 miles

---

[52] (Leon 2004, pers. comm.)
[53] (Lane 1955, 114)

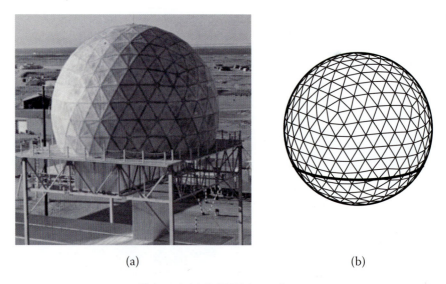

(a)   (b)

Figure 2.14. DEW Line radome.

per hour, the government built 58 radar stations across a 3,000-mile stretch of Alaska and northern Canada between 1955 and 1957. The Air Force required radar enclosures to be knocked down, flown to sites, and set up in 20-hour windows of predictable good weather.[54] They had to withstand a 210-mile-per-hour wind safety factor and not use any metal, since metal reflects the radar's beam. Fuller's fiberglass-reinforced plastic domes were feasible solutions.

In 1954, William Ahern, Bernard Kirschenbaum, Sherry Proctor, and William Wainwright of Geodesics, Inc., an office affiliated with Fuller in Cambridge, Massachusetts, designed a 55-foot-diameter 3/4 sphere that stood 40 feet high.[55] It was made of fiberglass-reinforced plastic and was based on prototypes developed earlier at the Massachusetts Institute of Technology. This dome weighed 12,000 pounds and was tested to withstand 150 mile-per-hour winds. The structure consisted of 363 glass-reinforced plastic components (polyester resin) that resembled giant diamonds and circle cake pans. Cake pan flanges were molded with epoxy resin molds and pans were bolted together to form the sphere and all parts were interchangeable. An outer perimeter edge of both diamond and circular pans were recessed to receive rubber gaskets, which sealed the joints between the diamonds and circular pieces.[56] A single radar dome, or *radome* as they were called, could be assembled in only 14 hours. At the time, radomes were the largest plastic structures ever built. Figure 2.14 shows a variation of the Geodesics, Inc. dome in service along the DEW Line in Alaska[57] alongside the icosahedral reference geometry that was used.

DEW Line radomes used the Ford breakdown method again, but instead of dividing the icosahedron's edge into four segments—a four-frequency subdivision—DEW Line domes such as the one shown in Figure 2.14(a) divided them into six. Other frequencies were also used. The dome support geometry was also different. Unlike the hemispheric Ford

[54] (Marks 1973, 61)
[55] The Geodesics, Inc. Cambridge, Massachusetts, office would later reincorporate and become Geometrics, Inc.
[56] (Geodesics, Inc. 1955)
[57] FOX-Main Radome 1962, photo courtesy of Brian Jeffrey, www.VE3UU.com.

Rotunda dome, DEW Line radomes were 3/4 domes, thus the support structure's ground plane intersects the sphere as a lesser circle, not a great circle, as can be seen in the diagrammatic view of a six-frequency grid in Figure 2.14(b). Bill Wainwright later modified this grid to eliminate the need for the numerous small nonstandard panels around the base. His modified grid is called the parallel truncatable grid and it greatly reduced fabrication and erection costs of radomes.

DEW Line domes were strong, weather resistant, and remained in service for many years, demonstrating, once again, the adaptability of geodesics to extreme conditions, innovative construction, and new materials, such as fiberglass.

Radome spherical breakdowns would evolve very differently than methods used for traditional domes. Two types of radomes became commonplace: air-supported radomes and self-supported radomes. Air-supported radomes are simple structures with fewer frame members. Additional support comes from having a pressurized interior, something like a slightly inflated ball. Self-supporting radomes, such as those used on DEW Line, have more structural frame elements because their frame supports the entire weight of the structure.[58] Self-supporting radomes were becoming larger and larger, and radar itself was changing. The higher frequency of subdivision and the inherent regularity of geodesic geometry were interfering with the radar's signal. A new breakdown approach was needed to eliminate long sequences of chords that approximated large great circle arcs. To overcome the interference problem, designers developed "randomized" layouts. So-called random domes are somewhat of a misnomer, as many random domes are actually based on highly regular spherical grids similar to the ones we have already seen. But instead of defining chords between closest neighboring grid points (the usual practice), triangles or diamonds are defined by selecting other combinations of vertices. Some domes also mixed isosceles and equilateral panels in a set pattern. Either way, the near-equilateral pattern we associate with geodesic domes is partially or totally eliminated, yet enough symmetry remains in the overall pattern that sets of identical panels are still used to build the dome. Figure 2.15 shows two approaches to randomization. Despite their appearance, they can be prefabricated from

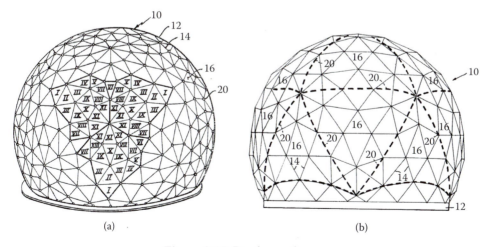

(a)   (b)

Figure 2.15. Random radomes.

[58] (Ahern 1968, 1)

standard, if irregular, components. A 1950s patent by Albert Cohen, Figure 2.15(a), uses 11 different panels for a 150-foot-diameter radome. Another design from the 1960s, Figure 2.15(b), by Bill Ahern and Bill Wainwright (who were part of Fuller's early prototype work at the Massachusetts Institute of Technology in the early 1950s) is based on four different panels.[59] In 1977, Ahern and Birgit Mathe would again refine this layout. They discovered that for moderately sized domes, only four panel sizes are required, whether or not the interior surface was the same or different from the exterior surface. Their layout was ingenious and an amazing demonstration of spherical trigonometry.[60]

In the early days of radome development, designs patented by companies or individuals not licensed to use Fuller's basic geodesic dome patent 2,682,235 presented potential intellectual property violations. Their "randomness" somewhat disguised their underlying geometry and staff at Fuller's Geodesics, Inc. office reviewed patent claims to see if the radome actually fell within Fuller's patent or not.[61]

Randomized radomes are highly sophisticated geodesic applications. Despite their appearance, many use a basic polyhedral framework and generate their grids by intersecting great circles. In most cases, the result is still a highly symmetrical subdivision of the sphere.

2.12   Kaiser's Domes

Donald Richter had met Fuller at the Institute of Design in Chicago and did calculations for the supine dome at Black Mountain as well as the first large-scale dome in Montreal in 1950. These projects, and several others like them, led Richter to develop a simple gridding technique he called the alternate grid. We will explain how it works in a later chapter. He was now working as a designer at Kaiser Aluminum and Chemical Company, and once again, a chance encounter with a geodesic dome model would alter the course of geodesics.

As the story goes, the famous industrialist Henry Kaiser was passing through his company's offices and noticed a model of a dome in Richter's office.[62] His first impression was that it was one of Kaiser's new products. But when the industrialist asked Richter what it was, Richter explained that it was a geodesic dome. Richter then proceeded to point out the virtues of the dome, i.e., high strength-to-weight ratio, suitability for prefabrication, and lightweight aluminum as an ideal material to build them. Kaiser, just as Colonel Lane had been, was completely sold on the idea. He saw enormous potential in producing aluminum domes, and immediately engaged company lawyers to make Kaiser Aluminum one of the first licensees of Fuller's geodesic patent.[63]

Kaiser was one of the earliest and biggest boosters of the Hawaiian tourist industry. He asked Richter to design a dome that would cover the auditorium for his new Hawaiian Village Hotel in Honolulu. Richter's design was an ingenious 49.5-foot-high, 149-foot-diameter dome that combined lightweight curved aluminum panels with aluminum struts. It was shipped to the resort and reported to have been erected in just 20 hours in 1957. Figure 2.16 shows Richter's basic design.[64]

Richter, like Duncan Stuart, had a keen sense of 3D geometry and the mathematical skills for the required calculations. His geodesic design was unusual in many ways. In most

[59] Figure (a) (Cohen et al. 1961, sheet 2/4, fig. 2); Figure (b) (Ahern and Wainwright 1968, sheet 1/4, fig. 1)
[60] (Ahern and Mathe 1977, sheet 1/5, fig. 1)
[61] (Howard 1958)
[62] (Fuller and Zung 2002, 36)
[63] Richter recounts "Working with Buckminster Fuller" in (Fuller 1985, 381–394).
[64] (Richter 1962, sheet 4/8, fig. 5A)

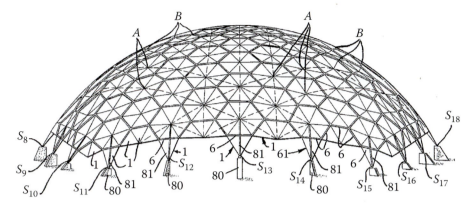

Figure 2.16. Elevation view of the basic Kaiser dome.

geodesic domes, struts alone distribute the compression and tension loads from the dome's own weight and external sources (wind, snow, rain, etc.) to the dome's foundations. But Richter also transferred loads through a membrane made of aluminum sheets, the same ones that weatherized the dome. The Kaiser dome and another dome, the Union Tank Car dome, which we will see next, both used the same structural membrane technique. But unlike Union Tank Car's inner layer of steel plates, outer layer of pipes, and in-between layer of rod bracing, Richter's design bent diamond-shaped sheets of aluminum and added a crossbar, also of aluminum, to form a 3D unit that resembled a tetrahedron. A typical unit is shown in Figure 2.17(a).[65]

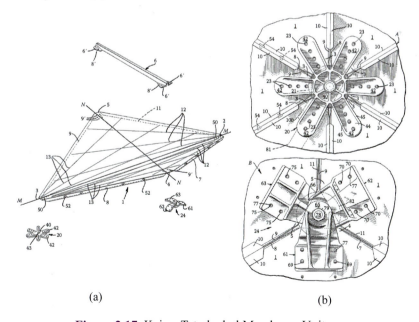

(a)   (b)

Figure 2.17. Kaiser Tetrahedral Membrane Unit.

---

[65] (Richter 1962, sheets 2 and 6/8, figs. 3A, 10, and 11)

The sheets of each diamond panel were lightly impressed with creases radiating from the vertices at the ends of the long diamond axes when sheets were cut and curving. These creases did not stretch or weaken the metal. Instead, they stabilized it, preventing the so-called "oil canning" effect that sometimes happens to sheet metal when it flexes or warps after a temperature change or the application of other loads. An aluminum crossbar spanning the narrow axis of the diamond is riveted to the sheet metal and completes the tetrahedral pattern. When the panels are assembled to make the dome, cast aluminum hubs, as shown in Figure 2.17(b), join the corners of adjacent panels. The fact that panel edges were bent, making them even stiffer, and riveting adjacent panels together weatherized them as well. A construction worker could easily handle a panel without using any equipment. The combination of metal membrane and lattice of struts formed an extremely rigid, weathertight integral structural system when the dome was completed.

2.13   Union Tank Car

By far, the most technically complex and highest risk geodesic project Fuller and the Synergetics team, lead by T. C. Howard, faced in the 1950s was the Union Tank Car Company dome built in Baton Rouge, Louisiana. The dome was 120 feet at its zenith and 384 feet in diameter, and would cover 110,000 square feet. A 200-foot paint tunnel, 20 feet high and 40 feet wide, was also part of the structure. When completed in 1958, it was the world's largest dome and free-span structure.[66] Figure 2.18(a)[67] gives an overall view of the massive structure.

The dome was approximately a one-quarter sphere, constructed of 321, 11-gauge (.125-in or 1/8-in) steel plates with an outer truss of 4-inch-diameter pipe struts braced with 3/4-inch-diameter tension rods. The plates were edge-welded to weatherize the dome's interior. Together, the outer and inner layers acted as a huge double-layer spheri-

(a)   (b)

Figure 2.18. Union Tank Car Company dome.

---

[66] The domes at St. Peter's Basilica in Rome, Pantheon of Ancient Rome and the Duomo in Florence are 136, 142, and 130 feet in diameter.
[67] Photo courtesy of RIBA Library Photographs Collection.

cal truss system approximately 4 feet deep. Figure 2.18(b)[68] shows how the plates, pipe struts, and tension rods form a double-layer truss structure.

What is particularly remarkable about the dome is that all fabrication and subassembly was done on site. Large sections were built on the ground. Steel plates, pipes, and tension rods were preassembled on fixtures to ensure accurate layup. Some fixtures are visible in the bottom left of Figure 2.18(a). The sections, some weighing as much as 4,000 pounds, were then hoisted into place by mobile cranes. Some cranes are visible in the background of (a). The structure was self-scaffolding; that is, throughout its construction, it always formed a structurally stable ring and did not call for any bracing. Temporary ground supports positioned subassemblies at the moment subassemblies were joined to alreadycompleted sections. The general construction sequence was from the ground up. Subassemblies were continuously added to the free edge as the dome progressively reached its final 384-foot span. This structure was a triumph of design and construction excellence for Synergetics, Inc. T. C. Howard was the principle architect. It was the first major dome to use the triacon subdivision method to create a double layer of hexagonal-octagonal grids, as seen in (b). The inner layer was made of steel plates, with the outer layer being constructed of compression pipes. Steel tension rods were placed between the outer and inner dome layers. The two layers, combined with the tension rods, formed an octahedral pattern. Recognizing the enormous strength and light weight this configuration offered, Howard would use variations of this octahedral design on future projects.

More than 20 domes were built in the 1950s, and many more followed in the 1960s. We have only touched on a couple of key ones long enough to see the origins of the alternate and triacon subdivision techniques; they would join other techniques in a formal classification system presented in Chapter 8.

The Union Tank Car project was not only the most challenging dome at the time, it was the first dome project to use a computer to calculate the geometry. In the spring of 1957, the hexagon wheel octahedral truss conceptual geometry for the dome had been approved by the client[69] and design detail work at Synergetics, Inc. had begun. At this point, every calculation was performed by hand. Synergetics had developed standard office forms for computing spherical triangles and they had implemented a system of double checks. They also made detailed 1-inch to 1-foot scale paper and wood models of critical jigs and assembly models to verify dimensions. By the summer of that year, Synergetics' staff was under tremendous pressure to accelerate design work and provide the shop drawings needed for start construction. It was at this point that the first computer program for spherical design was written; CAD had made its first appearance.

2.14   Covering Every Angle

Today's PCs and pocket calculators make short work of geodesic calculations, but calculations were a major challenge in the 1950s. Every one of the early geodesics domes was calculated by hand—one triangle at time! Office forms, developed by staff at the Fuller Research Foundation, Skybreak, Geodesics, or Synergetics, ensured that correct procedures were followed when using spherical trigonometry. A completed set of calculation forms for a modest-sized dome might be a stack of paper an inch thick. Forms included

---

[68] Photo courtesy of Tai Ray-Jones.

[69] The hexagonal wheel octahedral truss design was first proposed in 1956 but rejected for use in Kaiser Aluminum Company domes (Fitzgibbon 1957).

classic equations such as the law of cosines or sines or *Napier's rule* and a circle diagram of triangle parts (techniques explained in Appendix A). Forms also indicated where calculated values were filled in, added, subtracted, or multiplied (adding logarithms). Often, the form indicated the number of digits required for a calculation. To add to the tedium, most angles were represented in sexagesimal (base 60) notation; that is, degrees, minutes, and seconds. Seconds were sometimes carried out to six or seven decimal places.

Computers (a job title at the time) would follow the sequence of steps on forms developed by Stuart, Richter, T. C. Howard, Joe Stinborn, Richard Lewontin, and others. A typical form, when completed, would list the triangle's reference number, its central and surface angles, chord and arc factors, and sometimes its area. Some forms indicated how to check the results. This was particularly important because angles from one triangle were often used to start the calculations for an adjacent triangle, thus, any error that crept in would simply spread from triangle to triangle. In some cases, calculations were repeated by another person using different methods when possible. Often, the alternative method checks took more time than the original calculations that tended to use the most efficient method. Stuart was particularly careful in his calculations. He relied on high precision 15-place US Government Printing Office sine tables for trigonometry (see Figure 2.19).[70] He also computed his own supplementary tables to nine decimal places for converting angles in sexagesimal to decimal degrees and vice versa.[71]

By the late 1950s, dome projects were becoming very large and more numerous. They were capturing the public's imagination and numerous companies wanted to use the technology. It's said that it took Don Richter more than a year to work out the calculations for Fuller's Black Mountain 31 great circle model. The Union Tank Car project with its double truss dome would involve significantly more calculation than any previous dome project and it had to be done quickly. The success of geodesics not only depended on proving they were safe, but that they could be built to required tolerances. Fuller was vulnerable now. A project failure, especially in these start-up years, could discredit the entire geodesic concept. On the other hand, continued success would lead to more and more patent licensing,

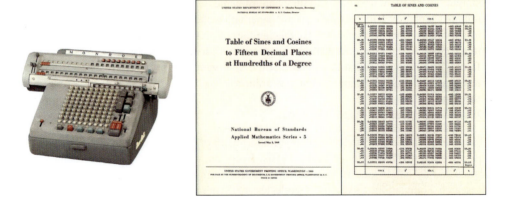

Figure 2.19. Monro-Matic calculator and 15-place sine and cosine tables.

---

[70] (US Department of Commerce 1948)
[71] (Stuart 1951)

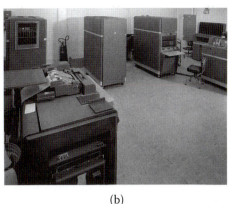

(a)   (b)

Figure 2.20. IBM 650 Magnetic Drum Calculator.

and this required design methodologies other companies could follow. Fast and accurate computation was now essential; better tools and procedures were needed.

High-end Post and Keuffel and Esser slide rules and the CRC standard mathematical tables and formulae[72] were used for some geodesic work, but slide rule accuracy, typically two significant digits or a maximum of three, was not adequate. In 1957, the Synergetics office started using Friden and Monro-Matic electromechanical calculators to compute spherical geometry. Some machines weighed more than 30 pounds.[73] Figure 2.19 shows the calculator and an often-used US Government 15-place sine and cosine tables used for calculation.[74] For sure, the calculator and reference tables sped things up and improved precision. However, the bulk of calculations still fell to human computers. The real breakthrough took place only when computation passed from human to electronic computers.

In 1955, NC State received one of the first IBM mainframes, the 650 Magnetic Drum Calculator. A typical installation is shown in Figure 2.20.[75] The operator in (a) is coding instruction with console switches. In the background is a computer card reader and punching unit. In (b), a tabulating machine for listing punched cards onto office print forms is shown in the foreground. Initially the 650 was used by farm management experts in the agricultural school to find the most profitable combinations of crops and livestock for a particular farm. But designers in Fuller's offices wanted to have it programmed to make geodesic reference tables. The idea was to have a desktop reference for spherical breakdowns, where the values of angles, chords, and coordinates could be looked up when needed. This was similar to the way engineers used the CRC standard mathematical tables.[76]

The IBM 650 was the first mass-produced digital computer sold to nongovernmental customers. It was inexpensive (only half a million dollars), small (it fit in a single room), and user-friendly (it was programmed in decimal rather than binary).[77] Most important, it

---

[72] (Chemical Rubber Company 1955)

[73] (Fitzgibbon October 1957). The Monro-Matic (Model 8N-213 shown) was fully automatic with ten keyboards, 11+10 counter, and 21-place accumulator. Photograph courtesy of John Wolff's Web Museum.

[74] (US Department of Commerce 1949)

[75] Photographs courtesy of IBM Corporate Archives.

[76] (Chemical Rubber Company 1953)

[77] Art Miller, professor emeritus of mathematics and computer science at Mount Allison University, made this characterization.

understood an English-like programming language called FORTRANSIT (FORTRAN Internal Translator), a subset of what was later called FORTRAN (FORmula TRANslation). The first geodesic computer program produced tables for Stuart's triacon breakdown.[78] It is based on the subdivision of a single spherical right triangle, the one shown in Figure 2.9. The angles between each grid member vary somewhat within this triangle, but once they were calculated, all other triangles were the same. Chord and arc lengths depended on the radius of the final sphere, but chord and arc factors can be calculated just once for a unit radius sphere. Actual arcs and chord lengths are found by multiplying each factor by the final dome's radius.

Perhaps the most tedious calculations involved grid point coordinates. Every point has unique coordinates, and even modest spherical subdivisions involve scores of angles and hundreds of coordinates. Having the computer calculate them would be a major step forward, improving accuracy and eliminating tedious, time-consuming work.

In the summer of 1957, Richard Lewontin was in charge of writing a FORTRAN program for the IBM 650 to do just that.[79] His program could be considered the first application of computers aiding design in architecture. For a given subdivision frequency, his program computed the angles, chord, and arc lengths and coordinates for all parts of the triacon right triangle for a spherical icosahedron and punched the results onto paper cards. Tabulating machines listed the card's numeric values on large paper forms. All angles were listed in decimal and sexagesimal (base 60) along with their sine and cosine values. Under considerable pressure to compute all panel dimensions and prefabrication fixtures for the Union Tank Car dome project, Lewontin's program initially produced tables for just 24- and 36-frequency subdivisions. Thanks to the symmetry of the triacon grid, the combination of these two tables also contained the angles and coordinates for 2-, 4-, 6-, 8-, 12-, and 18-frequency grids as well; they are subgrids within the 24- or 36-frequency grids.[80] Finding subgrid angles in a table designed for a larger grid is less convenient because it involves careful look up and additional hand calculations, but computing only a few high frequency tables kept computing costs down and produced the highest priority tables first. Lewontin's program had another advantage: it computed tables for a unit-radius sphere. One table could be used for many different-sized domes, even for domes where there were multiple layers of grids and trussing such as the Union Tank Car. Angles do not change, no matter how large or small the dome is, but chord and arc lengths, as well as vertex coordinates, do change, as the radius of the dome changes and must be rescaled to the actual radius of the sphere being built. Lewontin's program was so successful that by the fall of 1957, tables were being produced for a wide range of frequencies for both icosahedral and octahedral subdivisions and the Synergetics office was offering them or custom computer runs to other Fuller companies as proprietary and copyrighted data.[81]

The Cartesian coordinates for grid points presented another challenge. Their values depended on the sphere's radius and orientation in space and every vertex has different coordinates. This problem was solved by standardizing the orientations domes could take, based on which part of the spherical icosahedron reference polyhedron was uppermost.

---

[78] In 1954, John Backus headed the IBM team of researchers at the Watson Scientific Laboratory that invented FORTRAN, or FORmula TRANslation. FORTRAN was the first high level programming language and eliminated the tedium of programming in machine language. The FORTRAN language was commercially released in 1957.

[79] (Fitzgibbon 1957)

[80] A subgrid is sometimes called a harmonic and they are mathematically related (common factors) to the overall grid.

[81] Fitzgibbon letter to W. M. Wainwright (Geometrics, Inc. 1957).

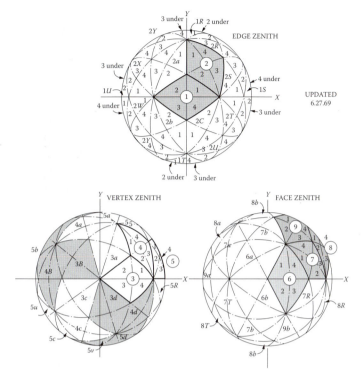

Figure 2.21. Triacon diamond and quadrant locations for spherical icosahedron.

Figure 2.21 shows edge-, vertex-, and face-zenith orientations for the icosahedron's basic triacon diamonds, as seen from above (positive z-axis). Due to the symmetry of the icosahedron, it turns out that all grid coordinates for a particular orientation can be derived from the coordinates of just the points within the view's shaded diamonds in Figure 2.21. For example, if the dome were edge-zenith, the topmost diagram in the figure, only the grid coordinates within diamonds 1 and 2 would be needed. All the other coordinates could be derived from these coordinates by applying a simple rotation equation or by manipulating the signs of their x-, y-, and/or z-component. Coordinates can be listed for a unit radius sphere and then multiplied by the radius of the actual sphere being built. This was a huge help in reducing calculations, and table books were prepared for just these diamonds. Today, PCs can calculate all coordinates in every diamond for these zenith orientations and any other in seconds. Scientific calculators today can even handle modest-sized domes. Appendix E describes how this is accomplished.[82]

The most difficult spherical calculations involve truncations. These are cases where grid members are cut by the ground plane or a structural support. Where to cut the sphere was an important decision based on factors such as required floor area, maximum height, or volume. Manufacturing and construction were also considerations. Most cuts would require a special band of panels or struts at the ground or supporting pylons. It is usually desirable to have triangular panels at the base so that their faces are always flat surfaces.

---

[82] (Kenner 1976)

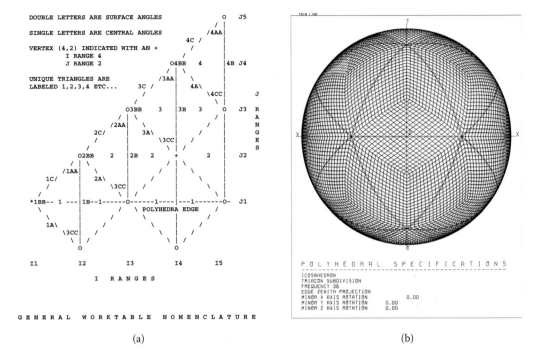

Figure 2.22. SPHERES geodesic dome computer program and graphic display.

Minimizing the number of special connectors at the base was also important, since these directly influence construction cost and complexity. Today's computers and programmable scientific calculators can find optimal solutions very quickly, but this was not the case in the 1950s.

The author, in collaboration with Duncan Stuart, developed a comprehensive triacon subdivision program called SPHERES in 1968. This FORTRAN IV program produced triacon work tables and computer graphics for the five Platonic solids with triangular, diamond, or hexagonal grids in single or multiple layers for very dense grids. For a given frequency and Platonic solid, all central and surface angles, chord and arc factors, triangle areas, and dihedral angles were found. All grid coordinates for all grid locations on a full sphere were computed. Figure 2.22(a) shows the spherical LCD right triangle guide printout for the naming conventions used for angles, chords, and grid vertices. The program also outputs data for graphics plotters. Figure 2.22(b) shows a diamond pattern for a 36-frequency icosahedral grid. The icosahedron reference edges are shown, though they are not part of the final subdivision grid. Frequency is the number of segments into which the grid divides the polyhedron's edge.[83]

Computer subdivision tables were essential to all large projects in the late 1950s and much of the 1960s. Bernard Taylor of Synergetics, Inc. wrote a program in 1958, the year

---

[83] SPHERES was initially developed for the IBM 1410 computer at the University of Detroit in 1968 and later run on an IBM 360 mainframe computer at Massachusetts Institute of Technology with CalComp drum plotter and Plot-10 graphics library support. SPHERES may have been the first geodesics computer graphics program. Note the use of I- and J-range references for each grid vertex in Figure 2.22(a), a holdover of the FORTRAN programming language where I and J were commonly used as integer DO loop indices (Popko 1968).

after Lewontin's program, to calculate the geometry of the American Society of Metals dome built in 1958. Other programs followed in the 1960s. Each new program would compute additional geometric information for unit spheres such as *dihedral angles* (the angle at the edge of two planar triangles), triangle areas, part counts, and calculation checks. These made design and cost estimating easier. Program output formats had improved as well; tables were easier to read but they did not eliminate the need for hand calculations. Unit radius spherical geometry still had to be converted to the geometry of the dome's actual radius. Subdivision tables remained in use until the early 1970s, only to be replaced by programs running on time-share computers and later on minicomputers. By 1975, computing had become inexpensive enough that any spherical grid could be computed "on the fly" when a project needed it. What's more, the exact value of arc lengths, chords, and vertex coordinates were now computed, thus eliminating the hand calculations that were required when scaling unit radius spherical values to the actual radius of a project sphere. Furthermore, computers now offered graphic output![84] Drum and flatbed plotters, driven by the coordinate output of geodesic programs, could make basic *orthographic projection* engineering drawings or perspective renderings from any viewpoint. Generalized spherical table books were now obsolete!

Today, personal computers make quick work of geodesic calculations and graphics, and even modest-sized domes can be worked out with programmable scientific calculators, though large grids can be somewhat tedious.[85] The IBM 650 was the beginning of computerizing spherical subdivisions and provided a base for optimizing layouts that simply could not be done by manual techniques. The computer largely eliminated tedious manual calculations and allowed spherical work to be highly automated. For companies licensing Fuller's geodesic patent, standard calculation techniques were a tremendous help in quality control and project management efforts. We will return to computation and spherical subdivision when we discuss computer techniques in Chapter 10.

2.15 Summary

The mid-1940s through the 1950s were launching years for geodesics. Synergetic geometry evolved from theory to practice with its first government and commercial successes. The Ford Rotunda, Union Tank Car, and Kaiser domes were winning architectural accolades. The Marine Corps demonstrations were successful. The Defense Department was to become a long-term patron of Fuller. The licensing success of the geodesic patent also demonstrated that other companies were willing to invest, and the variety of projects further demonstrated the soundness of the geodesic concept. Projects such as Union Tank Car and Kaiser domes were widely respected by engineers and architects alike. Design practices and spherical methodologies improved steadily and CAD was born from necessity.

Thus far, we have only touched on Fuller's start-up years, keeping our focus on the birth of the geodesic dome and the first spherical subdivision techniques. What, we may ask, have we inherited from these early years?

Surprisingly, the single most important legacy has nothing to do with domes at all: our most important inheritance from Fuller is a way of thinking about design, problem solving,

---

[84] Under contract to the National Aeronautics and Space Administration, Joseph Clinton prepared technical report "Advanced Structural Geometry Studies" detailing various spherical subdivision techniques. Each system was programmed in a computer; see (Clinton 1971).

[85] (Kenner 1976, Appendix 2, *Calculator Routines*, 166)

and "doing more with less," to use his expression. Fuller's synergetic geometry gives us a design framework rooted in both nature and science. Domes are just one by-product of synergetics thinking—an important one for sure—but a by-product nonetheless. Thousands of students and professionals in pursuits other than domes have benefited from geodesics and gynergetic geometry. The gift of Fuller and his associates is a creative process and a design imperative for conservation and service to mankind from which we continue to benefit.

In the context of *Divided Spheres*, our heritage is more specific: we inherit a framework for understanding the sphere, its symmetrical properties, and how to orderly subdivide it with great circles. The *alternate* and *triacon* subdivision methods we use today are direct descendants of Fuller's early work. The usefulness of these techniques can be seen in the diversity of geodesic applications today. In the next chapter, we sample some historic predecessors and modern applications.

Additional Resources

Buckminster Fuller Institute (BFI) is dedicated to accelerating the development and deployment of solutions that radically advance human well being and the health of our planet's ecosystems by facilitating convergence across the disciplines of art, science, design and technology, http://www.bfi.org/.

Chu, Hsiao-yun and Roberto G. Trujillo. *New Views on R. Buckminster Fuller*. Stanford, CA: Stanford University Press, 2009.

Edmondson, Amy C. *A Fuller Explanation: The Synergetic Geometry of R. Buckminster Fuller*. Design Science Collection, Boston: Birkhäuser, 1987. Reprinted *Back-In-Action* book series (Pueblo, Colorado: Emergent World Press, 2007).

Fuller, R. Buckminster. "3 Way Geodesic Grid B Calculations on Present." From *Noah's Ark II* notebook, 1948, Department of Special Collections and University Archives, Stanford University Libraries.

Fuller, R. Buckminster and E. J. Applewhite. *Synergetics: Explorations in the Geometry of Thinking*. New York: Macmillan Publishing Co., 1979.

Fuller, R. Buckminster and James Ward. *The Artifacts of R. Buckminster Fuller: A Comprehensive Collection of His Designs and Drawings*. 4 vols. New York: Garland, 1985.

Kenner, Hugh. *Bucky: A Guided Tour of Buckminster Fuller*. New York: Morrow, 1973.

Marks, Robert W. and R. Buckminster Fuller. *The Dymaxion World of Buckminster Fuller*. Garden City, NY: Anchor Books, 1973.

The Synergetics Collaborative (SNEC), a non-profit organization dedicated to bringing together a diverse group of people with an interest in Buckminster Fuller's to educate and support research and understanding of the many facets of synergetics, its methods, and principles, http://www. synergeticscollaborative.org/.

Zung, Thomas Tse Kwai. *Buckminster Fuller: Anthology for the New Millennium*. New York: St. Martin's Press, 2001.

3 Putting Spheres to Work

Buckminster Fuller's work in geodesic domes and spherical subdivisions was influenced by the work of others, just as his work is influencing the work of future generations. Today's geodesic designers benefit from the work of mathematicians and scientists in many diverse fields. We owe many analytic techniques to work carried out in descriptive geometry, biology, astronomy, climate, physics, and recreational equipment. This book touches on just a few of the more important sources and applications.

3.1 Tammes Problem

Pollen seems an unlikely place to look for examples of spherical subdivision. Nevertheless, we owe many useful concepts to early research in pollen grains. In 1930, the Dutch biologist P. M. L. Tammes published a landmark study *On the origin of number and arrangement of the places of exit on the surface of pollen-grains*.[1] This work bridges biology and mathematics, in particular, an applied area of discrete geometry called *circle packing*.

---

[1] (Tammes 1930, 1–84)

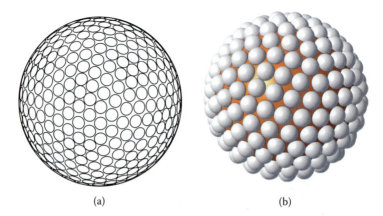

(a)   (b)

Figure 3.1. Circle and sphere packing.

Tammes was interested in the number and locations of exit points in pollen grains and he compared many types. He found 4, 6, 8, and 12 common; 5 never occurs; and 7, 9, and 10 occur less frequently. Very rarely does 11 occur.[2] He discovered that the distances between exit points were consistent within a type and their distribution on the grain was highly regular. These recurring numbers are no accident; they are optimal solutions to point distributions on the sphere. The similarity of pollen exits to circle packing lies in finding the optimal arrangement of a given set of points on the sphere. To measure how evenly distributed the points are, they are all surrounded by congruent circles large enough for two closest ones to be tangent but not overlap. Other circles will not be tangent, and the challenge is to make them as even as possible and minimize the total uncovered area outside the circles. Today, this optimization problem is called *Tammes problem* in his honor.

The best distribution of points occurs when the minimum distance between all points and their neighbors are maximized, or made as large as possible. A sphere surrounding each point offers a similar indicator—the closest spheres "kiss," an expression borrowed from billiards describing two balls that just touch. Examples of sphere and circle packing are shown in the random point distributions in Figure 3.1. Although their *packing* is quite good, you can see enough open space around some spheres and circles and realize the point distributions could be improved further.

Circle packing has a large body of literature and includes work by Conway, Goldberg, Tóth, Tarnai, Gáspár, Fowler, Robinson, and many other talented mathematicians (see the bibliography for references). Two- and three-dimensional circle packing finds practical applications in planning the locations of cellular phone towers, geo-synchronous satellite orbits, sports equipment, pharmaceuticals, and the development of new materials.

Although we are not involved in pollen research in this book, we can make use of circle and sphere packing to evaluate our own subdivisions. For example, the circles on the surface of the sphere can be seen as small slices taken from the sphere that resemble contact lenses. They are called *spherical caps*. A density factor—the ratio of the area of all caps to the total area of the sphere—measures what fraction of the sphere's surface is covered by the caps.[3] If two subdivision techniques distribute the same number of points on the sphere,

---

[2] (Aste and Weaire 2000, 108)

[3] All caps on the sphere have the same and maximum radius without causing overlaps. See (Tarnai 2002, 289).

the subdivision with a higher density factor (larger radius caps) has the better distribution of the two. In later chapters, we show basic methods for subdividing spheres and use sphere and circle packing as density factors to compare our results. We will also see how density factors are applied to sports equipment and how golf ball designers optimize circular dimple patterns to improve the ball's aerodynamics.

3.2 Spherical Viruses

In the mid-1950s, Francis Crick and James Watson (of DNA fame)[4] developed a theory about spherical viruses that hypothesized they had cubic symmetry, symmetry similar to the axes of a 3D cube. They also believed that the only way for these viruses to build their outer protein shell (the capsid) was to use a common molecular building block over and over again. These protein subunits would necessarily have to be positioned and packed so each had a nearly identical environment. Whatever their geometry, it was highly regular and symmetrical.

Building on their theory, X-ray crystallographer Donald Caspar, pictured on the right, analyzed spherical viruses and found in addition to cubic symmetry, that they also exhibited five-fold rotational symmetry.[5] This meant the virus would likely have the symmetry of an icosahedron, which has both *cubic symmetry* as well as *five-fold symmetry*. The icosahedron has 12 vertices, 30 edges, and 20 faces. If you look straight at any one of its vertices toward the center of the solid, you can rotate the icosahedron around your line of sight in increments of one-fifth of a circle and each view looks the same. This is an example of five-fold symmetry (around the vertex). This finding was also consistent with Crick and Watson's theory. Their theory further suggested that viruses with cubic symmetry might be built from aggregations of smaller asymmetrical pieces in multiples of 12 and a maximum of 60 identical subunits. Figure 3.2 shows conceptual models of two spherical viruses with cubic symmetry. A model of the Human Immunodeficiency Virus (HIV or AIDS virus) shown in (a) shows the external appearance

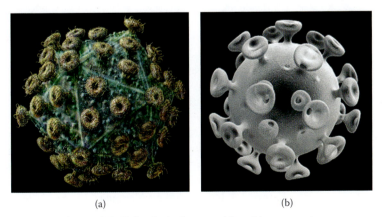

(a)   (b)

Figure 3.2. Spherical viruses with cubic symmetry.

---

[4] Francis H. C. Crick, James D. Watson, and Maurice Wilkins were jointly awarded the 1962 Nobel Prize for Physiology of Medicine "for their discoveries concerning the molecular structure of nucleic acids and its significance for information transfer in living material" (www.nobelprize.org).
[5] Image courtesy of Donald Caspar.

of the virus. The virus is coated (enveloped) in host cell membrane, which is drawn as a bluish green semitransparent layer in which various membrane proteins can be seen floating. The viral knobs (golden projections at the viral surface) insert into the matrix. These knobs allow the virus to attach to cells. The AIDS model (b) also has icosahedra symmetry (see Section 6.3.2 for a description of this type of symmetry).[6]

In the late 1950s, Caspar began collaborating with Aaron Klug,[7] who was involved in polio research. Robert Marks, an associate of Buckminster Fuller, noticed this work and introduced Klug to Fuller in 1959. Later, through Mark's writings,[8] Klug recognized similarities between his research data and the geometry of geodesic domes. Many domes use spherical icosahedra as their design base, and structural members are often arrayed in patterns of hexagons and pentagons.

Together Caspar and Klug formulated a new concept about spherical viruses they called *quasi-equivalence*. They visualized common viral subunits as something similar to a dome made of quasi-equivalent (nearly equal) triangles. They also envisioned common viral subunits clustered in an arrangement of pentamers and hexamers (pentagonal and hexagonal areas on the sphere defined by five or six triangles clustered around a central point).

Most subdivisions of spheres try to create grids of nearly equal (quasi-equivalent) triangles. Groupings of triangles define larger shapes such as hexagons and pentagons, and the total number of triangles in a subdivided sphere (with polyhedral base) follows a specific rule dictated by a *triangulation number*. By using different parameters in the triangulation number formula, a wide variety of *triangular grids* and grid orientations can be made over the icosahedron (or other 3D solids). Some parameters produce grids that are parallel to the edges of the icosahedron's equilateral faces, while other parameters skew the grid at an angle or make them perpendicular to the edges. Today, we classify a subdivision type by the orientation of the grid to the faces of the 3D solid being subdivided. They are called Class I, II, and III.[9] Chapter 8 describes subdivision techniques that produce grids in each of these classes.

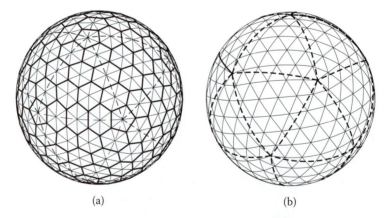

(a)   (b)

Figure 3.3. Icosahedral grid layouts.

---

[6] (a) Image © Russell Kightley 2009; (b) image © TurboSquid 2009.

[7] Aaron Klug won the 1982 Nobel Prize in Chemistry for his development of crystallographic electron microscopy and his structural definitions of nucleic acid-protein complexes (www.nobelprize.org).

[8] (Marks 1973)

[9] Caspar and Klug named their classes P = 1, P = 2, and P = 3 and they roughly correspond to the modern names Class I, II, and III.

Figure 3.3 shows different geodesic grids with triangular, hexagonal, and pentagonal patterns at different orientations to an underlying icosahedron (shown with dashed lines). In Figure 3.3(a), the highlighted edges of the hexagons and pentagons cross perpendicular to the icosahedron's edges (shown dotted). This grid orientation is called Class II. The grid in (b) is skewed and not parallel or perpendicular to the edges of the icosahedron and is called a Class III grid. Caspar and Klug experimented with combinations such as these when they analyzed their X-ray data.

The Caspar and Klug quasi-equivalence theory was presented in a landmark conference paper in 1962.[10] Its major idea is to describe the ways large numbers of identical protein subunits can build closed containers, such as virus capsids by a self-assembly process. Their fundamental concepts about the geometry of certain viruses impacted virology for years to come. They discovered that icosahedral symmetry is a preferred structure of certain isometric virus particles and the triangulation number shows only certain numbers of hexamers and pentamers are possible. Their work developed the notion of "skewed" spherical grids, which today are referred to as Class III subdivisions. These grids were totally different than those developed by Fuller. For sure, Fuller's work inspired Caspar and Klug, but once their paper was published, it was incorporated into the literature in a way that did not link it to Fuller.[11] Although our interests are in underlying geometric principles behind spherical subdivision, we can't help but marvel at how Caspar and Klug built up their theory of quasi-equivalence and made the connection between very big structures, such as geodesic domes, and the smallest of structures in viruses.

3.3   Celestial Catalogs

We still treat the sky as if it were one very large sphere surrounding the earth. This viewpoint, although not true, is practical and serves astronomers today just as it did the ancients. Astronomers with ground-based and space-based telescopes are continually surveying the sky and capturing unimaginable quantities of data recording the positions, magnitudes (brightness), and other characteristics of celestial objects, not to mention all the photographic records that stream back constantly. Catalogs today have more than 500 million such data records, and a billion will soon be catalogued. And this does not count the constant stream of images from space probes and orbiting satellites. Storing this huge quantity of data and making it readily accessible is an enormous archival challenge.

Researchers Peter Kunszt, Alexander Szalay, and Ani Thakar at Johns Hopkins University devised the Hierarchical Triangular Mesh (HTM) system[12] as one way to systematically subdivide the celestial sphere into small triangular reference areas, with each area acting as an efficient index to store and retrieve the astronomical data recorded within it.[13] It is used extensively for the Sloan Digital Sky Survey, which will map one-quarter of the entire sky in detail, determining the positions and absolute brightness of hundreds of millions of celestial objects. It will also measure the distances to more than a million galaxies and quasars.[14] Figure 3.4 shows how the sky can be divided into hierarchical sets

---

[10] (Caspar and Klug 1962, 1–24)

[11] (Urner 1991)

[12] (Szalay et al. 2005)

[13] Subdividing triangles by constructing a new one within it was a method used by Fuller in the 1950s. The technique was called the alternate method because subtriangles were created by adding chords between the alternate sides of a triangle.

[14] Sloan Digital Sky Survey, Johns Hopkins University.

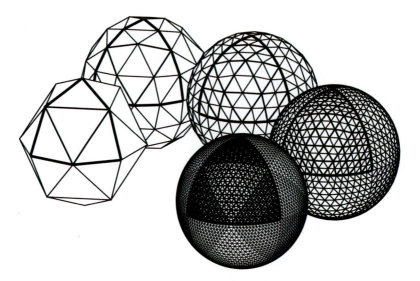

Figure 3.4. Hierarchical Triangular Mesh examples.

of triangles, each set making a progressively finer and finer grid. All five spheres in Figure 3.4 are oriented the same way to make it easier to compare them. At the furthest left of the figure is the simplest subdivision, in which a single highlighted face is divided into four triangles by creating a triangle in the center whose vertices are formed by the midpoints of the surrounding triangle's edge. Each of the four resulting triangles can be subdivided again using the same process. This recursive method, a subdivision of a subdivision of a subdivision, can make meshes as fine as needed, as demonstrated in the other spheres in the figure. If one area of the sky has more catalog entries, localized and finer meshes, just for that area, are also possible.

The spherical coordinate system, used by astronomers, records observational data (star information) with two coordinates: right ascension and declination. The system easily converts to latitude/longitude systems used by geographers or to an individual triangle in the grid called a *trixel*. A number identifies the trixel's exact location within a hierarchy and the portion of the sky it covers. An individual triangle is part of a cluster of four, and those four are one of four larger ones. The mesh can be continuously subdivided to meet the needs of growing catalogs. Forerunners of the HTM system include the Quaternary Subdivision System developed for cataloging and analyzing terrestrial data.[15] It is based on the same subdivision principle as HTM, but uses a different reference scheme for searching a particular triangle. In Chapter 8, we explain how to do this type of subdivision. We call the subdivision method *Mid-arcs*.

3.4   Sudbury Neutrino Observatory

The Sudbury Neutrino Observatory (SNO) is unlike any other observatory on Earth. Instead of sitting atop a dry mountaintop in Chile or Hawaii making observations through clear skies, the observatory is located 1.3 miles (2,070 meters) underground in an active

---

[15] (Dutton 1996, 8B.15–28)

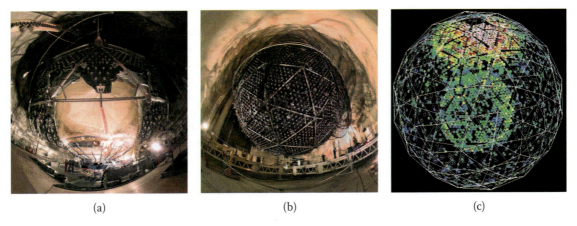

(a)   (b)   (c)

Figure 3.5. SNO geodesic sphere.

Canadian nickel mine.[16] The two kilometers of rock above this ultraclean facility help block false signals from cosmic rays and other radiation from space. Only a minimum amount of cosmic radiation can reach SNO and interfere with its observations.

Particle physics is a basic science in which researchers look for the smallest or most fundamental components of matter and the forces that govern their interactions. SNO, a collaborative research project between government agencies and universities, is designed to measure one of the smallest particles known: neutrinos. Neutrinos are subatomic particles so small and so fast, they simply pass through the earth unimpeded by matter. Our sun and other stars emit neutrinos as they generate and release energy. By detecting neutrinos from the sun, researchers learn about the physics of how the sun generates energy. SNO researchers have already discovered that the sun's neutrinos change from one type of neutrino to another on their journey from the sun to the earth. This discovery solved a 30-year-old scientific problem and confirmed earlier theories about neutrinos' mass and energy.

SNO's outer structure, referred to as the Photomultiplier Tube Support Structure (PSUP), is a 60-feet-diameter (18 meter) geodesic sphere made of hexagonal and pentagonal stainless-steel modular sections. The surface of the sphere supports arrays of highly sensitive photo detectors, 9,600 in all. Inside the PSUP is a second sphere, 39.4 feet in diameter (12 meters). This sphere is the world's largest acrylic vessel with 2-inch thick walls (5 cm); it holds 1,000 tons of heavy water.[17] Figure 3.5(a) shows the partially completed PSUP with a view of the acrylic vessel inside. The PSUP's photo detectors are mounted between its structural sections. Every detector aims towards the center of the sphere, where they can record light traces from neutrinos passing through SNO at any angle. The completed PSUP is shown in Figure 3.5(b).[18] Highly purified water fills the space between the PSUP and the acrylic sphere as well as the cavern where the instrument is suspended.

When a neutrino passes through the observatory, it interacts with the heavy water contained in the central spherical vessel, causing a slight flash of cone-shaped light. The photo

---

[16] INCO's Creighton Nickel Mine in Sudbury, Ontario, Canada.

[17] Heavy water is water in which deuterium, an isotope of hydrogen, replaces hydrogen atoms and H_2O becomes D_2O. The deuterium in heavy water enables the detection of all three types of neutrinos.

[18] PSUP photographs (a) and (b) courtesy of Lawrence Berkeley National Laboratories, Roy Kaltschmidt photographer.

detectors on the PSUP record the time, location, cone shape, and number of photons emitted by the flash. This data tells researchers what the neutrino is and how it behaves. Figure 3.5(c) shows a computer graphic simulation of SNO detecting neutrinos.[19] Some of the detectors are recording the passage of neutrinos (the small red cones in the image).

SNO's original mission is now complete. However, the facility continues to gather data, and new experiments are underway to advance solar physics. SNO is also capable of detecting a supernova, an exploding star, within our galaxy if one occurs while the detector is online. Supernovas release neutrinos earlier than photons (visible light); thus, it is possible to have some advance warning that a supernova will be visible with an optical or radio telescope.

3.5   Climate Models and Weather Prediction

It's impossible to ignore daily and seasonal weather, but more subtle and long-term changes are also becoming apparent. NASA says that over the past 40 years, the global temperature has increased 0.5°C. Mountain glaciers all over the world have receded. The average thickness of the arctic sea ice has decreased 40 percent, and sea levels have risen around the world. Moreover, NASA reports that the growing season now starts seven days earlier and ends two to four days later than it did 40 years ago, and that seven of the century's ten warmest years occurred in the 1990s.[20] Any one of these developments might not be a cause for alarm, but when seen together, their impact is considerable. The changes have already affected global agriculture, water supply, coastal habitation, energy generation, and energy demand. The effects are not the same for everyone. Some countries and societies are impacted more than others; some will adapt to change better than others.

Essentially, *climate modeling* is the use of computers to simulate the cause-and-effect relationships of the physical, chemical, and biological components of climate. Atmospheric general circulation models are simulation programs with dynamic cores that integrate equations for variables such as surface pressure, water vapor, solar gain, temperature, and fluid motion in spatial mediums, such as the atmosphere and oceans, land surface, and areas covered by ice. Although the techniques used can be applied to local weather forecasting, the emphasis is on large-scale global patterns where huge volumes of air and areas of landmass and oceans interact. Enormous amounts of spatial data, gathered over time or forecasted for future dates, make climate models computer-intense. They demand the highest processing speeds and memory possible and are typically run on the largest super computers.

Climate models need a systematic way to reference all areas of the planet, from the depths of oceans to the atmosphere above the oceans and landmasses. The design and choice of a reference grid is one of the earliest decisions made by programmers of climate models because all data references and time-space calculations depend on it. The grid facilitates calculations and makes it easier to deal with 3D dynamics, data storage/retrieval, and computer memory. In essence, the grid is the way spatial data is aggregated and disaggregated and is the link between data, the climate model, and the computer they run on.

Initially, climate models used latitude/longitude grids, because they were familiar to geographers, and the coordinate system was primarily used to classify data and refer to loca-

---

[19] Image © courtesy of MIT Laboratory for Nuclear Science.

[20] National Aeronautics and Space Administration, World Book at NASA—Global Warming, 2010, http://www.nasa.gov/worldbook/ global_warming_worldbook.html.

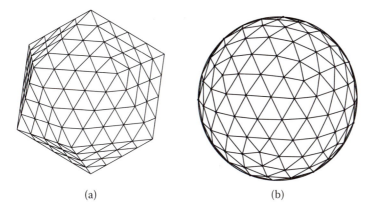

(a)   (b)

Figure 3.6. Icosahedral tessellation.

tions on Earth. But as climate models became more sophisticated and simulations for small areas became possible, latitude/longitude grids presented drawbacks. The convergence of meridians at the north and south poles creates more cells near the poles than at the equator. In addition, the poles had to be treated differently because the latitude terms 90° north and 90° south lose significance and any time-based simulations near the poles required smaller steps than at the equator to get stable results. Other grids were explored, since the polar regions greatly influence climate.

One solution to the "pole problem" has been to use geodesic grids resembling the same patterns as geodesic domes. Geodesic grids were first considered in the 1960s.[21] There are a number of ways to generate these grids. One way is to subdivide the icosahedron's equilateral faces into a series of smaller equilateral triangles and then project the resulting grid onto the surface of a circumscribing sphere. The projection process uniformly distributes the points and changes the shape of the triangles slightly, but the result is a grid of quasi-isotropic nearly equal triangles. The grid is also called a *tessellation*. Figure 3.6 shows the subdivided planar icosahedron (a) and its spherical projection (b). The scale and orientation of both figures are the same for comparison purposes.

Climate modelers David Randall and Ross Heikes at the Center for Atmospheric Sciences at Colorado State University use geodesic grids similar to the ones we have been discussing, but with an important enhancement. They use the triangular grid to derive a hexagonal grid, and it is the hexagonal grid on which they base their climate model. Why this extra step? In two dimensions, only three regular (all sides equal) tessellations—triangles, squares, and hexagons—cover all areas without gaps or overlaps.

Figure 3.7 shows all three grid systems surrounding a central cell shown shaded. For triangular and square grids, you'll notice that the cells neighboring the shaded one are either across one of its edges or across one of its points. Randall and Heikes call these "wall-point neighbors."[22] But notice that for hexagonal grids, all cells surrounding the shaded one are only across edges; there are no point neighbors. By eliminating point neighbors, you greatly simplify calculations when computing the differences between cells. All cells can be treated in the same way.

---

[21] (Sadourny, Arakawa, and Mintz 1968)
[22] Illustration based on (Randall 2002, 35, fig. 5)

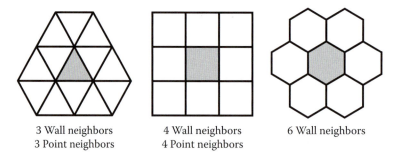

3 Wall neighbors
3 Point neighbors

4 Wall neighbors
4 Point neighbors

6 Wall neighbors

Figure 3.7. Tessellation wall-point neighbors.

Cells in a spherical tessellation have wall-point relationships, and it's possible to convert a triangular tessellation with wall-point neighbors into a hexagonal one with only wall neighbors simply by defining each point's Voronoi cell.[23] *Voronoi cells* are polygons that define an area (of points) that is closer to a designated point than to any other. If the original points are evenly spaced, or nearly so, the Voronoi cells will resemble a grid of hexagons. Randall and Heikes illustrate this in Figure 3.8(a). The Voronoi cells in green define the points (areas) closest to each of the points in the following triangular grid. The spherical versions of both grids are shown in (b).[24]

When a Voronoi tessellation is made of a fully subdivided icosahedron, such as the one shown in Figure 3.9, the result is mostly hexagonal cells, but pentagonal cells also result. One pentagonal cell surrounds each of the icosahedron's 12 vertices. Some are pointed out in the figure. Each hexagonal and pentagonal cell is quasi-isotropic because their areas are nearly equal. Both spheres in Figure 3.9 are oriented and scaled the same for comparison.

When it comes to climate modeling, hexagonal geodesic grids offer many benefits over other grid systems. They are largely isotropic and do not suffer from oversampling near the poles like longitude/latitude grids. The poles are no different than any other area in the hexagonal grid. All cells in the grid are equal wall-neighbors. There are no point neighbors,

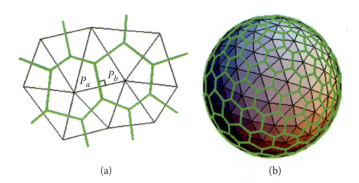

(a)   (b)

Figure 3.8. Hexagonal Voronoi cells from triangular tessellations.

---

[23] Voronoi diagrams are named after Russian mathematician, Georgy Feodosevich Voronoy (1868–1908). However, they were recognized much earlier by René Descartes (1596–1650) and used by Lejeune Dirichlet (1805–1859). Voronoi tessellations are also called Dirichlet tessellations.

[24] Climate model color illustrations in this section courtesy of Ross Heikes.

Pentagonal Voronoi cells surround each
icosahedral vertex.

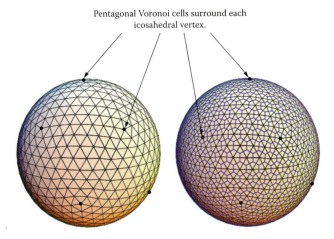

Figure 3.9. Hexagonal Voronoi cells of a triangular tessellation.

as in other grid systems. Hexagonal geodesic grids can also accommodate different resolutions where the atmosphere might be represented by a coarse hexagonal grid, and land-surface areas represented by a finer grid as seen at the right. Randall says coupling the atmosphere and ocean models with multiple hexagonal grids simplifies modeling because the atmospheric state is "interpolated down" from large hexagonal cells to smaller ones below and fluxes are "averaged up" from small cells to larger ones. Hexagonal geodesic grids can also be warped or morphed to meet local needs. Although hexagonal geodesic grids are more complicated to implement than rectangular longitude/latitude grids, their benefits far outweigh the additional work needed to set them up and use them in climate models. Figure 3.10 shows typical climate modeling output based on a geodesic grid.

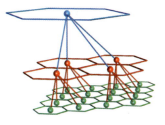

Climate modeling is an intense field of research today. Most efforts involve cooperative agreements with government agencies, such as the US Department of Energy, and

Figure 3.10. Geodesic grid climate model.

designs that take advantage of the latest super computers at laboratories across the country. Randall estimates that simulating a century's worth of climate with a full-coupled ocean-atmosphere model will require several weeks of round-the-clock supercomputer time.

3.6 Cartography

The essence of cartography is the transformation of information from a 3D spherical surface, the world, to a flat 2D surface, a map. This transformation is achieved through a spherical reference system and is often shown as a grid—tools and techniques also used by climate modelers and astronomers. In previous sections we have shown a number of cartographic examples, including Buckminster Fuller's Dymaxion maps, Cahill and Waterman butterfly maps, the Hierarchical Triangular Mesh (HTM) for celestial catalogs,[25] and Randall/Heikes' icosahedral climate model grids. Collectively, these systems (as well as many more not included in this book) are called discrete global grids (DGGs).

Cartographers Kevin Sahr, Denis White, and A. Jon Kimerling describe DGGs as a set of spherical regions dividing up the earth's (or a general sphere's) surface. Within each of these regions, there is a single point that can be used to assign data associated with the region. For spherical grids that define only regions, the *centroids* of the regions can be used as the single associated point for the region. Conversely, if a spherical system defines only points, the Voronoi regions of those points become the regions.[26] Sahr, White, and Kimerling point out the following common characteristics of all DGGs:

- a base regular polyhedron (i.e., Fuller used the cuboctahedron for the Dymaxion map),

- a fixed orientation of the base regular polyhedron relative to the earth (i.e., Cahill's vertex-zenith octahedron orientation),

- a hierarchical spatial partitioning method defined symmetrically on a face (or set of faces) of the base regular polyhedron (i.e., HTM's nested and reference system of triangular areas),

- a method for transforming that planar partition to the corresponding spherical/ellipsoidal surface (i.e., Randall/Heikes Voronoi region grid), and

- a method for assigning points to grid cells (i.e., centroids of regions).

Climate modelers would likely add one more DGG design characteristic to the list: DDGs provide a way to transform volume data to a spherical surface area or a summary statistic to a single point. Recall that the Randall/Heikes climate model uses a series of layered hexagonal-pentagonal grids that provide a mechanism for aggregating or disaggregating area or volume data in one grid to another overlaying or underlying grid.

In Chapter 8 we will describe six schemas for subdividing spheres. Each results in a triangular mesh that can be grouped into squares, pentagons, hexagons, or diamond shapes, depending on the foundation polyhedron used, the intensity of the grid, and the subdivision technique used. Except for the two butterfly maps, shown in Figure 2.4, the cartographic examples above use previously of these schemas.

---

[25] HTM is a variation of the Hierarchical Coordinate System; see (Dutton 1985).
[26] (Sahr, White, and Kimerling 2003, 121–123)

While climate modelers are using honeycomb grids to analyze macroscale atmospheric problems, supercomputer architects are looking to use those same grids for microprocessor networks performing those analyses.

3.7   Honeycombs for Supercomputers

Some of the largest supercomputers in the world are dedicated to global weather forecasting and climate modeling. Supercomputers being used for these purposes are often tasked with the discretization of partial differential equations (PDEs). Solving billions of PDEs is a highly repetitive and computationally intense endeavor. With each new generation of supercomputers, the number of computational clusters or processors increases, along with their processing speed. Connecting these processors and maintaining efficient communication between them has become a formidable challenge. Honeycomb networks of processors can reduce bottlenecks between processors and improve PDE discretization.

There are two main ways to interconnect a computer's processors: switched and switchless. In switched systems, messages sent from one processor to another pass through a hierarchy of switches that enable the communications path between them. This approach has worked well in the past. But as processor speed increases, the number of messages being passed from cluster to cluster also increases and soon the switch or hierarchy of switches is overwhelmed. The switch becomes a bottleneck, in effect, hindering further performance improvements.

The latest generation of supercomputers is switchless and employ many processors working in parallel. In a switchless system, computational clusters communicate with one another through a set of paths and nodes much like a *network*. The network provides gateways for directing communications. Graphically speaking, a network might resemble a Cartesian grid with computational clusters located at grid points, and the grid itself as a set of communications pathways connecting every processor with every other one. A lucky processor message might pass through only one or two links to get to its destination processor. Other messages might have to pass through many links to get to their destination. The challenge in processor design is to keep the number of internode links as few as possible, while making the links as short as possible. In most network layouts, there will be cluster nodes on the edge or ends of the network system, and they need a path that loops back and returns to the network. This usually means connecting this cluster back to the network with a "long wire," a path much longer than the usual ones. As the number of clusters increases, long wires are almost unavoidable. These long wires slow up any communication they carry. The goal of system designers is to minimize or completely eliminate long wires altogether.

Inspired by weather forecasting and climate modeling, researchers Joseph Cessna and Thomas Bewley have developed a new logical and physical approach to solving processor communication problems in supercomputers. The approach is highly adaptable to spherical applications, particularly those dedicated to PDE processing.[27] In effect, Cessna and Bewley have applied spherical grids, similar to those already used by climate modelers, to supercomputer architecture.

In the previous section, we showed examples of honeycomb (hexagonal) grids used in climate modeling. The Voronoi cells, derived from an initial triangular grid covering the earth, produce a honeycomb grid. A significant feature of a Voronoi grid is that every cell

---

[27] (Cessna and Bewley 2009)

Figure 3.11. Honeycomb processor grids.

(section of earth, or volume of atmosphere) is separated from its neighboring cells only by its edges; there are neighboring cells adjacent to any of the cells' vertices (see Figure 3.7). Most processor activity in climate modeling and weather forecasting involve adjacent cells, so arranging computational clusters like a Voronoi's honeycomb network is appropriate.

Cessna and Bewley found that computational clusters and the communication networks between them can be arranged in an analogous way. A honeycomb vertex is a computational cluster location. A honeycomb grid is a communications path. The number of networked clusters in any grid can be regulated according to the problem to which the supercomputer is applied.

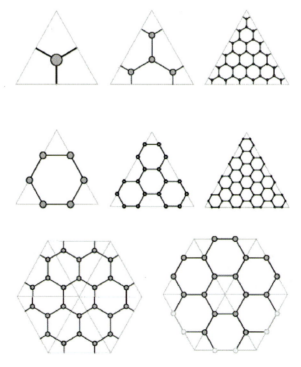

Figure 3.12. Typical processor and network arrangements.

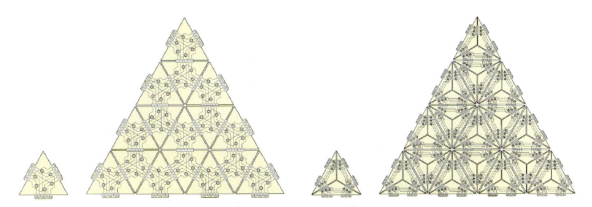

Figure 3.13. Examples of computation clusters and networks.

Figure 3.11 shows a series of increasingly dense honeycomb grids, demonstrating that the number of processors can be varied according to the modeling requirements.[28] In other words, there isn't a one-to-one correspondence between a climate model honeycomb grid and the number of computational clusters and communication paths in the supercomputer. What is the same, however, is the hexagonal grid and the way the discretization of partial differential equations is performed. Cessna and Bewley found that honeycomb arrangements of computational clusters reduced the number of communication paths, all but eliminating the need for long wires between clusters, and optimizing the discretization of PDEs used for spherical applications.

Figure 3.12 shows typical honeycomb graphs over tessellations of equilateral triangles and hexagons, the two types of grids commonly used to cover a sphere (or Earth). The choice of how many computational clusters that could be used in a computer simulation can be regulated to match analysis requirements. Figure 3.13 shows sample clusters and their communications paths in layouts that can cover the surface of the sphere.

The honeycomb grid approach offers many advantages when arranging networks of computational clusters in supercomputers. Applications such as climate modeling and weather forecasting can optimize PDE discretization for local area simulation with this type of architecture.

3.8   Fish Farming

From the 1950s to the early 2000s, global demand for fish products doubled as world population growth dramatically increased the number of people eating seafood. Skyrocketing demand combined with overfishing, climate change, and the adverse effects of pollution on water habitats are negatively impacting food supplies as well as economies that depend on fish products. Some of the highest valued species are now threatened, including cobia, grouper, amberjack, pompano, cod, halibut, and haddock.

Today more than 40 percent of the world's seafood comes from aquaculture, or fish farms. Aquaculture contributes to the global supply of fish, crustaceans, mollusks, and other aquatic animals, and production is substantial. With growth rates as high as 8 percent per year, aquaculture is outpacing all other animal food-producing sectors. Today, China

---

[28] Images courtesy of Joseph Cessna and Thomas Bewley.

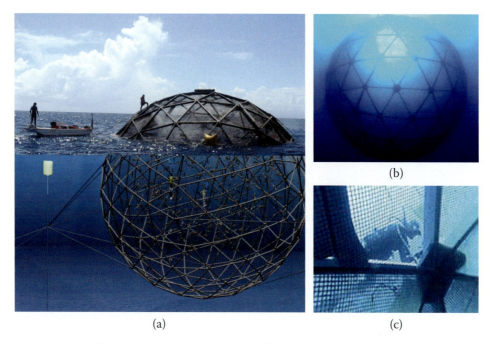

(b)

(a)   (c)

Figure 3.14. Ocean Farm Technologies' Aquapod fish pen.

produces almost 70 percent of the total quantity of aquaculture products, accounting for more than 50 percent of the total sales worldwide.[29] At this pace, aquaculture is expected to surpass beef production by 2011. In the United States, about 80 percent of the seafood consumed is imported from aquaculture farms in other countries.[30]

Most aquaculture today is based on land or near shore. However, recent design innovations could extend aquaculture to offshore deepwater sites, increasing possibilities for would-be fish farmers lacking protected near-shore sites or near-shore aqua-environments that are suitable in terms of water temperature, quality, and/or depth. These innovations would also be a boon to fish farmers facing shoreline zoning, protected shallow-water habitats, or other restrictions. The ability to farm offshore will greatly increase the number of locations that can host fish farms. Fish pens can be located wherever water oxygen levels, quality, and temperatures are most suitable. These conditions optimize fish growth and allow adequate separation between farms. Submersible pens protect fish and pens from storms and surface wave action, as well as potential collisions with ships. Fish pens must incorporate feeders and fish transfer/harvest mechanisms, and allow for remotely controlled monitoring and feeding, if they are to be economically viable. New high-tech fish farms now offer such possibilities.

Ocean Farm Technologies, Inc. has developed a fish pen using geodesic spherical design principles. Their patented design, called the Aquapod, is a 64-feet-diameter (19.7 meters) sphere with struts made of fiberglass-reinforced polyethylene (80 percent recycled content). Figure 3.14(a) shows a composite photo and an artist's drawing of an Aquapod viewed above and below water, while (b) shows an Aquapod completely submerged.

---

[29] (Food and Agriculture Organization [FAO] of the United Nations 2007, 16–17)
[30] (National Oceanic and Atmospheric Administration [NOAA] 2008, 4)

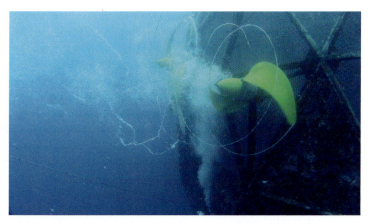

Figure 3.15. Prototype of a self-propelled Aquapod.

Wire-mesh netting covering the inside of the sphere makes the Aquapod both escape- and predator-proof, as shown in Figure 3.14(c).

The entire sphere, when assembled, is close to being neutrally buoyant in salt water. Fixed buoyancy can be set with closed cell foam, and buoyancy can be controlled with inflatable tubes mounted on the frame struts. Several inflatable tubes are visible above the water in (a). The sphere is anchored or moored in place but may occasionally be moved to other locations to maintain water quality.

Why a geodesic sphere? Geometrically speaking, a sphere encloses the most volume with the least amount of surface area. Submerged, the sphere is subjected to uniform water pressure over its entire surface. A sphere is the best shape for ensuring that these pressure loads are distributed evenly when the pen is stationary or moves about. Geodesic layouts can be prefabricated from just a few standard components. In addition, they are easy to assemble on shore and repair at sea. In Chapter 8, we describe several layouts that minimize the number of different parts needed to build a sphere.

In 2008, Ocean Farm Technologies, Inc. worked with researchers at Massachusetts Institute of Technology Offshore Aquaculture Engineering Center to develop a self-propelled design. Figure 3.15 shows a prototype configuration.[31] Pens can be repositioned to maintain water quality or temperature. A self-propelled pen could also eliminate the high cost of towboats that typically move cages from site to site today.

3.9 Virtual Reality

Virtual reality: it sounds like a contradiction in terms or, at best, pure escapism from the unpleasantness of daily life. In some ways, virtual reality is just that, though there are virtual realities that deliberately replicate stress and life-threatening situations. Aircraft flight simulators, for example, train pilots to handle emergencies.

Some real-life tasks are simply too difficult, unsafe, or costly to carry out without significant preparation or simulated practice. Examples of such tasks would include testing the first prototype of an advanced spacecraft, performing a first-ever difficult surgery, disarming a bomb, retrofitting new equipment in a nuclear reactor, or training for an assault on

---

[31] All Aquapod photos courtesy of Ocean Farm Technologies, Inc., Searsmont, Maine.

a terrorist's camp. In such cases, virtual reality offers safe ways to solve problems, develop skills, and gather data regarding the resources needed to accomplish the task in the best way possible.

Technically speaking, virtual reality is a computer generated 3D environment in which the user has complete freedom of motion and can view and interact with this environment in real time. In virtual-reality applications, a computer uses special devices or sensors to measure the user's movements and line-of-sight in order to send back visual, audio, and tactile feedback that give him the feeling he is actually in the world he is experiencing. Most systems today only offer limited movement and, in most cases, the user is actually stationary and uses devices like 3D mice, joysticks, or kinetic controllers (paddles, rackets, golf clubs, etc.) to move forward, backward, or change direction in the virtual world.

Three-dimensional computer games are a basic form of virtual reality. Few players will admit how addictive they are. Home entertainment gaming computers like Sony's PlayStation3, Nintendo's Wii, or Microsoft's Xbox offer photorealistic graphics and real-time animation of virtual worlds in which players engage in activities ranging from flying airplanes to playing golf, musical instruments, or first-person-shooter games. In 2008, game console software sales totaled $8.9 billion with 189.0 million units sold.[32] These are impressive figures, but the future market for virtual reality is even brighter because new applications are going far beyond home entertainment. For example, some new systems focus on military, law enforcement, SWAT training, construction, and other dangerous occupations. Others focus on health and exercise or virtual travel and tourism. What is new is the use of immersive systems that make the user part of the virtual reality program. Users can attach sensors to their bodies, use 3D computer mice, stand on tactile pads that measure their position and weight, and wear special cyber-gloves that follow any movement of the hand. Advanced head mount displays give users full 3D stereo views as they look around their virtual world. For sure, the combined effects of 3D graphics, tactile and motion sensors, and locomotion bring virtual reality much closer to real reality.

Until now, the missing sensation in virtual reality has been locomotion—letting people move freely about their virtual world in a natural way. Virtusphere, Inc. has solved this problem with a patented technology that literally immerses the user inside his or her virtual world.[33] The user is inside a Virtusphere, a ten-foot hollow sphere supported by a special platform of rotating wheels. The sphere, shown in Figure 3.16(a), is free to rotate in any direction but remains in place on its fixed support platform. The sphere is assembled from only a couple of different prefabricated spherical sections. The user enters the sphere through a circular hatch, which closes behind him. At this point, he is free to walk or run around inside the sphere in any direction or at any speed he wishes. The sphere is an ideal form for omnidirectional walking or running. Sensors mounted inside the structural frame's great circle arcs give complete coverage of the user as the sphere rotates. The user wears a wireless head-mounted display with gyroscopes, which tracks his head movements while he walks or runs inside the sphere, as shown in (b).[34] Special sensors track the combined rotation of the sphere (as the user walks or runs) and changes in the user's line of sight. The computer inputs this sensor data and outputs constant updates to the user's head-mounted stereo display. In the figure, the soldier can aim his weapon and fire at objects in his virtual

---

[32] NPD Group (formerly National Purchase Diary) market reports on PC and video games. See (NPD Group, Inc. 2009).

[33] (Latypov and Latypov 2003)

[34] All Virtusphere photos courtesy of Virtusphere, Inc.

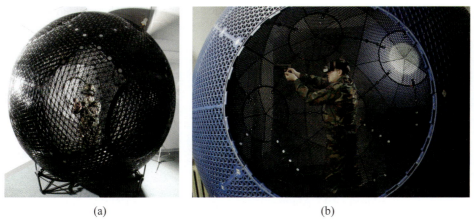

(a)   (b)

Figure 3.16. Assault training in Virtusphere.

world while on the run. What is unique about Virtusphere is the fact that the player can walk or run inside the sphere, and his movement will translate to movement within his virtual world. Imagine running down a street and looking in any direction, or standing on the edge of a steel beam high up a skyscraper under construction. Imagine interacting and communicating with another Virtusphere user who is elsewhere in your same virtual world. The combined effects of dynamic stereo vision, locomotion detection, and tactile feedback (your feet running, for example) add up to a truly startling experience.

Virtusphere's inventor, Nurakhmed Latypov, says that virtual reality gives a user a chance to feel, almost in real fashion, as if he or she is in some other world. In this world, users can see one another, talk, and interact in a 3D environment, and can even appreciate colors and sounds. Such immersion creates the impressive illusion of being in that other world.[35]

Our only question now is when can we get one for our home entertainment system?

3.10 Modeling Spheres

In the 1970s, everyone recognized Buckminster Fuller's geodesic domes. Nevertheless, few people understood the basic principles of spherical subdivision. Magnus Wenninger's classic book, *Spherical Models*, changed this and made the subject accessible, easy to understand, and fun by getting readers to make models.[36] He showed that the only tools you need to make complex, beautiful spherical models are a simple compass, a protractor, a straight edge, some paper and glue, a little patience, and a few back-of-the-envelope calculations. He literally put the principles of spherical subdivision at the reader's fingertips.

Wenninger starts modelers off with models of the basic Platonic solids (the tetrahedron, octahedron, cube, dodecahedron, and icosahedron). From there, he progressively builds the reader's projective geometry skills, so he or she can tackle increasingly complex models. Using various Platonic frameworks, Wenninger goes on to show how to make models of all three types, or classes, of spherical grids. These same classes are used today in a broad range of applications, from arranging dimples on a golf ball, to developing vaccines against viruses, to studying how climate change affects the earth.

---

[35] (Latypov and Latypov 2003, 1)
[36] (Wenninger 1979); republished by Dover with a new appendix in 1999.

Figure 3.17. Magnus Wenninger, polyhedral modeler.

Spherical polyhedra, like their planar versions, exhibit many forms of symmetry that can be used by modelers. Small spherical sections can be prefabricated in paper, wood, or sheet metal, and then duplicated and assembled to complete a full sphere. Today's CAD modelers employ the same techniques, only in a digital form. Three-dimensional sketches of spherical models use projective geometry methods. Digital computer models of geodesic arcs are analogous to circular paper bands, and mathematical constraints join 3D digital geometry, just as glue bonds paper strips together. Building paper models develops the same skills employed by CAD modelers.

Today, Wenninger is building models of complex spherical subdivisions that resemble basket weaves. In these designs, spherical patterns interlock and pass under and over one another, much like woven fabric. He still relies on the same projective geometry techniques he wrote about in *Spherical Models* (with the Dover mathematics appendix), but he also uses modern spherical calculator programs and personal computer polyhedral modelers like *Stella*[37] to make templates for more complex models. We devote two chapters to spherical modeling and CAD techniques. Figure 3.17 shows Wenninger completing a basket weave spherical model.[38] Several more are under construction on his modest workbench and his bookshelves display a small fraction of the thousands of polyhedron and spherical models he has made over the years. Wenninger[39] is author of the two other classic books on model making, *Polyhedron Models* and *Dual Models*.[40] A complete listing of his publications can be found in the bibliography.

3.11   Dividing Golf Balls

Despite images of players strolling along lush fairways and enjoying leisurely cart rides from hole to hole, golf is an intense and competitive game. Professional golfers can swing a club at 130 miles per hour with enough force to send a ball more than one-sixth of a mile at speeds in excess of 160 miles per hour. Golf's record-setting tournaments are a testament to the game's well-prepared athletes who devote countless hours of practice to their sport.

---

[37] *Stella* is a personal computer polyhedral modeler developed by Robert Webb. See Chapter 11 for details.
[38] Photos by Edward Popko.
[39] Father Magnus Wenninger is a monk of Saint John's Abbey, a community of Benedictine monks at Collegeville, Minnesota, USA.
[40] *Polyhedron Models* 1971, latest reprint 1996. *Dual Models* 1983, reprinted 2003.

But players at all levels also benefit from specialized high-tech equipment, including clubs and the golf ball itself.

Almost all golf balls have a surface with a large number of depressions, called dimples. All else being equal, the flying distance of a ball with dimples can be almost twice that of a smooth ball. Dimples deliver distance by generating small-scale turbulence around the ball's surface. This turbulence, somewhat counterintuitively, reduces drag on the ball because it makes the wake (drag) behind the ball much smaller. A ball with a rough surface simply flies better than a smooth one. Golf ball dimples represent one of the most creative and diverse applications of spherical subdivision principles.

The United States Golf Association (USGA) defines the maximum weight (1.62 oz.), minimum diameter (1.68 in), and spherical symmetry requirements of competition-level golf balls.[41] Spherical symmetry requires that a ball be designed and manufactured to behave as though it were symmetrical to enhance its control and predictability of flight; both are safety considerations. In other words, the ball must perform exactly the same in flight, no matter how it is oriented when placed on the tee. Designers meet those requirements in an amazing variety of ways, covering the ball's surface (8.87 in²) with between 200 and 1,000 dimples arranged in everything from simple bilateral patterns to complex polyhedral symmetries. Golf balls are wonderful examples of optimizing design tradeoffs where manufacturing techniques, aerodynamics, and design overlap. Amazing creativity is displayed in these 8.87 square inches, an area somewhat larger than a business card. If a golfer does not recognize the difference when he plays, he will when he pays; high-performance, professional-quality balls can cost four times as much as a knock-about ball.

In 1931, John Vernon Pugh disclosed in British Patent Provisional Specification Serial No. 377,354 that by using an icosahedral lattice for defining dimple patterns on golf balls, it is possible to make a geometrically symmetrical golf ball.[42] The lattice he developed divided the sphere into 20 equilateral spherical triangles, thus defining a spherical icosahedron. Pugh detailed how to plot the icosahedron onto a sphere accurately.

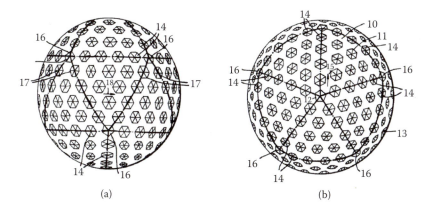

(a)   (b)

Figure 3.18. Early icosahedral dimple arrangement.

[41] United States Golf Association and the Royal & Ancient Golf Club, Equipment Testing, Appendix III—The Ball, www.usga.org/equipment/guide/book/appendix3ball.html.
[42] (Nesbit and Stiefel 1991)

Figure 3.19. Golf ball patent 4,560,168.

In the United States, most balls had a relatively simple dimple layout that placed "parallel" rows of dimples in an octahedral layout. Things changed quickly when Frank Martin and Thaddeus Pietraszek patented the first icosahedral layout with modern dimples. Figure 3.18(a) shows their 1978 hex-pent dimple arrangement.[43] This was a pivotal patent. It focused attention on the icosahedron, and it showed the limitless design possibilities of dimples. Within a year, Frank Martin had teamed up with Terence Melvin and others to secure yet another icosahedral patent. Although the essence of their patent claim focuses on ball coverings, the skin of the ball, their patents show circular dimple arrangements based on a spherical icosahedron.

The goal of ball designers is to design a pattern of dimples that improves the ball's flight characteristics (mostly flight distance) and directional control. By analyzing wind tunnel and robotic hitting machine data, designers concluded that the best way to achieve these goals was to increase the number of symmetry axes on the ball, and spherical icosahedra offered the best polyhedral base to work from. Despite many designs that use other bases, the icosahedron, or a small variation of it, remains by far the most common today.

In the mid-1950s, the icosahedral framework revolutionized dimple layouts. The icosahedron has six axes of vertex rotation, in contrast to other approaches that offer fewer. As a result, the additional axes increased flight stability. The ball's initial orientation on the ground, when launched, became less important. In 1985, Steven Aoyama was awarded a landmark patent that established the icosahedron as the framework of choice. Figure 3.19 demonstrates the flexibility of Aoyama's schema in a series of different dimple layouts.[44] In his patent[45] he subdivides icosahedral faces into four smaller triangles: a central equilateral

[43] (Martin and Pietraszek 1978, sheet 1/2, figs. 2–3)

[44] (Aoyama 1985, sheets 4, 5, 7, and 9 or 11, figs. 10A/B, 11A/B, 12A/B, and 14A/B)

[45] (Aoyama, op.cit.1985)

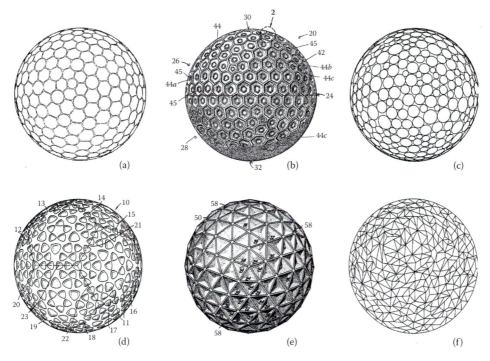

Figure 3.20. Samples of golf ball spherical subdivisions.

surrounded by three congruent *isosceles triangles* (see Mid-arcs schema in Chapter 8). In effect, this subdivision creates a variation of the icosahedron called the *icosidodecahedron*. These 3D forms are covered in detail in Chapter 6. The Mid-arcs schema adds six great circles to the sphere and divides each of the icosahedron's 20 faces into four smaller triangles. The additional great circles and smaller triangular areas define a totally new layout schema.

Aoyama's patent drawings show how the density of dimples can be regulated by simply changing the size and mix of dimples within the four working areas. The upper row of vertex-zenith views shows entire balls covered, with the dimple design shown below them. From left to right, three different-size dimples cover the first ball with 240 dimples, whereas 6, 10, and 13 sizes create 312, 692, and 912 dimples on the others. With 16 different sizes (not shown in the figure), the ball is covered with 1,212 dimples. Three of the four configurations in the figure show dimples along the icosahedron's face edges, but no dimples cross any of the six faces that subdivide great circles. These are kept dimple-free for manufacturing flexibility. Unlike our earlier examples of spherical caps, there may be more than one radius dimple.

Dimple research and development is as competitive as ever, often pitting hobbyists against international sporting equipment companies. More than 150 US patents have been awarded for dimple designs since Aoyama's work. The dimple wars are intense, and the battlegrounds shift from time to time, sometimes focusing on the overall layout, other times on the design of individual dimples and how they pack. As of this writing, more than 40 patents are pending based on application submissions dating back to 2003. Some are incremental refinements of older designs; others develop new designs like zero-ground balls

(balls that minimize their contact with the ground), dimples within dimples, noncircular dimples, or random dimple patterns. Figure 3.20 is a small sampling of patented dimple designs.[46] We will explore golf ball dimples further in the next chapter when we discuss their polyhedra and symmetry. Golf balls may be small, but you will surely appreciate how creative their designers subdivide spherical polyhedra to solving difficult geometric and manufacturing problems, keeping improved flight, control, and player scores in mind.

3.12   Spherical Throwable Panoramic Camera

If you can toss a ball, you might be the next Ansel Adams. The throwable panoramic camera is an amazing invention by Jonas Pfeil.[47] Throw this ball-shaped camera into the air and you will have a spherical picture centered on the ball when it comes down. It is an ingenious application of spherical design.

The camera is a little larger than a grapefruit, 7.6 inches in diameter, and weighs about a pound and a half. It sports 36 fixed-focus two-megapixel mobile phone cameras protected by an exterior shell of cushioning scales. Inside the ball are an accelerometer to compute the time the toss reaches its apex and a computer to trigger the 36 cameras at that precise moment. After downloading, computer software combines the 36 images into a seamless panoramic image.

Unlike wide-screen cameras and projection systems such as IMAX, the throwable's images are truly spherical; there are no blind spots and they can be shown in planetarium-like settings or simply viewed on a PC or pad display with pan/zoom in any direction. Figure 3.21 shows a couple of examples.[48]

What is really novel about the camera is the way it takes pictures. The accelerometer measures the launch acceleration and the onboard program integrates it to determine the launch velocity and the rise time to the throwable's highest point. When the rise time is reached (apex of the throw), camera motion is zero and the program triggers the cameras producing a blur-less set of panoramic images.

(a)   (b)

Figure 3.21. Sample views from the throwable panoramic camera.

---

[46] Patent illustrations from (a) (Kashima 2008, sheet 2/3, fig, 4); (b) (Simonds 2006, sheet 1/12, fig. 1); (c) (Maehara 2001, sheet 6/6, fig. 10); (d) (Machin 1995, sheet 5/14, fig. 5); (e) (Gobush 1991, sheet 12/13, fig. 4); and (f) (McGuire 2002, sheet 3/8, fig. 2).
[47] Jonas Pfeil first conceived of the camera as a diploma thesis at the Technical University of Berlin entitled "Throwable Camera Array for Capturing Spherical Panoramas" in 2010. He has applied for a patent.
[48] All images courtesy of Jonas Pfeil.

Pfeil highlights the importance of the number and spherical location of the cameras:

> The cameras are positioned on the ball's surface by numerically optimizing their orientations and ensure full panoramic coverage. Also the smallest amount of overlap between neighboring cameras is maximized to facilitate blending between the images.[49]

Each toss produces 36 JPEG format images, which are combined after downloading to a PC or tablet by way of standard USB connection. The camera uses rechargeable batteries, much like a cell phone. Editing and viewing software allows complete 360° pan/zoom and printing of panoramas. A commercial version of the camera is expected soon.

3.13   Hoberman's MiniSphere

Toys and games embody many spherical principles, and they are a playful way to see spherical subdivisions and 3D geometry in action. Charles Hoberman's expandable *Mini-Sphere* and Mark White's *Code World* are two examples. Both are embodiments of patented technologies for totally different applications—one for expandable roof construction systems, and the other for genetic coding.

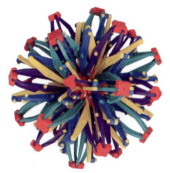

The *MiniSphere* is a series of interconnected plastic trusses that can expand and contract, causing its spherical form to greatly increase or decrease in size without changing its basic geometry. One version is shown at the right in its folded-up configuration. It's a fascinating, dynamic toy that one can watch for hours, wondering how it works without locking itself up or distorting its shape as it changes size. In some large exhibition versions, a motor facilitates expansions and contractions.

Hoberman, who was interested in expandable truss systems, developed several patented designs that could expand and collapse. These designs offer high levels of prefabrication, lightweight elements, ease of transportation, and the ability to span large spaces with minimum support.[50] The *Mini-Sphere* is an embodiment of this patented truss design.

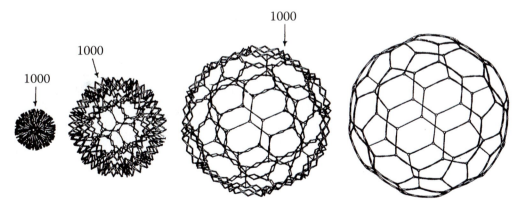

Figure 3.22. Radial expansion/retraction truss structures.

---

[49] (Pfeil 2012, pers. comm.)
[50] See Hoberman's reversibly expandable doubly-curved truss structure (USPTO Patent 4,942,700) and radial expansion/retraction truss structures (USPTO Patent 5,024,031) for examples of unfolding structures and toy transformations.

The core of this invention is a set of trusses, each made of a great circle "loop-assembly." A loop is made of many smaller scissor-pairs of lightweight tubular elements jointed to one another or to small hubs. A loop can expand or fold up, changing its size but not its basic geometry. In one version, three mutually orthogonal closed loops are arranged, each forming a great circle on their respective Cartesian *xy*-, *yz*- and *zx*-planes. The three loops intersect each other at six places and are joined via small hub connectors. In this arrangement, even though one loop-assembly passes through (intersects) another, they do not inhibit each other's synchronous unfolding or folding, as they are expanded or collapsed together. Their intersections always maintaining their orthogonal alignment. Smaller scissor-pair trusses fill the spaces between the three main closed loop-assemblies. These trusses are very flexible, and many combinations are possible. Figure 3.22, from Hoberman's patent, shows a typical expansion sequence.[51] One of the loop-assemblies made of scissor-pairs is labeled 1000, and you can follow its successive expansion. At its most expanded position, the scissor-pairs are so extended they look like a double-thick strut. Folding is simply the reverse sequence.

3.14   Rafiki's Code World

Rafiki's Code World is a very different spherical experience that proves you can learn subtle geometrical principles while having fun. This copyrighted logic game challenges players to define a sequence of spherical moves to match color combinations drawn from a deck of col-

or-coded challenge cards. This innovative game is not played on a flat board; rather, it is played on a sphere made up of three different spherical polyhedra. The main parts of this game are a brightly colored globe defined by two polyhedra surrounded by a glider, a third polyhedron (see figure at left).[52] The globe is a spherical icosahedron with its 20 color-coded triangular faces and circular insets. Across its faces is a continuous track or groove that defines the edges of another spherical polyhedron, the dodecahedron. These two polyhedra, the icosahedron and dodecahedron, are *duals*, which mean they have a symmetrical relationship. The 12 vertices of the icosahedron are centered midface to each of the dodecahedron's 12 faces and the dodecahedron's 20 vertices are centered within the icosahedron's 20 faces. Thus, one polyhedron has as many faces as the other has vertices and vice versa. You can easily verify these relationships in the globe. Note, also, that the 30 edges of the dodecahedron and icosahedron cross perpendicular to each other. Both polyhedra, therefore, have the same number of edges. We will take a closer look at polyhedra and their dual relationships in Chapter 6. We use them extensively when we subdivide spheres.

A glider, shaped like the edges of a spherical tetrahedron, surrounds the globe. The tetrahedron is one of the five Platonic solids. It has four large equilateral faces. Its six edges are represented in the model by wide green arcs spanning four vertex hubs colored red, blue, green, and yellow (not visible). Although the glider is free to move about the globe, pins under the hubs maintain the glider's orientation within the dodecahedral groove. Game inventor Mark White calls the entire unit, glider and globe, a "stone," short for the sorcerer's stone (a legendary tool of alchemists) that figures prominently in the game. Each time the glider is repositioned, a new configuration of four faces and four circle colors are mapped (decoded) by the hubs. There are 120 possible color configurations. The challenge

---

[51] (Hoberman 1990, sheet 7/12, figs. 13–16)
[52] Image courtesy of Mark White.

of *Rafiki's Code World* is to solve various *Code World* puzzles with a sequence of glider moves. Puzzles are drawn from a deck of cards, which shows combination pairs of colored triangle-circles or just circles. White says that puzzles have an infinite number of valid solutions; some sequences are shorter than others and herein lies the spatial challenge.

From a purely polyhedral point of view, *Rafiki's Code World* game is a wonderful kinematic introduction to dual polyhedral relationships and color mapping problems. But, like Hoberman's MiniSphere, Mark White's *Code World* is more than a game; it's an embodiment of another totally different invention, one based on genetic mapping. White sees a relationship between symbol-sets (analogous to the colored faces in *Code World*) and sequences of the amino acids in a protein (the way the glider's position spans particular colored faces). His patent application describes the model this way: it is really a spherical slide rule for biologists.[53]

3.15   Art and Expression

Subdivided spheres appear to defy gravity; they seem to want to be in space like stars, or under water like bubbles. And when they are held up, they feel as if columns of light alone could support them. Most of our spherical subdivisions are highly symmetrical—thus, balance is maintained in what's kept and what's left out. It's like your hand. Your fingers, along with the spaces between them, make up something larger.

Artists often equate mass and monumentality. However, subdivided spheres point out contradictions in this equation. Subdivided spheres can be massive, yet light, solid, airy, space-filling, or full of holes.

Here we sample the work of George Hart, Duncan Stuart, Vladimir Bulatov, Bathsheba Grossman, Rinus Roelofs, Francesco de Comité, and Nick Sayers.[54] No explanation is needed—they are here to delight and to educate. For the spherist, subdivision art is where surfaces, edges, and points meet light, texture, material, and you, the viewer.

Spaghetti Code
Aluminum, 2 meters,
2004 © George W. Hart

5-D Hypertetrahedron
Pencil and illustration board, 20 in. × 30 in.,
undated, estate of Duncan Stuart

---

[53] (Mark White 2003).

[54] Used by permission of the Estate of Duncan Stuart; image courtesy of George Hart; used by permission of Vladimir Bulatov, Abstract Creations, Corvallis, Oregon, USA; image courtesy of Bathsheba Sculpture LLC; image courtesy of Rinus Roelofs; image courtesy of Francesco De Comité; image courtesy of Nick Sayers.

Dodecahedron XII
Stainless steel and bronze, diameter 2 in.,
2007 © Vladimir Bulatov

Quintron
Direct-metal painting in steel, diameter 4 in.,
2009 © Bathsheba Grossman

Interwoven Sphere
Metal tubing, 60 elements,
2008 © Rinus Roelofs

Hamiltonian Circuits on a Dodecahedron
Computer 3D rendering,
2009 © Francesco De Comité

Coke Bottle Lampshade
58 plastic Coke bottles slotted together, diameter 48 cm.,
2010 © Nick Sayers

Additional Resources

Science and Nature

Caspar, Donald L. D. and A. Klug. "Physical Principles in The Construction of Regular
Viruses." *Cold Spring Harbor Symposia on Quantitative Biology* 27 (1962): 1–24.

Conway, John H., Heidi Burgiel, and Chaim Goodman-Strauss. *The Symmetries of Things*.
Wellesley, MA: A K Peters, Ltd., 2008.

Mathematics, Geodesics, and Cartography

Clinton, Joseph D. "A Limited and Biased View of Historical Insights for Tessellating a
Sphere." *Space Structures 5* ed. G. A. R. Parke and P. Disney. London: Thomas Telford
(2002): 423–431.

Ringler, Todd D., Ross P. Heikes, and David A. Randall. "Climate Modeling with Spherical
Geodesic Grids." *Computing in Science and Engineering IEEE Computational Science
and Engineering* 4.5 (2002): 32–41.

Sahr, Kevin, Denis White, and A. Jon Kimerling. "Geodesic Discrete Global Grid Systems."
Cartography and Geographic Information Science 30.2 (2003): 121–134.

Szalay, Alexander S. et al. "Indexing the Sphere with the Hierarchical Triangular Mesh," Microsoft Corp. 2005. Technical report MSR-TR-2005-123, 2005.

Tarnai, T. "Polymorphism in Multisymmetric Close Packings of Equal Spheres on a Spherical Surface." *Structural Chemistry* 13.3-4 (2002): 289–295.

Polyhedra

Cromwell, Peter R. *Polyhedra.* Cambridge [Eng.], Cambridge: Cambridge University Press, 1997.

Cundy, H. M. and A. P. Rollett. *Mathematical Models.* Oxford: Oxford University Press, 1961.

Hart, George W. and Henri Picciotto. *Zome Geometry: Hands On Learning with Zome Models.* Emeryville, CA: Key Curriculum Press, 2001.

Holden, Alan. *Shapes, Space and Symmetry.* New York: Columbia University Press, 1971.

Wenninger, Magnus J. *Polyhedron Models.* Cambridge [Eng.]; New York: Cambridge University Press, 1971.

The Sphere

Aste, Tomaso and Denis L. Weaire. *The Pursuit of Perfect Packing.* Philadelphia, PA: Institute of Physics Publishing, 2000.

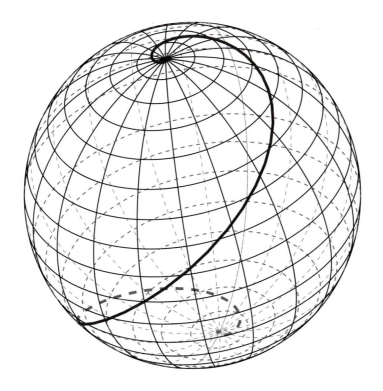

④ Circular Reasoning

W e live on a small part of a large sphere. Most of the time our senses tell us that it is flat. We perceive our world as Euclidean, with straight lines and distances measured in feet or meters. We see parallel lines in roads and railroad tracks. Objects come in various sizes and we can enlarge or reduce any image, preserving all the angles. Our daily life reinforces these perceptions.

But on a sphere, none of these impressions are true. The shortest path joining two points on a sphere is called a geodesic and it is an arc of a great circle defined by the intersection of the sphere and a plane through its center and distances are measured in arc length. Geodesic arcs are the "straight lines" of spheres. On a sphere, every pair of distinct great circles intersect, not once but twice, at points opposite one another called antipodal points. There isn't even a unique shortest path between these opposite points, either. We see this on a globe where there are an infinite number of geodesics (longitudinal lines are arcs of great circles) that pass through the north and south poles; *all* of them are the "shortest path" between these two antipodal points.

There are other differences between Euclidean geometry and spherical geometry. There are no "rectangles" with straight sides and four right angles, either. And somewhat

Euclid's Postulates

Euclid (about 300 BC) wrote the first set of axioms for ordinary Euclidean geometry. Euclid did not invent geometry, but he did invent rigor and publicized it. *Euclid's Elements* consisted of 13 books containing 465 propositions (theorems and proofs). These propositions were based on five postulates:

1. A straight line segment can be drawn joining any two points.

2. Any straight line segment can be extended indefinitely in a straight line.

3. Given any straight line segment, a circle can be drawn having the segment as radius and one endpoint as center.

4. All right angles are congruent.

5. If two lines are drawn that intersect a third in such a way that the sum of the inner angles on one side is less than two right angles, then the two lines inevitably must intersect each other on that side if extended far enough (also called the parallel postulate).

surprisingly, there are polygons with just two "straight" sides. Triangles are different too. When the three corresponding angles of one triangle are equal and in the same order as another triangle, the two triangles are congruent. But you cannot "enlarge" or "shrink" a spherical triangle and preserve its angles as you can with similar triangles in Euclidean geometry. There are no two ways about it: spherical geometry is simply non-Euclidean.

No geometry on a sphere can satisfy all of Euclid's five postulates (see side bar) for plane geometry, so any geometry we define is "non-Euclidean" by nature. Non-Euclidean geometry is one of many geometric systems where some of Euclid's first four postulates hold but his fifth one, the so-called parallel postulate, does not. Euclid's fifth postulate, restated in a simpler way, says that through a point not on a line, there is no more than one line parallel to the first line. In Euclidean geometry, parallel lines—lines in the same plane that do not intersect—are fundamental. However, on a sphere, the parallel postulate does not hold, either because there are no lines parallel to the given line through the point, or there will be more than one line through the point that does not meet the original line.

Mathematicians since Euclid realize many self-consistent geometries are possible where one or more of Euclid's postulates do not hold. Spherical geometry is one of them.

In this chapter, we explore the properties of spheres and see where we can employ Euclidean principles and where we can't. As the chapter title says, we use "circular reasoning." We start with arcs of great circles joining nonantipodal pairs of points and proceed to construct spherical polygons on the surface of the sphere. Since spherical triangles are the easiest to use when we want to subdivide a sphere, we concentrate on properties of triangles, such as shape, symmetry, and area.

Spheres are homogeneous; that is, a region about every point looks like a region about every other point, in every direction we look. To compensate for this sameness, we introduce coordinate systems so we can locate specific points and distinguish one region from another. The earliest coordinate systems go back to astronomers studying the celestial sphere, or geographers mapping the surface of the earth, many of whom were also astronomers. As in plane geometry, the choice of coordinate systems is very important for the sphere. Often a problem can be stated and solved much more easily in one coordinate system than another. A basic task is to determine which system to use in which situation and how to convert results found in one system so they can be used in another.

We begin with the fundamental notions on which everything else is based; namely, points and circles.

4.1 Lesser and Great Circles

Any plane cutting through a sphere intersects the sphere in a circle. If the plane goes through the center of the sphere, the circle is called a great circle; and if not, the circle is a small circle, also called a lesser circle. A family of parallel planes in space will cut the sphere in one great circle and infinitely many lesser circles and some of these circles are particularly useful in geodesic applications. In Figure 4.1(a), every circle is a lesser circle defined by a plane that intersects the sphere but does not pass through the sphere's center. In the figure, the plane defining each lesser circle is perpendicular to one of three Cartesian axes. Some circles have such small radii that they appear as points, but the intersection of the sphere and a plane is a point only when the plane is tangent to the sphere at the point so it does not cut through the sphere.

A slice of a sphere also creates something else besides the great or lesser circle, namely two spherical caps. The spherical caps of a great circle are hemispheres or two half-spheres. For a lesser circle, one spherical cap is smaller than a hemisphere and shaped somewhat like a contact lens. A spherical cap has a center on the sphere and that is different from the center of the disc in the plane that cuts out the circle on the sphere. If we choose a collection of points on the sphere and consider a spherical cap centered on each one, and if all the caps have the same size and the largest possible radius without causing any caps to overlap, we can compare the total area of the caps and the area of the sphere to get a measure of how well distributed the points are on the sphere (see Figure 4.1(b)). The dashed caps are on the other side of the sphere. The radii of the cap's lesser circles are all the same. Some are tangent to others, but none overlap. Thus, for this particular number and distribution of points on a sphere, the total area of the caps as a percentage of the total area of the sphere tells us how uniformly the points are distributed. In Figure 4.1(b), the caps cover about 71 percent of the surface of the sphere. This important use of lesser circles and spherical caps will be explored again later in the chapter.

Great circles are the geodesics of the sphere; they are the essential tool of spherical subdivision. Unlike lesser circles and spherical caps, a great circle is a circle on the surface

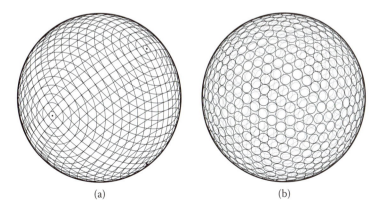

(a)   (b)

Figure 4.1. Lesser circles and spherical caps.

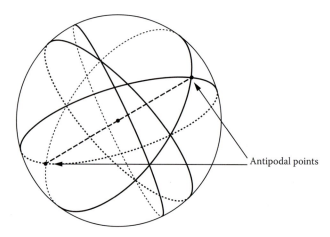

Antipodal points

Figure 4.2. Random great circles.

of a sphere having the same diameter as the sphere; they are the largest possible circle on a sphere. A great circle shares its center with the center of the sphere. Any two points on a sphere, plus the sphere's origin, uniquely define the plane of a great circle—so long as the two points on the surface of the sphere are not coincident or on exact opposite sides of the sphere. A sphere's great circles are all equal in diameter, curvature, and planar area.

An infinite number of great circles can be drawn on any sphere (a few of them are shown in Figure 4.2) by randomly selected planes, each passing through the sphere's center. Notice that every one of the great circles intersects every other one and the line where they intersect will intersect the sphere's surface at two points opposite each other (antipodes).[1] One pair of great circles, their line of intersection and intersecting antipodes are indicated in the figure.

In Figure 4.2, it is easy to see each great circle divides the sphere into two symmetric parts called hemispheres and that great circles are the largest circle possible. The amount of *curvature* of a great or lesser circle can be measured. It is equal to the reciprocal of the circle's radius (the smaller the reciprocal, the "flatter" the circle's curve). Great circles have the least curvature of any circle on a sphere while small circles have higher curvature on the same sphere because they bend more sharply. For example, if the radius of the sphere in Figure 4.1(b) is equal to 1, the curvature of the great circles will also be equal to 1 whereas the curvature of the cap's lesser circles with radius of .063 (each cap has the same radius in the figure) is 15.9. It is clear, from this example, that arcs of great circles are the only possible choice for locally shortest curves in the plane of the circle and curves with this property are the straight lines of spheres and the shortest distance between any two nonantipodal points. This distance is often called the *great circle distance*.

Great circles have another relationship with their spheres. The area of a sphere is given by the classic formula:

$$\text{area of sphere} = 4\pi r^2.$$

---

[1] Antipodal points have the same Cartesian coordinates, but opposite signs for each component. For example, on a unit sphere, point (–0.5, 0.5, 0.707) is opposite (0.5, –0.5, –0.707).

Notice in the equation that the sphere's area is four times the planar area of one of its great circles. There is a handy way to remember this: you need twice as much paint for the cathedral dome as you do for the floor under it.

4.2   Geodesic Subdivision

Great circles and their geodesic arcs are essential in spherical subdivision. Great circles are used to subdivide a sphere, and great circles are the result of spherical subdivision. We need to take a closer look at their properties to understand this dual role.

We have said that geodesics are the straightest lines on a sphere and they are arcs of great circles. In common usage, a geodesic is the shortest distance between two points on a sphere. But from a mathematical perspective, there are actually two geodesics between a pair of points that are not anitpodal: a short path and a long path. Figure 4.3(a) shows both of them—the geodesic short path is solid, while the long path is dashed—between two points. In spherical subdivision, we almost always use the short path geodesic.

The situation is a little different when the points are antipodal. In this case, an infinite number of geodesics pass through them. Both long and short paths have the same arc distance and none are unique in length. Figure 4.3(b) shows a few of them and illustrates two facts again: every great circle bisects every other great circle, and every great circle intersects every other great circle at antipodal points. In (b), every great circle intersects at the same antipodes, but this is certainly not the case in Figure 4.2, where every pair of intersecting great circles creates different antipodes.

In spherical subdivision, we regularly intersect geodesic arcs to derive points that define a polygon or grid on the sphere's surface. We will develop these techniques in detail later, but here, we can get a feeling for the general subdivision process. In Figure 4.4(a), points a and f, c and h, and the sphere's origin define two great circles that intersect at antipodes 1 and 2 (Appendix C explains the simple procedure for doing this). We can say that points 1 and 2 are derived from, or are the result of, intersecting two great circles.

We can derive points to make very complex grids and spherical subdivisions by intersecting different combinations of great circle arcs. Figure 4.4(b) outlines how this is done. The process starts by defining a series of reference points around the edges of a spherical

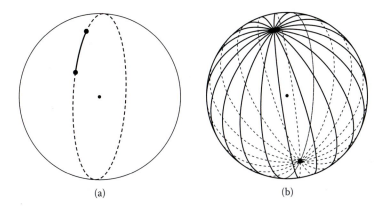

(a)   (b)

Figure 4.3. Long- and short-path geodesics.

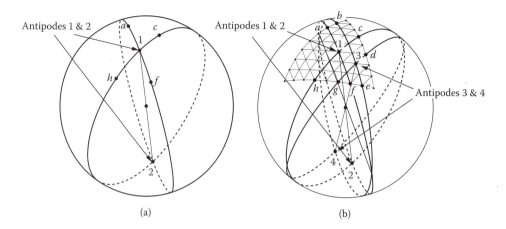

Figure 4.4. Defining points on a sphere by intersecting great circles.

triangle. In this example, there are 24 edge reference points including the three triangle vertices. The triangle's edges themselves are geodesic arcs between the triangle's vertices. The reference points are represented by small dots around the edge of the large spherical triangle in (b), and some are labeled *a* through *h*. We will discuss how these reference points are defined in a later chapter. For now, we will focus on using them to derive new points defined by intersecting geodesics.

By selectively picking pairs of edge reference points, for example, points *a* and *f* with points *c* and *h*, or *b* and *e* with *d* and *g*, we define four geodesic arcs that intersect at antipodal points 1 and 2, 3 and 4. Points 1 and 3 fall within the spherical triangle and define useful vertices on a subdivision grid. Points 2 and 4 fall on the opposite side of the sphere and are ignored. We can repeat this process—select pairs of edge reference points, define their geodesic arc, and select one of their intersection antipodes—over and over until all the subdivision vertex points within the triangle are defined.

To review geodesic subdivision again from a slightly different perspective, in the previous example, we started with 24 reference points around the edge of a large spherical triangle, as shown in Figure 4.4(b). Three of those points are the vertices of the large spherical triangle we want to subdivide. Notice that these 24 reference points define sets of eight great circles; each great circle in a set runs somewhat "parallel" to the edges of the large spherical triangle. By intersecting combinations of one great circle from each of the three sets, we create 42 antipodal points, 21 of those points falling within the large spherical triangle we are subdividing; the other 21 fall on the opposite side of the sphere and are ignored. The 45 total vertices shown for this subdivided triangle consist of 24 edge reference points plus 21 points derived from the intersection of great circles.

Once the vertices of the grid are defined, the grid members themselves can be either short geodesic arcs, such as the ones shown in Figure 4.4(b), or straight line chords between grid points. The former results in a grid of small spherical triangles; the latter results in planar triangles. If we want to subdivide the whole sphere, we simply make copies of the subdivided large spherical triangle and then position them to cover the rest of the sphere without gaps or overlaps.

We have just given the briefest of introductions to subdividing spheres. This method demonstrates a basic technique for using great circles to generate points that make up a subdivision grid. These techniques have several variations, and we will devote an entire chapter to them. Some variations start by subdividing a different shaped triangle, an equilateral triangle, or perhaps a right triangle. The number and spacing of the edge reference points around it may also vary. Other variations change the number of great circle intersections needed to derive a single subdivision point.

Regardless of these variations, the technique of deriving points by intersecting great circles or geodesic arcs remains the same. In Chapter 8, we use this important technique in several ways, and in Chapter 9, we compare their effects in subdivision. Depending on your application, one variation may result in a more suitable distribution of points than another and we will examine some ways of telling which are best for a given situation.

4.3   Circle Poles

Every great and lesser circle has two poles, each equidistant from every point on the circle's circumference. The poles are points on the sphere and define an axis that is perpendicular to the plane of the circle, passing through the circle's center and the sphere's origin. Figure 4.5 shows a great (a) and lesser (b) circle, their poles, axes, and points on the circumference.

Polar distance is the arc distance on the sphere between the point and either the north or south pole. All points on the equator are the same distance from the north or south pole, measured in degrees or *radians* (90° or π/2 radians) as shown in Figure 4.5(a). For the lesser circle shown (b), the polar distances are somewhere between 0° and 180°.

The *angular radius* of a circle (lesser or greater) is the interior angle between the circle's closest *pole* and any point on its circumference. This angle is labeled θ in Figure 4.5(a) and (b). Note that the angular radius of any great circle is always 90°. The angle between the two points in the sphere measured from the origin is called the *central angle*.

One last relationship between great and lesser circles is worth mentioning. Pairs of points on a sphere that are not coincident or antipodal, together with the origin of the

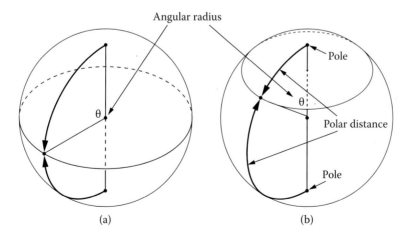

Figure 4.5. Circle poles and polar distances.

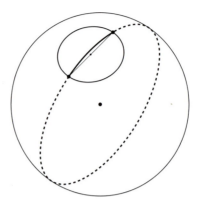

Figure 4.6. Points define great and lesser circles.

sphere, actually define two circles. In geodesics, we are usually interested in the shortest path between the pair, but the same two points also define the circumference of a lesser circle, as shown in Figure 4.6. The center of this circle is inside the sphere and at the middle of the chord between the two points.[2] The plane of the lesser circle is perpendicular to the plane of the great circle, defined by two points and the origin of the sphere. If those two points are antipodal, the lesser circle becomes a great circle, and an infinite number of them pass through these opposite points, as shown earlier in Figure 4.3(b).

4.4   Arc and Chord Factors

Distances between points on the surface of the sphere are measured in one of two ways: either along the surface following the great circle arc between them, or a straight line that cuts "through" the sphere as a chord.

The *chord distance* is the direct straight line distance between any two points on a sphere. The chord's length depends upon the radius of the sphere and is directly proportional to it. *Chord factors* express this length relative to both the radius of the sphere and the central angle between the two points measured at the origin. Chord factors are very useful in geodesics. If the sphere's radius is one, a unit sphere, the chord factor can be scaled up or down for any real-world application. That is, a chord's factor can be multiplied by the true radius of the intended sphere to arrive at the exact dimension of a required implementation. For any sphere with radius r, we can express the general relationship of a chord's length to its central angle θ as

$$\text{chord factor} = 2r \sin(\theta/2).$$

Arc factors are conceptually related to chord factors. Instead of measuring the direct straight-line distance between points, arc factors measure the arc length (geodesic distance) between the points in radians. The arc factor is simply the radian measure of the central angle of the arc. One radian is the same length as the sphere's radius "wrapped" around the

---

[2] The line between these two points is a chord. The perpendicular distance from the mid point of the chord (indicated with a dot) to the great circle is called the *sagitta*, and to the center of the great circle is called the *apothem*. Both are somewhat antiquated terms but useful in spherical work.

circumference of the sphere or one of its great circles. The number of radians or degrees in a great circle is

$$1 \text{ radian} = 360°/2\pi \text{ or } 180°/\pi = 57.2958°.$$

Thus, the number of radians in a degree is

$$1° = \pi/180° \text{ radians.}$$

An arc factor provides two pieces of information around the arc at the same time: it defines the size of the angle and the length of the arc it makes on the surface of the sphere. The arc factor is simply the radian measure of the central angle θ between two points on the unit sphere. The formula is

$$\text{arc factor} = \pi/180°\theta r,$$

where θ is the central angle between two points.

4.5   Where Are We?

The sphere is an omnidirectional object; every view is the same. There is no top, bottom, left, or right side to a sphere unless we impose one. To answer the question, "Where are we?" we need a reference system, and several important ones are used in geodesics. The ones we consider here are all based on an origin and axis system.

We start with the oldest and most intuitive system, the altitude-azimuth system, introduced by ancient astronomers. From there, we look earthbound and see what the familiar latitude-longitude system tells us about the sphere. Neither of these systems considers the radius of the sphere. The stars are so far away, the earth's radius is simply ignored. So we will look at two other systems, spherical and Cartesian systems, where distance and radius do play a role.

Why do we need so many systems? For a unit sphere, any one these systems will do, if we make certain assumptions. But in practice, it often is easier to solve certain geometry problems by using one system rather than another. It is common practice to solve part of a problem in one system, and then convert the result to another system for the final resolution. Reference systems help us to find where we are, just as the ancient astronomers figured out their location by looking at the sky.

4.6   Altitude-Azimuth Coordinates

The altitude-azimuth coordinate system, also called the horizontal coordinate system, is a 2D reference system that astronomers and celestial navigators have used for centuries to locate bodies in the sky. The sky is treated as if it were a huge spherical shell surrounding the earth, with the observer inside, at the center. Altitude-azimuth coordinates are easy to understand and are widely used because they are based completely on references relative to the observer or a reference point at the center of the sphere.

The references in this system are the observer's horizon plane (the earth or sea), *zenith* and *nadir* (points directly overhead and beneath the observer), and the *meridian plane*

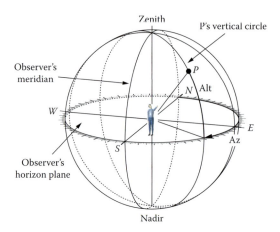

Figure 4.7. Altitude-azimuth coordinate system.

passing through the observer's location north and south and the zenith-nadir axis. Figure 4.7 shows the general arrangement of these references relative to an observer at the center. Point P on the sphere (i.e., a star, planet) is located by two angles: its altitude (Alt) and *azimuth* (Az).

Referring to Figure 4.7, the altitude is the angle from the observer's horizon to point P and can be any value from 0° (the body is on the horizon) to 90° (the body is at the observer's zenith or directly overhead). It is measured along the arc of a great circle called a *vertical circle*. The observer's elevation is not relevant.[3] He or she can stand on the deck of a ship or atop a mountain and will still achieve the same result.

The azimuth angle is measured from the observer's meridian. Two conventions are in common use. In North America, azimuths are measured from the north, increasing in a clockwise direction. East is 90°, south is 180°, and so on. In Europe and in celestial navigation, azimuth is measured from the south, with west being taken as positive. The azimuth of the eastern point on the horizon is −90° and the azimuth of the west is 90°.

Altitude-azimuth coordinates do not include the distance to the point. Distances to the stars are so great that these small discrepancies are not significant.[4]

Although widely used, the altitude-azimuth coordinate system does have some drawbacks. When applied to astronomy, the main drawback is the fact that its reference frame moves with the earth. It is difficult to track celestial bodies without having accurate time and location information, because their coordinates are constantly changing, a result of the earth's second-by-second rotation around its axis, and its yearly revolution around the sun. Altitude-azimuth coordinates also are specific to a particular location on Earth, but a measurement can be made, the time noted, and the result easily converted to any other coordinate system. For the same date, time, and point in the sky, coordinates will be different for

---

[3] In celestial navigation, a sextant measures the altitude of celestial bodies above the horizon, which might be the sea or Earth's land horizon. In these cases, the user's height must be factored in. The result is a small angle called dip of the horizon, which must be subtracted from the observed angle above the horizon. Otherwise, the sextant angle reading is too large, which will have the effect, if not corrected, of placing the observer further from his actual line of position.

[4] For objects closer to the earth, such as the moon, *parallax* must be considered.

observers at other locations, if the body is even observable from all of them. Oftentimes, other systems are used and then converted to altitude-azimuth coordinates for a particular time and observer location.

4.7   Latitude and Longitude Coordinates

Undoubtedly, the most familiar coordinate system is the one using latitude and longitude. All points on the sphere can be located with latitude-longitude coordinates. Unlike the altitude-azimuth coordinate system, this system combines lesser circle references with greater circles. Here we describe the general features of latitude-longitude coordinates and conversions to the Cartesian system.

Before we start, we should point out that terrestrial latitude-longitude coordinates describe positions on a sphere. The earth is not a perfect sphere, as it is slightly flat at the poles and bulges a bit at the equator. Technically speaking, the earth is a spheroid or an ellipsoid. For discussions of coordinates, however, we will treat the earth as if it were a perfect sphere.

The concept of latitude has been known for centuries, as astronomers recorded everyday observations of the sun's movement throughout the year and the altitude angle of Polaris, the North Star. Even without corrections for atmospheric distortion and for Polaris offset from the true north celestial pole,[5] latitude can be determined to within a few miles just by measuring the North Star's altitude above the horizon.

Referring to Figure 4.8, the latitude of a point, usually denoted by the Greek letter *phi* (φ), is the angle from the equator to the point. Latitude circles are lesser circles defined by points of equal latitudes. The equator, latitude 0°, is the only one that is a great circle.

Longitude references, denoted by the Greek letter lambda (λ), are great circles through the north and south poles and are also called meridians of longitude. The line connecting

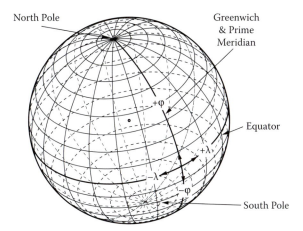

Figure 4.8. Latitude and longitude system.

---

[5] The star Polaris is not exactly on the earth's north polar axis. Instead, it is offset by a small amount and circles the true point by an arc distance of ¾°, thus tracing a small circle around the north celestial pole of about 1½° diameter. In celestial navigation, a small correction angle must be applied to a sextant's measured angle when Polaris sights are taken. The amount of this correction depends on the time the sighting is taken. Twice each day, there is no correction. At two other times, a maximum positive and negative correction is required.

the two poles is called the axis. Longitude measures angles relative to the Greenwich Meridian[6] and increases 0–180° east or west. There are no common sign usages, except for time-zone differences where east longitudes are positive.

4.8   Spherical Trips

Because latitude-longitude coordinates are so familiar, we can take some mental trips around the world to review what we have learned about spheres and see other characteristics of spherical triangles.

We take a companion on our first trip and begin at the north pole. The first thing we notice is that every direction is south. Remember, there are an infinite number of great circles (longitude meridians) through two antipodal points (north and south poles). We decide to walk south, but each of us sets off in a different direction. After a long walk, we both meet up again at the south pole. We arrive at about the same time, because we traveled at approximately the same rate and our overall great circle distances were the same.

The only difference between our two trips is how far apart we were at any time. This distance is called the *separation angle*—the angle, measured from the origin of the sphere, between two meridians at a given latitude. If we each took the same great circle but walked in opposite directions from the north pole, the most we could be separated is 180°. The fact that our separate courses started and ended at common points also means that our paths defined the sides of a spherical polygon—a two-sided one. A two-sided polygon, called a lune or *digon*, is only possible on a sphere; they are always something less than a hemisphere. If we had made our trip by traveling the same great circle, but had walked in opposite directions, our paths would have defined a hemisphere.

Our second trip involves three courses. Figure 4.9 shows our trip plan. Starting again at the north pole, we decide to walk south to the equator by way of longitude 90° west. Along the way, we pass through New Orleans, Louisiana. When we reach the equator, we turn

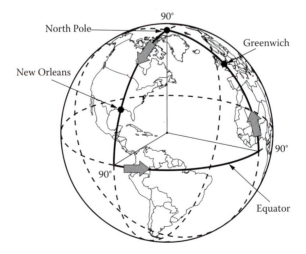

Figure 4.9. Spherical triangle with three right angles.

---

[6] The Greenwich Meridian has not always been the prime meridian. Throughout history, it has changed many times, favoring the dominant maritime nation at the time.

left 90° and proceed to walk one quarter of the way around the equator (or 90°) until we get to the prime meridian, longitude 0°. We then make another 90° left turn and walk due north, past Greenwich (near London) until we arrive again at the north pole. We have just walked a triangular course and arrived back where we started. We have also demonstrated that spherical triangles can have three 90° angles. The north pole is the only place on Earth where you could travel south, then east, and then north and return to your starting point. From the south pole, you could accomplish the same thing by traveling north, then east or west one quarter of the way around the equator, and then south again to make a similar trip.

Our last trip to the equator tells us something else about spherical triangles—any two triangles where two angles and a side are equal are not congruent on a sphere even though they are in plane Euclidean geometry.

We can see this with a slight variation in a third walking tour. Once again, we leave from the north pole and walk to the equator. But instead of walking along the equator exactly one quarter of the way around the sphere, we decide to walk a little less (or more) and then make the same left 90° turn to get back to the north pole. This route still forms a triangle, but this one only has two 90° angles instead of three. The distances from the north pole to the equator on the two polar legs of the trip remain the same as before, but the equator side between the two 90° angles is different. Unlike in Euclidean geometry, this new triangle is not congruent to one we made in the second trip, even though they both have an equal side (north pole to equator) and two equal angles (90° angles at the equator).

Our last two trips each had three courses and formed the edges of a large spherical triangle. But what if we take a trip with four courses and try to form a square or rectangle? Let's say we start at mid-latitude 45° north longitude. This trip might shake your faith in the utility of the compass. We start by walking 500 miles due east, then turn left 90° and walk north for another 500 miles. We make yet another 90° left turn and walk west another 500 miles, then make another turn left and walk due south another 500 miles. We have just walked a long, "squarish" route, where each course was 500 miles and each turn was 90°. But we did not end up where we started from! How did this happen? The reason is that although the distances were equal on the north-south legs along arcs of longitude (great circles), the two east-west legs did not cover the same number of degrees of latitude. We traveled more degrees at the higher latitude than at the lower one.

4.9   Loxodromes

Latitude and longitude are so familiar we can be fooled when navigating. Consider the use of a compass. A compass performs a very specific function; it always points in the same direction—north.[7] Think of it this way: when you change directions, the compass still points north. In effect, you are turning under the compass; the compass continues pointing in the same direction. The value of a compass is that it allows you to maintain a course over a long distance by ensuring that you cross every meridian (longitude) at the same angle. This course is called a *rhumb line* or a *loxodrome*, from the Greek words *loxos* (oblique) and *dromos* (running, from *dramein*: to run).

A loxodrome is a line on a sphere of constant bearing that cuts across all meridians at any constant angle except a right one. A loxodrome is also called a *spherical spiral* and it

---

[7] A compass is influenced by many factors. The magnetic north pole is not exactly located at the geographic north pole. Electrical currents (nearby wires), and large ferrous metal sources near the compass (engines, pipes, etc.) influence which way the compass needle points.

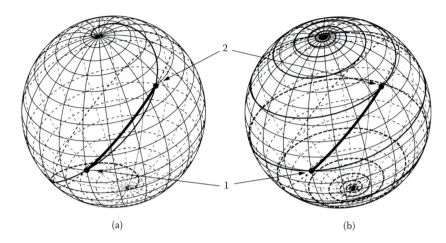

Figure 4.10. Loxodrome courses 56° and 84.2° with a great circle route.

displays as a logarithmic spiral on a polar projection. If the course is not due north-south or east-west, a loxodromic spiral will eventually spiral to either the north or south pole and then endlessly loop that pole in smaller and smaller circles.

A loxodrome displays as a straight line on a Mercator map. When plotted on any other map projection, it is a curved line. A loxodrome is a great circle, only if it is along the equator or a meridian. A rhumb line course between two points that is not along the equator or a meridian is always longer than the *great circle route* between the same two points. And there are an infinite number of rhumb lines between any two nonpolar points that are not on the same latitude or longitude. They only differ in the number of times they encircle the earth.

Figure 4.10 shows rhumb lines or loxodromes between two sample points, 1 and 2, and how they compare to a direct great circle route, shown with a bold line. In Figure 4.10(a), the loxodrome takes the most direct route from point 1 to point 2 by following compass course 56°. In (b), the loxodrome follows a different course of 84.2°. It still leads to point 2, but has to circle the earth one time to do so. In both illustrations, the great circle route between the two points is the shortest route possible. Notice, too, if you miss your destination on any rhumb course other than along the equator, you will eventually end up at the north or south pole.

Can we ever use a compass to follow a great circle route, if using a compass to navigate any direction other than due north-south or east-west puts us on a rhumb line or loxodrome? The answer is a qualified yes. You can approximate the great circle route by taking a series of short rhumb line courses. You would travel a particular compass course for a while before changing to another course, and then another, until you arrived at your destination. To do this, the navigator divides the overall track between his starting and ending points into a series of small segments. Between each segment is a waypoint, a location where he or she will change course slightly. The length of a course segment is about the distance the navigator expects to travel in a day, if by ship, or in an hour or so, if by plane, depending on the plane's speed.

On a ship, the course change might take place at the same time every day. For instance, navigators might change at local noon, the time when navigators traditionally used their sextants to determine their latitude by the sun's noon-time meridian passage. Today, airplanes and ships are equipped with autopilots, and course corrections are made automatically from time to time in order to stay close to a theoretic great circle course. The more rhumb lines or legs the course is divided into, the closer individual rhumb lines will approximate the great circle route.

When a great circle and rhumb line course are shown together on a Mercator map, the great circle route appears as a curved line, while the rhumb line is straight. On a sphere, they appear just the opposite, the great circle being the shortest "straight line" course, while the rhumb line is the longer curved course between points unless the points are directly north or south of each other or along the equator.

To give an example of just how dramatic the difference between navigating a rhumb line and great circle course can be, consider a trip between London, England, and Anchorage, Alaska. The latitude-longitude of London's Heathrow International Airport is 51° 28′ 15.72″N, 0° 27′ 30.98″W. Anchorage International Airport is located at 61° 10′ 14.69″N, 149° 59′ 33.66″W. The great circle distance between them, the most direct route, passes over Greenland and near the north pole, and is 4,475 miles (7,201 km). The rhumb line bearing to Anchorage from London is about 276° (almost due west by compass) and the distance is 5,731 miles (9,223 km). Clearly, the rhumb line distance is longer, almost 28 percent longer. These differences cannot be ignored by airlines or shipping companies. The time and fuel-cost implications are considerable.

4.10 Separation Angle

We have described the geodesic between any two points on a sphere as an arc of a great circle expressed in degrees. This same angle has different names, depending on the context of its use. In spherical geometry, it is an interior angle or arc angle. In astronomy, it is called a *separation angle* and it is the angle measured between celestial objects. On Earth, it is sometimes used synonymously with the great circle distance between pairs of latitude-longitude points. For example, in our first spherical trip previously, two explorers walked in opposite directions from the north pole on their way to the south pole. Their distance apart at any given moment was a function of their latitude-longitude positions. Using φ, λ, and the sign conventions already discussed to represent latitude-longitude of two points (indicated by subscripts), the general formula for separation angle is

$$\cos(\text{separation angle}) = \sin(\varphi_2) \sin(\varphi_1) + \cos(\varphi_2) \cos(\varphi_1) \cos(\lambda_2 - \lambda_1).$$

A degree of the great circle arc and the distance traveled on the earth's surface are related. One degree equals 60 minutes, and each minute equals one nautical mile, or 1,676 feet. This is one reason that ocean and air navigation use nautical miles, not statute miles, to measure distance. Along the same great-circle meridian (changing only latitude), one degree equals a nautical mile. But a minute of longitude along a line of latitude is a nautical mile only at the equator.

The above equation is quite useful, but it assumes the earth is a perfect sphere. Some caution must therefore be exercised when finding the separation angle for very short arcs or for arcs very near the poles because, in trigonometry, the sin function is sensitive to rounding errors for values close to 0°.

4.11 Latitude Sailing

Approximating great circle routes with rhumb line segments is a modern-day technique. It is only possible when you know your longitude, your distances, and when to change course.

Before the world's land masses and relative directions and distances were accurately established, explorers would combine coastal and offshore sailing, known as latitude sailing, to reach their destination. Instead of going directly to their destination, as we would today, earlier explorers would sail north-south or along familiar coasts until they reached the latitude they desired. Then they would sail due east or west using a compass orsun and star sightings to maintain the direction until they arrived at their desired port. Latitude sailing was therefore also called "easting" or "westing." Navigators would check their latitude by taking regular sightings of Polaris, the North Star, or by measuring the elevation of the sun at their local noon, when thesun was due north or south of their ship. The angle told them their latitude and was measured with a wide variety of instruments, including the *astrolabe*, quadrant, cross staff, and sextant.

A cross staff is shown in Figure 4.11.[8] For general sailing, only minor corrections in the angle were needed when sighting Polaris. Sights of the sun and moon, however, required a seasonal and daily correction. Nautical almanacs produced by observatories provide these correction factors.

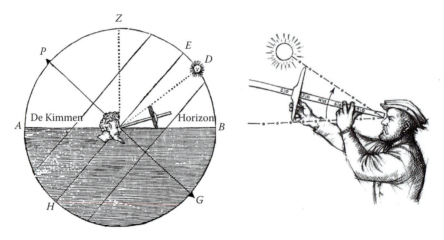

Figure 4.11. A cross staff measures the horizon angle to the sun.

4.12 Longitude

The Chinese used observations of the positions of Jupiter's moons to establish longitude on land hundreds of years before Europeans could accurately measure it at sea. This preceded the invention of a clock that could withstand the rigors of sea conditions: temperature, humidity, and a continuous rocking motion. The first such clock was designed by John Harrison (1693–1776).[9] The most advanced scientific instrument of its day, it revolutionized

---

[8] Image caption in Robert McGhee, *Canada Rediscovered* (Libre Expression and Canadian Museum of Civilization, 1991): "Pilot calculating the height of the sun above the horizon using a Jacob's staff (cross-staff), 16th century."

[9] In 1707, the British fleet was returning from war when beset by fog as it neared the English coast. A navigational miscalculation led to the grounding and sinking of the entire fleet off the Isles of Scilly. The cost of life was huge

navigation and *geodesy*. How does time, as measured by a clock, relate to longitude? The earth rotates through 360° of longitude in 24 hours, or 15° per hour. A celestial object will cross two meridians at different times, and the time difference converted to degrees is the difference in longitude between the two places.

In practice, a clock on board a ship would be set to the local time of a reference location, such as the Old Royal Observatory in Greenwich, England. The Royal Observatory at Greenwich would prepare celestial tables, giving the exact time that the stars, planets, moon, and especially the sun would pass the Greenwich Meridian[10] (longitude 0°) for every second of every day of the year.

At sea, a mariner would use a sextant or other device to observe one of these bodies and record the exact time it crossed the meridian (due north or south) of his ship. This was the moment at which the body was highest in the sky. The exact moment is hard to know, because the body appears to dwell at its high point for a short period. But averaging sites before and after meridian passage led to fairly accurate timing. Consulting his almanac, the mariner would note the time difference between the body crossing his meridian and Greenwich. This is the time it took the earth to rotate under the body the distance between the two places. Since the earth rotates 360° every 24 hours or 15° per hour, it is a simple matter to compute longitude. But again, accurate timing and an almanac with the times of meridian passage for celestial bodies were required.

Here's an example. If the sun crosses the meridian at Greenwich, England, at noon Coordinated Universal Time (UTC), but doesn't cross the ship's meridian until 16:19 UTC, it took 4 hours 19 minutes for the earth to rotate that much. The earth rotates 15° in one hour (360° in 24 hours); thus, the ship is at 64° 48.0′ west longitude, the same as Bermuda. It is also possible to get a good value for the longitude of a new place by observing lunar eclipses and, with the appropriate tables, Jupiter's moons. The clock method is by far the easiest and most accurate.

4.13   Spherical Coordinates

Spherical coordinates, also called spherical polar coordinates, are naturally suited to geodesic work and share some characteristics with latitude-longitude coordinates. Unlike the previously discussed systems, spherical coordinates include the sphere's radius. A point can be accurately plotted in three dimensions and located on the surface of a sphere, if you know the sphere's radius (a length) as well as two polar coordinate angles.

The radius, usually represented by lowercase Greek letter *rho* (ρ), is the distance from the origin to the point. This book uses unit radius spheres, so ρ is always equal to 1. The first polar angle, represented by (φ), is the geodesic arc along the great circle from the positive *z*-axis to the point. This arc is also called the zenith distance or colatitude and the angle's value range is $0 \leq \varphi \leq 180°$. This is the same polar angle described earlier in

---

and the English navy was seriously weakened. In response to the tragedy and public outcry, an Act of Parliament established the Board of Longitude to offer a reward of £20,000 prize for a method or contrivance that could determine longitude within 30 nautical miles (56 km). John Harrison, a clockmaker from Foulby, West Yorkshire, England, spent more than 30 years perfecting a series of clocks called H1 to H4. In 1736, Harrison demonstrated H1, a clock unlike pendulum designs, which did not rely on gravity to maintain its rate. It was the most revolutionary scientific instrument of its time. Harrison used H1 to solicit funding from the board for further refinements. H4, his last design, resembled a large pocket watch. He finally was awarded the full prize from a begrudging Board but only through the intervention of King George III.

[10] Today, tables are prepared and published in cooperation with the US Naval Observatory, Washington, DC.

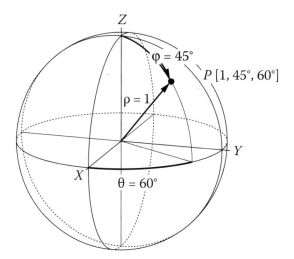

Figure 4.12. Spherical coordinate system.

this chapter in Section 4.3. The azimuth, the other polar angle, is represented by (θ). The azimuth is the angle between the x-axis and the point's great circle intersection with the xy-plane. This is the same angle in the azimuth-altitude system, except it is measured from the x-axis instead of the north reference pole. Theta's value range is $0 \leq \theta \leq 360°$, and the positive direction is counterclockwise around the z-axis.

Figure 4.12 shows the spherical coordinates of a point P, where (ρ, φ, θ) coordinates are (1, 45°, 60°). Notice that the azimuth angle (θ) will be undefined when (φ) is 0° or 180°. The effect is to define an infinite number of great circles passing through the z-axis, as illustrated earlier in Figure 4.3(b).

Some caution is needed when using the spherical coordinate system. The (ρ, φ, θ) convention used in this book follows the standard notation of mathematicians. However, in fields like physics, angles (φ) and (θ) are reversed. To add to the confusion, polar angle (φ) is sometimes measured from the xy-plane rather than the $+z$-axis, and azimuth angle (θ) is sometimes measured from the x-axis 0 to $\pm90°$ rather than counterclockwise 0 to 180°.[11] Scientific calculators with spherical coordinate functions have the same drawbacks. This book and the algorithms shown in Appendix C use the mathematics convention (ρ, φ, θ), with the angular specifications described above.

4.14   Cartesian Coordinates

Without a doubt, the Cartesian system is the most familiar coordinate system. A point's position in 3D space, relative to the origin of an orthogonal (perpendicular) axis system, is specified by three distances (positive, zero, or negative) in the x, y, and z directions. The Cartesian origin and some sample points, along with their coordinates, are shown in Figure 4.13.

---

[11] For standards and conventions, see report ISO 31-11 Quantities and units—Part 11: Mathematical signs and symbols for use in the physical sciences and technology, International Standards Organization, 1992.

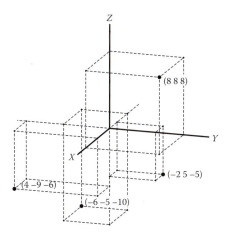

Figure 4.13. Cartesian coordinate system.

Some very useful relationships exist between these points. The distance of a point from the origin is simply

$$\text{distance} = \sqrt{x^2 + y^2 + z^2} \, .$$

The distance between any two points is

$$\text{distance} = \sqrt{(x_1 - x_2)^2 + (y_1 - y_2)^2 + (z_1 - z_2)^2} \, .$$

Many transformations can be performed on Cartesian coordinates. Key ones described in Appendix C include scaling, translation, and rotation. *Scaling* simply moves points closer to or farther from the origin by multiplying all coordinates by a constant value. The directions of points relative to one another remain the same. Translation moves one or more points to a new location. If more than one point is being translated, the move does not change the angular relationships of any of these points to one another. In a sense, the points are like passengers seated on a bus—they all move together, but their positions with respect to each other remain the same.

Rotation is very important in geodesics. In a rotation, one or more points spin around an axis such as children rotating on a carousel. Their distances to the rotation axis remain the same; only their positions around the axis change. Just like in translation, rotations do not change the relationship of the rotated points to one another. See Appendix E for a complete explanation of rotations used in geodesic subdivisions.

Although Cartesian coordinates can define any point in 3D space, they are not independent when the points lie on the surface of a sphere. They must be related by the equation $r^2 = x^2 + y^2 + z^2$, where r is the radius of the sphere. A simple relationship between the Cartesian and spherical coordinate systems is illustrated in Figure 4.14. It is a simple matter to convert from one system to another, and converting back and forth is common practice in subdivision work. Referring to the earlier spherical coordinate example, where P's coordinates were $(1, 45°, 60°)$; the Cartesian value is approximately $(0.3536, 0.6124, 0.7071)$.

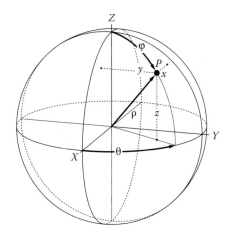

Figure 4.14. Cartesian and spherical coordinates.

Converting from spherical to Cartesian coordinate systems is quite easy. Scientific calculators can do the conversion just by changing the coordinate specification mode. The equations are

$$x = \rho \sin\varphi \cos\theta \quad \rho = \sqrt{x^2 + y^2 + z^2},$$

$$y = \rho \sin\varphi \sin\theta \quad \tan\varphi = \left(\frac{\sqrt{x^2 + y^2}}{z} \right), \text{ and}$$

$$z = \rho \cos\varphi \quad \tan\theta = \frac{y}{x}.$$

4.15 ρ, φ, λ Coordinates

The (ρ, φ, λ) coordinate system is not a new coordinate system, it is simply a variation on the latitude-longitude system discussed earlier. Converting latitude-longitude coordinates to Cartesian coordinates involves intersecting great circles and lesser circles. The equator is usually treated as the Cartesian xy-reference plane. In the (ρ, φ, λ) system, the xz-plane acts as the equatorial reference plane. This simple change in orientation makes it easy to find the Cartesian coordinates of *triacon* grid points, a spherical subdivision technique we explain later in Chapter 8. The (ρ, φ, λ) coordinate system is handy in other geometric situations as well.

The three-part coordinate (ρ, φ, λ) specifies a radius (ρ) and two angles $(\varphi$ and $\lambda)$. Since we work with a unit sphere, radius $\rho = 1$. The "latitude" angle φ is measured from the "equator," which is the zy-plane and is in the range of 0 to $\pm 90°$. In a similar fashion, the "longitude" angle is measured relative to the zy-plane from the $+z$-axis. Its range is also 0 to $\pm 90°$. Figure 4.15 shows the basic setup. A point P at $(1, 0°, 0°)$ is at the zenith or $+z$-axis. In the figure, the great and lesser circle references are shown every 5°. Small dots show points at great and small circle intersections every 5°, and the point P at coordinates $(1, 45°, 45°)$

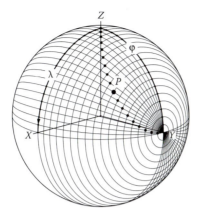

Figure 4.15. (ρ, φ, λ) coordinate system.

is emphasized. It is interesting to note the loxodrome appearance of the intersection points, where both α and β increments were equal. We will make use of the (ρ, φ, λ) coordinate system in Chapter 8, when we develop the triacon subdivision schema.

To derive the Cartesian coordinates for a sphere of radius ρ at the intersection of a great and lesser circle with z-axis polar angles φ and λ:

$$x = \rho \sin \lambda \cos \varphi,$$

$$y = \rho \sin \varphi,$$

$$z = \rho \cos \lambda \cos \varphi.$$

Now that we know how to define points in space and on a sphere, let's look at how we can define spherical polygons by creating edges between points.

4.16 Spherical Polygons

Spherical polygons have two or more sides, always defined by the arcs of great circles. In this section, we look at the most common ones: lunes, quadrilaterals, and regular polygons since any spherical polygon can be broken down into two or more back-to-back spherical triangles. We will look at spherical and planar triangles in great detail, since the subdivision schemas we develop later also define triangles.

4.16.1 Lunes

Lunes, also called digons or bigons, are two-sided polygons defined by the intersection of two great circles. Lunes are unique to spheres and do not exist in plane geometry. Anyone who has seen a beach ball is familiar with lunes, even if they do not realize it. Technically, a beach ball is a *hosohedron*, a spherical polyhedron made completely of lunes.

A lune is less than a hemisphere, and the great circles that define one lune actually define another congruent one on the opposite side of the sphere. Figure 4.16(a) shows

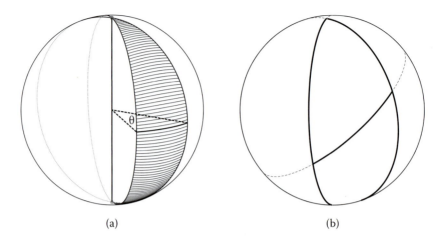

(a)   (b)

Figure 4.16. Lune, lunar angle, and colunar.

both of them, one is shaded, and its *lunar angle*, theta (θ), is indicated. This same angle is also called the dihedral angle or *angular separation*, depending on the context of where it is used. A key point to remember is that the angle where the great circles intersect is the dihedral angle between them; it is measured on a plane perpendicular to the line of intersection of the two great circles defining the lune. If another great circle, such as the one shown in (b), divides a lune into two congruent spherical triangles, the two triangles are said to be *colunar*.

A lune demonstrates another useful spherical concept. For a dihedral angle of θ, the area of a lune is the proportion of the sphere covered by the lune. The surface area of a sphere is $4\pi r^2$, so the area of a lune is $\theta/2\pi \times 4\pi r^2$ or $2r^2\theta$, where θ is expressed in radians. For a unit sphere, a lune's area is twice the lunar angle. If the lunar angle is measured in radians, the area is expressed in steradians or square radians. If the lunar angle is measured in degrees, the resulting area is spherical degrees. These units are explained shortly.

4.16.2 Quadrilaterals

Quadrilaterals are four-sided polygons and can be rectangles, squares, trapezoids, or rhombuses. A complex spherical subdivision problem can often be made easier by using quadrilaterals or by dividing the quadrilateral into two back-to-back triangles by adding a diagonal arc or chord.

Like other spherical polygons, the sides of quadrilaterals are always arcs of great circles. None of a quadrilateral's sides are parallel. Although an infinite number of quadrilaterals are possible, only a few of them with special angles can cover an entire sphere without gaps or overlaps.

Figure 4.17(a) shows quadrilateral faces of a spherical cube. Each face angle is 120° (not 90°), while the sides are 70.5288°. Figure 4.17(b) is a spherical icosahedron subdivided by rhombic quadrilaterals. The long axis of each rhombus along the sphere's surface corresponds to an edge of the spherical icosahedron (dashed). For the icosahedron, 30 large rhombus quadrilaterals (accented in the figure) cover the sphere. The resulting polyhedron is called a rhombic triacontahedron. Each accented quadrilateral is congruent

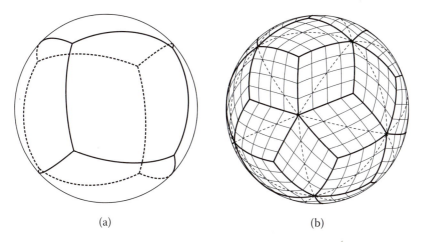

(a)   (b)

Figure 4.17. Spherical square and rhombic quadrilaterals.

and covers the entire sphere without overlaps or gaps. Its angles are 120° and 72°, and all four sides are 37.377°. The quadrilaterals within the large accented ones are not all congruent, and none of them individually could cover the entire sphere. The guidelines for spherical quadrilaterals are the following:

- *Rectangle.* Four equal angles, two opposite sides are equal.

- *Square.* Four equal angles, four equal sides.

- *Rhombus.* Two opposite pairs of equal angles, four equal sides.

- *Trapezoid.* Two pairs of equal and adjacent angles; only two opposite sides are equal (share equal angles).

Some caution must be exercised when working with quadrilaterals. Geographers often refer to areas bounded by two different latitudes and longitudes as spherical quadrilaterals. Strictly speaking, these areas are not spherical quadrilaterals because latitude references, except for the equator, are lesser circles. The sides of all spherical polygons, quadrilateral or otherwise, are always arcs of great circles.

4.16.3   Other Polygons

In addition to lunes, triangles, and quadrilaterals, many other regular convex spherical polygons exist. A *convex polygon* is one where any line or arc between its vertices defines a side or lies totally inside the polygon. The vertices of regular spherical polygons lie in a plane in 3D space. They are also the vertices of a regular Euclidean polygon inscribed in a sphere.

We can demonstrate this by showing the infinite variety of sizes that even a single polygon can take. Figure 4.18(a) shows a set of spherical pentagons. Within a single pentagon, its angles are equal and its sides are as well. But the angles of any two pentagons are not the same. The smaller a spherical *pentagon* becomes, the more its angles approach (but never

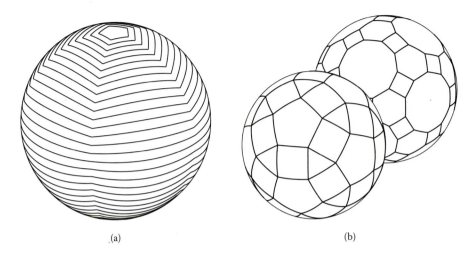

(a)   (b)

Figure 4.18. Spherical polygons.

equal) a regular planar pentagon. The larger the pentagon is, the more it approaches a great circle.

Figure 4.18(b) shows two spherical Archimedean polyhedra with vertices centrally projected onto the surface of a surrounding sphere. They provide additional examples of regular spherical polygons such as hexagons and dodecagons.

No matter how many sides a spherical polygon has, each side and each angle is less than 180°, and the sum of the sides is always less than 360°. Later, we will discuss the concept of spherical excess and the area of triangles. Excess is essentially the sum of the interior angles of the spherical polygon less the sum of the interior angles of its planar version. The difference is its area. We will see that excess applies to spherical triangles and any spherical polygon because spherical polygons can be subdivided into two or more spherical triangles.

4.16.4   Caps and Zones

A plane passing through a sphere defines either a lesser or greater circle. The plane also slices off a spherical cap. Lesser circles define small caps that resemble contact lenses, while a great circle defines a hemisphere, the largest possible cap.

Figure 4.19 shows a lesser and great circle cap superimposed on each other, with both centered over the same point (the larger dot) on the surface of the sphere. Both caps have the same polar axis. The two radii for the lesser circle, along with its height, are indicated along with the formulas to calculate them.

Caps play many roles in spherical work. One of the most useful is analyzing the distribution of points on a sphere. When caps are centered over every point on the sphere and their angular radii are all the same and as large as possible without causing any caps to overlap, the area of all the packed caps can be compared to the sphere's total area, indicating their packing density. If two spheres distribute the same number of points, the one with the higher density distributes points more evenly. If we call this *density factor D*,

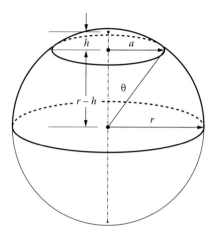

Figure 4.19. Spherical caps.

then the relationship is

$$\text{density} = \frac{n}{2}(1 - \cos\theta),$$

$$\text{surface area}_{cap} = 2\pi rh = \pi\left(a^2 + h^2\right),$$

where n is the number of points and θ is the angular radius of the cap.[12] The more even n points are distributed on the surface of the sphere, the larger the common radius of all caps will be, and thus the greater the final density.

In order to find the density, we have to know the area of a spherical cap cut out by a plane at distance h from the tangent plane to the sphere at the center of the cap. By a special property of 2D spheres, already known in some form to Archimedes, the most brilliant mathematician-scientist of antiquity, this area is given by $2\pi rh$, where r is the radius of the sphere.

Figure 4.20 shows the largest possible caps for each of the spherical Platonic solids. In clockwise order: (a) tetrahedron, (b) dodecahedron, (c) icosahedron, (d) octahedron, and (e) cube. We can show how to compute the density and measure the distribution of points by using the six spherical caps centered at the endpoints of the three Cartesian coordinate axes of the spherical octahedron shown in Figure 4.20(d). By looking at the slice of this sphere by a coordinate plane, we can see that there are six nonoverlapping spherical caps with radius $r/\sqrt{2}$ so h equals $r - r/\sqrt{2}$ and the area of each cap is $2\pi r(r - r/\sqrt{2}) = 2\pi r^2(1 - 1/\sqrt{2})$. The total area of the six caps is then $12\pi r^2(1 - 1/\sqrt{2})$ while the area of the whole sphere is $4\pi r^2$. The ratio of the area of the caps to the area of the sphere is then $3(1 - 1/\sqrt{2}) = 3(\sqrt{2} - 1)/\sqrt{2} = (3/2)(\sqrt{2} - 1)(\sqrt{2}) = (3/2)(2 - \sqrt{2})$, about $(3/2)(.5858) = .8787$ so a little more than 7/8 of the sphere.

The highest possible density is 1.0, a case where caps are centered over two antipodal points and each cap covers an entire hemisphere. As expected, these two hemispherical caps cover the

---

[12] (Tarnai 2002, 289)

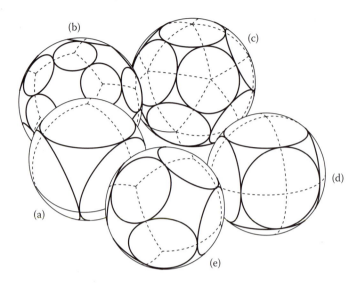

Figure 4.20. Spherical Platonic caps.

Spherical Platonic Caps						
Polyhedron	Angular Radius	Circle Radius	Cap Area	Total Caps	Total Area	Density
Tetrahedron	54.7356	0.8165	2.6556	4	10.6223	0.8453
Octahedron	45.0000	0.7071	1.8403	6	11.0418	0.8787
Cube	35.2644	0.5774	1.1530	8	9.2239	0.7340
Dodecahedron	20.9052	0.3568	0.4136	20	8.2721	0.6583
Icosahedron	31.7175	0.5257	0.9384	12	11.2607	0.8961

Table 4.1. Spherical Platonic caps.

entire sphere with no spaces or gaps. Table 4.1 summarizes their angular radius, circle radius, and surface area on a unit radius sphere. The total number of caps and the resulting density are also tabulated. Note that the caps covering the icosahedron have the highest density; they cover the highest percentage of the sphere's overall surface area. We will revisit spherical caps in the next chapter when we look at various point distribution strategies.

4.16.5 Gores

Gores are a series of 2D polygons, each polygon defined on its own plane, that when tiled together form a continuous strip approximating a 3D surface. The points distributed on a sphere can be connected with chords to define a covering mesh of triangles. A sequence of adjacent triangles creates a gore. Some series can wrap all the way around a sphere, such as the one shown in Figure 4.21(a). It is easy to imagine cutting a gore out of paper, creasing it where two triangles share an edge, and folding it to make a faceted model of the sphere.

Although triangles are the most flexible way to approximate a spherical surface, gores are not limited to triangles; any planar polygon with every vertex on the surface of the

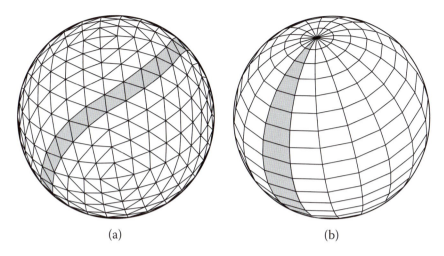

(a)   (b)

Figure 4.21. Triangular and quadrilateral gores.

sphere will suffice. Figure 4.21(b) shows a gore made of quadrilaterals and a triangle at the pole. We make use of gores in Chapter 10 to assemble complete spheres from sequences of triangles or in replicating geometry in other parts of the sphere.

4.16.6 Spherical Triangles

Triangles play a dual role in spherical subdivision: they are the result of subdivision, and they are used to subdivide. You have already noticed that points on a sphere can often be connected to form triangles, and these points can define either spherical or planar versions. Here we take a look at the properties of spherical triangles and what relationships they have to planar ones defined by the same points on a sphere.

We have already seen that a spherical triangle is formed by the intersections of three great circles. Figure 4.22 shows the details. Unlike plane triangles, the sides of a spherical triangle are measured in angles, and they are always less than a semicircle. The sum of the three angles of a spherical triangle always exceeds 180°, but is always less than 540°. A spherical triangle can also have two or even three 90° angles. We saw several examples before in Section 4.8.

Surface angles are the "corners" of a spherical triangle. They occur where two great circles intersect. All spherical triangles have three surface angles and, although there is no hard and fast rule, we will adopt the convention of using uppercase letters like A, B, and C to refer to them. The sides opposite these angles are traditionally given lowercase letters like a, b, and c. Figure 4.22 labels a spherical triangle this way. Surface angles are measured on a plane tangent to the sphere at the point of intersection.

The sides of a spherical triangle are segments of great circle arcs. In reality, their "length" is arc-length and they are measured by a central angle, also called an *arc angle* or *interior angle*, and usually denoted in lowercase. These terms are often used interchangeably in spherical trigonometry. Figure 4.22 locates central angles a, b, and c. It is easy to see how this angle measures the portion of the great circle arc that forms a side of spherical triangles.

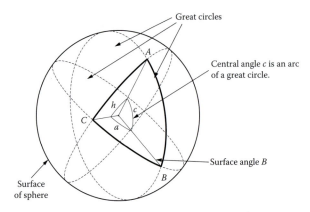

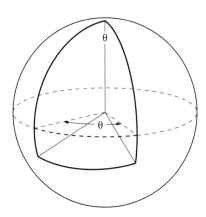

Figure 4.22. Spherical triangle *ABC*.

Figure 4.23. Dihedral angle.

Spherical and Planar Triangles Compared	
Spherical Triangles	**Planar Triangles**
Triangles lie on the surface of a sphere	Triangles lie on a plane
Angles are formed by the intersection of two great circles	Angles are formed by the intersection of two lines
Any angle is less than 180°	Same
Sides are arcs and measured in degrees or radians	Sides are straight and measured in length
Each side is less than the sum of the other two sides	Same
The sum of the sides is less than 360°	N/A
Any side is less than 180°	N/A
If two sides are equal, the angles opposite them are equal	Same
If two sides are unequal, the angles opposite these sides are unequal, and the greater side lies opposite the greater angle	Same
Sum of the surface angles is always <u>more</u> than 180° but <u>less</u> than 540°. The sum greater than 180° is called spherical excess	Sum of the angles is always 180°
Can have one, two or even three 90° surface angles	Can have, at most, one 90° angle
If two surface angles are equal, then the sides opposite those angles are also equal.	Same
Area depends on the arc-length of the sides and the radius of the sphere	Area depends on the length of the sides
There are no similar triangles on a sphere that are not congruent	Similar triangles have the same angles but different length sides
Two triangles are congruent, if they have the same angles and sides in relationships: ASA, SSS or SAS	Same
Angle-Angle-Angle, AAA, specifies a unique triangle	Angle-Angle-Angle specifies an infinite number of similar triangles

Table 4.2. Spherical and planar triangles compared.

We have already encountered dihedral angles when we covered lunes. The dihedral angle is the angle between the planes of two intersecting great circles, and it's the same angle as the surface angle of the three vertices A, B, and C in our spherical triangle (Figure 4.23).

Spherical triangles have many of the same characteristics as their planar cousins, but there are some important differences. Table 4.2 compares them. Like plane triangles, if two sides of a spherical triangle are equal, the angles opposite them are equal. If two sides of a spherical triangle are unequal, the angles opposite these sides are unequal, and the greater side lies opposite the greater angle.

But there is another important difference. The sides of a spherical triangle are entirely determined by its three surface angles. Two spherical triangles are congruent only if their corresponding surface angles are congruent. This is not true in planar triangles, where specifying the triangle's three apex angles defines an infinite number of similar triangles. And unlike planar triangles, the sum of the three sides (central angles) of a spherical triangle is less than 360° degrees, while the sum of the surface angles is always greater than 180°, but less than 540°.

4.16.7   Congruent and Symmetrical Triangles

In Euclidean geometry, triangles are congruent if the combinations of their angle (A) and side (S) parts are the same: *ASA*, *SSS*, *SAS*. These relationships are the same for spherical triangles plus one more: *AAA*. *Congruent spherical triangles* have all of their angles and sides equal and both triangles perfectly coincide. If two non-equilateral triangles have equal angles and sides, but in reverse order, they are said to be *symmetric triangles*. Symmetric triangles are equal, part for part, but they cannot be made to coincide because their respective sets of parts are arranged in opposite orders. Each triangle is the opposite hand of the other. Let's look at some examples of congruent and symmetrical triangles.

The same three great circles that form one spherical triangle actually form eight triangles. How do we get eight for one? This is quite a bargain! It is not hard to see how this happens. The three great circles that form any spherical triangle intersect each other at six points, three antipodal pairs. Every time one spherical triangle is formed, the planes of the great circles that define it have also defined seven other triangles. The simplest example

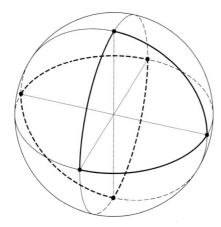

Figure 4.24. Orthogonal great circles define eight equilateral triangles.

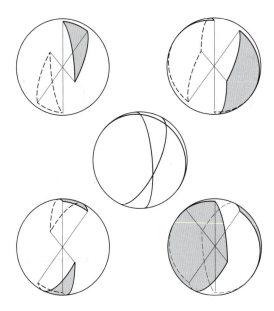

Figure 4.25. Three great circles define eight spherical triangles.

is when three great circles are orthogonal or 90° to each other. Notice in Figure 4.24, all eight triangles that result are congruent equilateral triangles and every surface angle is 90°, something that is only possible with spherical triangles (this is also the geometry of the spherical octahedron). If the great circles are not orthogonal, four pairs of symmetrical triangles result. Figure 4.25 illustrates this point. Notice that each of the four pair of triangles have antipodal vertices and are symmetrical but not congruent to one another.

4.16.8 Nothing Similar

A surprising difference between planar and spherical triangles is there are no similar triangles on a sphere. Triangles can be acute, obtuse, right, and equilateral. They can be congruent. But they cannot be similar; that is, they cannot have the same surface angles but proportionately larger or smaller sizes. At first, this seems impossible because similar triangles are so fundamental to plane geometry. But on a sphere, three angles always specify a unique triangle. An angle-angle-angle (*AAA*) triangle specification defines a unique spherical triangle, whereas on a plane, it defines an infinite number of similar triangles, each with the same angles but sides of different lengths. The following three examples demonstrate this essential concept.

Figure 4.26 shows an isosceles triangle formed by two meridians, each perpendicular to the equator. A smaller triangle is inscribed in the larger one. Side *a* is the arc of a great circle between the two medians, a requirement for spherical triangles. Arc *b* is from a lesser circle and does not qualify as a component of a spherical triangle, even though it looks like it is parallel to side *c*. Although both triangles have a common apex angle, the other two face angles are clearly not the same.

Let's look at similar triangles from another perspective. In plane geometry, we can cluster four equilateral triangles together to make a larger one. We can also subdivide a

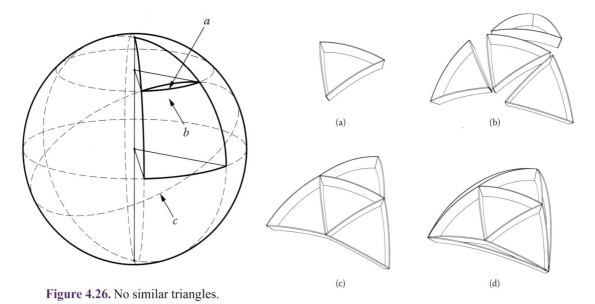

Figure 4.26. No similar triangles.

(a)

(b)

(c)

(d)

Figure 4.27. Appending equilateral triangles.

large equilateral triangle, making four smaller, congruent ones. In this example, and the next one, we show that neither of these two constructions is possible on a sphere.

In Figure 4.27, an equilateral triangle (a) is replicated three times and the copies appended to each side of the original triangle in (b) and (c). The resulting configuration of four triangles (c) is not, however, a larger equilateral triangle with the same angles. Instead, a new triangle with still larger angles is created, as indicated by the outermost sides in (d).

In our third example, it is easy to make a triangle within a triangle by creating sides between pairs of mid-edge or medial points. The new triangle is sometimes called a *medial triangle*. In plane geometry, the medial triangle of an equilateral triangle results

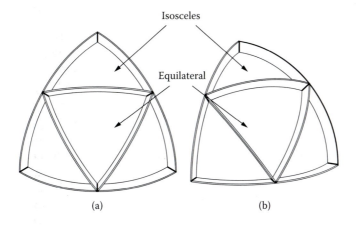

Isosceles

Equilateral

(a)

(b)

Figure 4.28. Subdividing a spherical equilateral triangle.

in four smaller, congruent equilateral triangles, and the sides of the medial are parallel to the sides of the original triangle. But none of this is true on a sphere, and Figure 4.28 shows why.

Two views of the same equilateral spherical triangle and its medial triangle are illustrated. The result is a single equilateral triangle in the middle, surrounded by three isosceles triangles. Figure 4.28(a) clearly shows that the central triangle is equilateral, while (b) views a typical apex isoceles triangle that surrounds it. The original equilateral angles are all 90°. The newly created medial triangle in the center is equilateral and its vertex angles are each 70.5°. The three surrounding isosceles triangles are congruent and have vertex angles of 90°, 54.7°, and 54.7°. The edges of the original equilateral triangle are each 90°, while the interior equilateral triangle has 60° edges. None of the edges are parallel to one another.

These three examples clearly demonstrate that there are no similar triangles on a sphere that are not congruent. Every spherical triangle is either unique or congruent, and three vertex angles are all that is needed to specify one. Next, we will look at spherical triangles with special properties.

4.16.9   Schwarz Triangles

Subdividing a sphere often involves subdividing just a small subarea and then replicating the geometry to cover the rest of the sphere. This subarea must be a special spherical polygon if it is to cover, or tile, the rest of the sphere's surface without causing overlaps or leaving gaps.

Hermann Amandus Schwarz's (1843–1921) work includes differential geometry and spherical surfaces. He discovered special spherical right triangles that could be replicated to cover a sphere through *reflections* and rotations without leaving any gaps or causing overlaps. Today, we call these triangles Schwarz triangles, and they are very useful in spherical subdivision because we can subdivide just this one triangle and then replicate the subdivision layout to cover the rest of the sphere. There is an additional benefit with Schwarz triangles: many are right triangles, and right spherical or planar triangles are the easiest to work with in trigonometry. Buckminster Fuller called Schwarz triangles "least common denominator" (LCD) triangles because he thought of them as the smallest triangle that can represent the whole sphere.

Schwarz triangles are different for each of the spherical Platonic solids. We list their angles here and show them in Figure 4.29 (clockwise from the top in the same order). The faces of each spherical Platonic solid are accented.

- 60°, 36°, 90° Icosahedron

- 60°, 45°, 90° Octahedron

- 45°, 60°, 90° Cube

- 60°, 60°, 90° Tetrahedron

- 36°, 60°, 90° Dodecahedron

The faces of each Platonic solid shown in Figure 4.29 have 6, 8, or 10 right Schwarz triangles. Half of the triangles are left- and half are right-handed. The tetrahedron has

60°, 36°, 90° Icosahedron
60°, 45°, 90° Octahedron
45°, 60°, 90° Cube
60°, 60°, 90° Tetrahedron
36°, 60°, 90° Dodecahedron

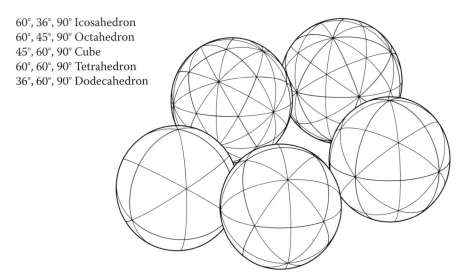

Figure 4.29. Platonic Schwarz triangles.

12 left- and 12 right-handed triangles, while the cube, octahedron, dodecahedron, and icosahedron have 24, 24, 60, and 60 left- and right-handed Schwarz triangles, respectively. In a later chapter, we describe the triacon subdivision method, which is based entirely on right Schwarz triangles. Because of this, the triacon method can subdivide all the spherical Platonic solids shown in Figure 4.29.

Although there are an infinite number of different right spherical triangles, only a few of them can cover the surface without gaps or overlaps. There are two ways to know if a triangle can do this: determine its area or calculate its vertex sums.

For a triangle to cover the entire sphere, its area must divide evenly into the total area of the sphere. Another way to know if a spherical triangle can tile a sphere is to examine its three vertex angles. For a triangle to tile an even multiple of times, each of its angles must add up to 360°, as they are arranged surrounding a common vertex point.[13] For example, in the icosahedral Schwarz triangle, the number of times any of their three angles share a common vertex on the sphere are 6 × 60°, 10 × 36,° and 4 × 90° and each adds up to 360°.

Figure 4.30 shows some additional spherical right triangles that pass these two tests and can perfectly cover the whole sphere. But note that several of these Schwarz triangles have more than one right angle. From back to front in clockwise order, they are

- 90°, 90°, 90° (Spherical Octahedron);

- 90°, 90°, 45°;

- 90°, 120°, 60°; and

- 90°, 120°, 90°.

Within this new group, only the spherical octahedron is suitable for spherical subdivision. Although all of them can cover the sphere, their *tiling* arrangements are not suited for

---

[13] (Dawson 2008)

90°, 90°, 90° (Spherical Octahedron)
90°, 90°, 45°
90°, 120°, 60°
90°, 120°, 90°

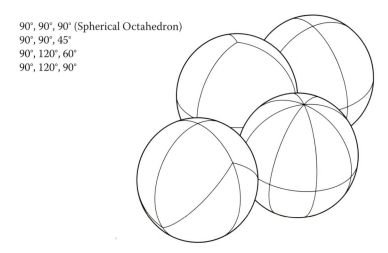

Figure 4.30. Other right triangle tiles.

subdivision because they do not have the necessary edge-to-edge or vertex-to-vertex symmetry we need. For the subdivisions demonstrated in this book, we use the smaller Schwarz triangles that subdivide the five spherical Platonic solids shown in Figure 4.29. We will use Schwarz triangles later in Chapter 8 to help generate small-area grids and again in Chapter 10 when we use subdivided Schwarz triangles to assemble 3D spheres with computer-aided design.

4.16.10   Area and Excess

Thus far, we have seen how great-circle arcs define the angles and sides of spherical polygons. Here we see how those same angles define the polygon's area as well. Later, we will use an amazing property called spherical excess, which relates the sum of a spherical triangle's angles to its area. But first, we need to see how area is measured.

4.16.11   Steradians

Steradians are used in many applications of spheres.[14] While the area of any spherical polygon can be measured in square units such as inches, feet, meters, and so forth, a more general way to express spherical area is to use units relative to the sphere's radius. While radians measure angles relative to the radius of a circle, steradians measure area relative to the radius of a sphere. Steradians can be thought of as "square radians" and they are abbreviated (sr). Steradians are part of the International System of Units (SI) and many high-end scientific calculators include this unit.

The number of steradians in a sphere is equal to the surface area of the unit sphere (or any other sphere—the size of the sphere does not matter). An entire unit sphere, therefore, measures 4π steradians, or about 12.57 steradians in all. One steradian is about 8 percent of the sphere's total surface area. To get a visual idea of a single steradian, look at Figure 4.31,

---

[14] Steradians are use in many areas of science and physics. For example, in the calculations of radiant intensity, the intensity of electromagnetic radiation (how brightly a light bulb shines) is measured in watts per steradian. Astronomers using radio telescopes use steradians to measure the density of radiant sky sources (stars or other celestial objects).

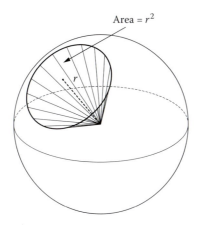

$Area = r^2$

Figure 4.31. Steradians and solid angles.

which shows a spherical cap whose surface area (not the cap's lesser circle area) is 1 steradian. If this is a unit sphere (radius = 1.0), then the area of the cap is r^2. The choice of a spherical cap is completely arbitrary; the surface area could be any shape.

4.16.12   Solid Angles

The *solid angle*, abbreviated (ω or Ω), is the name of the area projected onto the surface of a unit sphere and is measured in steradians. The solid angle of the spherical cap in Figure 4.31 is 1 sr. Spherical polygons that have the same area have the same solid angle regardless of their shape. Solid angle is simply the ratio of the polygon's area to the square of the sphere's radius. Again, the area's shape does not matter. The maximum solid angle possible on a sphere is the total area of a unit sphere itself, which is 4π or 12.5664 sr. The general formula for a solid angle is

$$\Omega = \text{area/radius}^2.$$

4.16.13   Spherical Degrees and Square Degrees

Steradians are one way to express area, but it can be expressed in *spherical degrees* or *square degrees* as well. Neither of these are SI units, but they are frequently encountered in spherical work and are easy to work with.

   Just as degrees are used to measure parts of a circle, square degrees are used to measure parts of a sphere. A spherical degree (sd) is 1/720th of the area of the whole sphere. Each degree can be visualized as a spherical triangle with angles 90°, 90°, and 1° or one-half of a 1° lune. The sphere in Figure 4.32(a) is divided into 720 areas, each equal to one spherical degree, 360 in each hemisphere. One triangle of one spherical degree is accented. One spherical degree equals $(4\pi r^2)/720$ or $(\pi r^2)/180$ sr, which is about .0175 sr.

   Square degrees, on the other hand, are units based on a small spherical square with one degree of arc on a side. Astronomers often use square degrees to refer to areas on the celestial sphere. For example, a NASA image of the sky might show a field of view

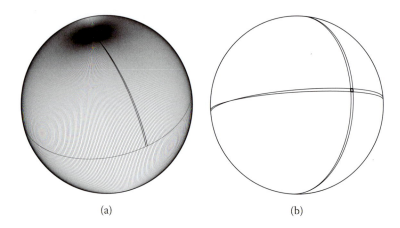

(a)   (b)

Figure 4.32. Spherical degrees and degrees squared.

of six square degrees. Square degrees are abbreviated deg² to distinguish them from other units.

You can think of a square degree as a small postage stamp on the surface of a large sphere; one is shown in Figure 4.32(b). Remembering that there are $180/\pi$ degrees in a radian, you can figure out that the number of square degrees in a full sphere is $4\pi(180/\pi)^2$, or about 41,252.96 deg². Therefore, one square degree equals $(\pi/180)^2$ steradians, or about 0.0003046 sr. To review:

1 radian	$180°/\pi$ degrees or about 57.295780°
1 steradian	$(180°/\pi)^2$ square degrees or 3,282.8064 square degrees
Area of the whole sphere	4π steradians or about 41,252.9612 square degrees
Solid angle (in steradians)	Ω = area/radius²
Solid angle of entire unit sphere	4π steradians

Armed with ways of expressing spherical area, let us look at how we find the area of spherical polygons. Area is related to excess, one of the most useful properties of spherical polygons.

4.17   Excess and Defect

A major difference between planar and spherical triangles is the fact that the sum of their angles is different. The interior angles of planar triangles always add up to 180°, but on a sphere the sum is always greater than 180° but less than 540° (or between π and 3π radians). There is a direct relationship between the area of a triangle and the sum of its interior angles. The amount the sum exceeds 180° or π is called the *spherical excess* and is denoted E (or delta or Δ).

Spherical excess is very useful because it is also the area of the triangle on a unit sphere. If a spherical triangle (or any convex polygon for that matter) is very small relative to the

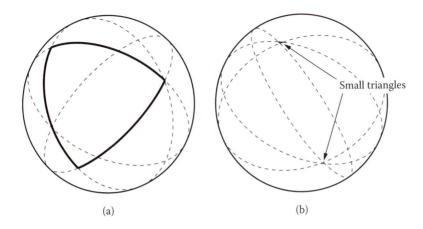

(a)   (b)

Figure 4.33. Large and small triangle spherical excess.

size of the sphere, its excess will be small and the polygon begins to behave like a regular planar polygon. On the other hand, if the spherical triangle approaches the largest size possible, like an equilateral where each vertex angle is just under 180°, then the excess approaches half the area of the total sphere. We will illustrate these two extremes next.

Spherical excess was first described by Thomas Harriot (c. 1560–1621), but Harriot did not publish his work. Excess was independently rediscovered by French-born Dutch mathematician Albert Girard (1595–1632) and excess is sometimes referred to as the Harriot-Girard theorem. The theorem states that excess, E, is the difference between the sum of the angles of a spherical triangle and the sum of the angles of a plane triangle π (radians) or 180° and is measured in radians or degrees accordingly. The formula is

$$E = (A + B + C) - \pi.$$

where A, B, and C are measured in radians. The area of a triangle is simply the relationship area $= Er^2$, expressed in steradians or spherical degrees depending on the units used. For a unit sphere, excess is the area of the triangle.

Figure 4.33 illustrates the difference in spherical excess between large and small triangles. The large triangle (*a*) has angles 90°, 90°, and 90°. Its excess is 90°. In steradians, this excess is 1.5708 or $\pi/2$. Since this particular triangle is $1/8^{th}$ the area of a unit sphere and eight of them equal the whole area of the sphere, this confirms that excess (E) is the area of a sphere: 8(1.5708 sr) = 12.566 sr, the total area of a sphere.

The two small triangles in Figure 4.33(b) both have angles 45.05°, 45.05°, and 90°. Their excess is 0.1°, which is quite small. When triangles are small relative to the size of the sphere, the sum of their angles approach 180° and they behave similarly to planar triangles.

Spherical defect is a concept similar to excess. The defect of a spherical triangle is the difference between 2π radians and the sum of the triangle's side arc lengths. It is usually denoted D. The formula is $D = 2\pi - (a + b + c)$ in radians or $360° - (a + b + c)$ in degrees. The triangles shown in Figure 4.33 and summarized in Table 4.3 illustrate that the larger a triangle becomes, the closer its spherical defect begins to approach 0. On the other hand, the defect of a smaller triangle, relative to the size of the sphere, approaches 360°.

	Spherical Defect						
	Surface Angles			Interior Angles			
Triangle	A°	B°	C°	a°	b°	c°	Defect°
Large	90.0000	90.0000	90.0000	90.0000	90.0000	90.0000	90.0000
Small	45.0500	45.0500	90.0000	3.3842	3.3842	4.7845	348.4472

Table 4.3. Spherical defect.

Although triangles are the most common polygon formed when subdividing spheres, squares, pentagons, hexagons, and others shapes can result, and their area is also based on spherical excess. The area, S, of a spherical polygon with n sides and sum θ of its interior angles on a sphere of radius r is

$$S = [\theta - (n - 2)180°] \, r^2.$$

A simple example shows the relationship of excess; the sides and sum of its angles follow the same rules we have been discussing. The four angles of a spherical cube's face, shown earlier in Figure 4.17(a), are each 120° and total 480°. Thus the area of the face on a unit radius sphere is $480° - (4 - 2)180° = 120$ sd or 2.09 sr. We can easily check this. The cube has six equal spherical faces; thus 6(2.09 sr) = 12.566 sr, the area of a full sphere.

4.17.1   Visualizing Excess

The preceding equation for spherical excess is extremely useful in spherical subdivision work. Albert Girard used the area of lunes and their proportional relationships to the sphere's area to derive the relationship and, in so doing, he proved that a triangle's area *is* proportional to its spherical excess. His logic sharpens our circular reasoning by synthesizing the relationship of lunes, triangles, and their areas, giving us an extremely useful formula for geodesic work.

We said earlier that a lune is a spherical polygon with just two angles (both equal), and two antipodal vertices (with both sides formed by 180° arcs of great circles). A lune's area is $2r^2\theta$, thus it is directly proportional to its lunar angle (θ). For a unit sphere, a lune's area is simply twice its lunar angle expressed in radians.

In this example, we define a spherical triangle with three orthogonal great circles. In other words, the great circles are positioned at 90° angles to each other and arranged similarly to the Cartesian coordinate system reference planes. Figure 4.34(a) shows this arrangement.[15] As expected, eight spherical triangles are defined (four pairs in which each pair is symmetric).

To illustrate Girard's theorem, we will concentrate on the surface of just two triangles, the yellow pair. Figure 4.34(b) pulls the sphere apart to make it easier to see these two triangles, which are on opposite sides of the sphere. Notice that every vertex in one yellow triangle is antipodal to every vertex of its symmetrical counterpart. Arrows show their relationship. Next, notice that the surface of each yellow triangle is defined by a set of three overlapping lunes colored red, green, and blue.

---

[15] This arrangement of great circles also defines a spherical octahedron.

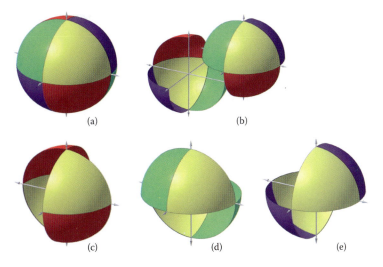

(a)   (b)

(c)   (d)   (e)

Figure 4.34. Lunes and spherical excess.

When great circles define a lune, they define two identical lunes that share the same two vertices (points where the great circles intersect). Figure 4.34(c), (d), and (e) show these pairs of identical lunes (i.e., those of the same color) and how they define the surface and vertices of their respective yellow triangles. It is easy to see that for each lune, one lunar vertex defines one of the triangle's vertices, while the other lunar vertex defines a vertex in the symmetric triangle. Notice, too, that half of every lune's surface covers one of the two yellow triangles' surfaces.

We can now summarize the relationships between the areas of the six lunes, the overall sphere, and both triangles; they lead directly to the excess formula and a spherical triangle's area. Simply stated: the sum of all lunar areas (there are six) is equal to the area of the sphere, plus four times the area of the two triangles. The "four" in the equation is explained by the fact that there are two extra overlaps in two symmetric triangles. Since pairs of lunes (red ($_r$), green ($_g$), and blue ($_b$)) have the same area and we know the area of the whole sphere is $4\pi r^2$, we can express their area relationships and the area of a single triangle (T) such as

$$2r^2\theta_r + 2r^2\theta_r + 2r^2\theta_g + 2r^2\theta_g + 2r^2\theta_b + 2r^2\theta_b = 4\pi r^2 + 4_{\text{area}}(T).$$

Girard's theorem for excess is a simplification of the preceding equation.[16] While excess is expressed in degrees or radians, area is expressed in steradians, square degrees, or degrees squared. Since the areas of the lunes are proportional to their lunar angle, which is also an angle of the triangle, the excess of a spherical triangle is directly related to the lunes that define it. Girard's theorem on spherical excess and the area of a triangle beautifully summarizes the relationship between spherical triangles and lunes.

We have established the relationship between a triangle's vertex angles, excess, and its area. We have also seen that when spherical triangles (or any polygon) are relatively small

---

[16] For complete derivation, see Eric W. Weisstein, "Girard's Spherical Excess Formula," MathWorld—A Wolfram Web Resource, http://mathworld.wolfram.com/GirardsSphericalExcessFormula.html.

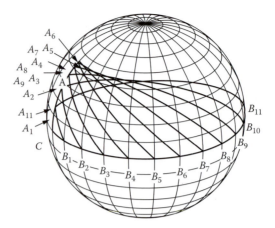

Figure 4.35. Excess in successive right triangles.

compared to the size of the sphere, they behave almost like regular 2D triangles. But large spherical triangles behave very differently than plane triangles.

Examining a diagram based on one by James Wertz and Wiley Larson,[17] we can see how excess changes as triangles get larger. Figure 4.35 shows a series of 11 right spherical triangles with a latitude-longitude reference grid of 10° and 15° degrees, respectively. The 11 triangles become successively larger from left to right. They all share the same right angle vertex on the left, where $C = 90°$ and each triangle has the same angle $B = 45°$. All triangles partially overlap and their base sides are all coincident to the equator. The smallest triangle's base side is 15°, and each successive triangle increases its base side by 15° (indicated by the reference longitudes). The largest triangle's base somewhat wraps around the sphere and is 165°.

Our example shows 11 right triangles with angle-side-angle (*ASA*) specifications. They each have the same vertex angle C, a side that is a multiple of 15°, and the same vertex angle B. Although B is the same angle, we have subscripted them 1 to 11 to make their correspondence with their A angles 1 to 11 more evident. In two dimensions, these 11 triangles would all be similar triangles but on a sphere, each triangle is quite different because their vertex angles A_{1-11} are different; this also affects the two nonbase side lengths of each triangle.

Table 4.4 lists the characteristics of these 11 triangles on a unit sphere. Keep in mind that every triangle has the same C and B angles, only angle A changes from triangle to triangle (a consequence of the different length side between C and B). What we notice from the table is the three vertex angles, A, B, and C, for the smallest triangle (first row entry) are almost like a regular 2D isosceles triangle. This is confirmed by the fact that there is almost no excess at all because their sum is near 180°. Figure 4.35 confirms the smallest triangle on the left. It's also the least "distorted" one.

In contrast, vertex A for the largest triangle (last row) is almost three times larger than vertex A for the smallest triangle. The triangle's excess is quite large too; as a result, it looks quite distorted. This series of triangles demonstrates that as the triangles get larger, their excess increases, their area increases, and they behave less like planar triangles. This

---

[17] (Wertz 1999, p.172, fig. 508)

Characteristics of Spherical Right Triangles										
A°	a°	B°	b°	C°	c°	A°+B°+C°	E°	Area (sr)	Area (sd)	Area (d²)
46.920	15.0	45.0	14.511	90.0	20.754	181.920	1.92048	0.03352	1.920	110.036
52.239	30.0	45.0	26.565	90.0	39.232	187.239	7.23876	0.12634	7.239	414.750
60.000	45.0	45.0	35.264	90.0	54.736	195.000	15.00000	0.26180	15.000	859.437
69.295	60.0	45.0	40.893	90.0	67.792	204.295	24.29519	0.42403	24.295	1392.012
79.455	75.0	45.0	44.007	90.0	79.271	214.455	34.45471	0.60135	34.455	1974.109
90.000	90.0	45.0	45.000	90.0	90.000	225.000	45.00000	0.78540	45.000	2578.310
100.545	105.0	45.0	44.007	90.0	100.729	235.545	55.54529	0.96945	55.545	3182.511
110.705	120.0	45.0	40.893	90.0	112.208	245.705	65.70481	1.14677	65.705	3764.608
120.000	135.0	45.0	35.264	90.0	125.264	255.000	75.00000	1.30900	75.000	4297.183
127.761	150.0	45.0	26.565	90.0	140.768	262.761	82.76124	1.44446	82.761	4741.870
133.080	165.0	45.0	14.511	90.0	159.246	268.080	88.07952	1.53728	88.080	5046.585

Table 4.4. Characteristics of spherical right triangles.

series of 11 triangles demonstrates in yet another way that there are no similar triangles on a sphere.

4.17.2   Centering in on Triangles

Spherical subdivision creates points on the surface of the sphere that can be triangulated. When planar triangles are constructed from point-to-point chords, their centers are very useful in deriving other geometry or in comparing one set of triangles to another. The four most important centers of planar triangles used in subdivision are the centroid, *incenter*, *orthocenter*, and *circumcenter*. They are defined at various intersections of angle bisectors or the edge midpoints opposite an angle (*medians*). Figure 4.36 shows how each center is

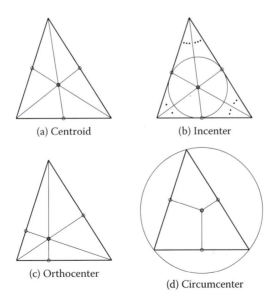

(a) Centroid   (b) Incenter

(c) Orthocenter   (d) Circumcenter

Figure 4.36. Planar triangle's centroid, incenter, orthocenter, and circumcenter.

Centers of Planar Triangles					
Center	**Intersection of**	**Equilateral**	**Acute**	**Right**	**Obtuse**
Centroid	medians	mid-triangle	inside	inside	inside
Incenter	angle bisectors	mid-triangle	inside	inside	inside
Orthocenter	altitudes	mid-triangle	inside	on right angle vertex	outside
Circumcenter	perpendicular bisectors of sides	mid-triangle	inside	mid-point of hypotenuse	outside

Table 4.5. Centers of planar triangles.

defined for planar triangles. Here we highlight their main characteristics. Table 4.5 provides a quick comparison as well.

Centroid

- The point of intersection of the medians of the triangle.

- The point always falls inside the triangle.

- The areas of all six subtriangles are equal.

- The center of gravity (COG) of the triangle (triangle will balance at its centroid as well as balance along any of its medians); the COG is also called the *barycenter* of the triangle.

- Divides each median, the line from a vertex to the midpoint of the opposite side, in a ratio of 2:1.

- The ratio of the length of sides to one another remains the same, if the triangle is scaled about the centroid.

- A triangle continuously scaled down converges on its centroid.

- The point's coordinates are the average of the vertex coordinates:

$$xc = (x^1 + x^2 + x^3)/3,$$

$$yc = (y^1 + y^2 + y^3)/3,$$

$$zc = (z^1 + z^2 + z^3)/3.$$

Incenter

- The point of intersection of the angle bisector of the triangle.

- The point always falls inside the triangle.

- The point is equidistant from the sides.

- The center of the triangle's *incircle*, the largest circle that can be inscribed and touch all three sides.

- Tangent points of the incircle and triangle sides are the same as angle bisector intersections only when the triangle is equilateral.

Orthocenter

- An altitude is the line that passes through a vertex and is perpendicular to the opposite side.

- Point of intersection of the altitudes of the triangle.

- Similar triangles are formed.

- Orthocenter for *acute triangles* falls inside the triangle and outside for *obtuse triangles.*

- For right triangles, the orthocenter is the vertex, which is the right angle.

Circumcenter

- The circumcenter is the center of the triangle's circumscribing circle, the circle that passes through all three of the triangle's vertices.

- The point is equidistant from the vertices of the triangle.

- The point of intersection of the perpendicular bisectors of the sides of the triangle.

- The circumcenter of acute triangles fall inside the triangle and outside for obtuse triangles.

- For a right triangle, the circumcenter is at the midpoint of the hypotenuse.

- The radius of a circumcenter given the sides of a triangle a, b, and c:

$$r = abc / \sqrt{(a+b+c)(b+c-a)(c+a-b)(a+b-c)}.$$

Figure 4.37 shows a simple application of incenters in spherical work. A circle is drawn inside each of the planar triangles. The circle's radius is 80 percent of the distance between the incenter and the edge of its triangle. The effect is to give a visual cue about the relative size and shapes of triangles. In these side-by-side comparisons, it is immediately clear that subdivision (a) is more uniform than (b) and that within (b) there is a central equilateral triangle (emphasized) whose vertices are at the midpoint of the outermost triangle. This triangle is much bigger than the other three. We will use incenter circles a great deal in Chapters 8 and 10.

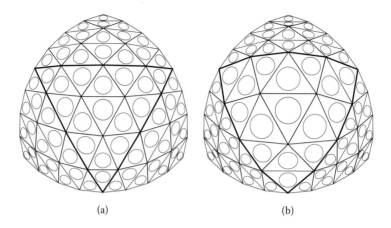

Figure 4.37. Incenter circles illustrate triangle shapes and relative sizes.

4.17.7   Euler Line

We have already seen the orthocenter, circumcenter, and centroid of triangles, where they are the intersections of the altitudes, perpendicular bisectors of the edges, and medians of a triangle. Leonhard Euler (1707–1783), the prolific Swiss mathematician and physicist, observed that in all planar triangles, except an equilateral, the orthocenter, centroid, and circumcenter all lie on an imaginary line (in that order). In his honor, this line is called the *Euler line*. Other centers of a triangle lie on this line also, but these three are the most important for our applications. Note also that a triangle's centroid is always located between the circumcenter and the orthocenter.

Figure 4.38 shows the Euler line segment between the circumcenter and orthocenter for acute (a) and isosceles (b) triangles, similar to the ones created by the points on subdivided spheres. The line segment between the outermost orthocenter and circumcenter is accented. A good feature about this segment is that its length is quite sensitive to the ratio of the triangle's sides. A slight change in the sides causes a significant change in the length of the line segment, and this is what makes it so useful in geodesics. It gives us a way to compare

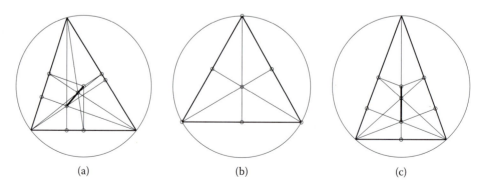

Figure 4.38. Euler line segments for (a) acute, (b) equilateral, and (c) isosceles triangles.

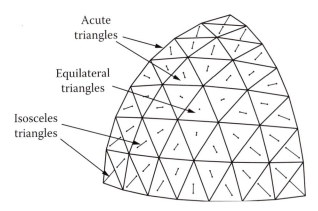

Figure 4.39. Euler line segments for a subdivision triangle.

the triangles produced by two subdivision schemas. Note, too, that there is no Euler line for the equilateral triangle (c). This is because all three centers are coincident.

We can use the length and direction of the Euler line segment as a measure of how triangles produced from spherical subdivision deviate from perfect equilaterals. If the schema produces all equilateral triangles (only the vertices of the tetrahedron, octahedron, and icosahedron on their circumsphere do this), the total length of the Euler line segments would be zero. But spherical subdivision produces acute and isosceles triangles as well, so the variation of this small length and its direction on the face of the triangle tells us how much the triangle deviates from equilateral.

Figure 4.39 shows the Euler line segments for every subdivision triangle of a polyhedral face and shows examples of equilateral, isosceles, and acute triangles. Despite the fact that the overall equilateral spherical triangle's apex angles are all 90°, there are no perfectly planar right triangles. With just a quick glance, you can tell what kinds of triangles resulted from the subdivision.

First, it is clear that only one triangle is equilateral—the one in the middle, because it is the only one without an Euler line. By inspecting the figure, it is also clear there are many isosceles triangles because the Euler line is perpendicular to one of the triangle's edges. If the triangle is acute, the Euler line will not be perpendicular to any edge. Some are labeled in the figure. The subdivision triangles that share one of the polyhedron's face vertices are almost right triangles because their circumcenter point, one of the end points of the Euler line segment, is almost tangent to the triangle's hypotenuse at its midpoint.

We will make use of Euler lines in the next chapter to see which schemas produce the most uniform spherical triangles, an important criterion in some geodesic applications.

4.17.8 Surface Normals

A surface normal is a vector, a mathematical entity that has direction and magnitude in 3D space. The normal to a surface is a direction perpendicular to that surface. If the surface is a plane, the *normal vector* is perpendicular to the plane and everything that lies on it. If the surface is curved or spherical, the normal direction is perpendicular to a plane tangent to the surface at that point. There is only one normal to a plane, but the surface of a sphere has an infinite number of normals. If we visualized the normals to a sphere as arrows, they would all point outward from or inward towards the origin.

Normals are very useful in geodesics calculations, even though they are rarely listed or displayed. Instead, normals are used in various computations. In images generated by computer graphics, they are used to determine whether a viewer can see the faces of an object just by inspecting the normals to the object's faces. If the angle between the viewer's eye direction and the normal (a perpendicular) to the object's face is more than 90°, the viewer can see just the edge of the object or nothing at all. If the angle is less than 90°, the viewer can see the face, although additional checks must be done to see if it is obscured by some other object between it and the viewer's eye.

Another use (unique to spherical subdivision) is to use normals to compare face orientations. If a subdivision results in perfectly equilateral triangles, their normals would have a uniform pattern and the angle between adjacent normals would be the same. But most spherical subdivisions result in a mix of isosceles, acute, or obtuse faces, and the pattern of their normals is also mixed. The angles between normals can be analyzed and displayed with stereographic projection, a topic we will cover later. The analysis tells us how uniform the subdivision faces are.

We do not draw normals when evaluating subdivisions, but we will show them here as arrows to make it clear that a normal is a direction and the direction is perpendicular to each plane of a face that results from a subdivision of the sphere. Figure 4.40(a) shows the normals of a set of subdivision triangles. The base of each arrow is located at the point where the normal's direction line intersects perpendicular to the plane of the triangle and passes through the origin of the sphere. Most of their bases are not in the exact center of their respective triangle, an indication that the triangle is not equilateral but slightly isosceles or acute.

Figure 4.40(b) is a larger section of the same subdivided sphere. The cutaway image exposes the sphere's origin, which is represented by a small black dot. Every normal radiates outward from the origin and is perpendicular to the plane of its respective triangle. The

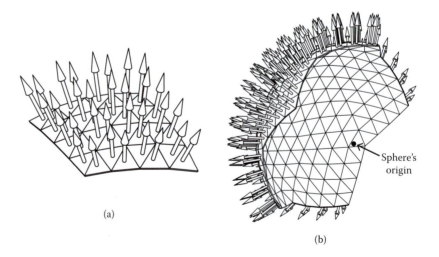

(a)

(b)

Sphere's origin

Figure 4.40. Surface normals for the planar surfaces of a subdivided sphere.

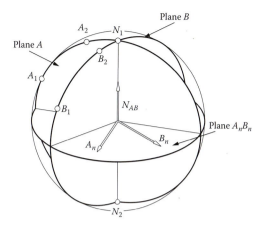

Figure 4.41. Defining points where great circles intersect.

angle between any normal and its neighbors is fairly uniform in these figures, suggesting that the subdivision schema produces uniformly oriented faces. Figure 4.40(b) also shows examples of normals and hidden lines. Notice that where normals point away from your view, you cannot see the outside face of the triangle; instead, you see its inside face.

In addition to hidden line computations and evaluating face orientations, normals are very important to the subdivision process itself. Earlier we described how points on a sphere are derived from intersecting great circles. To find the two antipodal points, where two noncoincident great circles intersect, we use normals of the great circle's planes. Figure 4.41 shows how this works. Points A_1 and A_2 define a great circle, and its normal A_n is shown as an arrow perpendicular to the plane of the great circle. Likewise for points B_1 and B_2 and normal B_n to their plane. Notice that these two normals, A_n and B_n now define a third plane labeled A_nB_n. As you might have guessed, this new plane has a normal, too, and it is labeled N_{AB} in the figure. This third normal also defines the line of intersection of the original two great circle planes. Although the line is infinite in length, it pierces the surface of the sphere at two antipodal points, N_1 and N_2, and these are the points we are looking for. Points N_1 and N_2 are the points of intersection of great circles, A and B, on the surface of the sphere.

To review, we have created a normal to the normals of two great circles and defined the line of intersection of these two great circles. The intersection line pierces the surface of the sphere at two antipodal points defining two points on the sphere where the two great circles intersect. In most subdivision cases, we will use one of the two points as a subdivision point and ignore the other. There are applications, however, where we use both of them. We will review this technique again in Chapter 8 and show the computation details in Appendix C.

4.18   Summary

We have come full circle in our reasoning. Our tool kit includes primitive geometry such as circles, points, arcs, and chords between points. We are able to locate them in three dimensions in several different ways, and we can use these points to define convex polygons, some of which can cover the surface of a sphere without gaps or overlaps. Spherical and planar triangles are particularly important, and warrant details about their angles and sides, as well as properties such as excess, area, centers, and orientation.

Evenly distributing points around the circumference of a circle is easy, but doing so over the surface of a sphere is not. In the next chapter, we define this challenge with a series of geometric objectives. We need to be precise about what "even" means and have a way of knowing how close we come to achieving an even point distribution on a sphere's surface.

Additional Resources

Aste, Tomaso and Denis L. Weaire. *The Pursuit of Perfect Packing.* Philadelphia, PA: Institute of Physics Publication, 2000.

Critchlow, Keith. *Order in Space: A Design Source Book.* New York: Viking Press, 1969.

Henderson, David W. and Daina Taimiṇa. *Experiencing Geometry: Euclidean and Non-Euclidean with History.* Upper Saddle River, NJ: Pearson Prentice Hall, 2005.

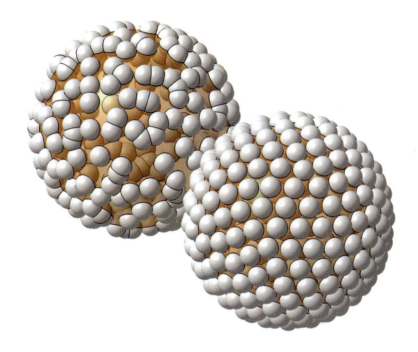

5 Distributing Points

In this chapter, we will look at three different but interrelated measures for point distributions on a sphere: covering, packing, and volume. Each provides a metric that allows us to compare one point distribution with another and to know which is more even. We use these measures in Chapter 9 to compare the results of different subdivision schemas.

It is easy to evenly distribute points around the circumference of a circle to any desired precision. But doing the same on the surface of the sphere is very difficult. In this section, we see there are several different perspectives on what an "even" point distribution means. We are particularly interested in techniques that can compare the uniformity of one distribution to another.

The essence of spherical subdivision is to define point distributions that meet application needs. The main one we focus on is making every point as equidistant and equiangular as possible to the points immediately around it. Other criteria might include minimum variations in chord lengths or in face areas.

In the strictest sense, the greatest number of points that can be equidistant and equiangular to every other one is four, and these four points are the vertices of a tetrahedron. But we use a less restrictive interpretation in spherical subdivisions and accept distributions that are "symmetrical" points on a sphere, where the distance from a point to its neighbors is equal, such as the vertices of the spherical Platonic solids, or nearly equal, like subdivided spherical Platonic solids. Sometimes mathematicians say points are evenly distributed, when the minimum distance between points is maximized.

The point sets that result from our spherical subdivision schemas are not arbitrary at all. The number of points and their distribution depend entirely on the spherical polyhedron we start with, the intensity of subdivision, and the particular subdivision schema used. We are interested in arbitrary point sets because several techniques have been developed to measure how even their distributions are, and we can use these same techniques to evaluate the results of our own not-so-arbitrary subdivisions.

5.1   Covering

The covering problem is sometimes called the Tammes problem named for the 1930s Dutch botanist P. M. L. Tammes, who studied pores on pollen grains. We briefly mentioned Tammes' work in Chapter 3. Tammes noticed that pores are distributed on spherical pollen grains in highly symmetric ways and their arrangement was essentially one of packing the largest possible equal circles on a sphere.[1] Tammes posed three questions:

- What is the largest diameter of *n* equal circles that can be placed on the surface of the sphere without overlap?

- How can they be arranged to achieve this maximum?

- Is the arrangement unique?

To meet these conditions, the minimum distance between the centers of the circles must be as large as possible and all distances (circle diameters) must be the same.[2] For a certain small number of circles, such as 2, 3, 4, 5, 6, 12, or 24, there are exact solutions for the Tammes problem. For other distributions of *n* circles, most are numeric (computer) solutions, though they remain unproven *conjectures.*

Circle packing has a large and rich body of literature. In 1964, Fejes Tóth[3] defined the conditions for packing equal circles on a sphere. Theoretic work by H. S. M. Coxeter and applied research by Tibor Tarnai and Zsolt Gáspár also determined the limits of packing equal circles on a sphere. Circle packing remains a key area of research because optimal packing addresses practical challenges such as locating cell phone repeaters, optimizing satellite orbits, deriving error-correcting codes, or developing antiviral drugs.

There are two broad approaches to distributing *n* points on a sphere—unconstrained and constrained. In unconstrained distributions, the placement of circles is subject only to tangent conditions where a maximum radius is achieved. Any distribution mechanism whatsoever that places them on the surface of the sphere maximizes their angular radius and does not cause any overlaps. Some ingenious techniques have been developed to find good distributions. One technique randomly distributes *n* points in three dimensions and constrains them to lay on the surface of the sphere. From there, the two closest points are moved a little farther apart and the point set is then reexamined. The two closest points are found and moved apart, and the distribution of points is periodically tested for evenness.

---

[1] (Tammes 1930)
[2] (Radin 2006)
[3] (Tarnai and Wenninger 1985)

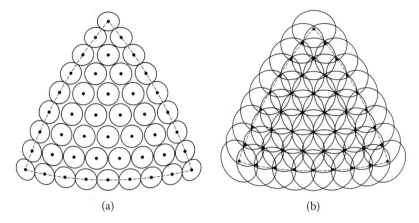

(a)   (b)

Figure 5.1. Tangent and overlapping spherical caps.

When there is no measurable improvement in the evenness of point distribution, the testing of pairs stop. In another technique, n points are mathematically modeled as n like-charged particles. Each particle is given the same electrical charge. They are treated as if each is attached to the end of an imaginary spring with the same elastic pull. The other end of the spring is fixed to the origin of the sphere.

Particles with the same charge will repulse each other and when all n points are released, they expand outward from the origin, trying to get away from their neighbors, which are trying to do the same thing. The springs pull them back, keeping the particles from flying off into space. After a number of expansion-contraction-repositioning cycles, the particles settle down and reach an equilibrium state on the surface of an imaginary sphere. The distribution is stable, because particles are equidistant from the center (equal spring force) and maximally separated from their closest neighbors (electric repulsive forces). The angular radius of a spherical cap centered over each particle is now as large as it will get.[4] We saw an example of this earlier in Chapter 3. These are just some of the many approaches to distributing unconstrained points on a sphere. Except for some relatively low number of points, however, it has been impossible to prove that a particular arrangement is unique, or the best one, or that a particular distribution technique consistently produces better results than another.

In constrained distributions, some type of point-group symmetry is imposed during the distribution process. For example, two or more circles may be clustered sidebyside or they may have a predefined relationship (i.e., they are always placed two at a time as antipodal pairs or four at a time as tetrahedral quartets).[5] In these last two cases, n is a step function and will always distribute multiples of two or four points on the sphere's surface.

Another form of constrained distribution is *polyhedral packing*. Circles are centered at the vertices of a polyhedron inscribed in a sphere. Tibor Tarnai has investigated polyhedral packing extensively and we list some of his papers in the additional resources at the end of the chapter.[6] The spherical caps for the Platonic solids, shown in Chapter 4, are the most

---

[4] Paul Burke maintains a very useful website with many examples of distributing points on a sphere. See http://local.wasp.uwa.edu.au/~pbourke/geometry/spherepoints.

[5] (Fowler, Tarnai, and Gáspár 2002)

[6] (Tarnai 1980)

basic examples of polyhedral packing. In the subdivision schemas we develop later on, we constrain points to icosahedral symmetries. Circle packings with icosahedral symmetry have been tabulated for large numbers of circles.[7]

Figure 5.1(a) shows circle packings (caps) for points on a spherical equilateral triangle. All caps have the same radius and the radius is as large as possible without causing overlaps. In this distribution, the distances between any two adjacent edge points of the triangle (dashed) are the same. Thus, every circle over the triangle's edge is tangent. However, the distances between points within the face are close, but not the same. Thus, there is some space between them.

Figure 5.1(b) shows a variation on the packing theme. The points in (b) have the same distribution as (a), and all caps have the same radius. But this time, the cap radius in (b) is determined by the amount of overlap needed to eliminate all gaps and spaces. In other words, in Figure 5.1(b), there are no *interstices*, spaces between each cap, such as in (a). The effect is to tile over the entire surface of the sphere. Tibor Tarnai and Magnus Wenninger have investigated these spherical circle-coverings and shown interesting results where points are defined when the cap's lesser circles intersect (see the bibliography for references). They take on a new hexagonal-pentagonal pattern.[8] When the caps overlap like this, the sphere is "shingled" with roofing plates, as it were. Several architects have explored the idea for roofing geodesic domes.[9] They find, however, the caps need more than the minimum radius to ensure enough overlap to avoid rain infiltration.[10] The total area of overlapping caps is not a useful metric because of overlap, but the common cap radius that just eliminates interstices is a good index. For our subdivision evaluations, we use the nonoverlapping caps when we compare one subdivision schema with another. We include overlapping caps here just for completeness.

What do the caps tell us about point distributions? First, it is easy to see the distribution pattern because the spaces stand out. In Chapter 9 we describe stereographic projection, which is a method for making 2D stereograms of points, arcs, and circles on a sphere that can be measured. Second, we can find the percentage of the sphere's area covered by caps. If two subdivision schemas distribute the same number of points, the one with caps covering more of the sphere's surface has a better distribution of points than the other.

The number of points and their distribution created by our subdivision schemas are not arbitrary, and they are not a continuous function. The total number of points depends on the number of polyhedral faces or edges, the intensity of subdivision, and the particular schema used to subdivide. In all schemas, the maximum radius of a circle or the cap is determined by the arc distance between the two closest vertices in the entire subdivision. Figure 5.2 shows two spherical icosahedra with caps centered at every subdivision grid point. The subdivision in Figure 5.2(a) distributes 642 points on the surface of the sphere. A cap that is centered at every point as large as possible and has no overlapping will have an angular radius of 3.96°. A single cap is about .015 steradians and all of them together have a density of .77 or cover a little more than three quarters of the sphere's surface. This particular example has a very good point distribution. The example in Figure 5.2(b)

---

[7] (Hardin, Sloane, and Smith 2008)

[8] (Tarnai 1993)

[9] Leonard Spunt developed a patented modular dome structure that essentially uses circular units, somewhat like spherical caps, to replace struts or chord-like structural members when building a dome. The benefit is that fewer different-sized units are needed than traditional strut systems. See (Spunt 1976).

[10] (Spunt 1976, 237)

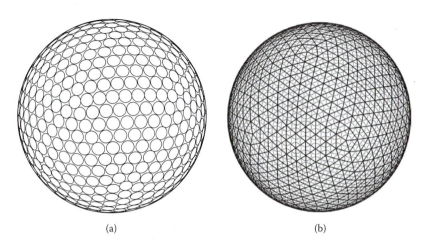

(a)   (b)

Figure 5.2. Spherical caps.

shows both the subdivision grid and spherical caps at every grid vertex. This particular grid distributes 752 points and its caps cover about 71 percent of the sphere's surface. In Chapter 9 we will use the density of spherical caps to compare one subdivision with another. The goal is to determine which subdivision schema produces the most uniform distribution for the same number of subdivision grid points.

5.2 Packing

Sphere packing is a variation of the Tammes covering problem and arranges identical spheres around points on the surface of a sphere. Sphere packing focuses on maximizing the minimal distance between points. This seems like double talk, but it basically means that for any point distribution, the goal is to make the distance between points as large and nearly equal as possible. One researcher states this objective somewhat philosophically: "How can you best stay away from other people in a spherical universe?"[11] In sphere packing, the objective is to find the largest common diameter that a set of equal spheres can have and be placed on another sphere without interference. The centers of the spheres all lie on the surface of the sphere being evaluated.

Let us take the simplest packing case in which just three spheres touch each other. Figure 5.3 shows that their centers will be equidistant and the angular separation between one another will be 60°, forming an equilateral triangle between the sphere's centers. We can imagine a fourth sphere cradled on top of these three, with all four centers equidistant and with the same angular separation. This is the maximum number of points in three dimensions that can be both equidistant and equiangular to one another; they also define the vertices of the tetrahedron.

Within limits, we can place identical spheres around a central sphere where packed spheres will kiss, but not interfere with one another. *Kissing*, a term borrowed from billiards, refers to balls that just touch but do not interfere with each other. The *kissing number* is the largest possible number of equal size spheres that can touch a given sphere in three dimensions. Finding that number of spheres is known as the kissing number problem. The

[11] (Bagchi 1997)

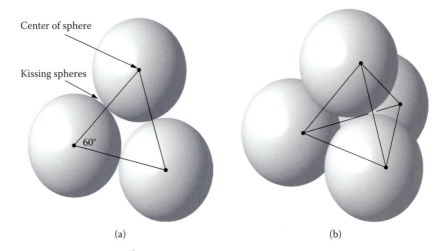

Center of sphere

Kissing spheres

60°

(a)   (b)

Figure 5.3. Angular separation of three tangent spheres.

goal is to minimize the gaps between the balls and not have any ball interfere with another. Figure 5.4 shows that the maximum kissing number in three dimensions is 12— that is, 12 spheres surround a central one, with all having the same radius. The centers of the outer 12 spheres form the vertices of an icosahedron.

The kissing number problem in three dimensions can also be described in terms of the angular separation. This angle between the centers of kissing, or near-kissing spheres, will be at least 60°, as we saw in Figure 5.3. Circle packing and the kissing number problem are two views of the same thing.

We can adapt the principles of packing and kissing spheres to spherical subdivision by centering spheres of the same radius at each point on the surface of the sphere that results from subdivision. These packed spheres will have the same radius, but it will be far less

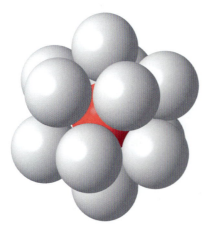

Figure 5.4. Twelve packed spheres.

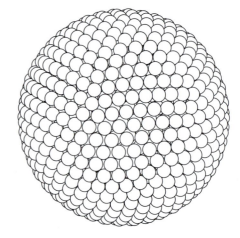

Figure 5.5. Distributed point sphere packing.

than the unit sphere's radius they surround. The largest radius of the packing spheres will be half the chord length between the two closest grid points in the subdivision. If their radius were larger, the spheres would interfere; if less, then the spaces between would increase. Figure 5.5 shows a typical sphere packing for a subdivided spherical icosahedron. All spheres have the same radius, and their centers are all on the surface of the unit sphere. In this particular schema, only the spheres that are immediately clustered around each of the 12 icosahedra vertices kiss. There are varying gaps between all other spheres.

If two subdivision schemas distribute the same number of points on the sphere, the one that has the larger radius packing spheres is the one whose points are more uniformly distributed. Thus, the metric of comparison of the two schemas is simply the packing sphere's radius.

5.2.1 200-Year-Old Kissing Puzzle

In three dimensions, the number of spheres that can be packed around a central one is not always obvious. It is easy to arrange 12 spheres to touch a central one, as shown in Figure 5.4. But it's not clear if you can add another. There is enough total space left over, but will one more fit? In the late 1600s, Isaac Newton and astronomer David Gregory argued this point. Both recognized there was enough space left over between the spheres to add another, but would an additional sphere kiss or interfere? Newton thought the kissing number remained at 12 and none could be added, but Gregory argued otherwise. The question was not resolved until 200 years later. In 1874, Newton was proven correct.[12]

5.3   Volume

The volume perspective sees the point distribution problem as one of maximizing the volume of an enclosing surface surrounding the points. This surface is called the *convex hull* and it is a mathematical concept that applies to *n* dimensional space where *n* can be 2, 3, 4, or more dimensions. In two dimensions, it is the minimum convex polygon that surrounds a set of points. The perimeter of this polygon is like a rubber band stretched tightly against the outermost points in a set of points. The polygon is always convex because a line between any two points on the perimeter is always inside the polygon or coincident to its perimeter.

In three dimensions, the convex hull is like shrink-wrapping around the extent of the outermost points. It is the minimum *convex polyhedron* that surrounds the points. The surface resembles a faceted polyhedron that perfectly contains the object or the set of points. A simple example is the Platonic solids. They are convex polyhedra with all points equidistant from the center; therefore, their convex hull is the same polyhedra. This is also the case for our subdivided spheres, where neighboring point sets form planar triangles.

Three-dimensional convex hulls are important in many applications. For example, in space management, they play a role in packing products and shipping cargo, activities in which the minimum volume needed to accommodate something is important (although this ignores the possibility of nesting objects one within another). In robotics, the convex hulls of objects indicate potential collision zones, where one object can interfere with another, like a robot passing through a maze of objects. In computer graphics, convex hulls are used to distinguish one object from another. The convex hulls of horses do not look like

---

[12] (Szpiro 2003)

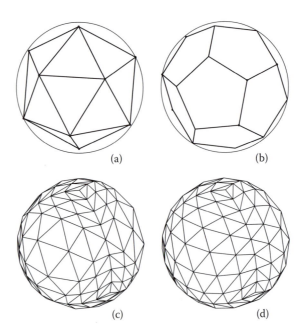

(a) (b)

(c) (d)

Figure 5.6. The number and distribution of points affect the convex hull.

the convex hulls of tea pots. Convex hulls can also be used to find the resting position of objects, or how an object moves or is to be picked up. In our application, the convex hull's volume provides a way to measure just how evenly points are distributed on the surface of a sphere.

The convex hull surface defines a volume and it is affected by the number and distribution of noncoincident points on the sphere's surface. Adding more points or spacing them more evenly will increase the volume. Figure 5.6 shows the effects of both. Figure 5.6(a) and (b) show how by increasing the number of points, it increases the convex hull. The icosahedron (a) and dodecahedron (b) evenly distribute 12 and 20 points on the surface of their respective unit circumsphere. The dodecahedron has almost 9.8 percent more surface area than the icosahedron, simply because it evenly distributes more points on the same unit sphere. When both are inscribed in a unit sphere, the dodecahedron's volume is 66.49 percent, whereas the icosahedron's volume is less at 60.54 percent of the sphere's volume.

In Figure 5.6(c) and (d), we see the effects of point distribution on the convex hull. Both subdivided spheres distribute 130 points on the sphere, but the points in schema (d) are more evenly distributed than (c), as demonstrated by the fact that there is less variation in the size of its triangles.. The convex hull of (d) is about 1.3 percent larger as a result. The convex hull of a fully subdivided sphere is simply the smallest convex polyhedron that completely encloses it and tends to look just like a regular faceted sphere, as shown in Figure 5.8. Efficient 3D convex hull algorithms remain a computer science research area, but our spherical subdivision applications greatly simplify calculations. In our applications, all points are on the surface of a sphere. Thus, every point defines part of the convex hull's surface. This is shown in Figure 5.7, where the convex hull for a subdivided spherical

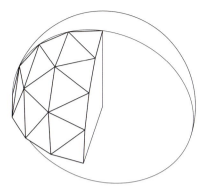

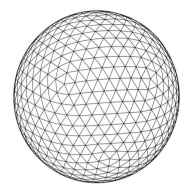

Figure 5.7. Convex hull segment.   **Figure 5.8.** Convex hull of a spherical subdivision.

polyhedron face is shown. A portion of the sphere's surface is shown, so that it is clear that the points are on the surface. The triangulated mesh that results from connecting the nearest neighboring points is simple to define, and every triangle becomes a surface patch in the convex hull. There are 16 triangular patches and 3 more surface patches along the sides to the origin of the sphere. The symmetry within our subdivisions reduces calculation as well. Most of our schemas are based on standard geometric subareas of spherical polyhedra such as an equilateral face or a Schwarz triangle. Figure 5.7 is a subdivision of a single face of a spherical octahedron. There are a total of eight octahedral faces subdivided like these in the whole sphere. Highly precise surface and volume calculations need only be performed for these standards and then multiplied by the number of instances to get the surface area and volume for the whole sphere. Later, in Chapter 9, we will compare the convex hull volumes of different subdivision schemas. The differences tell us which ones produce the best point distributions.

Figure 5.8 shows a complete convex hull surface for a higher order spherical icosahedron subdivision. The surface looks no different than a faceted version. Once again, high-precision calculations are performed on a standard single subdivided face and then multiplied by 20 to get the convex hull volume for the whole sphere.

5.4   Summary

Point distributions on spheres can be defined as covering, packing, or volume problems. Each approach tries to optimize the arrangement of points, maximizing or minimizing a different aspect of their distribution. Cap density tells us what percentage of the sphere's surface is covered by lesser circles (all with the same maximum radius) without causing overlaps. Likewise, the radius of the largest possible sphere centered at every point provides another metric for evaluating point distributions. Spheres, all with the same radius, kiss or just touch each other but do not interfere with each other. The radius is related to the minimum chord distance in the subdivision and indicates how uniformly points are spread out. The convex hull surface provides an area and volume metric for evaluating point distributions. Both increase as more points are placed on the surface of the sphere, and both increase as the same number of points is more evenly distributed.

In the next chapter, we will look at ways to use spherical polyhedra as geometric references to generate and distribute points.

Additional Resources

Aste, Tomaso and Denis L. Weaire. *The Pursuit of Perfect Packing.* Philadelphia, PA: Institute of Physics Publishing, 2000.

Fowler, P. W., T. Tarnai, and Zs. Gáspár. "From Circle Packing to Covering on a Sphere with Antipodal Constraints." *Proceedings of the Royal Society* (London) 458.2025 (2002): 2275–2287.

Szpiro, George. *Kepler's Conjecture: How Some of the Greatest Minds in History Helped Solve One of the Oldest Math Problems in the World.* Hoboken, NJ: John Wiley & Sons, 2003.

Tarnai, T., Zs. Gáspár, and Lidia Szalait. "Pentagon Packing Models for All-Pentamer Virus Structures." *Biophysical Journal* 69 (1995): 612–618.

Tarnai, T. and Magnus J. Wenninger. "Spherical Circle-Coverings and Geodesic Domes." *IASS International Congress* Moscow (1985): 5–21.

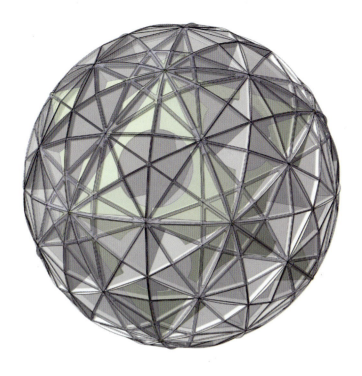

6 Polyhedral Frameworks

In this chapter, we cover the characteristics of Platonic solids and Archimedean solids and how to define their geometry precisely, a prerequisite for subdivision. We begin our discussion of polyhedra by defining their most common characteristics. We then detail the advantages each solid offers for spherical subdivision, show their relationships to each other, and highlight the symmetry properties we can exploit when we use them to subdivide spheres. In the next chapter, we use them to jump-start spherical subdivision.

For centuries, polyhedra have fascinated mathematicians, scientists, and artisans alike. They are one of the most intensely studied aspects of geometry. Their 3D faces appear in seemingly infinite arrangements of faceted and star shapes. Their geometry appears in everything from microscopic sea creatures to crystals, from virus cells to soccer balls. Their symmetries apply to mathematics, architecture, art, chemistry, biology, product design, and physics.

Two types of polyhedra, Platonic and Archimedean, are ideal frameworks for subdividing spheres. Platonic solids (we use the terms *solid* and *polyhedron* interchangeably) have faces made of only one type of regular polygon: equilateral triangles, squares, or pentagons. There are five Platonic polyhedra. *Archimedean solids* have faces made of two or more regular polygons and include hexagons, octagons, and dodecagons as well. There are many Archimedean solids, some quite complex. Both of these types of solids have planar and spherical versions and they are highly symmetrical. Spherical polyhedra give us a huge head start in the subdivision process because their vertices are already evenly distributed on the surface of a sphere that surrounds them and their regular face shapes; with equal edge length, edges give us small and manageable areas to work with as well. We can develop grids within these small symmetrical areas and then replicate the grid repeatedly to cover the rest of the sphere without causing overlaps or leaving any gaps. The same symmetry ensures that the total number of vertices, faces, and edges in a polyhedron will always have a predictable relationship. Spherical subdivision based on spherical Platonic and Archimedean solids are the most common approach, by far.

6.1   What Is a Polyhedron?

Polyhedra are mathematically sophisticated and appear in more than three dimensions. In solid geometry, a polyhedron is a 3D solid whose faces join at their edges. The word derives from the Greek *poly* (many) plus the Indo-European *hedron* (seat). There are many distinct families of polyhedra, the result of differences in their geometry and various mathematical properties. We focus on two types, the Platonic and Archimedean, because they are the most common polyhedral frameworks used in spherical subdivision. The five Platonic solids are regular polyhedra, which means their faces are the same regular type— equilateral triangles, squares, or pentagons. The 13 Archimedean solids are closely related to the Platonic solids, but have two or more regular faces and some have hexagons, octagons, and decagons.

We can define Platonic and Archimedean solids as 3D convex forms with no holes, depressions, protrusions, or penetrations. They are totally defined by their vertices, edges, and faces. All vertices lie on the surface of a surrounding sphere called the circumsphere, whose origin is the same as the center of the polyhedron. The edges of Platonic and Archimedean solids are equal-length straight edges between two neighboring vertices. On spherical versions, edges are equal-length geodesic arcs. The faces of planar polyhedra are regular polygons, and the plane of any one of them, if extended, does not pass through the interior volume of the polyhedron. An individual Platonic or Archimedean solid will have the same number and types of faces sequenced in the same way around every one of its vertices. Faces do not penetrate one another and every face joins another at an edge, leaving no gaps and causing no overlaps. For spherical polyhedra, the sum of the face angles surrounding any vertex is always 360°.

Figure 6.1 shows the planar and spherical version of a dodecahedron, a Platonic solid with 20 vertices, 30 edges, and 12 pentagonal faces. The planar face angles shown in (a) are all 72°, and the dihedral angle, the angle between faces measured perpendicular to their shared edge, is approximately 116.56°. The spherical version shown in (b) has face angles $\beta = 120°$. The arc angles of the spherical edges α are each approximately 41.81°.

Spherical Platonic and Archimedean solids can be derived from their planar versions very easily. All vertices of these solids are the same distance from the center of the polyhedron, thus they are all on the surface of a circumscribing sphere. The vertices on either end of a chord, a platonic edge, and the center of the solid define the plane of the arc, a geodesic arc, between neighboring points. In Figure 6.1(a), one such plane is shown as a small

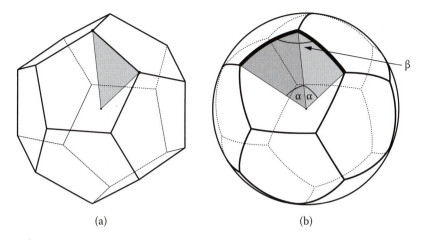

Figure 6.1. Planar and spherical polyhedra.

gray triangle defined by three points: the origin and two vertices.[1] Figure 6.1(b) shows the spherical version of the dodecahedron from the same viewpoint and scale as (a), and every edge in (a) has been transformed into a geodesic arc in (b). The planes of two arcs are emphasized in (b). You can think of these geodesic arcs as *central projections* of a planar polyhedron's edges. The projection is on the polyhedron's circumsphere, which acts like a curved projection screen. An imaginary point light source located at the origin produces a shadow of the chord, its arc, on the circumsphere screen.

In Figure 6.1(b), the two emphasized arcs demonstrate another fact about angles on a sphere. The surface angle β between the two arcs is the dihedral angle, between the great circle planes that define them (see Section 4.16.6). The arcs themselves are the sides of the spherical face and are measured by their central angles from the center or origin of the sphere. They are each labeled α. In any given Platonic or Archimedean polyhedron (planar or spherical version), all chords or arcs are the same length.

With this general introduction, let us take a closer look at the properties of these two families of polyhedra.

6.2   Platonic Solids

The five Platonic solids are *regular polyhedra*. Known since antiquity, regular polyhedra are polyhedra having the same size and type of regular convex face (with equal angles and length sides), and the same number of faces meeting at every vertex. Regular polyhedra are the only polyhedra that enclose a volume of space with a surface composed of only one kind of plane polygon (for example, squares, triangles, or pentagons) and feature one kind of polygon meeting at every vertex. Figure 6.2 shows all five Platonic solids. They are the tetrahedron (1), octahedron (2), cube[2] (3), dodecahedron (4), and icosahedron (5). In Figure 6.2, all solids have the same edge length, and the differences in their enclosed volumes are apparent.

---

[1] Three points define a plane.

[2] A cube is also called a *hexahedron*.

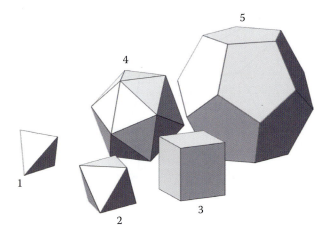

Figure 6.2. Platonic solids.

The Platonic solids are the most widely used frameworks for spherical subdivision:

- All vertices are evenly distributed and lie on the surface of a surrounding sphere called a circumsphere.

- Each Platonic solid is composed of equal surface angles, equal dihedral angles, equal chord length, and equal arc lengths.

- Every face, edge, and vertex has rotational and reflection symmetry.

- A single small Schwarz right triangle totally defines the entire sphere's subdivision.

- A subdivision of a Platonic solid also subdivides its dual, even though its dual's face type is very different. (Duals will be discussed shortly.)

- A spherical equilateral triangle can describe the surface of any Platonic solid, even if its faces are not equilateral.

- Except for the tetrahedron, the planes of faces on opposite sides of the polyhedron are parallel.

Each of these characteristics will be discussed in the remainder of this chapter.

One could say that the tetrahedron is the most equal of polyhedra. It has four vertices, six edges, and four equilateral faces. Not only are its edges equal and all of its faces the same, it is the only polyhedron in which every vertex is equidistant from every other one. It's also the only Platonic solid whose vertices have no antipode, which means no two vertices are on opposite sides of the solid or its circumsphere. The tetrahedron can be formed by truncating a cube with two cuts; that is, by slicing off pieces of the solid. Each cut starts along the diagonal of one of the cube's faces and slices through to the opposite side's vertices without passing through the center of the cube. In effect, the opposite vertices of one face and its nonantipodal companion on the cube's opposing face define the tetrahedron's vertices. We discuss truncation in more detail when we describe how some Archimedean solids can be created from Platonic solids.

The octahedron is a familiar form. Many people visualize it as two back-to-back square pyramids. It has 6 vertices, 12 edges, and 8 equilateral faces. Four faces meet at any vertex. Each of the octahedron's six vertices has an antipode and the three axes between antipodal pairs are mutually orthogonal. The octahedron plays a special role in spherical subdivision. Its spherical version is the only spherical polyhedron where its contiguous edges define three mutually orthogonal great circles. Cartographic applications often use octahedral bases when grid systems must have an equatorial and Greenwich great circle. The spherical octahedron also appears in applications where a Cartesian coordinate system is being modeled. From a trigonometric point of view, the faces of the spherical octahedron clearly demonstrate that a spherical triangle can have more than one 90° angle. The octahedron's faces each have six: three 90° surface angles and three 90° edges.

The cube, with its 8 vertices, 12 edges, and 6 faces, is the most familiar polyhedron. It is defined by three parallel pairs of square faces whose midface points form orthogonal axes. Pairs of opposing edges are parallel, and every vertex has an antipode. The cube has a special role in spherical subdivisions because it has subtle symmetry relationships with other polyhedra. These relationships are discussed later in the chapter.

The dodecahedron has 30 edges, 12 pentagonal faces, and 20 vertices, the most vertices of any Platonic solid. The dodecahedron also has the largest volume of any Platonic solid if all the solids have the same length edges. From a symmetry point of view, the dodecahedron's 12 pentagonal faces are 6 pairs of faces on opposite sides of the polyhedron. Opposing pairs of faces are rotated 36° relative to each other around an axis between their midface points. Every edge of the dodecahedron is parallel to another edge on the opposite side of the polyhedron.

The icosahedron is by far the most frequently used polyhedral foundation in spherical subdivision. The icosahedron has 12 vertices, 30 edges, and 20 equilateral faces, the most faces of any Platonic solid. In its spherical version, each face can be subdivided into 6 spherical right triangles, which means 120 (6 × 20 faces) of these right triangles tile together to cover the entire sphere. Subdivisions of one right triangle can define the subdivision of the whole sphere. The icosahedron has 15 pairs of opposing edges and its 12 vertices define 6 pairs of antipodal points. Each antipodal pair defines an axis. These symmetries are very important and described in great detail later in the chapter.

6.2.1 Platonic Duals

The dual of a polyhedron is another polyhedron resulting from exchanging vertices and faces. A vertex of the dual is located at the center of the original polyhedron's face. The duals of Platonic solids are unique; they are other Platonic solids. Figure 6.3 illustrates the dual relationship for all the planar and spherical Platonics. In the front row, from left to right, are the planar dodecahedron-icosahedron duals (a) and the octahedron-cube duals (b). The tetrahedron (c) is a dual to itself. The back row is the spherical version of the front row. The dual relationship between two polyhedra is especially evident in the planar versions, where you easily see that the faces of one correspond to the vertices of the other. In the spherical versions, the colors of the planar versions are retained to better see how their spherical faces and edges relate. In the spherical versions, you can easily see the edges (great circle arcs) of each dual cross perpendicular to the other and they cross each

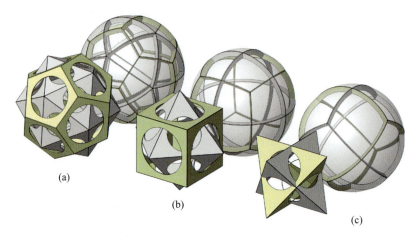

(a)

(b)

(c)

Figure 6.3. Planar and spherical platonic duals.

other at their midpoints. This is also the case in the planar versions, but its easier to see in the spherical version. Notice also that for any pair of duals, both have the same number of edges, but their numbers of faces and vertices are swapped. That is, if one polyhedron in the dual set, the octahedron-cube dual for example, has six vertices and eight faces (the octahedron), the other (the cube) will have eight vertices and six faces. Another consequence of the symmetry between duals is the spherical subdivision of one of them is also a subdivision of the other.

6.2.2   Shorthand for the Unpronounceable

The Platonic solids have simple names and their characteristics (vertices, edges, and faces) are easy enough to remember. But the Platonic solids are not the only polyhedra. In fact, there are hundreds. Many have catchy titles such as tetrahemihexahedron or the greatinverted-retrosnub-icosidodecahedron. Shorthand notations were inevitable to make it easier to refer to polyhedra and to see their similarities and differences at the same time. There are several notation systems. Most are based on the polyhedron's face types and summarize how faces meet at edges or cluster around vertices. With experience, one can visualize complex polyhedra simply by looking at their notation. In some systems, a rearrangement of notation describes their dual and their *symmetry group*, something we discuss shortly.

Ludwig Schläfli (1814–1895) worked mostly in geometry, arithmetic, and function theory, though some of his best work was in polyhedra, where he invented the notation named after him. *Schläfli symbols* describe all the Platonic and Archimedean polyhedra. We highlight this particular system for the Platonic solids because Coxeter, a renowned mathematician, extended this system to describe geodesic subdivisions of spherical polyhedra. The extended system is particularly useful when describing the spherical subdivision schemas. It is a useful notation to become familiar with.

Schläfli symbols take the form $\{p,q\}$ to describe regular polyhedra, with p the number of sides in a regular polygon face and q the number of those faces that meet at any vertex. A cube, for instance, has square faces each with four sides (p) and three of these faces meet

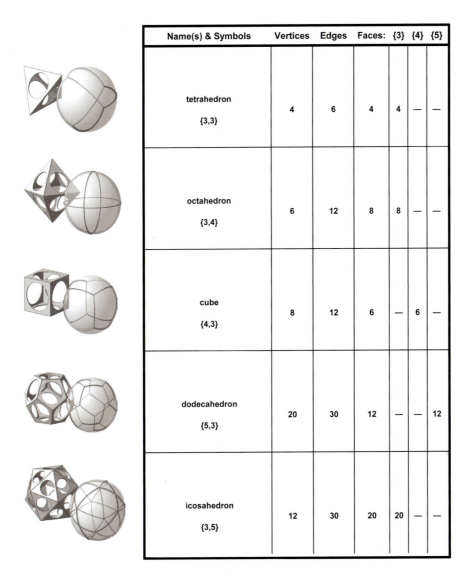

Name(s) & Symbols	Vertices	Edges	Faces:	{3}	{4}	{5}
tetrahedron {3,3}	4	6	4	4	—	—
octahedron {3,4}	6	12	8	8	—	—
cube {4,3}	8	12	6	—	6	—
dodecahedron {5,3}	20	30	12	—	—	12
icosahedron {3,5}	12	30	20	20	—	—

Table 6.1. Planar and spherical Platonics.

at each vertex (q). The Schläfli symbols for the Platonic solids (paired by dual) are:

Icosahedron	{3,5}	Dodecahedron	{5,3}
Octahedron	{3,4}	Cube	{4,3}
Tetrahedron	{3,3}		

One useful property of Schläfli symbols is that the reverse of $\{p,q\}$ or $\{q,p\}$ is the symbol for the dual of that polyhedron. In other words, if $\{p,q\}$ is a possible solid, so is $\{q,p\}$. You can verify this by looking at Table 6.1, which summarizes the vertices, edges, and faces for the five Platonic solids we have been discussing. The Schläfli symbols for each polyhedron is given under the polyhedron's name. Notice also that p and q values are simply reversed for duals.

6.2.3 Circumsphere and Insphere

We already mentioned that the vertices of the Platonic solids are tangent to a circumsphere whose origin is coincident with the center of the polyhedron, a point equidistant from each vertex. Less obvious, however, is the fact that another sphere, called the *insphere*, can be inscribed inside each Platonic solid and this sphere is tangent to the midpoint of each Platonic face. These points of tangency define the vertices of its dual.

Figure 6.4 illustrates this relationship by showing the circumsphere and insphere for a dodecahedron. The radius of the insphere, the *inradius*, is the perpendicular distance to the face from the center of the polyhedra. The view in Figure 6.4 shows both spheres and several vertices tangent to the circumsphere. Midface points tangent to the insphere define the vertices of the polyhedron's dual, the icosahedron. Both polyhedra share a number of systematic relationships that we will look into next.

The late twentieth-century geometrician H. S. M. Coxeter used the circumsphere-insphere tangent points to define regular polyhedra in a unique way. He states, "A polyhedron is regular if there exists three concentric spheres one of which contains all the vertices, one contains the midpoints of all the edges, and one that meets all the centers of all the faces."[3] The five Platonic solids satisfy these three conditions.

Dual Platonic solids share another subtle relationship. For the icosahedron-dodecahedron and cube-octahedron dual, if either pair's circumsphere has the same radius, they will also have the same radius insphere. This property means the ratio of their volumes is the same as the ratio of their surfaces for either pair of duals.

How can this happen? A polyhedron and its dual are different solids! The answer lies in the fact that when both polyhedra have the same circumsphere, the radius of *circumcircles* around their faces is also the same.

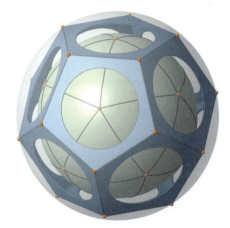

Figure 6.4. Polyhedra circumsphere and insphere.

---

[3] (Coxeter 1973)

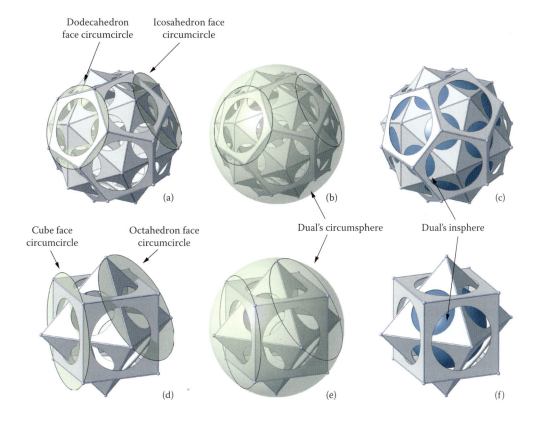

Dodecahedron
face circumcircle

Icosahedron face
circumcircle

(a)

(b)

(c)

Cube face
circumcircle

Octahedron face
circumcircle

Dual's circumsphere

Dual's insphere

(d)

(e)

(f)

Figure 6.5. Common circumsphere and insphere.

Figure 6.5(a) shows an icosahedron and a dodecahedron inscribed in the same circumsphere. The vertices of both polyhedra are the same distance from their shared center, and the circumcircle of one face from each polyhedron is shown in the figure. The circumcircles around the dodecahedron's pentagonal faces have the same radius as those around the icosahedron's triangular face. Since the circumference of these circumcircles pass through their respective face's vertices, which are already on the surface of the sphere, the circumcircles are actually lesser circles on the surface of the circumsphere. As shown in Figure 6.5(b), they form congruent spherical caps whose base planes are the same perpendicular distance from the origin of the sphere. Thus, both polyhedra have the same insphere radius.[4] The icosahedron's insphere is shown in blue in Figure 6.5(c). It is tangent to the midfaces of both the dodecahedron and icosahedron.

The exact same relationship applies to the octahedron and cube, as shown in Figure 6.5(d) to (f). The faces of both these polyhedra have the same radius circumcircles when they both have the same circumsphere, as shown in (d) and (e). Thus, the ratio of their two volumes is the same as the ratio of their two surface areas. You can verify this by looking at the volume and area ratios for the icosahedron-dodecahedron and cuboctahedron later in the chapter.

---

[4] Note that the polyhedral face circumcircles and resulting spherical caps illustrated here are not the same as the caps surrounding polyhedral vertices illustrated earlier in Chapter 4.

6.2.4 Vertex-Face-Edge Relationships

In 1639, René Descartes observed that the total number of a polyhedron's vertices, faces, and edges were systematically related. He noted his discovery in private correspondence and described it in a manuscript that remained unpublished until after Leonhard Euler independently rediscovered the same relationship and published it in 1752.[5] Today, this formula is called the Descartes-Euler polyhedral formula, but more commonly, Euler's law.[6] For convex polyhedra (without any holes), Euler's law states the number of vertices + faces = edges + 2.

Euler's law applies equally to the Platonic and Archimedean solids and to the highly subdivided spherical versions based on them. Figure 6.6 shows the tetrahedron, cube, and dodecahedron in the front row, and subdivided spherical versions behind them. Each polyhedron is the same scale and is oriented the same way as its spherical version. Although the Platonic and spherical versions (shown in the figure) differ considerably, Euler's law demonstrates that the relationship of their total vertices, faces, and edges remains the same. Table 6.2 confirms Euler's law by listing the vertices, faces, and edges for each of the polyhedra in Figure 6.6.

In the next chapter, we develop techniques for subdividing spheres. We make use of a very handy triangulation number called T to describe the triangulated mesh resulting from a subdivision. With T, it is easy to determine the number of faces, edges, and vertices in a highly subdivided polyhedron. These counts are consistent with Euler's law, no matter how intensely the sphere is subdivided. We will use both T and Euler's law to check our subdivision results.

Descartes made another important contribution in understanding the relationship between the faces, edges, and vertices of convex polyhedra. He noted that the sum of the face angles around any vertex, an amount he called the *closure deficit*, is always less than 360°. A simple example shows why this is the case. If five equilateral triangles cluster around the same vertex, they form a *pentagonal cap*, as seen in Figure 6.7(a). When each

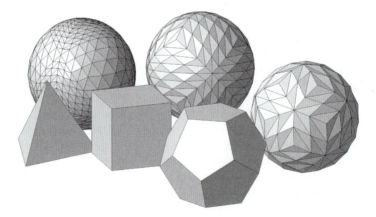

Figure 6.6. Vertex, face, and edge examples of Platonics and subdivided spherical Platonics.

[5] (Descartes, Adam, Tannery 1996, 265)
[6] See (Richeson 2008) for an excellent history of Euler's law and how this law created the field of topology.

Table 6.2. Examples of Euler's law for Platonics and subdivided spherical Platonics.

Euler's Law								
Polyhedra	Vertices	+	Faces	=	Edges	+	Constant	=	Sum
Tetrahedron	4	+	4	=	6	+	2	=	8
Subdivided Spherical Tetrahedron	130	+	256	=	384	+	2	=	386
Cube	8	+	6	=	12	+	2	=	14
Subdivided Spherical Cube	194	+	384	=	576	+	2	=	578
Dodecahedron	20	+	12	=	30	+	2	=	32
Subdivided Spherical Dodecahedron	122	+	240	=	360	+	2	=	362

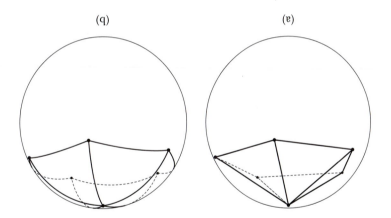

(a)

(b)

Figure 6.7. Five equilateral triangles define space.

vertex is on the surface of a sphere, the closure deficit is also measuring the sphere's curvature. In this example, the closure deficit is 60° or 360° − 5(60°). If the deficit is larger than this, the sphere gets smaller and the curvature is greater. Conversely, if the deficit is smaller, the size of the sphere gets larger and has less curvature. But there are limits. As the deficit approaches 0°, the polygons surrounding the vertex approach a plane. We see this happening if we try to position six equilateral triangles around the same vertex; the deficit would be 0° or 360° − 6(60°). They form a perfect plane, which has no curvature. As a result, no space is defined.

In Figure 6.7(b), we see the spherical version of the same vertex points shown in (a). If we extended the pentagonal cap analogy to every vertex that is not already completely surrounded, we will quickly recreate an icosahedron. This demonstrates that 20, the number of equilateral faces in an icosahedron, is the maximum number of faces possible for a regular polyhedron.

Descartes also observed the sum of all the closure deficits for all vertices in convex polyhedra is always 720°. We use this rather amazing observation later, when we check the results of our spherical subdivisions. Speaking of convex polyhedra (with no holes), Descartes stated: "If four plane right angles are multiplied by the number of solid angles and from the product are subtracted eight right angles, there remains the sum of all the plane angles, which exist on the surface of that polyhedron."[7]

What this means is the total face degrees in convex polyhedra equals four right angles (360°) times the number of vertices in the polyhedra less eight right angles (720°).

6.2.5 The Golden Section

The so-called *golden section* or *golden ratio* is an amazing geometrical construction that has been studied since antiquity. It has been related to human perceptions of balance and beauty and appears in nature, art, architecture, science, and mathematics. We can use the golden section to define very precise polyhedra.

The golden section is the result of asking a simple question: what is the result when a line is divided, so the ratio of the lengths of the divisions is the same as the ratio of the division to the length of the line?

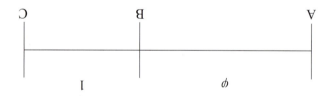

That is, divide a line segment with AC:AB :: AB:BC. In the language of ancient mathematicians, line AC is divided in an "extreme and mean ratio." The result is AB = 1.61803989... × BC.

This ratio is universally known as φ.

The value of φ is found from a simple algebraic statement of the original problem. If BC = 1, then,

$$\frac{\phi+1}{\phi} = \frac{\phi}{1}$$

or,

$$\phi^2 - \phi - 1 = 0.$$

Among the amazing properties of φ is this result:

$$\phi - 1 = \frac{1}{\phi},$$

so 1/1.61803989... = 0.61803989....

It is also interesting to note that the ratio of two successive terms of the Fibonacci series, where each term is the sum of the preceding two terms (1, 1, 2, 3, 5, 8, 13, 21,...) converges to φ.[8]

The symbol φ occurs often in the geometry of regular polygons. Perhaps the most dramatic illustration is the pentagon, from which the mystical *pentagram* derives, as shown

---

[7] From American Mathematical Society's commentary on Descartes' unpublished *Treatise* manuscript, 1999.

[8] See Mario Livio, *The Golden Ratio: The Story of Phi, the World's Most Astonishing Number* (New York: Broadway Books, 2002) for a wonderful review of phi.

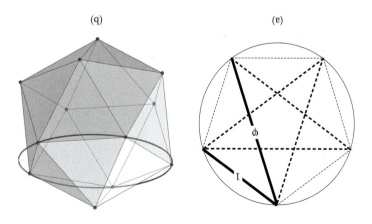

(a)   (b)

Figure 6.8. The pentagram's golden ratios.

in Figure 6.8. The ratio of the short and long lines (two are accented in bold) is $1:\phi$, the golden ratio.

6.2.6 Precise Platonics

Precise definitions of the Platonic solids are essential when subdividing their spherical versions. We can use unit and ϕ relationships to define the exact vertex coordinates of all five Platonic solids. The face angles of spherical Platonic solids are integral divisions of $360°$, and their sides are arc lengths based on trigonometric ratios of the square roots of 2, 3, and phi.

The coordinates of the tetrahedron, octahedron, and cube are simple:

Tetrahedron: $(+1, +1, +1)\ (-1, -1, +1)\ (-1, +1, -1)\ (+1, -1, -1)$
Octahedron: $(\pm 1, 0, 0)\ (0, \pm 1, 0)\ (0, 0, \pm 1)$
Cube: $(\pm 1, \pm 1, \pm 1)$

The coordinates of the dodecahedron and icosahedron are related to the golden ratio. The 12 icosahedron vertices are defined by the orthogonal intersection of 3 golden rectangles; that is, rectangles whose sides have the proportions of 1 to ϕ. Figure 6.9(a) and (b) shows the arrangement of the golden rectangles to the icosahedron's vertices and to its faces. The coordinates of the icosahedron's 12 vertices corresponding to the figure are

$$(0, \pm 1, \pm \phi),\ (\pm \phi, 0, \pm 1),\ (\pm 1, \pm \phi, 0).$$

Figure 6.9(c) and Figure 5.6(d) show the dodecahedron. The coordinates of its 20 vertices are also based on the golden ratio and defined by

$$\left(0, \pm \phi, \pm \frac{1}{\phi}\right), \left(\mp \phi, 0, \pm \frac{1}{\phi}\right), \left(\mp \frac{1}{\phi}, \pm \phi, 0\right), (\pm 1, \pm 1, \pm 1).$$

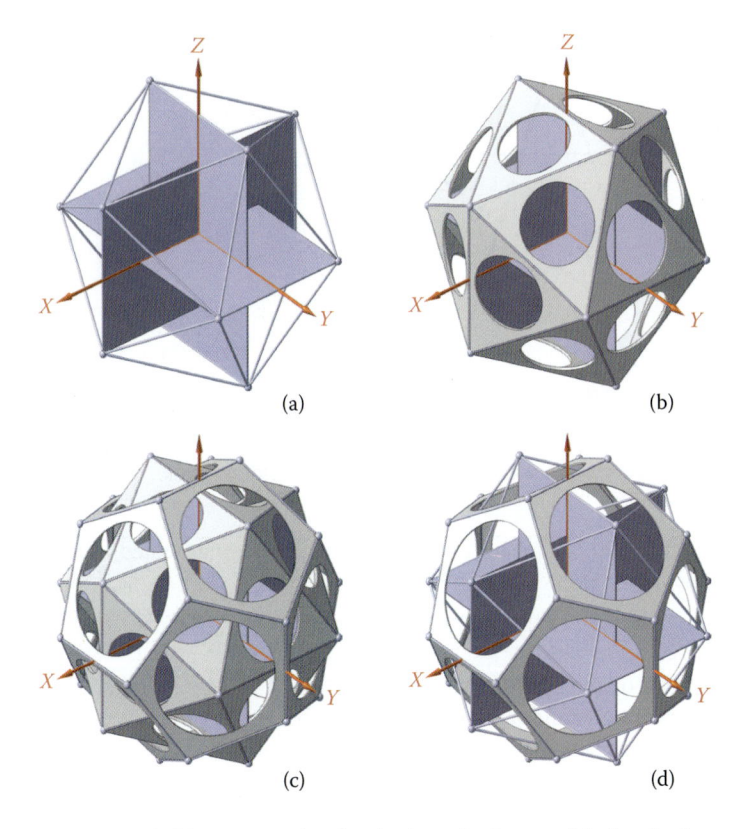

Figure 6.9. Golden rectangles for the icosahedron and dodecahedron.

The coordinates given previously for the Platonic solids are precise, but they do not create polyhedra with a unit radius circumsphere. To rescale vertex coordinates to be on the surface of the unit sphere, simply divide each of the coordinate's x, y, and z components by the absolute value of the coordinate:

$$x_{\text{unit sphere}} = \frac{x}{\sqrt{x^2 + y^2 + z^2}}$$

$$y_{\text{unit sphere}} = \frac{y}{\sqrt{x^2 + y^2 + z^2}}$$

$$z_{\text{unit sphere}} = \frac{z}{\sqrt{x^2 + y^2 + z^2}}$$

6.2.7   Platonic Summary

At this point, we have identified all the key properties needed to use any Platonic solid as a polyhedral framework for spherical subdivision. Table 6.3 provides more specific properties for each one, such as surface and central angles, vertex coordinates, and so

Planar Platonics in Unit Circumsphere					
Measure	Tetrahedron	Octahedron	Cube	Dodecahedron	Icosahedron
Symbol					
Schläfli Symbol	{3,3}	{3,4}	{4,3}	{5,3}	{3,5}
Face					
Edge Length	1.632993162	1.414213562	1.154700538	0.713644180	1.051462222
Circumcircle Radius	0.942809042	0.816496581	0.816496581	0.607061998	0.607061998
Incircle Radius	0.471404521	0.408248291	0.577350269	0.491123473	0.303530999
Vertex Angle	60.0	60.0	90.0	108.0	60.0
Dihedral Angle					
Adjacent Faces	70.528779366	109.471220634	90.0	116.565051177	138.189685104
Adjacent Faces (radians)	$\cos^{-1}\left(\frac{1}{3}\right)$	$\cos^{-1}\left(\frac{-1}{3}\right)$	$\cos^{-1}(0)$	$\cos^{-1}\left(\frac{-1}{\sqrt{5}}\right)$	$\cos^{-1}\left(\frac{-\sqrt{5}}{3}\right)$
Area					
Face r^2	1.154700538	0.866025404	1.333333333	0.876218520	0.478727069
Full r^2	4.618802154	6.928203230	8.0	10.514622242	9.574541383
Volume					
r^3	0.513200239	1.333333333	1.539600728	2.785163863	2.536150710
Vertex Coordinates	$(\pm1,+1, +1)$ $(-1, -1, +1)$ $(-1, +1, -1)$ $(+1, -1, -1)$	$(\pm1, 0, 0)$ $(0, \pm1, 0)$ $(0, 0, \pm1)$	$(\pm1, \pm1, \pm1)$	$(0, \pm\varphi, \pm1/\varphi)$ $(\pm1/\varphi, 0, \pm\varphi)$ $(\pm\varphi, \pm1/\varphi, 0)$ $(\pm1, \pm1, \pm1)$	$(0, \pm1, \pm\varphi)$ $(\pm\varphi, 0, \pm1)$ $(\pm1, \pm\varphi, 0)$
Reference Spheres					
Insphere Radius	0.333333333	0.577350269	0.577350269	0.794654473	0.794654473
Circumsphere Radius	1.0	1.0	1.0	1.0	1.0

Spherical Platonics in Unit Circumsphere					
Measure	Tetrahedron	Octahedron	Cube	Dodecahedron	Icosahedron
Face					
Angle	120.0	90.0	120.0	120.0	72.0
Central Angles					
Side	109.471220634	90.0	70.528779366	41.810314896	63.434948823
Side (radians)	$2\sin^{-1}\left(\frac{\sqrt{2}}{\sqrt{3}}\right)$	$2\sin^{-1}\left(\frac{\sqrt{2}}{2}\right)$	$2\sin^{-1}\left(\frac{1}{\sqrt{3}}\right)$	$2\sin^{-1}\left(\frac{1}{\sqrt{3}\,\phi}\right)$	$2\sin^{-1}\left(\frac{1}{\sqrt{\phi^2+1}}\right)$
Median	125.264389683	90.0	90.0	69.094842552	58.282525588
Arc Factor					
Side	1.910633236	1.579796327	1.230959417	0.729727656	1.107148718
Median	2.186276030	1.579796327	1.570796327	1.205932490	1.017221960
Chord Factor					
Side	1.632993162	1.414213562	1.154700538	0.713644180	1.051462224
Median	1.776147668	1.414213562	1.414313562	1.134176274	0.983929040
Area					
Face r^2	3.141592654	1.570796327	2.094395102	1.047197551	0.628318531

Note: 1) all angles in degrees unless noted

Table 6.3. Planar and spherical platonic summary.

forth. The most important facts we need when starting a subdivision are the precise vertex coordinates, face angles, and side angles of the Platonic solid we will use as our framework for spherical subdivision and Table 6.3 provides this information.

The central angles, values, or formulas can be used as is, but once again, the vertex coordinates, shown in the table, are based on unit and φ ratios and must be scaled to place vertices on the surface of a unit circumsphere. With the basics of each Platonic solid established, we now look into their symmetry. Symmetry shows us how to find common small areas where we can focus our subdivision efforts. Symmetry also shows us how the subdivision of a small area can be made to cover the entire sphere.

6.3   Symmetry

The symmetry of polyhedra is one reason we find them so visually attractive. But symmetry is more than beauty; there are mathematical symmetries at the core of the dual relationships we have seen, and can be exploited to define areas whose subdivision can be replicated over the remaining surface of the sphere.

There are many types of symmetry. Two types are particularly important in spherical work: rotation and reflection. A polyhedron has *rotational symmetry* if its appearance stays the same after being rotated around an axis by some integral fraction of a full rotation of 360°. Rotational symmetry is one of the most commonly used types of symmetry in spherical work. For example, if you hold a cube in your hand and orientate it so one side faces you and you keep that side pointing toward you while rotating the cube in 90° increments either clockwise or counterclockwise around an axis between your eye and the center of that side, the cube will still look the same each time. This is rotational symmetry. The axis of rotation is your line of sight to the center of the cube's side.

A polyhedron has *reflection symmetry* if one or more planes (mirrors) can be passed through the polyhedron's origin, and the image of faces, edges, or vertices on both sides of the plane are identical. For example, if you passed a plane through a pair of opposite edges of an icosahedron, the images on both sides of this plane would be identical. Since an icosahedron has 30 edges, or 15 opposite pairs, we can see that for edges alone there are already 15 reflection symmetries. Almost all polyhedra have this symmetry, but there are a few that don't. These exceptions are called *chiral* forms because the images on either side of any plane passed through their centers are different. We will discuss chiral forms later in the chapter.

The *identity symmetry,* sometimes called the "do nothing" symmetry, is equivalent to rotating a polyhedron 360° around any point in space; the polyhedron simply returns to its original position, without any change taking place. At first, identity symmetry seems useless because no apparent change has taken place. But the identity serves an important mathematical function analogous to multiplying a number by one (unity). We do not use the identity symmetry directly in subdividing spheres, but we include it whenever we enumerate a polyhedron's symmetries.

Figure 6.10 shows examples of rotation and reflection symmetry for a cube. In (a) through (c), a single rotation axis is shown through the centers of two opposing faces, edges, and vertices. All axes also pass through the center of the polyhedra, represented by a small sphere. The cube can be rotated any multiple of 90°, 120°, or 180°, respectively, around these three axes, and the cube will still look the same. This demonstrates rotational symmetry.

Figure 6.10(d) shows two reflection planes, one through mid-edge points, and the other through edges on opposite sides of the cube. These planes also pass through the cube's center. The cube's image on either side of either plane is identical and illustrates reflection symmetry.

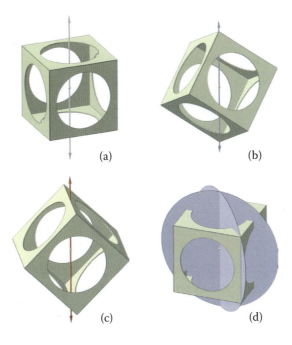

(a)

(b)

(c)

(d)

Figure 6.10. Rotation and reflection symmetry.

6.3.1 Symmetry Groups

A symmetry group is a collection of symmetry-preserving mathematical transformations that when applied to an object, such as a polyhedron, leave it invariant, that is, the object looks exactly the same in shape, position and orientation as it did before the transformation. Subdivisions of spherical polyhedra can retain some, if not all, of the polyhedron's planar symmetry.

Each of the five Platonic solids falls into one of three symmetry groups. The icosahedral group includes the icosahedron-dodecahedron dual. Sometimes the letter I is used when making mathematical references to the group. The octahedral group, designated O, includes the octahedron and cube. Finally, the tetrahedral group, T, includes just the tetrahedron, since it is a self-dual.

We will begin with icosahedral symmetry, which plays a central role in subdivision schemas, because its spherical faces subdivide the sphere into the most number of equal areas. This provides a big jump-start in the subdivision process. For this reason, icosahedral symmetry is the most frequently used spherical subdivision framework.

6.3.2 Icosahedral Symmetry

The icosahedron has the highest degree of symmetry of the five Platonic solids. Its 12 vertices define 6 pairs of antipodes and each antipodal pair defines a rotation axis. Figure 6.11(a) shows all six as blue-gray colored arrows. Each one passes through the center of the polyhedron and around any vertex; five equilateral triangles form a pentagonal cap. If the icosahedron is rotated around any one of these vertex axes any multiple of 72°, the appearance of these caps, or any other part for that matter, will not change.

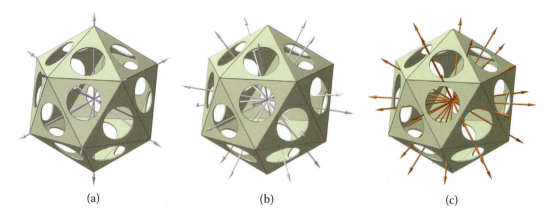

Figure 6.11. Icosahedral rotation symmetry.

The icosahedron has 20 faces. The plane of each is parallel to a face on the opposite side of the polyhedron. We can pass an axis through the midface points of opposing pairs of faces and see that each face in the pair is rotated 60° with respect to the other. White arrows in Figure 6.11(b) show all ten of them. The icosahedron can be rotated any multiple of 120° around these axes, and its appearance will not change.

The icosahedron has 15 pairs of opposing and parallel edges. A rotation axis can pass through the mid-edge points of each pair. Thus, for just its edges alone, there are 15 axes of rotation symmetry. The red arrows in Figure 6.11(c) show how the axes relate to pairs of edges. Around any one of these axes, the icosahedron can be rotated any multiple of 180°, and it will still look the same.

To fully describe icosahedral symmetry, we also need to include its reflection symmetries. The icosahedron has 15 planes of symmetry. Each one defines a great circle that passes through a pair of opposing edges and the center of the polyhedron. Figure 6.12 shows these planes in a series of views from the same perspective. A single plane is shown in Figure 6.12(a). It is tangent to four vertices and bisects four of the icosahedron's equilateral faces around their medians. Figure 6.12(b) and (c) show all 15 symmetry planes and the

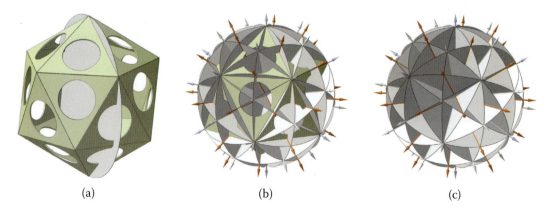

Figure 6.12. Icosahedral reflection symmetry.

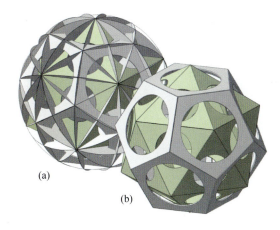

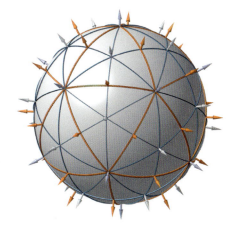

Figure 6.13. Icosahedron and dodecahedron dual reflection symmetry.

Figure 6.14. Spherical icosahedron and dodecahedron with Schwarz triangles.

great circles they create. Figure 6.12(c) shows the relationship of the reflection planes to the vertex, face, and edge rotation axes we saw earlier.

To summarize the icosahedron's symmetries, there are:

24 — 4 rotations of 72° around centers of 6 pairs of antipodal vertices
20 — 2 rotations of 60° around 10 pairs of opposite faces
15 — 1 rotation of 180° around centers of 15 pairs of opposing edges
 1 — identity symmetry

_____

60 symmetries.

Reflection symmetry doubles the number of symmetries, for a total of 120.

The dodecahedron and icosahedron are duals and share the same symmetries, though the numbers are rearranged a bit—2 rotations of 120° around the centers of 10 pairs of antipodal vertices, 4 rotations of 72° around 6 pairs of opposite faces, and 1 rotation of 180° around 15 pairs of opposite edges. The identity symmetry adds one more and reflection doubles the total, equaling the icosahedron's 120 symmetries. Figure 6.13 shows the icosahedron and dodecahedron's dual reflection symmetry with (a) and without (b) symmetry planes.

When we consider all of the icosahedron's rotation and reflection symmetries together, 15 great circles are defined in the polyhedron's spherical form. They are shown in Figure 6.14 and they subdivide the icosahedron into 120 Schwarz triangles. The red arcs are icosahedral edges; the blue-gray ones are dodecahedral edges. We will make good use of icosahedral Schwarz triangles later on when we develop different schemas for subdividing the sphere and then use CAD techniques to assemble copies of the triangle to make a full sphere.

There are many more great circle combinations for the spherical icosahedron, but none have reflection symmetry, beyond the 15 we have already identified. Figure 6.15 shows a configuration that adds 16 more great circles for a total of 31. Buckminster Fuller

Figure 6.15. Icosahedral 31 great circle symmetry.

discovered this set in the mid-1940s, and the framework greatly influenced his thinking about spheres and inspired his invention of the geodesic dome (see Chapter 2). The additional great circles are the result of adding two new sets. One set is from arcs spanning the mid-edge points of each icosahedral face. These arcs subdivide the icosahedron's face into four smaller triangles—one central equilateral surrounded by three isosceles triangles. Since the triangles on either side of the great circle's plane are not the same, these new great circles do not add any new reflection symmetries (see Section 4.16.8). A second set of great circles is added when arcs span from the mid-edge of one icosahedral face to the nearest non-shared mid-edge of a neighboring face. These great circles do not add reflection symmetries.[9] Note that while the spherical subdivision, shown in Figure 6.15, contains many more triangles than before, it can still be described by just one left- and one right-handed Schwarz triangle based on the original equilateral icosahedral face.

6.3.3 Octahedral Symmetry

The octahedron and its dual, the cube, have similar rotation and reflection symmetries as the icosahedron, but the octahedron has fewer of them, 48 in all. Figure 6.16(a) shows all of the vertex, face, and edge rotation axes together. Three rotational axes pass through three pairs of opposing vertices. The octahedron can be rotated any multiple of 90° around any of these three axes without changing its appearance. In similar fashion, there are four axes through the centers of opposite faces, and any rotation that is a multiple of 120° will not change its appearance. Finally, there are six axes through the midpoints of six pairs of opposing edges. A rotation of 180° around any of these three axes will not change the octahedron's appearance.

Figure 6.16(b) and (c) show the octahedron's rotation axis along with the nine reflection planes and the nine great circles they create. Figure 6.16(c) shows the relationship of the reflection planes to the vertex, face, and edge rotation axes. Another way to visualize these nine great circles in (c) is to see that three of them follow the edges of the octahedron and six follow the edges of its dual, the cube. We can summarize all of the octahedron's

---

[9] For a comprehensive analysis of icosahedral great circles, see Robert Gray's website at www.rwgrayprojects.com.

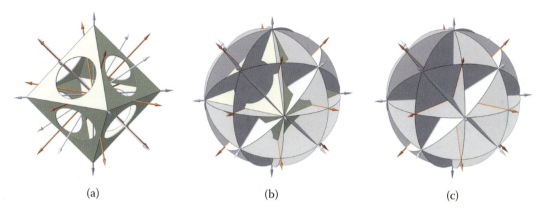

(a)   (b)   (c)

Figure 6.16. Octahedral rotation and reflection symmetry.

symmetries this way:

9 — three rotations of 90° around centers of three pairs of opposite vertices
8 — two rotations of 120° around four pairs of antipodal faces
6 — one rotation of 180° around centers of six pairs of opposing edges
1 — identity symmetry

24 total

Reflection symmetry doubles these symmetries, for a total of 48.

The octahedron and the cube are duals. As in the icosahedral symmetry we saw earlier, the cube retains the same symmetries, but the symmetries are applied in different ways— two rotations of 120° around the centers of four pairs of antipodal vertices, three rotations of 90° around three pairs of opposite faces, and one rotation of 180° around six pair of

Figure 6.17. Spherical octahedron with Schwarz triangles.

opposite edges. The identity adds one more symmetry and reflection doubles the total, equaling the octahedron's 48 symmetries.

Figure 6.17 shows a spherical octahedron with the rotation axes and reflection planes we have been discussing. Notice that the reflection planes define great circles that divide the octahedron into 48 Schwarz triangles, 24 left-handed and 24 right-handed. The octahedron-cube dual relationship is evident in this figure. The red great circles outline the octahedron's edges, while the blue-gray ones follow the cube's edges. Since both the octahedron and cube are defined by the same Schwarz triangles, any subdivision of one polyhedron is, in effect, a subdivision of the other.

6.3.4 Tetrahedral Symmetry

The tetrahedron is a self-dual with the fewest vertices, faces, and edges of any polyhedra. It's also the only Platonic solid in which the faces are opposed by a vertex rather than by another face. It is therefore not surprising that the tetrahedron, with fewer vertices, faces, and edges, has the fewest number of symmetries (24 total) of the three groups.

Figure 6.18(a) shows the relationships and all of the vertex, face, and edge rotation axes together. Referring to (a), the tetrahedron can be rotated two times, each 120° around each vertex-face axis, and one 180° rotation around the centers of three pair of opposite edges without changing its appearance. Figure 6.18(b) shows the tetrahedron's six planes of reflection symmetry, and (c) shows the relationship of reflection planes to rotation axis. Notice that each of the six symmetry planes, shown as great circles, pass tangent to an edge on one side of the tetrahedron and normal to the midpoint of an edge on the opposite side. The tetrahedron's rotation and reflection symmetries combine with the identity symmetry for a total of 12 symmetries in the following way:

8 — two rotations of 120° for each of four axis through a vertex and the middle of an opposing face
3 — one rotation of 180° around the center of three pairs of opposite edges
1 — identity symmetry
_____

12 total

Reflection doubles the count to 24.

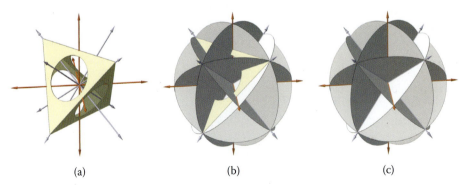

(a)   (b)   (c)

Figure 6.18. Tetrahedral rotation and reflection symmetry.

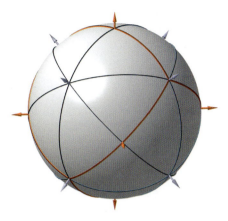

Figure 6.19. Spherical tetrahedron with Schwarz triangles.

Figure 6.19 shows a spherical tetrahedron and retains the rotation axes and reflection planes (great circles) we have been discussing. Combinations of great circles divide the tetrahedron into 24 Schwarz triangles, 12 left-handed and 12-right-handed. Notice the spherical cube pattern, a result of the tetrahedron-tetrahedron dual. It might seem surprising to see this pattern. However, a tetrahedron can be created from the diagonals of the cube's faces, and this is what is apparent in the figure.

6.3.5 Schwarz Triangles and Symmetry

In Chapter 4 we described the Schwarz triangle as the largest spherical right triangle that can tile a spherical polyhedron's face and cover the sphere without gaps or overlaps. Each pair of platonic duals or symmetry group has its own Schwarz triangle. But the story does not end there. Symmetry groups also reveal other relationships.

The icosahedron's Schwarz triangles define its dual, as expected; but somewhat unexpected is the fact that the same tiling also defines the spherical octahedron and tetrahedron. Figure 6.20 shows why this happens. Four identical spheres are shown with the same set of icosahedral Schwarz triangles. All views are from the same point and scale. Figure 6.20(a) is the full set of Schwarz tiling for the icosahedron. One face is highlighted in (b) and a dodecahedron face is highlighted in (c). But the same tiling also defines the octahedron's face, as shown in (d). How does this happen? Octahedral faces have three 90° angles and three 90° sides. The 90° angle of the icosahedral Schwarz triangle defines each octahedral face angle, and each octahedral face edge (also 90°) is defined by the three different sides of icosahedral Schwarz triangles. They add up to 90° (31.71747° + 20.90516° + 37.37737°). We also notice that the combination of Schwarz triangle edges that define octahedral edges form continuous great circles, each orthogonal to one another, a hallmark of the spherical octahedron. The point here is that although the octahedron has its own Schwarz triangle definition and it is larger in area than the icosahedron's (when both have the same circumsphere radius), we can use symmetry relations to define the octahedron with an icosahedral Schwarz triangle. In practice, however, we generally use only the Schwarz triangle for the particular polyhedral framework we want to use. This

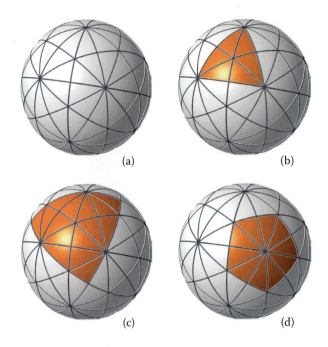

Figure 6.20. Icosahedral Schwarz triangle defines the dodecahedron and octahedron.

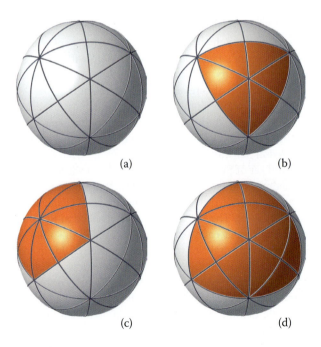

Figure 6.21. Octahedral Schwarz triangles define the cube and tetrahedron.

Plaltonic Schwarz Triangles					
Triangular Part	Tetrahedron	Octahedron	Cube	Dodecahedron	Icosahedron
Surface Angles					
A	60.0	60.0	45.0	36.0	60.0
B	60.0	45.0	60.0	60.0	36.0
C	90.0	90.0	90.0	90.0	90.0
Central Angles					
a	54.73561032	45.0	35.26438968	20.90515745	31.71747441
b	54.73561032	35.26438968	45.0	31.71747441	20.90515745
c	70.52877937	54.73561032	54.73561032	37.37736814	37.37736814
Arc Factors					
a	0.95531662	0.78539816	0.61547971	0.36486383	0.55357436
b	0.95531662	0.61547971	0.78539816	0.55357436	0.36486383
c	1.23095942	0.95531662	0.95531662	0.65235814	0.65235814
Chord Factors					
a	0.91940169	0.76536686	0.60581089	0.36284333	0.54653306
b	0.91940169	0.60581089	0.76536686	0.54653306	0.36284333
c	1.15470054	0.91940169	0.91940169	0.64085182	0.64085182
Area					
Stradians	0.52359878	0.26179939	0.26179939	0.10471976	0.10471976
Spherical Degrees	30.0	15.0	15.0	6.0	6.0

Note: all angles in degrees

Table 6.4. Platonic Schwarz triangles.

maximizes the symmetry we can exploit. It's evident from the octahedral face tiling in (d) that using the icosahedron's Schwarz triangle to define an octahedron's face instead of the octahedron's own Schwarz triangle has cost us some rotation and reflection symmetry.

We can examine the octahedron's Schwarz triangles and see similar relations to its dual, the cube, and to the tetrahedron. Figure 6.21 shows four spheres tiled with the same octahedral Schwarz triangles, as shown in (a). The view of each sphere is the same. Single octahedron and cube faces are shown in (b) and (c), and their dual symmetry is obvious. But since a tetrahedron can be defined by the face diagonals of a cube, it's no surprise that the octahedron-cube triangles also define the tetrahedron's face, as shown in (d). The spherical tetrahedron face angles are each 120° and its edges are 109.47122°. Two back-to-back octahedral triangles, each 60°, fit perfectly in each of the tetrahedron's angles, and the sum of two octahedral triangle hypotenuses (each 54.73561°) equals a tetrahedral edge. Unlike the octahedral face we previously defined with icosahedral Schwarz triangles, the tetrahedron's face defined here with octahedral Schwarz triangles has full rotational and reflection symmetry.

Given the relationships of Schwarz triangles between the three symmetry groups, one could argue that only two different pairs of Schwarz triangles, one for the icosahedron and another for the octahedron, are needed to define every spherical Platonic solid. But once again, the practice is to use the group's own Schwarz triangles because they are the largest right triangles that will tile their faces and still maintain full rotational and reflection symmetry. Table 6.4 summarizes the characteristics of the Platonic Schwarz triangles.

6.3.6   Deltahedra

Symmetry within the Platonic solids provides another tool for subdivision. Geometrists Martyn Cundy and A. P. Rollett proposed the term *deltahedra* to describe any polyhedra whose faces were all equilateral triangles.[10] The name reflects the Greek capital letter delta (Δ), which has the shape of an equilateral triangle. There are many different deltahedra. Three are Platonic solids, but no Archimedean solids are deltahedra. Figure 6.22 shows the three Platonic deltahedra: the tetrahedron (a), octahedron (b), and icosahedron (c). As we have already seen, they also define symmetry groups. What's important is the fact that any subdivision technique we develop for a general spherical equilateral will subdivide any deltahedra and its dual, even though the dual's face type is some other type. For example, subdividing an octahedron's equilateral face also subdivides a cube's square face. The same applies to the spherical icosahedron's equilateral face and the dodecahedron's pentagonal face. Although we can develop subdivision techniques for any regular spherical polygon that tile a sphere, we benefit most from techniques that subdivide equilaterals or a right Schwarz triangle. In Chapter 8, we will show how to do this.

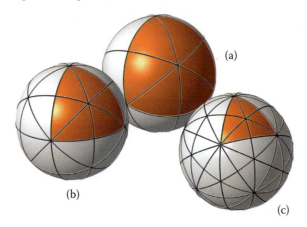

Figure 6.22. Deltahedra equilateral faces.

6.4   Archimedean Solids

The Archimedean solids, so named to honor Archimedes of Syracuse (287–212 BC), are the second-most important polyhedral frameworks for spherical subdivision. Known long before the mathematician's lifetime, these 13 solids are similar to Platonic solids in that they are convex polyhedra. But unlike Platonic solids, they have two or more types of regular polygon faces. For this reason, they are classified as *semiregular* polyhedra. Faces join at their edges, they do not penetrate one another, and they always sequence the same way around every vertex. Figure 6.23 shows all 13 of them with the same edge length.[11] Referring to the numbered solids in the figure, they are the truncated tetrahedron (1), truncated cube and truncated octahedron (2 and 3), lesser and greater rhombicuboctahedron (4 and 5), cuboctahedron (6), icosidodecahedron (7), snub cube and snub dodecahedron (8 and 9), truncated icosahedron

---

[10] (Cundy and Rollett 1961, 78).

[11] Two Archimedeans, the snub cube and snub dodecahedron, come in a left- and right-handed version. One could argue there are really 15 Archimedeans, not 13.

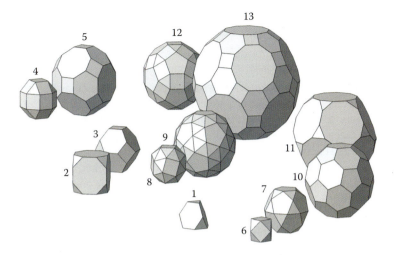

Figure 6.23. Archimedean solids.

and truncated dodecahedron (10 and 11), and finally, the lesser and greater rhombicosido-decahedron (12 and 13).[12]

There are important differences between the Platonic and Archimedean solids that affect their use in spherical work. The Platonic solids distribute points (vertices) evenly on the surface of their circumsphere. The distance between any vertex and its neighbors, and the angle between any two neighboring points and every dihedral angle for a solid, is the same. Platonic solids also have only one insphere that's tangent to the midface point of their faces. A pair of Schwarz triangles, left- and right-handed, can describe the entire surface of spherical versions.

Archimedean solids do not have every one of these properties. While their vertices are on the surface of their circumsphere and the distances between any vertex and its neighboring vertices are the same, the angle between adjacent edges at a vertex and the dihedral angle between faces are not always the same. Archimedean solids have two or more inspheres, each tangent to the midface points of one set of like faces but not tangent to the midface points of the other face types. The *midsphere* of an Archimedean solid does, however, intersect every edge of that solid at its midpoint. Multiple spherical triangles are needed to describe the surface of spherical versions. In some spherical applications, it is important to have a subdivision grid that has at least one sequence of grids that defines a complete great circle. Some subdivision techniques result in grids that form great circles, but two Archimedean solids, the icosidodecahedron and cuboctahedron, have sequences of edges that naturally form great circles in their spherical version. We will discuss them in a moment. The octahedron, a Platonic solid, is the only other polyhedron we have discussed having this property.

Archimedean solids have duals just like Platonic solids. They are called *Catalan solids*, and are quite beautiful in their own right. However, they do not offer any special advantages over Platonic or Archimedean solids as frameworks for spherical subdivision. We do not

---

[12] Most of the Archimedean solids were named by Johannes Kepler (1571–1630), the German mathematician, astronomer, and astrologer who formulated the laws of planetary motion.

cover them in this text. Instead, we include several excellent references in the additional resources at the end of this chapter.

6.4.1   Cundy-Rollett Symbols

Individual Archimedean solids can have two, three, or four different face types. The types of regular polygons and their order around each vertex of a solid is always the same for a given solid. Thus, an Archimedean solid can be fully specified by simply listing the order of its polygon faces around a vertex. Within an Archimedean solid, the vertex face orders are all the same and the solid is said to be *vertex-transitive*,[13] which means the face arrangement around any one of its vertices can be transformed to any other vertex position on the solid through some rotation or reflection symmetry. All Platonic and Archimedean solids are vertex-transitive.

Schläfli devised symbolic notation for Archimedean solids, as he did for Platonic solids. His Platonic symbols list the number of sides in the regular polygon and how many times they meet at a vertex. His symbol for a cube is {4,3}, which means four-sided polygons (squares) meet three at a time at a vertex. But Archimedean solids have more complex face orders than Platonic solids, and as a result, their Schläfli symbols are not as intuitive. Geometrists H. M. Cundy and A. P. Rollett, modified Schläfli's symbols to list the regular face types in their vertex order.[14] For example, the *Cundy-Rollett* symbol {3.5.3.5} means every vertex of this solid is surrounded by alternating triangles and pentagons. It could also be written as {3.5²}. This particular Archimedean solid is called the icosidodecahedron and we will see its details shortly. To distinguish Cundy-Rollett symbols from Schläfli symbols, we use periods between parameters in the former and commas in the latter.

6.4.2   Truncation

Truncation is the way some Platonic solids become semiregular Archimedean solids, and their names reflect this fact. When a Platonic solid is cut, new and regular (equal edge and

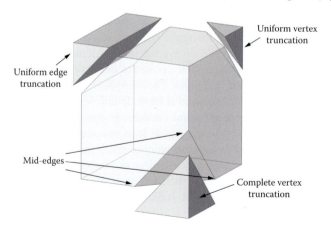

Figure 6.24. Vertex and edge truncations.

---

[13] A polyhedron is vertex-transitive if any vertex can be carried to any other by a symmetry operation (Cromwell 1997, 369).

[14] (Cundy and Rollett 1961, 84)

angle) polygonal faces result. *Uniform truncations* pass the cut plane less than halfway along the edges that meet at any vertex. *Complete truncations* cut every edge midway. Figure 6.24 shows the differences between the two. Some Archimedean solids require cutting a Platonic solid several times. There are a couple of Archimedean solids, called *snubs*, which cannot be made by simple vertex or edge cuts alone.

We introduce the Archimedean solids, both planar and spherical, by symmetry groups.

6.4.3   Archimedean Solids with Icosahedral Symmetry

The first six Archimedean solids are part of the icosahedral symmetry group and can be inscribed in either a dodecahedron or icosahedron. On average, this series distributes (not evenly) nearly twice as many vertices on the sphere as any other symmetry group. Table 6.5 shows the planar and spherical versions together from the same viewpoint scaled to the

Name(s) & Symbols	Vertices	Edges	Faces:	{3}	{4}	{5}	{6}	{8}	{10}
icosidodecahedron {5.3.5.3} or {5.3}²	30	60	32	20	—	12	—	—	—
truncated icosahedron {5.6.6} or {5.6²}	60	90	32	—	—	12	20	—	—
truncated dodecahedron {3.10.10} or {3.10²}	60	90	32	20	—	—	—	—	12
small or lesser rhombicosidodecahedron {3.4.5.4}	60	120	62	20	30	12	—	—	—
great rhombicosidodecahedron {4.6.10}	120	180	62		30	—	20	—	12
snub dodecahedron or snub icosidodecahedron {3.3.3.3.5} Two opposite-handed versions	60	150	92	80	—	12	—	—	—

Table 6.5. Archimedean solids with icosahedral symmetry.

same unit circumsphere. All six solids in this group can all be inscribed in a dodecahedron or icosahedron.

The icosidodecahedron combines the 12 pentagonal faces of a dodecahedron with the icosahedron's 20 equilateral faces; it can be created by complete vertex truncations of either a dodecahedron or an icosahedron. The result is alternating face types around each vertex. Polyhedra with two pairs of alternating regular face types at each vertex are called *quasi-regular*.[15] Perhaps the most important property of this solid, and the reason it is often used in spherical subdivision, is that its edges form six hemispheric planes, each of which is a *decagon* and will define a great circle in its spherical version. In Chapter 8 we show a subdivision technique called Mid-arcs. When applied to a spherical icosahedron, the technique creates the spherical icosidodecahedron framework.

The truncated icosahedron is widely recognized as the traditional design for soccer balls. In physics and chemistry, this solid resembles the newly discovered carbon 60 (C_{60}) molecule and has been named the "Fullerene" or "*Buckyball*" after Buckminster Fuller.[16] Both are shown on the left. The truncated icosahedron can be created by uniform vertex truncation of an icosahedron or dodecahedron. Thus, its faces are tangent to either the dodecahedron or the icosahedron. We will see variations of the truncated icosahedron in several spherical subdivision schemas. Much of what has been said about the truncated icosahedron applies equally to the truncated dodecahedron, which we will review in a moment. It can be created by uniform vertex truncation of the dodecahedron or icosahedron; thus, it can be inscribed in either solid.

The small rhombicosidodecahedron can be created by truncating the edges and vertices of other Archimedean solids. Each of its faces belongs to one of three circumscribing solids: the dodecahedron, the icosahedron, and the rhombic triacontahedron. The small rhombicosidodecahedron can also be made by rotationally expanding the faces of a dodecahedron along an axis perpendicular to their face centers. The great rhombicosidodecahedron is the most complex and the largest of the Archimedean solids when all polyhedra have the same edge lengths. It has more regular face types (squares, hexagons, and decagons) than any other Archimedean solid and, as a result, the most vertices and edges. The great rhombicosidodecahedron distributes twice as many vertices on the surface of the sphere of any Archimedean solid: 120 in all. When compared to other Archimedean solids with the same edge lengths, it is the largest by far. Like the others in this symmetry group, its faces also belong to a circumscribing dodecahedron and the icosahedron.

The snub dodecahedron, sometimes called the snub icosidodecahedron, places 12 pentagonal faces on the surface of a circumscribing dodecahedron. The snub dodecahedron and another Archimedean solid, the snub cube, are *chiral polyhedra*; that is, they do not have reflection symmetry. No plane passing through the center of the polyhedron produces equal images on both sides. There are two versions of this polyhedron: one is left-handed, while the other is right-handed. Neither version can be made by simple truncations of any

---

[15] Quasi-regular polyhedra are sometimes called combinational solids because they combine the faces of two other polyhedra.

[16] A new kind of carbon, C_{60}, was discovered in 1985 by scientists Robert Curl, Harold Kroto, and Richard Smalley. They were awarded the Nobel Prize in Chemistry in 1996 for discovering this new class of compounds. The molecular model resembles a truncated icosahedron and is named the Fullerene or Buckyball, in honor of Buckminster Fuller.

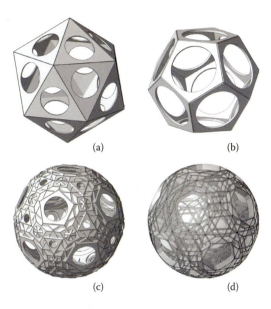

(a)   (b)

(c)   (d)

Figure 6.25. Superimposed Archimedean polyhedra with icosahedral symmetry.

Platonic solid. One of the spherical subdivisions schemas that we will develop in a later chapter creates an entire class of skewed subdivisions similar to the snub cube. We will look at skewed subdivisions in Chapter 8.

We can see the symmetry of this group by superimposing both their planar and spherical versions at the same scale. Figure 6.25(a) and (b) show the Platonic duals that define this symmetry group. They are positioned as duals—the dodecahedron's vertices in (b) correspond to the same position as the icosahedron's midface points in (a). Figure 6.25(c) and (d) show the Archimedean planar and spherical polyhedra in this symmetry group superimposed and from the same viewpoint as (a) and (b). The snub dodecahedron is omitted because it cannot be made by simple truncations. In (c) it's easy to see symmetrical patterns that result from the various vertex and edge truncations of the icosahedron and dodecahedron. Figure 6.25(d) is simply the spherical version of (c) and emphasizes the arc edges. The symmetry of all four figures is evident.

6.4.4   Archimedean Solids with Octahedral Symmetry

The next six Archimedean solids are related to the octahedron and its dual, the cube. They are the truncated octahedron and cube, cuboctahedron, the lesser and greater rhombicuboctahedron, and the snub cube. Table 6.6 shows them along with their basic characteristics. Every planar and spherical version is shown from the same viewpoint scaled to the same unit circumsphere. With the exception of the snub cube, five of the six solids can be formed by truncating the octahedron's vertices and/or edges to a certain depth. Each cut creates new face types and vertices. All six solids in this group can be inscribed in an octahedron or cube.

Referring to Table 6.6, the truncated octahedron and truncated cube are simple uniform vertex truncations of the vertices of an octahedron or cube. All cut planes are uniform vertex

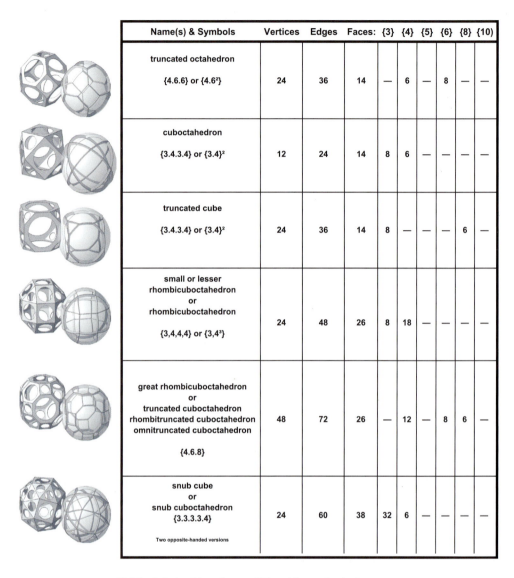

Name(s) & Symbols	Vertices	Edges	Faces: {3}	{4}	{5}	{6}	{8}	{10}
truncated octahedron {4.6.6} or {4.6²}	24	36	14	—	6	—	8	—	—
cuboctahedron {3.4.3.4} or {3.4}²	12	24	14	8	6	—	—	—	—
truncated cube {3.4.3.4} or {3.4}²	24	36	14	8	—	—	—	6	—
small or lesser rhombicuboctahedron or rhombicuboctahedron {3,4,4,4} or {3,4³}	24	48	26	8	18	—	—	—	—
great rhombicuboctahedron or truncated cuboctahedron rhombitruncated cuboctahedron omnitruncated cuboctahedron {4.6.8}	48	72	26	—	12	—	8	6	—
snub cube or snub cuboctahedron {3.3.3.3.4} Two opposite-handed versions	24	60	38	32	6	—	—	—	—

Table 6.6. Archimedean solids with octahedral symmetry.

truncations of the octahedron and cube at a distance between a vertex and neighboring edge midpoints. Sets of like faces (squares or hexagons) in the truncated solid are parallel to each other and axes normal to a pair of faces are mutually orthogonal, like Cartesian axes. The spherical forms of both solids are often used in applications, where x-, y-, and z-axes are needed, or in cartographic applications where great circles are required for the equator and Greenwich Meridian.

The cuboctahedron is unique among polyhedra. It can be formed several ways. First, it can be made by complete vertex truncations of either a cube or octahedron. It can also be made by combining the cube's six faces with the octahedron's eight faces or by combining

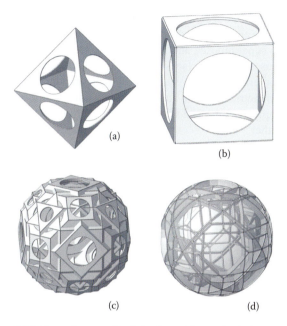

(a)

(b)

(c)

(d)

Figure 6.26. Superimposed polyhedra with octahedral symmetry.

eight tetrahedra, each sharing one point in the center of the polyhedron. No matter how the cuboctahedron is made, every one of its edges is equal and every vertex is equidistant to its neighbors, as well as to the center of the polyhedron. It's the only polyhedron like this. Another unusual property of the cuboctahedron is it has only two different face types, and each type surrounds the other at every vertex. Despite having two types of faces, every dihedral angle for the cuboctahedron is the same. The icosidodecahedron is the only other quasi-regular Archimedean solid. We will return to these two solids later in the chapter. They offer great circle advantages in spherical work.

The small rhombicuboctahedron is another unique polyhedron. Each of its faces belongs to one of three circumscribing solids: the cube, the octahedron, and the rhombic dodecahedron. What is unusual here is the same number of faces and types can be rearranged to make a different polyhedron that isn't a Platonic solid or an Archimedean solid. A polyhedron with this property is said to be *isomeric*. Few polyhedra have this property. We will take a closer look at this characteristic later in the chapter. The great rhombicuboctahedron can be made by truncating the cuboctahedron's vertices. Its faces belong to one of two circumscribing solids, the cube and the octahedron.

The snub cube, sometimes called the snub cuboctahedron, cannot be made by simple truncations of either an octahedron or a cube. Although its equilateral faces are skewed relative to the edges of the octahedron, its square faces lie on the surface of the circumscribing cube and eight of the triangular faces lie on the circumscribing octahedron.

Although each Archimedean solid in the Octahedral symmetry group is quite different, their group symmetry shows through when we superimpose them. The octahedron-cube dual that defines this symmetry group is shown in Figure 6.26(a) and (b). They are positioned as duals: the octahedron's vertices in (a) correspond to the cube's midface points in (b). The edges of both solids would be perpendicular to one another if both were overlaid. Figure 6.26(c) and (d) show the Archimedean planar and spherical polyhedra in this symmetry group superimposed and from the same viewpoint as (a) and (b). The snub cube is left out because its skewed faces obscure the others. It cannot be made by simple truncations like the others. The highly symmetrical patterns, clearly seen in Figure 6.26(c), are the result of symmetrical vertex and edge truncations of the octahedron and cube. Figure 6.26(d) is simply the spherical version of (c) and emphasizes the relationship of the arc edges to one another.

6.4.5   Archimedean Solids with Tetrahedral Symmetry

This is the smallest symmetry group and it contains only one solid, the truncated tetrahedron. It is the only polyhedron with just hexagonal and triangular faces. It can be created by making a uniform vertex truncation of a tetrahedron or by truncating all the vertices of a cube to different depths. The first technique is the most common. Either way, the result is a polyhedron bounded by four hexagons and four triangles, as shown in Table 6.7.

As far as we know, the truncated tetrahedron has not been used for spherical subdivisions in industrial products or architecture. It is included here to complete the Archimedean symmetry examples.

Name & Symbol	Vertices	Edges	Faces: {3}	{4}	{5}	{6}	{8}	{10}
truncated tetrahedron {3.6.6} or {3.6²}	12	18	8	4	—	—	4	—	—

Table 6.7. Archimedean solid with tetrahedral symmetry.

6.4.6   Chiral Polyhedra

Earlier in the chapter, we said two Archimedean solids, the snub cube and snub dodecahedron, were chiral polyhedra, but until now, we did not ascribe any significance to this. A chiral polyhedron is one that has two distinct forms or *enantiomorphs*, a *laevo* (left-handed) and *dextro* (right-handed) version. Your hands are nearby examples of enantiomorphs. No combination of rotations or translations of one hand will produce the shape of the other. In a sense, you have to turn one inside out to make the other. You can demonstrate this by turning a glove inside out. A chiral polyhedron cannot be made by simple truncations of a Platonic solid.

We care about chiral polyhedra because we will develop an entire family of subdivisions related to them. Class III subdivisions are chiral, and their lack of reflection symmetry means the geometry developed for a local area can only cover the remainder of the sphere, if it is rotated around one or more axes. We lose the convenience of reflecting geometry from one place on the sphere's surface to another place, something we can easily do in some cases by simply changing vertex coordinate signs.

Figure 6.27 shows the right- and left-handed versions of the snub cube (a) and (b) and the snub dodecahedron (c) and (d). Neither set of polyhedra can be created from a Platonic

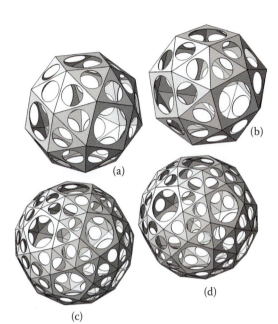

(a)

(b)

(c)

(d)

Figure 6.27. Snub cube and snub dodecahedron.

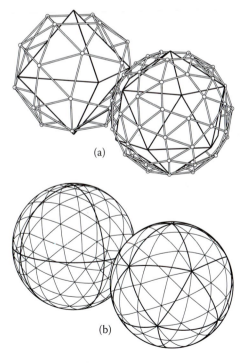

(a)

(b)

Figure 6.28. Chiral polyhedra and subdivisions.

solid by vertex or edge truncations. Both sets of figures have been scaled, so the most extreme vertices are tangent to a unit circumsphere. Try as you may, you cannot pass a plane through any of them (and their origin) and produce identical images on both sides.

Although none of the vertices of chiral polyhedra are antipodal pairs, they retain symmetric relationships with their Platonic octahedron and icosahedron cousins. In Figure 6.28(a), we see them superimposed—the octahedron and snub cube on the left, the icosahedron and snub dodecahedron on the right. Even though the snub's chiral form appears "twisted" relative to its Platonic counterpart, the systematic relationship between their faces, edges, and vertices is easy to see. The square faces of the snub cube and pentagonal faces of the snub dodecahedron are rotationally symmetric to the vertices of the octahedron and icosahedron.

Figure 6.28(b) shows a Class III spherical subdivision using an octahedral (left) and icosahedral foundation (right). The bold spherical Platonic edges in the figures are *not* part of the subdivision grid; they are displayed in the same orientation as the polyhedra in (a) to show their relationships to their Platonic frameworks. No matter how finely we subdivide the sphere using this schema, the vertices, edges, and faces produced by the grid will retain their symmetry relationship to their foundation polyhedron. Moreover, they will always remain chiral and will lack reflection symmetry.

6.4.7   Quasi-regular Polyhedra and Natural Great Circles

The cuboctahedron and icosidodecahedron are classified as semiregular and *quasi-regular polyhedra*. It turns out that these two are also the only Archimedean solids whose edges form natural equatorial planes that translate to great circles in their spherical versions. They are also quasi-regular polyhedra. The cuboctahedron {3.4.3.4} alternates equilateral and square faces around each vertex, while the icosidodecahedron {3.5.3.5} alternates equilateral triangles and pentagonal faces. Quasi-regular solids have equal dihedral angles, equal-length sides, and all vertices tangent to their circumsphere. These are also characteristics of the Platonic solids, but most important, they have continuous sets of edges that form planes through the center of the polyhedron and form great circles in their spherical versions. It's not surprising to see quasi-regular polyhedra used in spherical applications where at least one great circle is required. For example, a geodesic dome needs to sit evenly on the ground, and a mold for a golf ball needs to have a hemispheric parting line in order to mold it. The octahedron is the only Platonic solid, without further subdivision, that also has natural hemispheric planes. Figure 6.29 shows the two quasi-regular polyhedra and the octahedron (not classified as quasi-regular) in applications that exploit their great circles. In the figure, the polyhedra and application have approximately the same orientation and scale.

The cuboctahedron has four equatorial planes. Figure 6.29(a) shows it alongside a computer model of Buckminster Fuller's famous octet truss connector (b). This connector can join up to 12 lightweight tubular struts, forming a tetrahedral space frame structure with an enormous strength-to-weight ratio. What is especially innovative here is that each strut connector is cast with four "ears" that are offset in such a way that they overlap the ears of adjacent connectors while remaining tangent to, but not crossing, any of the cuboctahedron's great circle planes. One of the four planes is shown in Figure 6.29. We go into detail about this amazing design in Chapter 10.

The octahedron shows up in many spherical applications, not only because it offers great circles, but it also offers three mutually orthogonal hemispherical planes that natu-

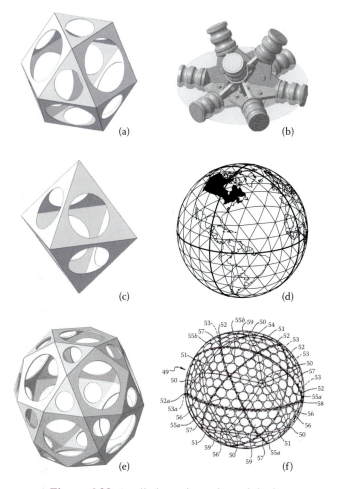

Figure 6.29. Applied quasi-regular polyhedra.

rally fit applications where a Cartesian *xyz*-reference system is needed. Figure 6.29(c) and (d) show its use in mapping where the equator and Greenwich references are required. The particular cartographic grid shown is called the *quaternary subdivision*, a meshing system that hierarchically subdivides equilateral triangles into four smaller ones by adding new edges that span between the mid-edge points of the original triangle.[17] In Chapter 8, we develop this schema in detail; it's called the Mid-arcs schema.

The icosidodecahedron offers six great circles; they are easily seen in Figure 6.29(e) and (f). This application shows the arrangement of dimples on a golf ball.[18] The designer needed at least one great circle plane as a parting line for the molding process. Using an icosidodecahedral foundation, he needed to develop two dimple arrangements to cover the entire ball, one for the triangle, another for a symmetrical part of the pentagonal face. In Hwang, the inventor of this particular design, chose to place six large dimples totally

---

[17] (Dutton 1991)
[18] (Hwang November 1996, sheet 5/9, fig. 8)

within each of the equilateral face. The icosidodecahedron's pentagonal faces are further subdivided into five slightly isosceles triangles. Three dimples are totally within this triangle, while the remaining ones straddle edges or center over the pentagon's midface point. Most golf balls are molded and, until recently, have required a hemisphere parting line, an equatorial line that is not crossed by dimples. Golf balls are one of the most creative and diverse applications of spherical subdivision. We discuss their spherical subdivision geometry shortly.

6.4.8   Waterman Polyhedra

We have discussed Platonic and Archimedean solids in detail because we will use them in the next chapter to jump-start the spherical subdivision process. For completeness, however, we want to mention the Waterman polyhedra because some are nearly spherical and already highly subdivided. Unlike spherical Platonic and Archimedean solids, Waterman polyhedra do not evenly distribute points on a sphere or even distribute points on the same radius sphere. Waterman polyhedra are important additions to a spherist's tool kit because they exhibit many of the properties we have been discussing, such as symmetry, great circles, antipodal points, and so forth, and they demonstrate a rich way to develop spherical geometry that can solve certain application problems.

In 1988, Steve Waterman began a series of investigations into gridded maps. We saw an example of his butterfly map projection alongside Cahill's map when we discussed Fuller's Dymaxion maps in Chapter 2. Waterman's projection studies led to the development of entirely new families of polyhedra and some resemble highly subdivided spheres. The way they are constructed is unique.

Waterman polyhedra begin with a 3D lattice of points. The points are the centers of stacked spheres arranged similarly to the way oranges are stacked in a grocery store. There are several ways to pack (stack) equal spheres to maximize the number in any given volume. In *cubic close packing (CCP)*, also called face-centered cubic (FCC) packing, an infinite number of tangent spheres are packed in layers or square grids and stacked, one layer on top of another. Each layer of spheres occupies the intergrid spaces of their adjacent layers. If all spheres have a diameter of $\sqrt{2}$, their centers will fall only on integer coordinate values, if one of the sphere's center occupies the origin of the Cartesian space at coordinates (0, 0, 0). Figure 6.30(a) shows how CCP spheres are arranged on alternating layers. The grid spacing is one unit in the x, y, and z (perpendicular to the page) direction. The solid spheres are on even z-planes, including the $z = 0$ plane. The spheres shown dashed are on odd z-planes. The centers of all the spheres, the lattice points in Waterman clusters of spheres, are marked with small dots on the grid. Figure 6.30(b) shows the section of the CCP stack indicated by (a). Both the grid in (a) and the stack in (b) are examples of a theoretical grid and stacking that extends infinitely in the x, y, and z directions. [19]

In CCP, 12 spheres surround every sphere in the pack, and their centers are equidistant from the center of the sphere they surround. If the centers of these 12 spheres are connected, they define the vertices of a cuboctahedron, Fuller's vector equilibrium (see Section 2.1). [20]

---

[19] It is interesting to note that Johann Carl Friedrich Gauss (1777–1855) proved that the densest possible packing of equal spheres is $d = \pi/(3\sqrt{2}) = 0.74048$, or 74 percent and that the space between spheres (not occupied) is 0.25951, or 25.9 percent. Coincidentally, this is the same density packing ratio of an inscribed sphere to a rhombic dodecahedron polyhedron. The rhombic dodecahedron also stacks as a polyhedron with coincident lattice points to the FCC.

[20] In synergetic geometry, Buckminster Fuller makes extensive use of CCP spheres and refers to some of their configurations as the *isotropic vector matrix* (IVM). See (Edmondson 1987, 127) for a good explanation of how

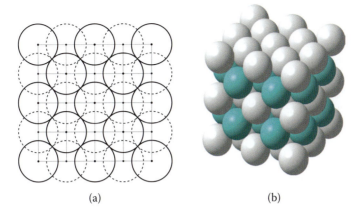

(a)   (b)

Figure 6.30. Cubic closest packing.

A Waterman polyhedron is defined by the convex hull of all the lattice points (the centers of stacked spheres) that are within a defined radius from the lattice's origin. Points outside this radius are ignored. A convex hull is the minimum convex polyhedra (faces, points, and edges) that completely surround the points. In a sense, the convex hull is like a shrink-wrap surrounding the most extended set of points.

The Waterman Polyhedra, called W1, are defined by a radius of $\sqrt{2} \times 1$. This radius selects only the 12 points surrounding the origin, and their convex hull is the same as the cuboctahedron or Fuller's VE. As the defined radius increases, more lattice points are selected and their convex hull, the Waterman polyhedron, becomes more and more complex and is made up of many more faces, points, and edges. When the defined radius has a root greater than 300, Waterman polyhedra resemble spherical subdivisions. Figure 6.31 shows two examples with selected lattice points (with spheres around them) and the resulting convex hull Waterman polyhedron.[21] These Waterman polyhedra are referred to by their square root numbers. Notice that a higher root number results in more points, edges, and faces on the sphere.

Waterman polyhedra exhibit high degrees of edge, face, and vertex symmetry. For some cases, it is possible to pass planes through the Waterman sphere stack and define a lattice of points that create a hemispherical or copula polyhedra that resemble a geodesic dome.[22] The FCC packing and positioning (one sphere at the origin of the Cartesian space) produces one family of Waterman polyhedra; however, other families result when the origin of the sphere cluster is repositioning to other symmetrical positions within the lattice. There are seven possible origins for FCC packings,[23] and corresponding Waterman Polyhedra can be created with sphere stacks having a $\sqrt{2}$ diameter, or any other consistent diameter. If their diameter is not $\sqrt{2}$, their centers will not necessarily fall on integer xyz-coordinates.

IVM extends Fuller's concept of vector equilibrium (VE) and how it relates to his octet truss design. Steven Waterman believes that Fuller was not aware that FCC (he called it CCP) had square root integer distance between any two-sphere centers.

[21] Images courtesy of Rene K. Mueller, SimplyDifferently.org.

[22] See Paul Bourke, "Waterman polyhedra data generator," http://local.wasp.uwa.edu.au/~pbourke/geometry/waterman/gen/, 2004.

[23] Seven possible point lattice origins: (1) sphere center, (2) spheres touch point, (3) spheres triangle center (there are two variations), (4) spheres forming tetrahedron, (5) spheres forming square-based pyramid, and (6) spheres.

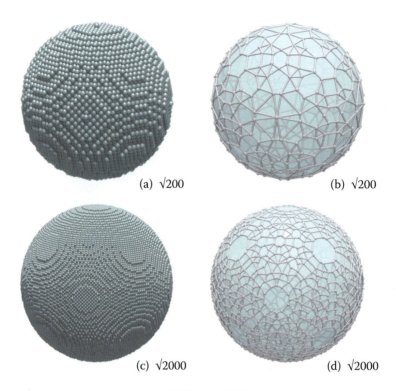

(a) √200   (b) √200

(c) √2000   (d) √2000

Figure 6.31. Waterman polyhedra $\sqrt{200}$ and $\sqrt{2000}$ point lattices and convex hulls.

Waterman polyhedra names are simple. They are labeled *Wn*, with *n* being the defined radius integer parameter.

The concept of Waterman polyhedra has been generalized to include point lattices created by packed (stacked) polyhedra in addition to spheres. Packs of polyhedra, restricted to certain integer *x,y,z* grids, will produce more than 60 new families of polyhedra. In these cases, the Waterman polyhedra naming convention is necessarily more complex because their names must account for the type of polyhedron in the stack, the origin location for the stack lattice, and the defined radius parameter. For more information on Waterman polyhedra, see Mark Newbold's website.[24]

To summarize, basic Waterman polyhedra are developed from Waterman spheres with diameters of $\sqrt{2}$ and packed in a CCP or FCC manner. Their centers are located at integer *x*-, *y*-, and *z*-coordinates, and one sphere in the pack is centered at the Cartesian origin, at coordinates (0, 0, 0). A defined radius from the origin, a multiple of $\sqrt{2n}$, selects a set of lattice points, the centers of spheres. A Waterman polyhedron is the convex hull of the selected set of lattice points. The number of points, edges, and faces in a Waterman polyhedron (the convex hull of the selected points) obey Euler's law, which states the number of vertices + faces = edges + 2. The number of vertices in a Waterman polyhedron is a fixed set. In other words, if one Waterman polyhedron has 2,000 points, another will not have an arbitrary 2,001 points. Chords between points define polygons, most of which

---

[24] http://dogfeathers.com/java/ccppoly.html.

are not regular (equilateral triangles, pentagons, hexagons, etc.). A general definition of Waterman polyhedra include FCC lattices created by both polyhedral and spherical packs. Different Cartesian origins for these lattices in combination with defined radii producing different families of Waterman polyhedra. Not all points of a Waterman polyhedron are on a circumscribing sphere, such as the Platonic and Archimedean solids already described; Waterman polyhedra exhibit high degrees of point, edge, and face symmetry. As the defined radius increases, the number of different face types, edge lengths, angles, and areas also increase. These may be limiting factors in some spherical applications.

6.4.9 Circlespheres and Atomic Models

In this chapter we have seen polyhedra enclose space, define great circles, and distribute points on their circumsphere. For the most part, polyhedra have been portrayed as static, highly symmetric bodies with predictable topological relationships among their faces, edges, and vertices. Artist Kenneth Snelson, who discovered the principle of *tensegrity* (tension/compression systems), found that a subset of Platonic and Archimedean solids possess adjacent face relationships and axial frameworks that can make new polyhedral forms he calls *circlespheres*. Snelson observed that certain polyhedra can have two-color face mappings; their faces can be colored one of two colors, and no face will be adjacent to another of the same color. Using this base, he then replaces two-color faces with dual-polarized (north/south) magnetic disks. This second form demonstrates how forces interact to create a stable body, while allowing systematic motion within the body. In so doing, the circlesphere acts as a model of the atom.

Figure 6.32 shows some of the polyhedra that can be two-color mapped. In these examples, four faces always meet at a vertex. Alongside each two-color polyhedron is their circlesphere version. Magnetic rings with dual-polarization replace certain faces of the polyhedron. Each lesser-circle ring is made of polarized material, one side of the ring is

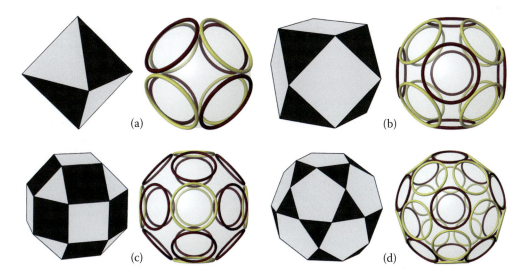

Figure 6.32. Circlespheres.

polarized north or south, while its other side is polarized in the opposite direction.[25] Figure 6.32 shows that the octahedron (a), cuboctahedron (b), lesser rhombicuboctahedron (c), and icosidodecahedron (d) meet the two-color and dual polarization criteria. In each case, no two faces or rings touch another face with the same color or polarization. In each circlesphere example, all lesser circles are on the same circumsphere, and all are the same diameter. What is unique here is the fact that every ring only touches other rings that have opposite magnetic polarization. Notice, too, that the points of tangency between rings suggest the mid-edge points of the polyhedron on which they are based.

6.4.10   Atomic Models

Snelson recognized that circlespheres could be exploited in yet another way. He could replace each magnetic ring with a magnetic disk. If the disks were mounted on axes that ran through the center of the polyhedron's faces and the middle of the polyhedron, the disks would form a spherical gear set—that is, turning one disk would cause all the others in the system to turn synchronously and at the same number of revolutions per minute.

Figure 6.33 shows examples of circlesphere disk configurations for the icosahedron (8 magnets), cuboctahedron (14 magnets), small rhombicuboctahedron (26 magnets), icosidodecahedron (32 magnets), and lesser rhombicuboctahedron (62 magnets).[26] Notice that for any of the configurations shown, it is possible to rotate one disk either clockwise or counterclockwise, which would cause all the other disks to rotate synchronously. If a north-polarized face is rotated clockwise, all other north-polarized faces on the circlesphere will rotate clockwise, while south-polarized faces will rotate in the opposite direction.

Snelson made another surprising discovery with circlespheres. He noticed that circlespheres behave much like the charged fields of atoms, and thus provide a visualization aid for how the atom works. Circlespheres with magnetic disks are kinetic models of forces

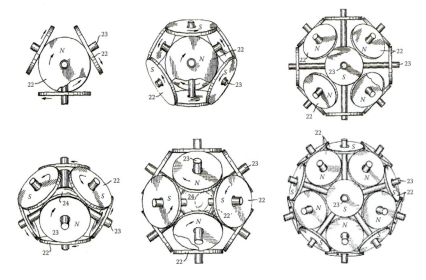

Figure 6.33. Models of atomic forms.

[25] Circlesphere images courtesy of Kenneth Snelson.
[26] (Snelson 1966)

(magnetism) and motion (rotary), and the number of magnets that can fully link together in circlespheres is uncannily close to the numerical sequences by which electrons fill "shells" or energy levels of atoms according to the periodic table of elements. In a sense, circlespheres had become models for atomic forms by representing the electronic structure of atoms and molecules with interacting magnetic rings and discs.

Snelson's circlespheres and magnetic polyhedra extend a subset of Platonic and Archimedean solids to new dimensions and offer new ways to visualize and model atomic structures.[27] It is somewhat ironic that Fuller's synergetic geometry, a design cosmology based on natural systems, geometric relationships, strength of materials (in compression and in tension), and charged particles, completely missed the synergetic system Snelson discovered in magnetic circlespheres and force fields that model the atom.

Additional Resources

Polyhedra General

Coxeter, H. S. M. *Introduction to Geometry.* New York: John Wiley & Sons, 1961.

——. *Regular Polytopes.* Mineola, NY: Dover Publications, Inc., 1973.

Cromwell, Peter R. *Polyhedra.* Cambridge [Eng.], Cambridge: Cambridge University Press, 1997.

Cundy, H. M. and A. P. Rollett. *Mathematical Models.* Oxford: Oxford University Press, 1961.

Holden, Alan. *Shapes, Space and Symmetry.* New York: Columbia University Press, 1971.

Pugh, Anthony. *Polyhedra: A Visual Approach.* Berkeley: University of California Press, 1976.

Richeson, David S. *Euler's Gem: The Polyhedron Formula and the Birth of Topology.* Princeton, NJ: Princeton University Press, 2008.

Wenninger, Magnus J. *Dual Models.* Cambridge [Eng.], New York: Cambridge University Press, 2003.

——. *Polyhedron Models.* Cambridge [Eng.], New York: Cambridge University Press, 1971.

Spherical Polyhedra

Wenninger, Magnus J. *Spherical Models.* Cambridge [Eng.], New York: Cambridge University Press, 1979.

Wenninger, Magnus J. and Peter W. Messer. "Patterns on the Spherical Surface." *International Journal of Space Structures* 11.1–2 (1996): 183–192.

---

[27] See (Hearney 2009, 110–118) for a detailed explanation of modeling atomic shells.

Polyhedra in Art, Design, and Nature

Loeb, Arthur L. *Space Structures: Their Harmony and Counterpoint.* Reading, MA: Addison Wesley Pub. Co., Advanced Book Program, 1976.

Pearce, Peter. *Structure in Nature Is a Strategy for Design.* Cambridge, MA: MIT Press, 1980.

Pearce, Peter and Susan Pearce. *Polyhedra Primer.* Palo Alto, CA: Dale Seymour Publications, 1978.

Posamentier, Alfred S. and Ingmar Lehmann. *The (Fabulous) Fibonacci Numbers.* New York: Prometheus Books, 2007.

Williams, Robert. *The Geometrical Foundation of Natural Structure: A Source Book of Design.* New York: Dover Publications, 1979.

Other Polyhedra

Hart, George W. and Henri Picciotto. *Zome Geometry: Hand-on Learning with Zome Models.* Emeryville, CA: Key Curriculum Press, 2001.

Richeson, David S. *Euler's Gem The Polyhedron Formula and the Birth of Topology.* Princeton, NJ: Princeton University Press, 2008.

7 Golf Ball Dimples

W e introduced golf ball dimples in Chapter 3 and pointed out how dimples have improved the distance and control golfers can hit a ball. Here we look at the basic ways designers distribute dimples on spherical polyhedra. We will look at patented designs and see how they exploit the symmetry and great circles of various polyhedral frameworks. We will see very creative arrangements of dimples that optimize the packing of dimples with different diameters. The approach, used in many designs, is to define an arrangement within a small working area, such as a Schwarz triangle, and then replicate the arrangement to cover the rest of the ball without causing overlaps or gaps. USGA regulations require that golf balls be spherically symmetric. While dimple shapes and sizes can vary a great deal, the more times a fixed pattern is replicated to cover the ball's surface the better, provided of course the dimples are large enough to be effective, and are evenly spaced to eliminate bald spots. Most designers develop their dimple patterns within standard design areas based on a polyhedral foundation.[1]

Most designers start with one of the five spherical Platonic solids or one of their 13 Archimedean relatives. Usually, the designer will subdivide the polyhedron's faces into

---

[1] (Solheim 1987, sheet 3/5, fig. 7)

smaller working areas, including Schwarz, equilateral, and isosceles triangles. The number of different working areas determines how many dimple patterns a designer needs to develop; each area requires its own pattern. For example, each of the icosahedron's 20 equilateral faces might be divided into six Schwarz triangles (three left-handed ones and three right-handed ones); a dimple pattern for one of these would be replicated 120 times, covering the entire ball through combinations of rotation or reflection. This is a good return for such a small designed area.

In the last chapter we described three symmetry groups that characterize the Platonic solids: the icosahedral (icosahedron and dodecahedron), the octahedral (octahedron and cube), and the tetrahedral groups. Polyhedral-based dimple patterns usually fall into one of these symmetry groups and we can use golf ball dimples to show how these symmetries are applied.

Why are these groups so important? There are two main reasons. First, the higher the number of rotation symmetries there are, the more spin axes a ball has and the more consistently the ball will fly. Of course, this assumes that the dimple pattern is uniform. Second is layout flexibility. Platonic solids share all of their symmetries with their Archimedean relatives. For example, a designer could develop a dimple pattern within the equilateral faces of the icosahedron or its Schwarz triangles. If he chooses an icosidodecahedron, he has equilateral and isosceles triangles to work with, all of which could be further subdivided. The icosahedron and icosidodecahedron are different polyhedra, but they belong to the same symmetry group; a symmetrical dimple arrangement in one will share symmetries with the other. The icosahedral group offers more choices than the other two groups.

7.1   Icosahedral Balls

The icosahedron and its dual, the dodecahedron, share the same 120 rotation and reflection symmetries around their vertices, edges and faces (see Section 6.3.1). The icosahedral group has the most symmetries of any group. For this reason, it is the framework of choice among ball designers.

Figure 7.1 shows a series of icosahedral group designs. We have already seen pure icosahedron examples, so here we start with its dual, the dodecahedron, which is also in the icosahedral group. Figure 7.1(a) shows a creative dimple arrangement for a spherical dodecahedron, which normally has no complete great circles. Brent Emerson's design arranges three different dimple sizes in such a way as to effectively subdivide each of the pentagonal faces into five diamonds, five isosceles triangles, and a center pentagon. In so doing, he creates ten dimple-free great circles.[2] Only one of the dodecahedron's 12 faces is shown in the figure. Note that the innermost pentagonal area, defined by five dimples labeled 60, surrounds the vertex of the dodecahedron's dual, the icosahedron. Figure 7.1(b) shows a design by In Hwang based on a spherical icosidodecahedron.[3] This Archimedean solid combines the 20 triangular faces of the icosahedron with the 12 pentagonal faces of the dodecahedron. They are clearly visible in the drawing, which emphasizes the six great circles formed between faces. The large triangles are equilateral, while the ones that subdivide the pentagonal face are isosceles.

Within these two types of triangles, the designer makes a straightforward dimple arrangement. Once again, the icosahedral symmetry is apparent. The center dimple in the

---

[2] (Emerson et al. 2003, sheet 4/8, fig. 4)
[3] (Hwang November 1996, sheet 5/9, fig. 8)

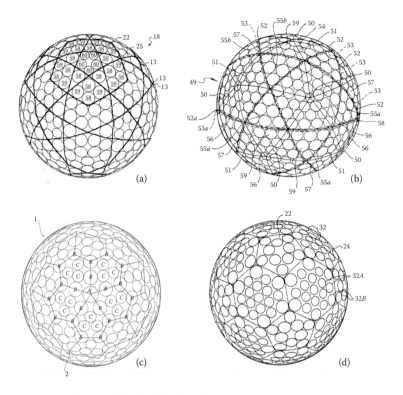

Figure 7.1. Golf balls with icosahedral symmetry.

middle of each pentagonal face is a vertex of the underlying icosahedron. In similar fashion (a), each of these vertices is surrounded by five equal-sized dimples. The ico-sidodecahedron is a popular design base because it offers all the symmetry axes of the icosahedron and has six natural great circles that can be used for parting lines, something a pure icosahedral base does not offer.[4]

Figure 7.1(c), is an unusual Catalan base. The 13 Catalan solids are duals of the 13 Archimedean solids,[5] and this one is called a pentagonal hexecontahedron.[6] It is the dual of the snub dodecahedron and has two distinct forms, which are *mirror images* (enantio-morphs) of each other. The distinct forms preclude reflection symmetries within a form, but some rotational symmetry remains. The dimple marked *A* is a vertex of the icosahedron and, once again, it is surrounded by five equal-diameter dimples marked *B*. We show this design to make the point symmetries, such as icosahedral, octahedral, or tetrahedral, are constant between a Platonic solid and its related Archimedean and Catalan solids.

Figure 7.1(d) uses another Archimedean base, the truncated icosahedron (resembles a soccer ball), in which 12 pentagonal and 20 hexagonal faces combine.[7] Dimple diameters

---

[4] This advantage is becoming less important as molding technology improves. Titleist's *Staggered Wave* golf ball molding parting line actually weaves in and out of the dimples that cross the hemispheric boundaries.

[5] The Catalan solids are named after Belgian mathematician, Eugène Catalan, who was the first to write about them in 1865.

[6] (Winfield and Aoyama 2003, sheet 1/6, fig. 1)

[7] (Mackey 1991, sheet 2/2, fig. 3)

do not vary much in this layout, which is one reason for the large land areas, the undimpled areas, between some clusters. Although this polyhedral base is quite different than previous ones, it still carries over some of the icosahedron's symmetry. The five equal-diameter dimples in the center of each pentagonal face surround an icosahedral vertex.

Commercial balls with icosahedral symmetry include Acushnet Co. Pinnacle *Platinum Distance*, Wilson *Ultra*, Ben Hogan *Hawk*, Dunlop Sports *DDH II*, Dicks Sporting Goods Maxfli *HT-9*, and Callaway/Spalding Top-Flight *Complete Distance*.

7.2   Octahedral Balls

Until the 1980s, octahedral layouts were very common. The octahedron's trirectangular work areas are easy to work with, and the base offers three orthogonal great circles for parting lines that facilitate molding and make it easier to make tools and dies. The octahedron and its dual, the cube, have only 48 rotation and reflection symmetries around their vertices, edges, and faces—less than half the number of the icosahedral group. Figure 7.2(a)[8] shows a classic dimple design, called the Atti pattern,[9] similar to the Class I subdivision schema we will see in the next chapter. Within each trirectangular face, dimples are arranged in three concentric "rings." Around the edges of each triangle, 21 dimples enclose an inner ring of 15 dimples. These 15 dimples, in turn, surround the six midface dimples.

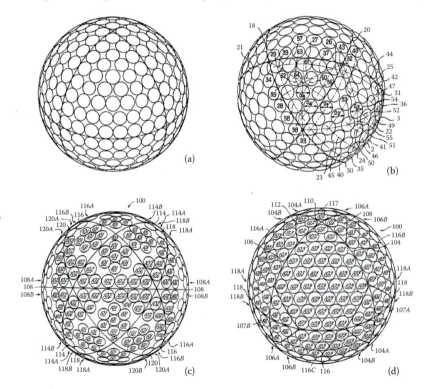

Figure 7.2. Golf balls with octahedral symmetry.

---

[8] (Morgan 2005, sheet 3/13, fig. 3)

[9] Named after Ralph Atti who received many patents for golf ball construction and manufacturing.

Each face will have 42 dimples; the entire ball has 336, covering more than 80 percent of its surface. The Archimedean solid cuboctahedron combines the cube's six quadrilateral faces with the octahedron's eight equilateral faces to form the polyhedral foundation in the design shown in Figure 7.2(b).[10] The designer further subdivides each equilateral into six Schwarz triangles and each quadrilateral face into four isosceles triangles. Dimples within the quadrilaterals have a grid-like arrangement.

The octahedron's dual, the spherical cube, is shown in Figure 7.2(c).[11] Each quadrilateral face is subdivided into four congruent isosceles work areas. Only one dimple pattern is required; it will be replicated 24 times over the ball. Dimple numbering indicates each triangle has seven 101-size and eight 102-size dimples, or 60 total for each of the cube's faces. The whole ball is covered by a total of 360 dimples. The octahedral vertices coincide wherever two great circles meet at 90°. One is labeled 103 in the drawing.

Figure 7.2(d)[12] shows the six squares and eight hexagonal faces that make up this solid. The truncations do not add any great circles to the octahedron's original three. The dimple numbers indicate their sizes. It is impressive that just two sizes can get this level of coverage and leave so little land area between them. Once again, the octahedral symmetry shows through; every vertex of the octahedron is surrounded by four equal-sized dimples.

These four examples illustrate the flexibility of the octahedral group. When great circles intersect at small angles, it is difficult to fill the corners without resorting to dimples too small to contribute to the ball's aerodynamics. If they are omitted, however, bare patches result. Notice how well dimples pack in the great circle corners in Figure 7.2(a), as compared to the other layouts.

Commercial golf balls with octahedral symmetry include the Slazenger *Raw Distance Practice*, Nike *Tour Accuracy DD*, Spalding *Super-Flight*, and MacGregor *Tourney*.

7.3   Tetrahedral Balls

Tetrahedral symmetry is not used often for golf balls, perhaps because it offers the fewest number of symmetries (24 in all), and its large equilateral faces have considerable spherical excess (see Chapter 4). We show some designs for completeness. The symmetry principles we have been discussing—repeatable design areas, great circle parting lines, and dimple arrangement—still apply. Figure 7.3 shows two golf ball designs based on the tetrahedron. Their equilateral faces are so large it is difficult to see their edges in these illustrations. We have, therefore, outlined one face in each design to make both the faces and their design areas easier to read. Both designs subdivide the tetrahedron's faces into six right Schwarz triangles. Dimples are arranged within these areas, and none cross any of the six great circles. One great circle serves as a parting line. In Figure 7.3(a),[13] two rows of small dimples labeled *S* parallel two sides of the Schwarz triangle; the longer run of these dimples is along the polyhedral face edge. The design in Figure 7.3(b)[14] also places same-sized dimples along the polyhedral edge and the face medians within each Schwarz triangle. Clusters of three small dimples in two of the Schwarz triangle's corners yields the distinctive dimple pattern in the middle of each of the tetrahedron's faces and around its vertices.

---

[10] (Morell 1990, sheet 1/4, fig. 2)

[11] (Gobush 20 Sept. 1988, sheet 6/10, fig. 11)

[12] (Gobush 23 Aug. 1988, sheet 4/6, fig. 6A)

[13] (Stiefel and Nesbitt 1999, sheet 1/2, fig. 2)

[14] (Yamada 1990, sheet 4/6, fig. 5).

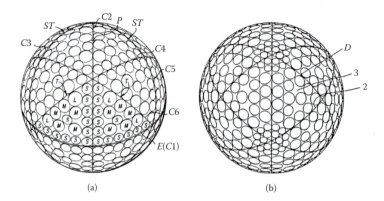

Figure 7.3. Golf balls with tetrahedral symmetry.

7.4   Bilateral Symmetry

Although we have emphasized dimple designs with polyhedral bases, we present one final dimple layout based on loxodromes. We described this unique spherical spiral earlier in Section 4.9. It is not immediately obvious that it could be used to distribute points on a sphere. A loxodrome design was first proposed for a golf ball in 1924 by Jarvis Hunt.[15] Here we show a modern, sophisticated use of this spiral, which produces *bilateral symmetry* instead of the multi-axis symmetry examples we have been discussing. Figure 7.4(a) shows a design by Donald Bunger and associates, with four left-handed and four right-handed spirals between the north and south poles of the golf ball.[16] These spirals define a single point at each pole, 24 in each hemisphere and 4 along the equator. Since none of the spirals are great circles, the four-sided areas formed by their intersections technically are not spherical polygons. The points where the spirals intersect each other are rotationally symmetric around the poles. Loxodrome spirals naturally continue to the lower half of the sphere, but the designer decided to break the curves at the equator and rotate the lower hemisphere's spirals 45° relative to the upper one; thus the equatorial plane is not a mirror

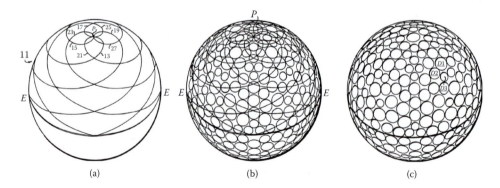

Figure 7.4. Golf balls with loxodrome design.

[15] (Hunt 1924)
[16] (Bunger et al. 1991, sheets 1/9, 4/9, and 5/9; figs. 2, 8, and 9, respectively)

plane. Great care is needed to find spirals that form useful intersections and design areas. If the spirals intersect at too shallow an angle, very small dimples result, and they have less of an aerodynamic effect. Note how the designer centers some dimples on the loxodrome or at their intersections, while others are arranged totally within four-sided areas. Figure 7.4(b) shows the relationship of the spirals to the dimple arrangement, while (c) shows the completed design, minus all reference lines. This particular design distributes 410 dimples of three sizes over the ball.

7.5   Subdivided Areas

At this point, we have a good sense of the three main symmetry groups a golf ball designer can use as a polyhedral foundation. Assuming the design is not a random layout, his choice of polyhedral foundation determines the regular spherical faces he will work with, either equilateral triangles, quadrilaterals, pentagons, or hexagons. In many cases, the designer will further subdivide the polyhedral face to gain additional great circles for parting lines. These subdivisions also may be used to define standard areas, so dimple arrangements developed within will cover the rest of the ball through rotation and reflection symmetries.

Archimedean solids and subdivided Platonic solids will have at least two, if not more, work areas. An icosidodecahedral base, for example, has pentagonal and equilateral triangular faces, and both may be further broken down into Schwarz triangles or nested equilaterals. If the ball is a hybrid, it may combine some standard work areas with entirely different ones. We have already seen hybrid cases where some of the regular faces of a spherical icosahedron are preserved, yet other areas are developed as bands of dimples around the equator.

Figure 7.5(a)[17] demonstrates an important subdivision, where nested equilateral triangles define the center points of dimples. The approach is quite different from the mid-arc subdivision of the equilateral triangle shown in (b),[18] which can also be viewed as a series of right Schwarz triangles. Figure 7.5(c) is an innovative subdivision of an equilateral triangle. The schema results in three congruent spherical quadrilaterals.[19] The subdivision, however, does not result in adding great circles to the polyhedral base. The pentagon within a pentagon, shown in Figure 7.5(d),[20] is similar in approach to the triangle within a triangle in (a). In Figure 7.5(e) the pentagonal area is subdivided, so two standard areas, a diamond and isosceles triangle, surround a central pentagon.[21] This calls for three dimple arrangements.

A number of designers use icosidodecahedral bases. Figure 7.5(f)[22] demonstrates one way to subdivide the 12 pentagons into isosceles work areas adjacent to the 20 equilateral triangles. Spherical quadrilaterals are not used often as work areas, but the arrangement in Figure 7.5(g)[23] shows that any number of triangular subareas could be defined, provided the patterns within are symmetrical. In Figure 7.5(h), the standard design area is an entire

---

[17] (Aoyama 1999, sheet 9/9, fig. 10)
[18] (Woo 1993, sheet 1/5, fig. 1)
[19] (Aoyama 1995, sheet 3/7, fig. 3)
[20] (Emerson et al. 2003, sheet 4/8, fig. 5)
[21] (Shaw and Haines 1979, sheet 3/3, fig. 3)
[22] (Hwang November 1996, sheet 4/9, fig. 7)
[23] (Aoyama 1990, sheet 5/8, fig. 8)

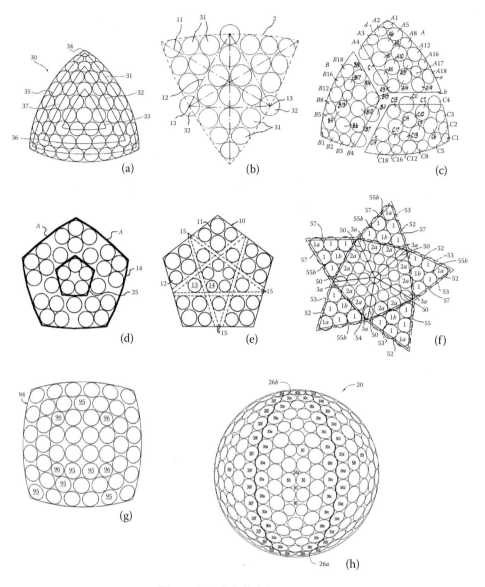

Figure 7.5. Subdivision areas.

lune, six of which cover the entire sphere.[24] Given the symmetry of dimple arrangement within, it would be possible to define a *birectangular* triangle as a subunit.

7.6   Dimple Graphics

Dimples made it easier for us to read the underlying spherical subdivision of golf balls. We are not designing golf balls in this book, but we can use graphic styles that look like

---

[24] (Veilleux, Simonds, and Shannon 2007, sheet 2/11, fig. 2)

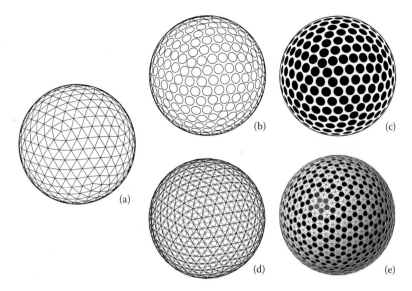

Figure 7.6. Grid accents, dimples, and holes.

dimples to make it easier to read the subdivision schemas we create. Each schema we present in the next chapter results in a triangular grid. We can place dimple-style graphics at grid vertices or in the middle of the grid's triangular faces. In the first case, we use circles or small balls centered at each grid point. In the second case, we place a hole centered at the triangle's centroid. Figure 7.6 shows five graphic styles for the same spherical grid. A typical subdivision grid in (a) is shown with circles at each vertex in (b) and shaded in (c). These last two styles resemble golf balls. Figure 7.6(d) shows the same subdivision, but this time, holes are located at the centers of each triangular face and then shaded in (e). Each style emphasizes a feature. Some make it easier to see how uniform the spacing is between grid points; others indicate differences in the triangles shape and size. We will make extensive use of these graphic styles in the next chapter.

7.7   Summary

We have defined what a polyhedron is, detailed the characteristics of the Platonic and Archimedean solids, and shown precise ways to define their planar and spherical geometry. Schläfli and Cundy-Rollett's modified Schläfli symbols provide simple and useful notations for specifying a polyhedron and seeing the similarities and differences between them. The most important polyhedral frameworks for spherical subdivision are the Platonic deltahedra—the tetrahedron, octahedron, and icosahedron. They have equilateral faces that break down into Schwarz triangles, and each defines a symmetry group. Any subdivision of a deltahedron's Schwarz or equilateral face triangle also subdivides its dual, even though the duals have a very different face type. While the Platonic solids, especially the icosahedron, are the most common frameworks for subdivision, we will see instances where quasi-regular Archimedean solids, the cuboctahedron and icosidodecahedron, solve special cases where a natural great circle of edges is needed. We have also seen that two of the Archimedean solids, the snub cube and snub dodecahedron, are chiral and do not have reflection symmetry. Chiral subdivisions are a recent and distinct class of spherical subdivisions. The rotational and reflection symmetry inherent in the icosahedral, octahedral, and tetrahedral groups gives us methods for replicating geometry from one part of the sphere to another. We can subdivide a small area, such as a Schwarz triangle, and then replicate it to cover the entire sphere without gaps or overlaps.

Golf ball dimples are comprehensive demonstrations of the principles of spherical subdivision. Dimple layouts are practical solutions to difficult geometry problems. Most golf ball dimple arrangements are developed from a polyhedral base. Designers exploit symmetry, great circles, and local working areas that can be replicated over the sphere to pack different-sized dimples. The variety and creativity of these spherical designs are amazing, not only in their creativity, but as solutions to difficult spatial problems.

In the next chapter, we explore various subdivision techniques based on the polyhedral frameworks and symmetry groups we have just covered.

Additional Resources

Stewart, Ian. "Crystallography of a Golf Ball." *Scientific American* (Feb. 1997): 96–98.

Tarnai, T. "Geodesic Domes and Golf Balls." *Space Structures* 4.2 (1993): 1176–1183.

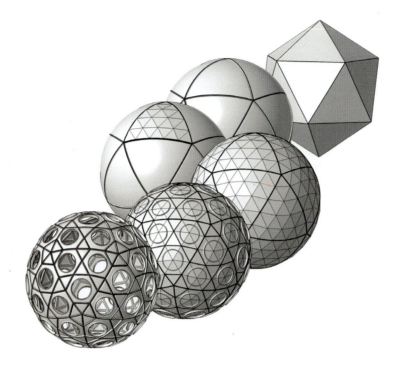

Subdivision Schemas

S pherical polyhedra provide a convenient starting point for subdividing spheres because they evenly distribute an initial set of points on the surface of the sphere. This chapter describes how to define more points between this initial set. When each point is connected by an arc or chord to the points immediately around it, a triangular grid results. The top four figures above show this sequence. Some spherical applications require design features, such as openings or panels or different face shapes. The geometry can become quite complex. The lower two figures above suggest some features, such as holes and panels that are based on the subdivision grid. These features are added with CAD programs and discussed in a later chapter.

In the mid-1960s Joseph Clinton developed a class system for categorizing spherical grids.[1] Each class refers to triangular grids that have the same basic orientation relative to the edges of the spherical polyhedron they subdivide. Class I grids run parallel to

---

[1] In 1966, Joseph Clinton and Mark Mabee, under NASA contract, undertook a study of spherical geometries and developed the spherical node identification and grid topological class system we use today. Their Classes I and II were first published in *Dome Book 2*; see (Clinton 1971). Magnus Wenninger added Class III for skewed grids in 1979 in his book *Spherical Models*; see (Wenninger 1979).

the polyhedron's edges. The subdivided spheres shown on the previous page are all Class I. Class II grids run perpendicular to the polyhedron's edges, while Class III grids are skewed and run at an angle to the edges. We will cover each class in a separate section in this chapter, highlighting techniques unique to each and showing examples of how the subdivision can be implemented.

The presentation of each subdivision schema in this chapter is identical. Each presentation starts with orienting one of the polyhedron's faces in a way that makes calculations easier. This face is called the *principal polyhedral triangle (PPT)*, and it serves as our subdivision workbench. Points are located on the PPT's edges, and great circle arcs span from one edge point to another. Subdivision grid points are defined where these arcs intersect over the face of the PPT or at some point near their intersection. We will illustrate each of these steps, noting the mathematical principles. We will also present the key formulas and algorithms in Appendix C.

We have tremendous flexibility in spherical subdivision. We can start with any number of spherical polyhedra and can develop different types of grids covering them. By selectively spanning points with either great circle arcs or chords, we can make triangles, diamonds, hexagons, pentagons, or other polygons. With so many choices, it is natural to ask which combination is best. Of course, the answer depends on the application. In the next chapter we present a series of metrics that make it easy to see their differences and similarities. In the end, you will be equipped to answer this question for your own applications.

8.1   Geodesic Notation

In Chapter 6 we saw examples of Schläfli symbols. They summarize polyhedral geometry with two numbers $\{p,q\}$, which are read as p-gons (pentagons, hexagons, etc.) that meet q times at a vertex. The Schläfli symbol for a dodecahedron, for example, is $\{5,3\}$, indicating that three pentagonal faces (five edges) meet at each vertex. This notation is extended to include symbols for grids that cover spherical polyhedra.[2] *Geodesic notation* is very useful because, at a glance, it states the type of tessellation, how many grid members intersect at a vertex, the intensity of the grid, and the grid's orientation with respect to the base polyhedron's edge.

The general form of geodesic notation is $\{p, q+\}_{b,c}$ where p indicates the tessellation shape and q is the valence (number of grids) that meet at a subdivided polyhedron vertex. When three subdivision grid members meet at a vertex, its valence is called *trivalent*, and when four, five, and six grid members meet at a vertex, they are called quadravalent, *pentavalent*, and hexavalent, respectively. All of our subdivision examples develop triangular grid, thus, $p = 3$. In an icosahedron, five grid members (pentavalent) meet at each polyhedral vertex. Thus, $q = 5$ (for tetrahedra or octahedra, $q = 3$ and 4, respectively). The "+" sign indicates that more than q grid members meet at other vertices; these cases occur at points other than the base polyhedron's vertices. For an icosahedron with a triangular tessellation grid, $\{p,q\}$ would be $\{3,5+\}$ or 3-gons (triangles) with 5 grid members meeting at each of the polyhedron's vertices. The + sign, indicated by 5+, means that at points other than the polyhedron's vertices, more than 5 grid members meet.

Perhaps the most useful aspect of geodesic notation is the b,c subscript because it indicates the frequency of the grid, how course or fine it is, and the orientation of the grid

---

[2] (Goldberg 1937, 106)

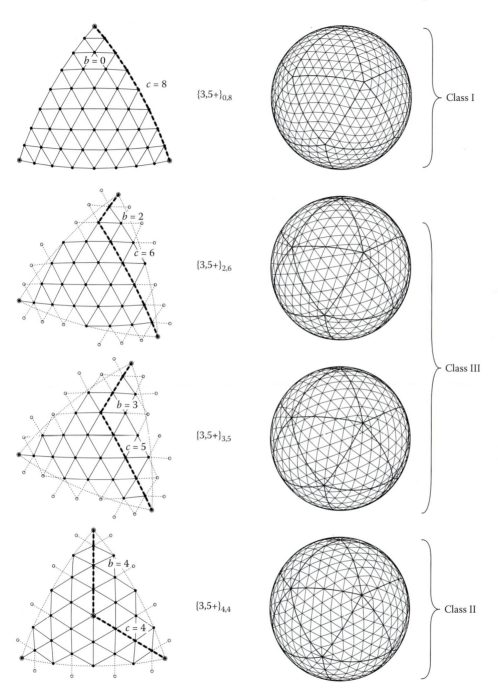

Figure 8.1. Examples of geodesic notations for Class I, II, and III tessellations.

with respect to the sides of the polyhedron's faces.[3] The frequency (designated by the superscript Greek letter *nu* or ^{v}) of a subdivision grid is the sum $b + c$, and both b and c can be any nonnegative integer.[4] Either b or c can be 0, but not both in the same subdivision.[5] Most of the examples in this chapter are frequency 8^v, thus, b and c will have values, such as 0,8 or 4,4 or 2,6, respectively.

Figure 8.1 shows three classes, or families, of grids for a single PPT—Class I, II, and III. A class is simply a group of subdivision methods or schemas that result in the same grid orientation with respect to the edges of the polyhedron's face. Within a class, there will be small variations in grid points within the PPT from one schema to another, but the general orientation of the grid remains the same. Each left-right pair of illustrations in Figure 8.1 shows one grid type covering a single PPT and covering a full sphere. We have said that Class I grids run parallel to the PPT's edges; a typical one is shown in the topmost pair in Figure 8.1. Class II grids are the other extreme and run perpendicular to the PPT's edges, as shown in the bottom pair. The Class III grids, shown in Figure 8.1, are placed between Class I and II in the figure to emphasize the fact that their grids are oriented somewhere in between Class I and Class II and are skewed with respect to the PPT's face.

Figure 8.1 also shows the geodesic notation and an example of a fully subdivided sphere. Each of these classes has the same {3,5+} specification because all the subdivisions shown are for a spherical icosahedron. Only their b,c values change because they indicate how the grid is oriented. To see how b,c coordinates work, start at any PPT apex and follow the shortest path of grid lines to reach the next PPT apex to the left of the direction you are traversing in the grid. In the figure, b directions are down and to the left, while c directions are down and to the right. If you start at the uppermost PPT vertex, count how many grids you traverse in the two directions; this is the b,c value that appears in the geodesic notation.

8.2   Triangulation Number

Goldberg used geodesic notation to investigate dual grids (grids based on other grids), and the symmetries of grids on a sphere. In the process, he found a direct relationship between b,c and the topology of a grid, the number of parts on a fully subdivided sphere. The relationship $T = b^2 + bc + c^2$, sometimes called the triangulation number, or simply T, is very useful.[6] It provides a coefficient that summarizes the total faces, edges, and vertices in a completely tessellated sphere.[7] Figure 8.2 shows how to find the triangulation number and the formulas that use it to enumerate the number of faces, edges, and vertices in a fully subdivided sphere. The example uses T to find the number of parts in a 4^v Class I spherical octahedron.

$$T = b^2 + bc + c^2,$$
$$\text{Faces} = T \times \text{number of polyhedra faces,}$$
$$\text{Edges} = T \times \text{number of polyhedra edges,}$$
$$\text{Vertices} = T \left((2 \times \text{valence}) / (6 - \text{valence}) \right) + 2.$$

[3] H. S. M. Coxeter used Michael Goldberg's formula of $T = b^2 + bc + c^2$ with geodesic notation $\{p,q\}_{b,c}$ to describe Buckminster Fuller's geodesic grids; see (Coxeter 1971).

[4] The Greek letter nu (v), a symbol for subdivision frequency, first appeared in early worksheets developed by Synergetics, Inc. and Geodesics, Inc. in the 1950s. It is still used in this context today.

[5] It is not often seen, but when either b or c is 1 and the other is 0, there is no tessellation and the notation describes an unsubdivided polyhedron. A basic tetrahedron polyhedron, for example, would read $\{3,3+\}_{0,1}$.

[6] (Goldberg 1937, 104–108)

[7] The triangulation number was important in defining the possible icosahedral surface lattices possible in the pioneering virus research of Donald Caspar and Aaron Klug. See Section 3.2.

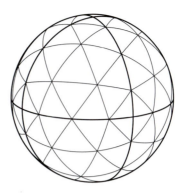

Figure 8.2. Triangulation number.

Example

Spherical octahedron:
Faces—8
Edges—12
Vertices—6

Geodesic notation: $\{3,4+\}_{0,4}$

Subdivided octahedron:
$T = 4^2 = 16$
Faces $= T(8) = 128$
Edges $= T(12) = 192$
Vertices $= T\,[(2(4)) / (6 - 4)] + 2 = 66$
Euler's law: vertices + faces = edges + 2 = 194

The triangulation number is also an indication of the possible patterns that can be made from the triangular grid. If the number is evenly divisible by three, you can make diamond and hexagonal-pentagonal patterns by selecting different combinations of points from the triangular grid. All Class II subdivisions can make these other patterns, but not every Class I or III combination can do so.

8.3   Frequency and Harmonics

In geodesic notation, the frequency of a tessellation, $b + c$, is the number of segments into which the grid divides the polyhedra edge. Some subdivision schemas can produce any even or odd frequency. Others are limited to even ones or to powers of 2. A grid may also contain subgrids, or harmonics. A harmonic is a subfrequency that is mathematically and geometrically related to another subfrequency. If a tessellation has harmonics, some of its grid points define other tessellation grids. These might be useful in some geodesic applications, for example, where grid overlays are used for trussing or harmonics that define smaller units within the PPT that facilitate construction or prefabrication.

The easiest way to visualize both frequency and harmonics is to see them side by side. Figure 8.3 shows an example of doubling frequencies for icosahedron Class I, II, and III tessellations. The frequencies in (a) and (b) are 2^v, 4^v, 8^v, and 16^v for Class I PPTs and Class II diamonds (described later). They are shown in clockwise order around a common icosahedra vertex. In a similar manner, (c) shows frequencies 3^v, 6^v, 12^v, and 24^v for Class III tessellations. Notice in each figure that the lower frequency grid is also part of the higher one. The lower frequency grid is a harmonic of the higher frequency grid.

Harmonics follow rules. All frequencies are positive integer numbers and the sum $b + c$. Class I and III can have even or odd frequencies. Class II and its harmonics are always even frequencies because the tessellation grid is symmetrical about the polyhedron's face vertical axis. A frequency can have a harmonic, if it is not a prime number. The numbers, 2, 3, 5, 7, 11, 13, 17, 19, 23, and 29 are prime, and therefore, any Class I subdivision with these frequencies will not have harmonics. Class III subdivisions are skewed, and both b and c specifications must be nonprime for there to be a harmonic.

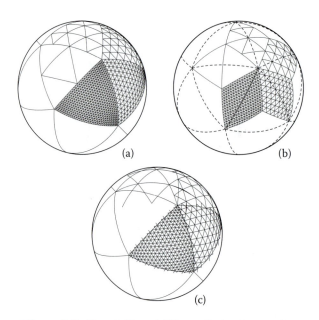

Figure 8.3. Class I, II, and III tessellation harmonics.

\multicolumn{7}{c}{**Tessellation Harmonics**}
ν	**Prime Factors**	\multicolumn{5}{c}{**Harmonic Frequencies**}
---	---	---	---	---	---	---
2	2 (Prime)					
3	3 (Prime)					
4	2,2	2^v				
5	5 (Prime)					
6	2,3	2^v	3^v			
7	7 (Prime)					
8	2,2,2	2^v	4^v			
9	3,3	3^v				
10	2,5	2^v	5^v			
11	11 (Prime)					
12	2,2,3	2^v	3^v	4^v	6^v	
13	13 (Prime)					
14	2,7	2^v	7^v			
15	3,5	3^v	5^v			
16	2,2,2,2	2^v	4^v	8^v		
17	17 (Prime)					
18	2,3,3	2^v	3^v	6^v	9^v	
19	19 (Prime)					
20	2,2,5	2^v	4^v	5^v	10^v	
21	3,7	3^v	7^v			
22	2,11	2^v	11^v			
23	23 (Prime)					
24	2,2,2,3	2^v	3^v	4^v	6^v	8^v 12^v
25	5,5	5^v				
26	2,13	2^v	13^v			
27	3,3,3	3^v	9^v			
28	2,2,7	2^v	4^v	14^v		
29	29 (Prime)					
30	2,3,5	2^v	3^v	5^v	6^v	10^v 15^v

Table 8.1. Tessellation harmonics.

Figure 8.3(c) shows Class III examples of frequencies 3^v, 6^v, 12^v, and 24^v. Their b,c specifications are doubled in each successive tessellation. Their geodesic notations are $\{3,5+\}_{1,2}$; $\{3,5+\}_{2,4}$; $\{3,5+\}_{4,8}$; and $\{3,5+\}_{8,16}$, respectively. Note that both b,c values double each time.

To determine if a frequency has one or more harmonics, first find its prime factors. The prime factors of a positive integer are the prime numbers that divide into it exactly, leaving no remainder. If the frequency has prime factors, the product of any combination of them is a subdivision grid harmonic, if it also satisfies the rules above. Table 8.1 shows the harmonics for frequencies 2^v–30^v. Column 1 is the base frequency, and column 2 lists its prime factors, if it is not a prime. Columns 3 onward are the frequency's harmonics. Referring to Table 8.1, a 24^v tessellation, for example, has prime factors 2, 2, 2, 3. Thus, a 24^v tessellation will have 2^v and 3^v tessellation harmonics as well as tessellations for the product of various combinations of these prime factors, six in all: 2^v, 3^v, 4^v, 6^v, 8^v, and 12^v. Keep in mind that for Class III subdivisions, both b and c must be nonprime for there to be harmonics.

To review, the frequency of a subdivision tessellation is $b + c$, or the number of times the grid divides up the polyhedron's edge. If the frequency is a prime number, the tessellation has no prime factors, and therefore, no harmonics. If the frequency is not prime, the product of any combination of its prime factors will be a harmonic grid in the overall tessellation.

8.4   Grid Symmetry

In Chapter 6, we defined symmetry as any combination of rotations and reflections that leave the polyhedron unchanged. Tessellation grids exhibit these symmetries. Most of the subdivisions developed in this chapter are subdivisions of the equilateral faces of the tetrahedron, octahedron, or icosahedron. Although these polyhedra are rotationally symmetric around their vertices, edges, and faces, the subdivision grids developed over them are not necessarily symmetric. Figure 8.4 shows some examples of three classes of grids we develop later in this chapter. The center of each face is marked.

The Class I and II grids, shown in Figure 8.4(a) and (b), are symmetric by reflection and by rotation. That is, the face can be rotated around its center (indicated by a ⊕ symbol) any multiple of 120°, and its appearance does not change. Likewise, the grid can be reflected around any of the principal triangle's altitudes or edges and remain unchanged. Class I and II tessellations are generally symmetric rotationally and by reflection, but Class III tessellations are not.

Class III grids run at an angle to the PPT's edges. As a result, their grids are *enantiomers*, which means the grid's reflection is an opposite-handed version.[8] You cannot pass a plane through a Class III subdivided sphere and produce equal images on either side. This is analogous to the images of your left and right hands. They are enantiomorphs—opposite-handed reflections of each other. Notice in Figure 8.4 that even after reversing the b,c coefficients for the Class I and II tessellations (a) and (b), the grid would still look the same. But when b,c is reversed in Class III, the grid becomes opposite-handed, as (c) and (d) demonstrate. Class III grids do have rotational symmetry, however.

---

[8] The Latin words *laevo* and *dextro* are sometimes used to designate the left- and right-handed image or to indicate a rotation direction.

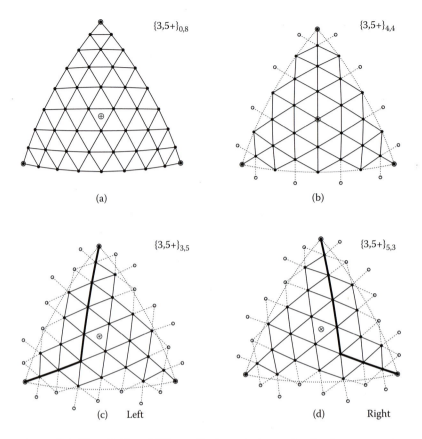

$\{3,5+\}_{0,8}$ $\{3,5+\}_{4,4}$

(a) (b)

$\{3,5+\}_{3,5}$ $\{3,5+\}_{5,3}$

(c)   Left (d)   Right

Figure 8.4. Class I, II, and III tessellation rotation and reflection symmetry.

To review: subdivisions of spherical polyhedra are grouped into three classes, each distinguished by the orientation of their subdivision grid with respect to the polyhedron's edges.[9] In Class I and II subdivisions, tessellation grids run parallel and perpendicular to the polyhedron's edges. In Class III, grid chords run at angles to the edges, never parallel or perpendicular to them. Class I and II subdivision schemas generally produce grids that are symmetric by rotation and reflection. However, Class III grids are only rotationally symmetric. Their reflections are not symmetric; they are opposite-handed.

For a subdivided spherical polyhedron, a geodesic notation has evolved that summarizes the type of tessellation grid, the number of grid members that meet at a typical polyhedron's vertex, its *valence*, and the coordinates of the tessellation's grid. Coordinates b,c define the orientation of the grid with respect to the polyhedron's edges, as well as the grid's frequency. The frequency of a subdivision grid is the sum $b + c$, or the number of times the grid subdivides a polyhedron's edge. Most Class I and III subdivision techniques can produce any even or odd frequency, but some produce frequencies that are only powers of 2. Class II frequencies are always even frequencies. A subdivision grid may contain harmonic grids (grids within grids), if the frequency is not a prime number. Harmonics, if

---

[9] (Wenninger 1979, 98–100)

present, offer more flexible implementations of geodesics in some applications. You can derive a triangulation number, T, from a subdivision's geodesic notation and use it to calculate the total number of faces, edges, and vertices in a fully subdivided sphere.

8.5   Class I: Alternates and Ford

This section shows the step-by-step process for subdividing a sphere with Class I tessellations. We demonstrate four different subdivision schemas. Figure 8.5 shows examples of Class I tessellations developed on several different polyhedral foundations. Although Class I subdivisions are typically used with deltahedra—polyhedra with only equilateral faces such as the octahedron, tetrahedron, and icosahedron—the schema can be used with isosceles areas as well. Figure 8.5 shows examples of subdivided deltahedra. From left to right they are icosahedron 8^v, icosahedron 4^v, tetrahedron 8^v, and octahedron 11^v.[10] Although some of their grids look very similar, there are subtle but important differences that may influence which Class I schema you use in your own applications. We highlight these differences in the next chapter and make side-by-side comparisons of all the schemas.

The earliest Class I schemas were developed by Buckminster Fuller, Don Richter, Jeffrey Lindsay, T. C. Howard, and Duncan Stuart in the late 1940s and early 1950s in Buckminster Fuller's various startup companies. Early Class I subdivisions were called *alternates* and *Ford* breakdowns. The name *alternate* most likely evolved to mean this subdivision was the alternate method to the triacon method, another popular subdivision technique at the time, or from the way the method subdivides alternating sides of a spherical triangle. The *Ford* moniker is likely a tribute to the grid developed for the 1953 Ford Rotunda dome project in Dearborn, Michigan.[11] Many geodesic domes have been built using Class I grids. Today, this classic grid finds new uses in product design, cartographic reference systems, computer graphics, scientific instruments, sports balls, and other industrial applications.

It is easy to recognize Class I subdivisions. Their three-way grid "parallels" the three edges of the polyhedron's triangular faces, as shown in Figure 8.5. As a result, the polyhedron's

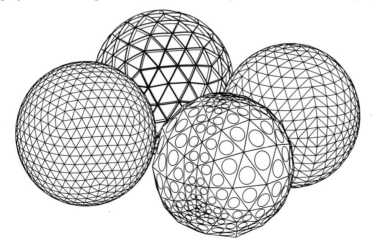

Figure 8.5. Examples of Class I subdivision schemas.

---

[10] Geodesic notations are: $\{3,5+\}_{0,8}$, $\{3,5+\}_{0,4}$, $\{3,3+\}_{0,8}$, and $\{3,4+\}_{0,11}$, respectively.
[11] See Section 2.8.

edges, though subdivided, remain part of the final grid. This makes Class I subdivisions easy to recognize, and makes it easy to see which polyhedron is being used as its base. None of the other classes have this characteristic. Class I can be any even or odd frequency, and one of its geodesic *b,c* specifications is always zero. The tessellation frequency is $b + c$, and the grid may contain harmonic grids, depending on their prime factors.

Each Class I subdivision schema we develop in this chapter begins by properly orienting the PPT. The PPT becomes our spherical workbench. Reference points are located along its edges. Pairs of reference points on opposite edges of the PPT define great circles that cross and intersect over the PPT's surface. Each great circle intersection defines two points. One point is within the PPT's face, while the other is on the opposite side of the sphere and is ignored. We repeat this process again and again, each time using a different combination of edge reference points and corresponding great circles. The process ends when this single PPT is completely tessellated. We copy the resulting PPT grid and replicate it to cover the rest of the sphere. The Class I schemas developed here differ only in the way the edge reference points are defined and how great circles between them intersect to define subdivision grid points. All Class I subdivision schemas start by orienting the PPT and defining reference points along its edges. We start with general considerations for the PPT.

8.5.1   Defining the Principal Triangle

A single face from a spherical deltahedron—a tetrahedron, octahedron, or icosahedron— serves as our subdivision workbench. Figure 8.6 shows three deltahedra above, each oriented vertex-zenith, along with one of their faces subdivided with a 4^v grid and its geodesic notation below.

It is evident that the icosahedron's PPT (far-right illustration in Figure 8.6) has the smallest face area and the most uniform grid of the three. Its chords, triangle face sizes,

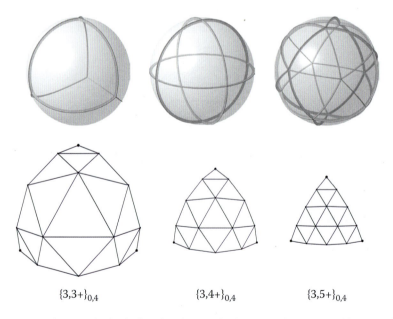

$\{3,3+\}_{0,4}$ $\qquad\qquad$ $\{3,4+\}_{0,4}$ $\qquad\qquad$ $\{3,5+\}_{0,4}$

Figure 8.6. Class I principal triangles for tetrahedron, octahedron, and icosahedron.

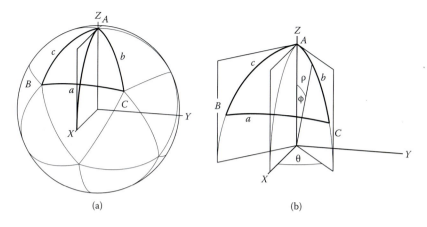

Figure 8.7. Principal triangle orientation and vertex coordinates.

and shapes show less variation than the other two. The spherical icosahedron is the best reference polyhedron of the three because its PPT has the smallest surface area of the Platonic solids and, of the three deltahedra, distributes the most evenly spaced vertices on its circumsphere. Its PPT has only one-twentieth of the area of a full sphere, whereas the tetrahedron and octahedron's PPTs are larger and cover one-fourth and one-eighth of the area of the sphere, respectively. The spherical icosahedron distributes 12 evenly spaced vertices on its circumsphere, while the tetrahedron and octahedron only distribute four and six vertices, respectively, on theirs. For these reasons, the icosahedron is the most frequently used reference polyhedron for spherical subdivision, by far. We will use the icosahedron in every subdivision schema we demonstrate and include an occasional tetrahedron or octahedron to make the point that the same techniques apply equally well to them.

With a spherical icosahedron as our base, the next task is to define precisely the geometry of one of its PPTs. We orient the PPT vertex-zenith and one of its faces symmetric to the xz-plane, as shown in Figure 8.7(a). This PPT orientation simplifies our calculations. We already know the coordinates of the PPT's zenith point, $(0, 0, 1)$, and we know that any point we calculate along edge b can be mirrored to edge c simply by reversing its y-coordinate sign. The vertex-zenith orientation does not affect any of the PPT's face or edge angles.

To find the coordinates of vertices B and C, it is easiest to first define them in the spherical coordinate system and then convert them to Cartesian values because we already have all the spherical information we need. Figure 8.7(a) shows the overall orientation, while (b) shows the relationship of the two coordinate systems to the PPT. Referring to Figure 8.7(b), the great circle planes that define edges b and c are tangent to the z-axis. The dihedral angle between these planes is simply the PPT's surface angle, or 72° (for the tetrahedron and octahedron, it is 120° and 90°, respectively). In the spherical coordinate system, a point is defined by three values: rho (ρ), pi (ϕ) and theta (θ), where ρ is the radius of the unit sphere, ϕ is the PPT's edge (angle), and θ is the PPT's face angle on the xy-plane, measured around the z-axis from the $+x$-axis (either clockwise or counterclockwise).

In Chapter 6, we used the golden ratio to precisely define the face and edge angles of the spherical Platonic solids. Using these angles, along with the radius of the unit sphere (1.0),

Deltahedra PPT Coordinates and Chord/Arc Factors							
	Vertex C Spherical Coordinates			Vertex C Cartesian Coordinates			Chord	Arc
Deltahdron	ρ	$\phi°$	$\theta°$	X	Y	Z	Factor	Factor
Tetrahedra	1.0	109.471220634	60.0	0.471405	0.816497	-0.333333	1.632993	1.910633
Octahedra	1.0	90.000000000	45.0	0.707107	0.707107	0.000000	1.414214	1.570796
Icosahedron	1.0	63.434948823	36.0	0.723607	0.525731	0.447214	1.051462	1.107149

Table 8.2. Deltahedra PPT coordinates and chord/arc factors.

we have all the information we need to define the spherical coordinates of PPT vertex *C*. Table 8.2 lists the spherical coordinates for apex *C*. The chord and arc factors are based on the sphere's radius and the PPT's edge (angle). We can find the Cartesian coordinates of vertex *C* by simply converting the spherical values to Cartesian values, using the following formulas. Most scientific calculators can do this automatically.[12]

$$x = r\cos\theta\sin\varphi,$$
$$y = r\sin\theta\sin\varphi,$$
$$z = r\cos\varphi.$$

To summarize the Cartesian coordinates of the icosahedron's PPT's apices are

$$A = (0.000000, 0.000000, 1.000000),$$

$$B = (0.723607, -0.525731, 0.447214), \text{ and}$$

$$C = (0.723607, 0.525731, 0.447214).$$

We have now established all the basic facts about the principal triangle: surface and edge angles, PPT vertex coordinates, and chord/arc factors. We have all the information we need to begin subdividing the PPT.

8.5.2   Edge Reference Points

Class I subdivision begins at the PPT's edges. A series of reference points is located along the edges, dividing the edges into segments equal in number to the subdivision frequency. Reference points have a regular pattern. Their spacing may be equal or unequal, but all three PPT edges will have the same pattern regardless of the spacing used.

There are two approaches to defining reference points: *Equal-chords* and *Equal-arcs*. Each approach can produce several different grids. For Equal-chords, the PPT's apex-to-apex chords (all three are the same length) are subdivided with equally spaced reference points. These reference points are on the PPT's edge chord; thus, they are inside the sphere. They must be projected to the surface of the sphere before they can be used as subdivision

---

[12] Two spherical coordinate conventions are in common use: (ρ, ϕ, θ) and (ρ, θ, ϕ). Be sure to use the convention consistent with your calculator.

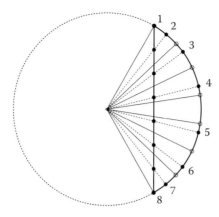

Figure 8.8. 7^v Equal-chords, Equal-arcs subdivision.

references. For Equal-arcs, the PPT's arc edge (not its chords) is evenly subdivided with points, which are used as references, as is. No projection is needed because these points are already on the surface of the sphere.

Figure 8.8 illustrates the difference between these two approaches with a 7^v subdivision of a tetrahedron's chord. We use an odd frequency so that reference points from the two approaches will not be superimposed on one another. This makes the figure easier to read. We also use the chord of a tetrahedron, the longest of any Platonic with a unit circumsphere, to make comparisons between Equal-chords and Equal-arcs more vivid. Referring to the figure, points 1 and 8 are the chord's end points on the surface of the circumscribing sphere represented in the cross section by the circumscribing circle.

For grids based on equal chord reference points, the chord between points 1 and 8 is divided into seven equal segments (7^v) with points located between them. The dashed lines show chord points projected from the center of the sphere to its surface. They are numbered 2 through 7 (1 and 8 are vertices of the tetrahedron's PPT edge and are already on the sphere's surface). By inspecting the figure, you can easily see the arc spacing between projected chord points is not equal. The difference between the largest and smallest arc length is more than 2:1 in this example. To put some numbers to this, the tetrahedron's overall edge arc is approximately 109.5°, and the seven arcs resulting from Equal-chords subdivision are 9.4°, 14.1°, 19.8°, 22.8°, 19.8°, 14.1°, and 9.4°, in that order. It is also evident that the arc lengths are symmetrical; that is, arc 1-2 equals 7-8, arc 2-3 equals 6-7, and so on.

The equal-arc points in Figure 8.8 are marked with small hollow circles on the surface of the sphere. They do not have to be projected. Each is spaced at 15.6°. Note that lines between them and the center of the sphere divide the chord into unequal segments. Appendix C lists simple formulas for dividing chords and arcs between any two points on the sphere into equal segments and projecting points to the surface of the sphere.

8.5.3   Intersecting Great Circles

Intersecting great circles is the most intuitive and easiest way to define points on the surface of the sphere. We make good use of this technique in our Class I schemas. Recall that a great circle is defined by three points: the origin and any two non-tangent, non-antipodal

Figure 8.9. Defining points by intersecting great circles.

points on the sphere. Figure 8.9 shows two great circles defined by two pair of points (red) and the origin (not shown). Two new antipodal points are created where these great circles intersect. One of them (brown) is visible in the figure; the other is on the opposite side of the sphere and cannot be seen.

The procedure for defining Class I grid points on the PPT's face is very simple. We use the origin and pairs of edge reference points on opposite edges to define the plane of each of the two great circles. The line of intersection between any two great circles pierces the surface of the sphere at two antipodal points. One of these points will be within the PPT's surface; we will ignore the other one, which is on the opposite side of the sphere. This new point will become a tessellation vertex in the final grid, or it will be used as an intermediate point to find a tessellation vertex, depending on which variation of grid we are making. We use this technique extensively in Class I subdivisions because it is extremely accurate, can be applied in many geometrical situations, and is easily programmed into computers.

Let's review the Class I technique. Class I tessellations run parallel to the edges of the spherical polyhedron, and the spherical polyhedron's edge remains a part of the final tessellation grid. The initial choice of polyhedron determines the size of the PPT to be subdivided. Only one PPT is subdivided and the tessellation is then replicated to cover the rest of the sphere. We position the polyhedron vertex-zenith with the selected PPT's edge symmetrical to the *xz*-plane. Its vertex angle, chord, and edge arc lengths are found along with the coordinates of each vertex. Next, we locate reference points along each PPT edge, which divides the edge into segments equal to the frequency. Two approaches to dividing the PPT edge are used: Equal-chords or Equal-arcs. The equal-chord approach produces unequal, but symmetric, points along the PPT's edges, while equal-arc produce equally spaced points and equal-length arc segments. All three edges of the PPT will have the same pattern of reference points. Based on these edge reference points, great circle planes are defined and intersected. The intersection of two great circles is a line that intersects the sphere's surface at two antipodal points. One of these points falls within the edges of the PPT's face and we ignore the other one. We can use at least four different combinations of edge reference points and great circle intersections to define a spherical grid. Each combination results in a slightly different grid.

8.5.4   Four Class I Schemas

Two approaches for defining PPT edge reference points—Equal-chords and Equal-arcs—have been described, along with a general technique for defining points on a sphere at the intersection of great circles. It is now time to combine these techniques and subdivide the PPT's face. Figure 8.10 shows how subdivision grid points result from different combinations of intersecting great circles. Each grid point will be a little different, depending on the choice of Equal-chords or Equal-arcs references and the intersection of great circles. An octahedron PPT is properly oriented with 4^v reference points located along its edges. The PPT in Figure 8.10(a) uses equal-chord reference points along its edges, and defines a new point within the PPT's face by intersecting three great circles. The planes of the three great circles are approximately parallel to a PPT edge.

In Figure 8.10(a), notice that all three great circles based on equal-chord edge reference points intersect at a single point. This means we could have intersected any two of them, and their intersection would have created the same tessellation point. We show three to emphasize the fact that for equal-chord reference points, all three will intersect at the same point. For reference purposes, we will call this subdivision schema Equal-chords. Figure 8.10(b) and (c) show two more schemas; both are based on equal-arc reference points. In (b), a grid point is defined by intersecting just two great circles. Two will always intersect and define a point within the PPT. As before, the planes of the great circles run approximately parallel to the PPT's edges.

Intersecting two great circles is certainly easier than intersecting three of them, but this simpler approach has one subtle, but significant disadvantage—the tessellation vertex

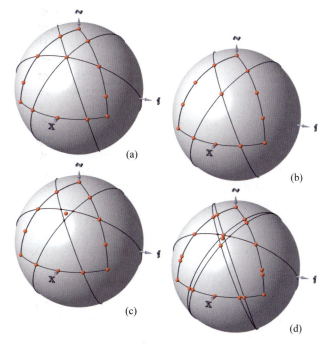

(a)

(b)

(c)

(d)

Figure 8.10. PPT edge points define vertices at intersecting great circles.

created is not rotationally symmetric around the center of the PPT. This asymmetry is too subtle to appear in Figure 8.10(b) with only one vertex, but it will become quite noticeable when we subdivide the whole PPT. This schema demonstrates how asymmetry occurs, even when all other references are symmetrical. There are applications where grid symmetry is not a concern and this technique is useful. We call this schema Equal-arcs (two great circles).

Despite the asymmetry problem just uncovered, equal-arc reference points yield superior results when we intersect three great circles, rather than just two. Figure 8.10(c) illustrates a surprising example of such situations. Unlike the great circles in the Equal-chords approach, three great circles based on equal-arc reference points do not intersect in a single line. They intersect in three lines instead, creating three pairs of antipodal points, three of which are on the PPT's face. These three points (brown) form a small triangle, which Duncan Stuart aptly called a *window*.[13] The planar version of the small window triangle is just inside the sphere; only its three vertices are on the surface.

Although the window appears to be a small equilateral triangle, most windows are acute triangles. We can use the triangle's centroid, a single point, to summarize the three window vertex points. We use the centroid because if the window was continuously scaled down until it was so small it approximated a single point, and if the ratio of its sides were preserved while doing this, it would get smaller and smaller and converge to its centroid. Thus, the centroid—the intersection point within the triangle of its three edge bisectors—is the single point that best summarizes the window's *apices*. The centroid, however, is not on the surface of the sphere, it is on the plane of the window triangle and must be projected to the surface before we can use it as a tessellation grid point. Figure 8.10(c) shows the final tessellation vertex point as a red vertex centered over the window. Although more computation is involved in this schema, it is well worth the effort. The vertices are uniformly distributed, rotationally symmetric about the PPT's center, and mirror symmetric around the altitude axis of the PPT's vertices. We call this schema Equal-arcs (three great circles). Figure 8.10(d) superimposes all three schemas to demonstrate the influence various combinations of edge reference points and great circles have on defining grid points. It is clear that each schema produces a different grid point for the same PPT and frequency.

There is another variation of Equal-arcs. It simply subdivides the equilateral PPT into four smaller spherical triangles by spanning great circle arcs between the mid-edge points of the PPT. We call this subdivision schema Mid-arcs and we will develop it later in the chapter.

We have outlined four different Class I tessellation schemas. The first defines vertices by intersecting three great circles based on selected pairs of equal-chord reference points. The second and third define vertices by intersecting two and then three great circles based on equal-arc reference points. The last schema is a recursive triangle within a triangle technique. We are now ready to apply them to subdividing a complete spherical icosahedron. Although our examples are icosahedral, they apply equally well to tetrahedra and octahedra. Only the characteristics (angles and coordinates) of their PPTs change. We will show completed examples of them at the end of the Class I section, though we do not develop their grids.

---

[13] The term "window" was first coined by Duncan Stuart in a technical report in 1952 but it did not pass into general usage until Joseph Clinton (1971, I-24), Shoji Sadao, Magnus Wenninger (1979, 93–95), and others popularized it in the literature.

8.5.5 Equal-Chords

The Equal-chords schema, as the name implies, starts with equal-chord reference points along the PPT's edge. The Equal-chords approach results in unequal arc lengths between edge reference points, when they are projected to the surface of the sphere. This is just noticeable along the PPT's edges in Figure 8.11(a) and (b). Pairs of reference points (and the origin) define great circle planes that are approximately parallel to the PPT's edges. Selected great circles intersections are shown in Figure 8.11(a) and their resulting tessellation vertices in (b). For example, great circle arcs $a1$-$b1$, $a6$-$c2$, $b3$-$c3$ intersect at a single line, creating two antipodal points. The ones that fall on the surface of the PPT are selected and shown in (b). Two other example vertices show how the process repeats itself. This process continues until every edge reference point combination has defined a PPT grid point. The completed grid, shown in Figure 8.11(b), is quite uniform. It is rotationally symmetric

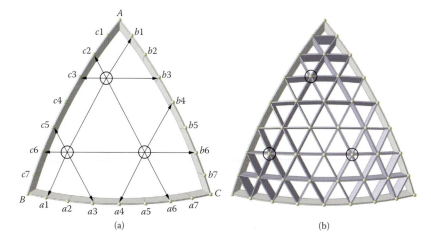

Figure 8.11. Equal-chords tessellation schema.

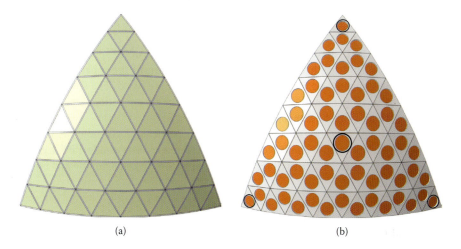

Figure 8.12. Equal-chords tessellated PPT.

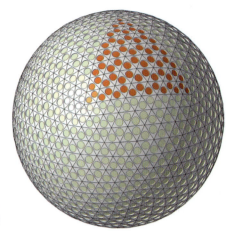

T: 64
Total Face: 1,280
Total Edges: 1,920
Total Vertices: 642

Figure 8.13. Equal-chords tessellated sphere.

about the PPT's center, its apices, and its mid-edge points. The grid is also mirror symmetric about the PPT's three altitudes. All grid points are on the surface of the sphere and define either planar or spherical triangles.

A great number of surface treatments are possible. In Figure 8.12(a), grid triangles are shown as simple planar surfaces. Panels are suggested in (b) with holes centered at each triangle's incenter. Their radius of the incircle is 80 percent of the incenter-to-edge distance; thus, each hole is equidistant to its triangle's three edges. Holes make it easier to compare face sizes and, to some extent, their shapes. An equilateral face would have a hole perfectly centered inside it, while the hole is more off-center if the triangle is skewed. Holes also tell us something about triangle sizes. The largest triangle in this schema is the equilateral triangle at the center of the PPT, while the smallest are found next to the three PPT apices; these four are highlighted in (b). Barely noticeable, the triangles along the edge increase in size slightly, as they approach mid-edge and move towards the center of the PPT. We will see this pattern in other schemas as well. We will take a closer look at face sizes and shapes in the next chapter.

The subdivided PPT, shown in Figure 8.12, is only one face of the spherical polyhedron. In Figure 8.13 the PPT has been replicated to cover the rest of the sphere. The PPT just developed is highlighted, and the total parts on the sphere are listed. The sphere is covered by rotating copies of the subdivided PPT to the other 19 face positions on the spherical icosahedron. We describe how this is done in Appendix E.

CAD users have other, but related, techniques for replicating geometry. A full sphere can be assembled by making multiple geometric references to a single PPT's geometry and projecting each reference to another place on the sphere. This technique and others are explained in Chapter 10.

8.5.6 Equal-Arcs (Two Great Circles)

We have said that two nontangent great circles always intersect at two antipodal points. When we use points along the edge of the PPT to define those intersecting great circles, one of the intersection points is always on the surface of the PPT. In Equal-chords, we used

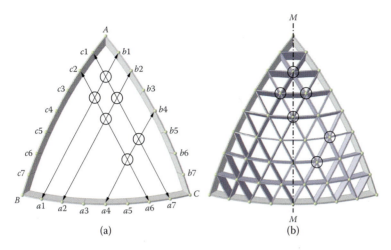

Figure 8.14. Equal-arcs (two great circles) schema.

three great circle intersections to define one tessellation vertex. We could have created the same tessellation by intersecting just two great circles. However, when equal arc reference points along the PPT's edge are used rather than equal chord reference points, the great circles intersect in unexpected ways. We will show this by first defining grid points with just two great circle intersections, and then with three. The differences between the two sets of points are surprising.

Figure 8.14 shows the icosahedron PPT with equal arc reference points along the edges. Their uniform spacing is evident. In this schema, a grid point on the face of the PPT is defined where two great circle arcs intersect. Figure 8.14(a) shows examples of intersecting great circles: c1-a7, c2-a6 with a1-b1, a2-b2, a4-b4.

Using just two great circles is an attractive option. The intersection is easy to calculate, and is a straightforward procedure. We repeat the process for each pair of great circles between reference points along edges c-a and b-a. As simple as this process is, there is an unexpected consequence, however, when you only use two great circles to define a grid point. The edge reference points in Figure 8.14(b) are labeled the same as (a). Notice that the grid points closest to edge a are closer to the edge than the corresponding grid points are to edges b and c. Every horizontal row of grid points seems somewhat displaced towards the bottom of the PPT. In fact, they are.

The biggest consequence of intersecting just two great circles (based on equal-arc reference points) is that the resulting grid points are not rotationally symmetric about the PPT's center. They are not symmetric about the PPT's edges or apices, either. There is only one axis where they are mirror symmetric, and that is the one PPT altitude line marked M-M in Figure 8.14(b). These symmetry artifacts are a direct result of using just two great circles to define a vertex. For equal chord reference points, three great circles will intersect at the same point as two great circles, but not when equal-arc reference points are used. For some applications, Equal-arcs (two great circles) asymmetry is not an issue, but for others this schema will not serve.[14]

---

[14] Rodolfo Aguilar made excellent use of this schema in "Study of the Stability of Framed, Triangulated Geodesic Domes under the Action of Concentrated Loads" (Aguilar 1964).

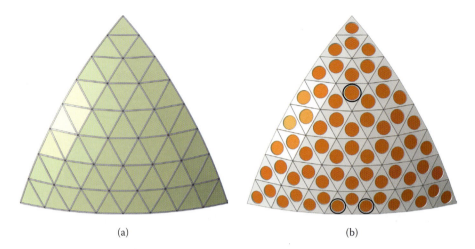

(a)   (b)

Figure 8.15. Equal-arcs (two great circles) tessellated PPT.

Figure 8.15 shows the surface treatments for a fully tessellated Equal-arcs (two great circles) PPT. They are the same ones used in Equal-chords. However, this time, the panel holes tell a different story. Just like the previous subdivision schema, each hole's radius is 80 percent of the distance between the incenter of the triangle and the closest edge of its triangle. First, the largest and smallest triangles are no longer in the center or at the PPT's apices, as they were in Equal-chords. The largest triangle now is above center, and the smallest ones are the two triangles along the PPT's bottom edge. They are circled in Figure 8.15(b).

Figure 8.16 shows a fully tessellated Equal-arcs (two great circles) sphere, along with the total part counts. This schema creates the same number of parts as Equal-chords. Although it appears quite uniform at this scale, an analysis of face size in the next chapter will show it is not as uniform as the previous Equal-chords schema. If rotational symmetry is not a requirement, Equal-arcs (two great circles) is an acceptable schema.

T: 64
Total Face: 1,280
Total Edges: 1,920
Total Vertices: 642

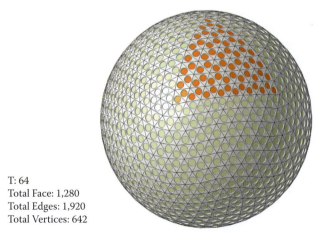

Figure 8.16. Equal-arcs (two great circles) tessellated sphere.

8.5.7   Equal-Arcs (Three Great Circles)

The previous example cast some doubts on the advantages of using equal-arc reference points to define great circle intersections. However, when three intersecting great circles are used to define grid points instead of two, the results are outstanding, yielding the most uniform tessellation gird of the three schemas thus far. Like the previous Equal-arcs (two great circles) schema, this one is not without its surprises, but they are easy to deal with.

In this schema, equal-arc reference points along the PPT's edges are selected so that pairs of them define three great circle arcs approximately parallel to their respective PPT edge. Unlike Equal-chords schemas where great circles reference equal-chord points, those based on equal-arc references do not intersect at a single line. Each pair of great circles intersects at a different line and defines its own pair of antipodal points. With three intersecting great circles, six antipodal points are created and three of them (one from each antipodal pair) fall on the surface of the PPT. Figure 8.17(a) illustrates the result.

In Figure 8.17(a), examples of four three-way intersections are circled. Taking just one of these examples, the great circle arcs between $c3$-$b3$, $c1$-$a7$ and $b2$-$a2$ intersect at three different places, creating three different points within the PPT. These points are on the surface of the sphere but they form a small window triangle rather than the single tessellation vertex we need. Figure 8.17(b) shows the 21 window triangles that result when the entire PPT is subdivided. Notice that the windows become larger and larger towards the center of the PPT; the smallest ones are the ones nearest the PPT's three apices.

Each window is actually a small triangle whose planar surface is just inside the sphere; only the apices of the window triangle are actually on the surface of the sphere. Figure 8.18(a) shows a detail of the window and its vertices. The red arcs represent the great circle arcs, and the small balls represent their intersection points. What is needed is a single point that summarizes the three window vertices. The triangle's centroid is the most appropriate choice because if the triangle were continuously scaled down while maintaining the proportions of its sides, its geometry would converge at a single point at its centroid. The centroid of a triangle is defined as the intersection of its medians. The average of its three

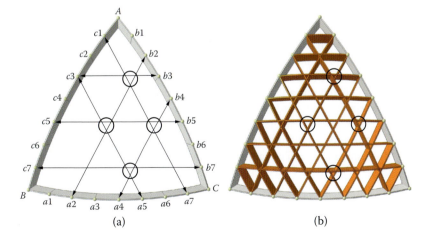

(a)   (b)

Figure 8.17. Equal-arcs (three great circles) schema.

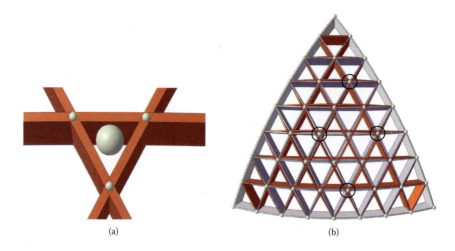

(a)   (b)

Figure 8.18. Equal-arcs windows and tessellation vertices.

vertex coordinates produces the same point we call C. To find the window's centroid point:

$$Cx = \frac{(x_1 + x_2 + x_3)}{3},$$

$$Cy = \frac{(y_1 + y_2 + y_3)}{3},$$

$$Cz = \frac{(z_1 + z_2 + z_3)}{3}.$$

We are almost done; only one small detail remains. The window's centroid is on the plane of the triangle, which is inside the sphere. The point must be projected from the origin to the surface of the sphere to create the final tessellation vertex we need. This is a simple matter of normalizing—creating a unit vector length—for the point. Appendix C explains this simple procedure. Figure 8.18(b) shows the tessellation vertices that result when all the centroids of the window triangles are found and projected to the sphere's surface. The original great circle arcs are shown in red, the final grid in blue-green. This figure highlights intersections that were used as examples in previous figures.

Figure 8.19 shows the completed PPT with the surface and panel treatments we have been using. The results are highly uniform, and are the best yet. The final grid resembles Equal-chords to some extent. The largest and smallest triangles are in the same positions as Equal-chords. The mid-PPT face triangle is also equilateral, and just as in Equal-chords, triangle areas increase as you move from the PPT apex to mid-edge, and from mid-edge to midface. The main differences between the Equal-chords and Equal-arcs (three great circles) schema are chord lengths and face angles. We compare the two schemas in the next chapter.

Figure 8.20 shows a complete sphere based on the PPT just created. Notice that its part counts are the same as the other schemas. The subdivision method did not change this fact,

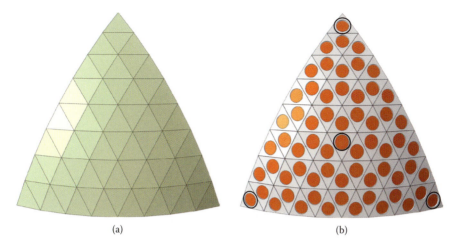

(a)   (b)

Figure 8.19. Equal-arcs (three great circles) tessellated PPT.

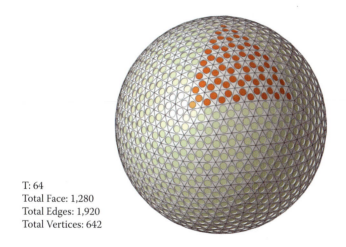

T: 64
Total Face: 1,280
Total Edges: 1,920
Total Vertices: 642

Figure 8.20. Equal-arcs (three great circles) tessellated sphere.

since the schemas all have the same *b,c* specification. Certainly, Equal-arcs (three great circles) involves more computation than either of the previous two schemas, but the results are worthwhile. Analysis will show that it has the best distribution of points and the least variation in chord lengths, angles, and face angles of any Class I schema.

8.5.8 Mid-Arcs

The earliest work with Mid-arcs was done by Jeffrey Lindsay in the early 1950s. At that time, the technique was referred to as the alternate method. We call the technique Mid-arcs because it is more descriptive of how the technique works. Deltahedra spherical faces are all equilateral. Mid-arcs takes advantage of this fact and subdivides a PPT by creating one mid-arc reference point on each of its edges. Each pair of mid-arc points defines a great

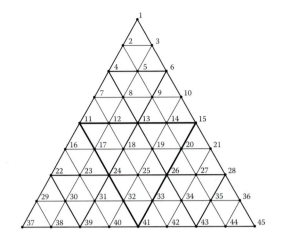

Figure 8.21. Mid-arcs vertex grid reference numbers.

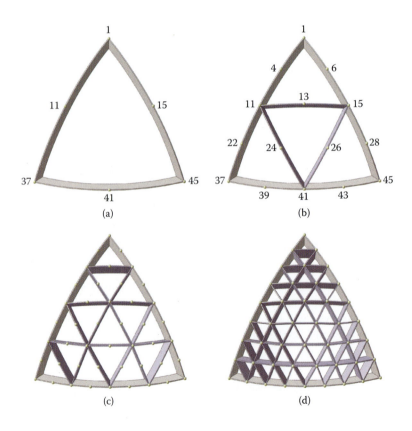

Figure 8.22. Mid-arcs tessellation schema.

circle arc across the triangle and these three arcs subdivide the triangle into four smaller ones. There is no need to intersect great circles. We use the same method of defining four new triangles within an existing one until we have subdivided the entire PPT. Each time we subdivide all the triangles in a PPT this way, we double the PPT's frequency. Mid-arcs is very easy to understand and to calculate. It is popular with cartographers and astronomers[15] who have devised ways to reference individual triangles on the sphere by taking advantage of the schema's hierarchical nesting of triangles within triangles.[16] Often, they subdivide a spherical octahedron rather than an icosahedron because the octahedron results in natural great circles that can be aligned with the earth's equator and Greenwich Meridian.

Figure 8.21 shows how Mid-arcs work. The pattern of triangles within triangles is evident. Each successive layer of triangles is based on the mid-arc points of an outer triangle. This recursive nesting of triangles within triangles is the essence of Mid-arcs.[17]

Figure 8.22 applies the schema to the same icosahedral PPT we have been using all along. We number the vertices from top to bottom, left to right so that Figure 8.21 and Figure 8.22 can be compared. As usual, subdivision begins at the PPT edges, where each has a single mid-arc reference point.[18] In Figure 21(a), they are numbered 11, 15, and 41. These points can be found by projecting the mid-chord point between 1 and 37 to the surface of the sphere or finding the middle of the arc between the PPT's apices. Either way, great circle arcs between these points define four new subtriangles (1-11-15, 11-37-41, 11-41-15, and 15-41-45), each of which has mid-arc points (4, 6, 13, 22, 24, 26, 28, 39, and 43); see Figure 8.22(b). These new mid-arc points define 12 more edges and 30 new vertex points when all edges are considered (c). This process repeats itself once again until a total of 42 tessellation vertices and 108 edges are defined. Figure 8.22(d) shows the complete tessellation. Keep in mind that all of these vertices and edges were derived from the apices of

T: 64
Total Face: 1,280
Total Edges: 1,920
Total Vertices: 642

Figure 8.23. Mid-arcs tessellated sphere.

---

[15] (Dutton 1996, 1991) and (Szalay et al. 2005).

[16] (Fekete 1990)

[17] Cartographers sometimes use quaternary subdivision meshing schemas that hierarchically subdivide equilateral triangles similar to Mid-arcs (Dutton 1996). See also Section 3.3.

[18] The edge midpoints of the icosahedron (or of the dodecahedron) are the vertices of the icosidodecahedron (20 triangular faces and 12 pentagonal faces). See Chapter 6 for examples.

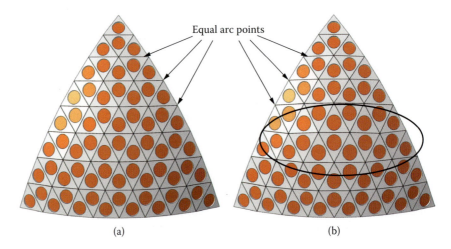

Equal arc points

(a)   (b)

Figure 8.24. Equal-arcs versus Mid-arcs PPT tessellations.

the original PPT and the mid-arc reference point along each edge. Calculations have to be precise; otherwise, errors will accumulate. Our example stops at 8ᵛ, but we could keep on subdividing each triangle indefinitely, doubling the frequency each time.

The Mid-arcs schema is straightforward and easy to calculate. However, there are some limitations. Unlike the other Class I schemas, Mid-arcs can only tessellate frequencies that are powers of 2 (i.e. 2ᵛ, 4ᵛ, 8ᵛ, 16ᵛ, 32ᵛ, etc.). Any even or odd frequency is possible with the other Class I schemas, but only even ones are possible with Mid-arcs. The second drawback concerns the tessellation faces themselves. Despite the fact that we start with an equilateral PPT face and place reference points only in the middle of each of its edges, the vast majority of subdivision triangles created in the final tessellation are not equilateral. Only those faces whose centroids are tangent to the PPT's own centroid are equilateral; all others are slightly isosceles or acute. In this example, only triangles 1-37-45, 11-15-41, 13-24-26, and 18-25-19 are equilateral.

A few numbers illustrate this last point. The icosahedron's PPT edges are each 63.43495°. The first new mid-arc triangle (11-15-41) is also equilateral with 36° edges (its centroid is tangent to the PPT's centroid), but the other three triangles created (1-11-15, 11-27-41, and 15-41-45) are slightly isosceles. Two of their edges are 31.71747°, and the other edge shared with the mid-PPT triangle is 36°. Although other Class I schemas have this same characteristic, Mid-arcs produce more variation in chord length and triangle area than others do. We will see this later in the next chapter. Figure 8.23 shows a full sphere subdivided by the Mid-arcs technique along with its part counts. The PPT we subdivided is highlighted.

Figure 8.24 illustrates another disadvantage of Mid-arcs tessellation (left) by comparing its grid side by side with an Equal-arcs (three great circles) tessellation for the same PPT. Although the PPT edge reference points are the same in both schemas, Mid-arcs faces increase in size more quickly as they approach the center of the PPT, and the chords change length faster as they approach the middle of the PPT. As a result, sequences of chords that cross the PPT are not as smooth in transition as they are with Equal-arcs. An example run of chords showing this difference is circled. Figure 8.25 makes the same comparison, but

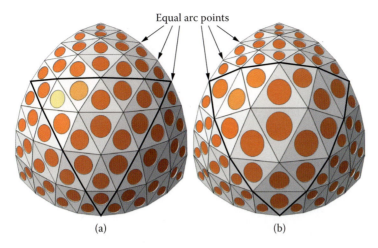

Equal arc points

(a)   (b)

Figure 8.25. Tetrahedral Equal-arcs versus Mid-arcs PPT tessellations.

uses tetrahedral PPTs instead. Tetrahedron PPTs are four times larger than an icosahedron's face (not to scale in the figure), and this makes it easier to see the differences between the two schemas. The first-level subdivision of the PPT is highlighted in each illustration, and the differences are pronounced. Mid-arcs central PPT face is noticeably larger and the faces near the PPT apices are smaller than they are in Equal-arcs tessellations. In Figure 8.25(b) the tetrahedron's central equilateral triangle (highlighted) has three times as much area as the ones at each apex. The tetrahedron's equilateral faces have significantly more spherical excess than the icosahedron's and highlight these differences.

Mid-arcs is the simplest subdivision technique presented thus far. The fact that frequencies are always powers of 2 and the repetitive definition of triangles within triangles offers a natural hierarchy that can be exploited to reference individual triangles on the sphere.[19] Mid-arcs also demonstrates a basic spherical principle—there are no similar triangles on a sphere. You cannot subdivide a spherical equilateral triangle around its mid-edge points and define four other equilaterals as you can in plane geometry.

8.5.9   Subdividing Other Deltahedra

We have developed four different Class I schemas in this chapter. Although each example subdivides a spherical icosahedron, the same methodology applies to tetrahedra and octahedra. The only difference is the initial PPT's characteristics (vertex angle, edge arc length, and vertex coordinates) used to start the process. Use the values in Table 8.2 to subdivide the others.

Figure 8.26 shows examples of $8v$ equal-arc schemas for the tetrahedron and octahedron (geodesic notation $\{3,3+\}_{0,8}$ and $\{3,4+\}_{0,8}$, respectively). They are shown in a partially disassembled state and with mixed surface treatments. Both of these polyhedra have larger PPT areas than the icosahedron for the same radius sphere. As a result, their tessellation triangles show much more variation in tessellation triangle area and shape, as the incenter circles clearly show. The tetrahedron shows the greatest variation of the two. These examples clearly demonstrate how important the choice of foundation polyhedra is when the

---

[19] (Fekete 1990)

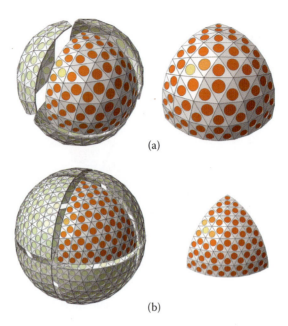

(a)

(b)

Figure 8.26. Equal-arc (three great circles) tessellations of tetrahedra and octahedra.

goal is to create the most uniform subdivision possible. It is no wonder the icosahedron is the most frequently used polyhedral base.

8.5.10 Summary

Class I schemas were among the first spherical polyhedra subdivision methods developed. They apply to any deltahedra. Except for Mid-arcs, which is limited to frequencies that are powers of 2, any even or odd frequency is possible and, depending on prime factors, may contain subgrids at harmonic frequencies. Class I subdivisions maintain the polyhedron's edge as part of the final tessellation grid, though the edge is segmented. The geodesic notation for Class I is distinctive: either the b or c coefficient is always 0.

Class I tessellations begin with Equal-chords or Equal-arcs reference points along the PPT's edges. Pairs of points (and the origin) define intersecting great circles that are approximately parallel to the PPT's edges. Grid points result where their line of intersection pierces the sphere's surface:

- *Equal-chords*. Based on equal-chord reference points, great circles parallel to the PPT's edges are intersected and their intersection points define points on the tessellation grid within the PPT's surface. Two intersect at the same line as three.

- *Equal-arcs (two great circles)*. Based on equal-arc reference points, two great circles are intersected. They intersect at a single line and define individual tessellation points on the PPT's surface. This procedure is computationally simple, but the resulting tessellation vertices are not rotationally symmetric and only mirror symmetric around one of the PPT's altitude axis.

- *Equal-arcs (three great circles)*. Based on equal-arc reference points. Three great circles are intersected, but they do not meet at a single line. Instead, they meet at three different lines, creating three points on the PPT. A small window triangle results and the window's centroid is projected to the surface of the sphere. The resulting grid is rotationally symmetric around the center of the PPT, and they are mirror symmetric around the PPT's three altitudes. Analysis will show that Equal-arcs, when developed with three great circles, produces a highly uniform grid, the best of the Class I schemas presented.

- *Mid-arcs*. Grids are generated by using the mid-arc points on a triangle's three edges to define four others within. The technique is simple to calculate, rotationally and reflection symmetric, and applicable to all deltahedra. The hierarchical nature of triangles nested within triangles offers convenient referencing systems for individual triangles. The grid is quite uniform, but not as good as Equal-arcs (three great circles).

All four schemas result in a three-way grid over the PPT, where arcs and lines approximately parallel the PPT's edges. If the triangulation number is evenly divisible by three, selected grid points can define hexagons, pentagons, or diamonds. The 8^v examples in this chapter have a triangulation number of 64, which is *not* evenly divisible by three; thus, grid points will not define uniform patterns of these other polygons. However, a wide variety of surface treatments is possible. The examples shown here are simple surfaces and panels with incenter holes. Chapter 10 shows other treatments.

8.6   Class II: Triacon

Class II subdivisions create grids that are perpendicular to the edges of the spherical polyhedra, rather than parallel as with Class I grids. Figure 8.27 shows examples of Class II tessellations. Notice that the tessellation grid is symmetrical to the polyhedron's edges and face medians. This means that Class II tessellations are always even frequencies. For this reason, Class II geodesic notation is easily recognized: b always equals c, and neither is equal to zero. Notice, too, that selective use of grid points produce triangles, diamonds, hexagons, or pentagons. In this section, we will continue to use a spherical icosahedron 8^v as our foundation polyhedron because we want to compare our results with the other schemas; its geodesic notation will be $\{3,5+\}_{4,4}$.

There are many ways to create Class II tessellations. The schema developed here is called *triacon* and was developed by Duncan Stuart in 1952 to make it easier to prefabricate geodesic domes by reducing the number of different parts needed.[20] In the next chapter we will see just how innovative Stuart's schema is and why it is widely used today.

The triacon is unique. It achieves a remarkably uniform distribution of points over the sphere and does so with the fewest number of different chord/arc lengths, face shapes, and face areas of any subdivision method for a given frequency. These are key traits when spheres are manufactured or assembled from piece parts. This accounts for the triacon's popularity with dome builders and why most large geodesic structures still use the triacon method. The triacon is based on the subdivision of a single Schwarz triangle, rather than an equilateral triangle as with Class I. When the Schwarz triangle is subdivided and the

---

[20] (Stuart 1952)

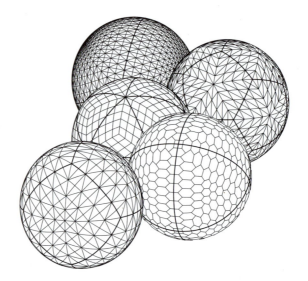

Figure 8.27. Examples of Class II subdivision schemas.

resulting geometry mirrored two times, it forms a diamond shape, which is positioned symmetric to the polyhedron's edge.[21] The fact that triacon is based on a diamond shape that is edge-symmetric to the polyhedron gives us an advantage; triacon can subdivide *any* of the five spherical Platonic solids, not just deltahedra. Another advantage of triacon is its triangulation numbers are always evenly divisible by three. This means its triangular tessellation grid can produce regular layouts of triangles, diamonds, hexagons, and pentagons, as we see in Figure 8.27. From back to front: tetrahedron 16^v, dodecahedron 8^v, icosahedron 10^v, octahedron 22^v, and cube 8^v.[22] The triacon does not create grid members coincident to the polyhedron's edges like Class I grids do. Instead, the polyhedron's edges are simply geometric references used to compute the grid. In Figure 8.27, the polyhedron's edges are shown only to highlight the underlying polyhedra and point out that selected grid points along the edge and face medians also define hemispheric great circles. They can be useful in certain geodesic applications. Class II grids always result in a grid point in the middle of the polyhedron's face. This is not always the case with Class I schemas, and only occasionally happens in Class III schemas. This can be a useful feature in some spherical applications.

8.6.1   Schwarz LCD Triangles

A finite number of Schwarz triangles cover a sphere without overlap or gaps. On a spherical icosahedron, a single Schwarz triangle covers one-sixth of a PPT's face or 120th of the total spherical icosahedron surface; 60 of them are left-handed and 60 are right-handed. There are 24 Schwarz triangles on a spherical tetrahedron, 48 on a cube, 48 on an octahedron, and 120 on a dodecahedron. The Schwarz triangle is the LCD we will subdivide.

---

[21] Triacon diamonds for an icosahedron resemble the faces of the rhombic triacontahedron polyhedron, the likely source for this nickname.

[22] From back to front, the geodesic notations are: $\{3,3+\}_{12,12}$, $\{3,5+\}_{4,4}$, $\{3,5+\}_{5,5}$, and $\{3,4+\}_{11,11}$.

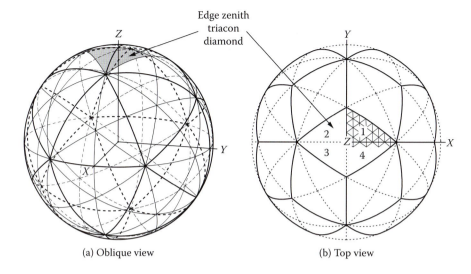

Figure 8.28. Edge-zenith triacon diamond.

We begin subdivision by orienting the spherical icosahedron edge-zenith and an edge coincident with the *xz*-plane. Figure 8.28(a) shows an oblique view of the edge-zenith diamond. A top view of the same edge-zenith diamond is shown in (b) where its four quadrants are numbered 1 to 4. A partial triacon grid is developed in quadrant 1 of the edge-zenith diamond. Orienting the spherical Platonic edge-zenith with the diamond axis coincident to the *xz*-plane makes calculations easier. First, the LCD triangle is a right triangle and easily solved with spherical trigonometry. Second, the grid coordinates we calculate for quadrant 1 are all positive and can be mirrored to the other three diamond quadrants, simply by changing their *x*- and/or *y*-coordinate sign; their *z* values never change. For example, changing only the sign of the *x*-coordinate, a point in quadrant 1 is reflected to a symmetrical position in quadrant 2. Changing the *y* sign reflects points in quadrants 1 and 2 to quadrants 4 and 2, respectively. This is a pretty good return on our subdivision effort. Calculate one quadrant, and the others will fall into place if you simply change the signs of various *x*- and *y*-coordinates.

8.6.2 How Frequent

Class II schemas create grids that are perpendicular to the polyhedron's edges, and the grid is symmetrical around the face's medians. Because of this symmetry, Class II frequencies are always even and their geodesic notations b,c are always positive and equal integers. If a Class II frequency is greater than 2^v, it may contain harmonics or subgrids just like Class I grids. Figure 8.29(a) shows a sphere with a series of triacon diamonds for 2^v, 4^v, 8^v, 16^v, and 32^v in clockwise order. Each diamond is made of four LCDs and the icosahedron's edges are shown in the figure as dashed lines. Note that the 32^v tessellation contains all the vertices of the others grids; thus, they are all harmonics of the 32^v grid. Figure 8.29(b) shows a single 16^v triacon diamond with different line weights for the 2^v, 4^v, and 8^v harmonics it contains.

To determine if a tessellation has harmonics, start by listing its prime factors. The tessellation will have a harmonic for even products of prime factor pairs. Table 8.3 lists the

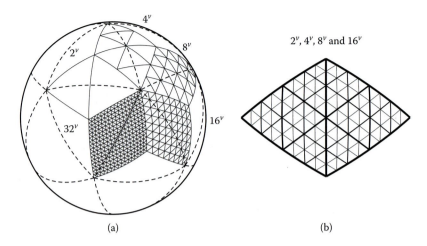

Figure 8.29. Triacon diamond subdivision harmonics.

ν	Prime Factors	Harmonic Frequencies				
2	2 (Prime)					
4	2,2	2^ν				
6	2,3	2^ν				
8	2,2,2	2^ν	4^ν			
10	2,5	2^ν				
12	2,2,3	2^ν	4^ν	6^ν		
14	2,7	2^ν				
16	2,2,2,2	2^ν	4^ν	8^ν		
18	2,3,3	2^ν	6^ν			
20	2,2,5	2^ν	4^ν	10^ν		
22	2,11	2^ν				
24	2,2,2,3	2^ν	4^ν	6^ν	8^ν	12^ν
26	2,13	2^ν				
28	2,2,7	2^ν	4^ν	14^ν		
30	2,3,5	2^ν	6^ν	10^ν		

Table 8.3. Triacon frequency harmonics.

harmonics for frequencies 2^ν through 30^ν. Notice that if you create geometric information for 16^ν through 30^ν tessellations, it will contain the same information for 2^ν though 14^ν tessellations, since these grids are also contained in the others.

8.6.3   A Quick Overview

The triacon subdivision method is one of the easiest to implement. Figure 8.30 shows the sequence in a series of top views (from the positive z-axis) of the LCD and evolving triacon diamond. The basic procedure is as follows:

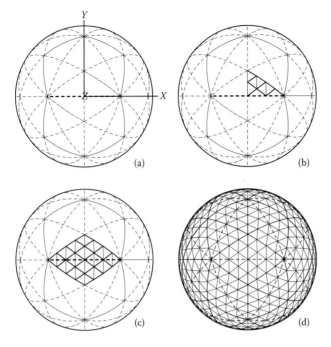

Figure 8.30. Triacon subdivision sequence.

- Orient the spherical Platonic solid (an icosahedron in our examples) with an edge at the zenith and coincident to the xz-plane, as shown in Figure 8.30(a). Four LCDs, two left- and two right-handed, form a diamond that is symmetric to the zenith-edge.

- Subdivide the LCD in quadrant 1 of the edge-zenith diamond, Figure 8.30(b). The LCD is a spherical right triangle, and the bases of the smaller right triangles are multiples of equal-arc subdivisions of the icosahedron's edge. Simple trigonometric relationships define every tessellation grid member; compute the coordinates of each of the tessellation vertices in the LCD, shown in Figure 8.30(b), and mirror the grid point coordinates to quadrants 2, 3, and 4 by changing the appropriate x or y signs of quadrant 1 vertices. This completes the definitions for the LCD and the edge-zenith diamond, Figure 8.30(c).

- Tessellate the remainder of the sphere by rotating copies of the edge-zenith diamond to all other polyhedral edge positions, Figure 8.30(d). Spherical icosahedra and dodecahedra will require 29 replications/rotations. Octahedra and cubes require 11 and tetrahedra require 5. Appendix E describes how this is done.

8.6.4   Establish Your Rights

The first step in triacon subdivision is to define the spherical polyhedron's LCD. Figure 8.31 shows the LCDs covering the spherical icosahedron and the way we label the

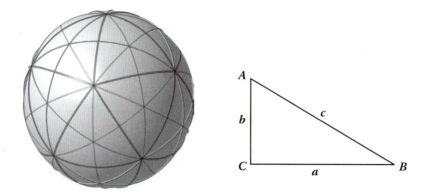

Figure 8.31. LCD right triangle.

parts of a typical right triangle. Whether or not the LCD is left- or right-handed does not matter in our calculations. The first step is to find LCD sides a, b, and c, given its three surface angles. Referring to the LCD triangle labels in Figure 8.31, the basic relationships are

$$\cos a = \cos A \, / \sin B,$$

$$\cos b = \cos B \, / \sin A, \text{ and}$$

$$\cos c = \cos a \times \cos b.$$

It's easy to establish the LCD's surface angles A, B, and C. Through symmetry and knowing that the sum of the surface angles around any point is always 360°, we know that $A = 60°$ and $B = 36°$, and, by definition, $C = 90°$. We now use the formulas in Figure 8.31 to find sides (central angles) a, b, and c. They are 31.7175°, 20.9052°, and 37.3774°, respectively. We only show a few decimal places here, but it is important to maintain the highest precision you can in your calculations.

We also need the arc and chord factors for the LCD's sides as well as the full PPT polyhedral edge and face meridian arc length. The arc factors for any of the LCD's sides are simply the radian value of their central angle, or $2\pi\theta/360$ where θ is the arc's central angle expressed in degrees. The chord factor between two points on the unit sphere is the straight-line distance between them and is simply $2r \sin(\theta/2)$ where r is the radius of the sphere and θ is the central angle between the two points. Given LCD sides, the polyhedron's edge and face medians are simply

$$\text{PPT polyhedron edge} = 2a, \text{ and}$$

$$\text{PPT face median} = b + c.$$

Table 8.4 summarizes the principal LCD parts for the icosahedron and the other four spherical Platonic solids. Angles are shown in degrees and arc/chord factors are for unit radius spheres.

Principal LCD Parts for Spherical Platonics					
Part	Tetrahedron	Octahedron	Cube	Dodecahedron	Icosahedron
Surface Angles					
A°	60.00000	60.00000	45.00000	36.00000	60.00000
B°	60.00000	45.00000	60.00000	60.00000	36.00000
C°	90.00000	90.00000	90.00000	90.00000	90.00000
Central Angles					
a°	54.73560	45.00000	35.26439	20.90517	31.71749
b°	54.73560	35.26439	45.00000	31.71749	20.90517
c°	70.52877	54.73510	54.73561	37.37738	37.37738
Arc Factors					
a	0.95531	0.78540	0.61548	0.36486	0.55257
b	0.95531	0.61548	0.78540	0.55357	0.36486
c	1.23096	0.95532	0.95531	0.65236	0.65236
Chord Factors					
a	0.91940	0.76537	0.60581	0.36284	0.54653
b	0.91940	0.60581	0.76537	0.54653	0.36284
c	1.15470	0.91940	0.91940	0.64089	0.64085
Full Edge	109.47118	90.00000	90.00000	41.81032	63.43350
Full Face Median	125.26434	90.00000	NA	69.09486	58.28254

Table 8.4. Principal LCD parts for spherical Platonic solids.

8.6.5   Subdividing the LCD

The triacon schema subdivides the LCD by nesting a series of right spherical triangles within it. Figure 8.32 shows each step for an eight frequency grid.

- *Step 1.* Define edge perpendiculars. Divide half the polyhedron's edge (arc) into $^v/2$ segments and drop perpendiculars to side a between every edge segment. The result is $^v/2$ nested right triangles, all sharing angle B. In this example, there are three nested right triangles within the overall LCD right triangle. We will label their angles and sides the same way we labeled the overall LCD triangle. Side a of each nested right triangle is an integral number of polyhedron edge segments. In the illustration, it appears that each of the nested triangles has the same angle A, but this is not the case. There are no similar triangles on a sphere. Each nested right triangle has a slightly different value for A, but angles B and C remain constant at $36°$ and $90°$, respectively. They are shown in Figure 8.32(a). Given a, B, and C for each nested triangle, we find parts A, b, and c using these relationships:

$$\cos A = \cos a \times \sin B,$$
$$\cos b = \cos B / \sin A,$$
$$\cos c = \cos a \times \cos b.$$

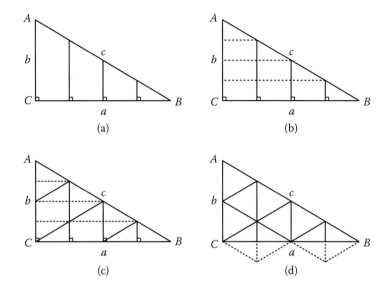

Figure 8.32. Triacon LCD tessellation sequence.

- *Step 2.* Subdivide LCD's side *b*. With all nested triangle parts solved, divide LCD's side *b* into arc segments equal to the *b* sides of each of the nested right triangles found in step 1. Each of the LCD side *b* segments is slightly different; there will be $^v/2$ segments in all. See Figure 8.32(b).

- *Step 3.* Strike diagonals. Triangulate the grid with diagonals by alternating nested triangle *c* arcs across the LCD. Each row of *c* arcs is the same length. The result is a series of triangles, each with their own *a, b,* and *c* arcs, as shown in Figure 8.32(c).

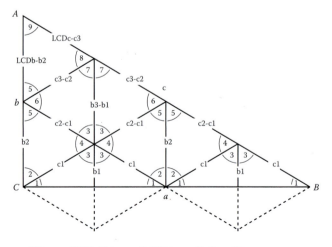

Figure 8.33. Triacon LCD tessellation 8^v summary.

- *Step 4*. Complete the triangulation. Diagonals are constructed with nested triangle *c* arcs to alternate points, as shown in Figure 8.32(d). When completed, all interior tessellation vertices have three-way intersections of two diagonals and a perpendicular to side *a*. Sides *a* and *c* of the overall LCD triangle will be subdivided $^v/2$ times, while *b* is subdivided $^v/4$ times.

Step 1 solves all six parts (three angles, three sides) of each nested right triangle within the LCD. Steps two through four replicate the parts across the LCD to create the three-way grid. The elegance of the triacon schema lies in the fact that parts *A*, *b*, and *c* of the nested triangles are replicated across the LCD to define the angles and sides of the other triangles in their respective rows. Figure 8.33 shows how the replication works. Starting at the smallest nested right triangle, the one immediately next to LCD vertex *B*, notice how its angles 1 and 3 and side *b*1 carry across the LCD to all other triangles in its row. Notice also that angle 2 is the complement of angle 1. Other angles are found this way, they are either complements or *supplements* of one or more nested triangle parts. For example, angle 4 is the supplement of two angle 3s. Some diagonal c arcs are found by factoring out c arcs from another nested triangle. The figure shows this easy step.

8.6.6   Grid Points

The triacon subdivision of the LCD has a special effect on the alignment of grid points within the LCD. Figure 8.34 shows that every grid point falls at the intersection of one great circle with one lesser circle, not two or three great circles shown in Class I schemas. Notice also that the planes of the lesser circles are parallel to the *xz*-plane of the zenith diamond (the great circle of the icosahedron's polyhedral edge), a result of replicating side *b* of each nested right triangle across the LCD during subdivision.

It is easy to find the Cartesian coordinates of grid points in the triacon diamond. Only two angles, *a* and *b*, are needed for each point and we already have their values when we

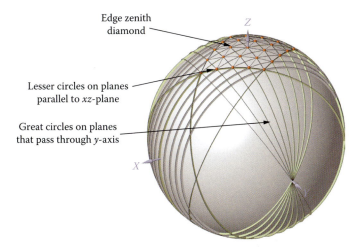

Figure 8.34. Triacon 8^v diamond great and lesser circle planes.

solved the parts of each nested right triangle in the previous step:[23]

$$x = \sin a \times \cos b,$$
$$y = \sin b,$$
$$z = \cos a \times \cos b.$$

Angle a is measured along the xy-plane and is positive, if it is on the $+x$-axis side of the plane; otherwise, it is negative. Angle b, measured on the yz-plane, is positive, when it is on the $+y$-axis side and negative, if on the $-y$ side. To find vertex coordinates for grid points in quadrants 2, 3, and 4, simply change the signs of the coordinates in quadrant 1. The z-coordinate for every point in the edge-zenith diamond will always be positive. Appendix C shows how and Appendix D lists the coordinates for the triacon diamond to higher precision and for all four quadrants of the diamond.

8.6.7 Completed Triacon

We have completely defined the triacon grid. Every triangle is solved and we have found the coordinates of every grid vertex. Figure 8.35 shows the triacon diamond and highlights the LCD in quadrant one (b). The overall diamond tessellation is quite uniform, and only four different triangles are needed (there are left- and right-handed versions of each).

Figure 8.36 shows a fully subdivided sphere and the part totals for an 8^v subdivision of an icosahedron. Notice that the total number of faces, edges, and vertices is far fewer than it is for any of the Class I schemas: 960, 1,440, and 482, respectively, versus 1,280, 1,920, and 642. The completely subdivided sphere is the result of replicating the single tessellated diamond over the remaining 29 icosahedral edges. The same graphics conventions used before are used here. The initial triacon diamond in its edge-zenith position is highlighted.

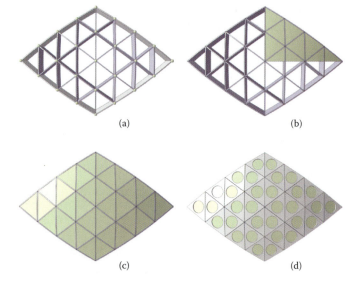

(a)   (b)

(c)   (d)

Figure 8.35. Triacon tessellation schema.

---

[23] These equations are similar to the general spherical coordinate system (ρ, φ, λ) described in Chapter 4.

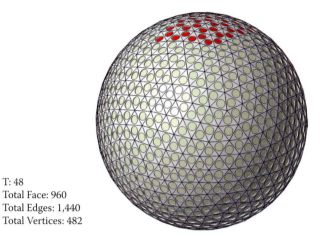

T: 48
Total Face: 960
Total Edges: 1,440
Total Vertices: 482

Figure 8.36. Triacon tessellated sphere.

8.6.8   Subdividing Other Polyhedra

The triacon subdivision is the only schema that can subdivide every Platonic solid. The triacon can do this because its basic subdivision unit is a diamond symmetrical with the polyhedron's edge. Figure 8.37 shows all five spherical Platonic solids with 8^v triacon tessellation. A single subdivided edge-zenith diamond and all polyhedral edges, shown in red, are superimposed on a fully tessellated sphere. From top down, left to right they are tetrahedron (a), octahedron (b), cube (c), dodecahedron (d), and icosahedron (e). The subdivision technique is identical in each case; only the initial LCD triangle characteristics differ.

Notice the difference in tessellation face orientations on each Platonic solid's grid. The octahedron and icosahedron, (b) and (e), have uniform orientations and appear almost like faceted spheres. By comparison, the tetrahedron's faces appear to be quite angular to one another. Also notice that, even at the small scale of these illustrations, it is clear that the icosahedron has the most uniform subdivision grid of the five solids. Since uniform face orientations and uniform grid are important in some geodesic applications, we will analyze them in the next chapter.

8.6.9   Summary

The Class II (triacon) subdivision method results in grids that are perpendicular to the edges of a spherical polyhedron's faces and symmetrical to face medians. As a result

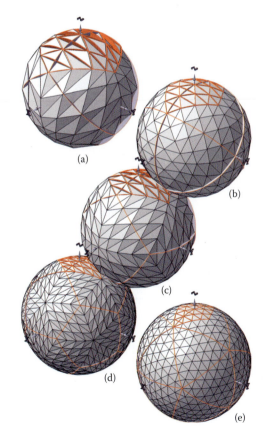

Figure 8.37. Triacon reference diamonds for Platonics.

of this symmetry, frequencies are always even and tessellations will contain harmonics for the products of any of the frequency's prime factors that are not themselves prime frequencies. Triacon subdivisions have a distinctive geodesic notation: b,c specifications are always equal, and neither is ever zero.

The principal subdivision unit is a spherical right Schwarz triangle, also referred to as an LCD. Four LCD triangles, two left-handed and two right-handed, define a diamond that is symmetric to each edge of the polyhedron. Due to the edge symmetry of the triacon diamond, this subdivision schema can be applied to every spherical Platonic solid. The subdivision process is identical for any Platonic solid:

- Select the spherical Platonic solid and desired subdivision frequency.

- Orient one edge at the zenith and coincident with the xz-plane.

- Define the basic LCD triangle characteristics (angles and edges).

- Compute surface and interior angles for a series of right spherical triangles within the LCD.

- Compute arc and chord factors for the same.

- Replicate the b and c arcs across the LCD to complete the grid.

- Complete triangle angles and edges by finding complements and supplements.

- Establish the naming convention for tessellation vertices based on a naming hierarchy of the diamond reference, the quadrant within the diamond, and the row/column of the grid point.

- Compute xyz's of all vertices in the first diamond quadrant and mirror them to quadrants 2, 3, and 4 by changing the signs of their x- and y-components.

- Replicate the completed edge-zenith diamond to all other parts of the sphere.

It is worth mentioning that there are other Class II schemas. The triacon, developed by Duncan Stuart, divides the LCD's base (tangent to the polyhedral edge) into equal arc segments; see Figure 8.32(a). However, it is also possible to create Class II grids by evenly dividing side b or c of the LCD.[24] At first, you might think the results would be the same as triacon, but remember, there are no similar triangles on a sphere. Any tessellation based on equal-arc subdivision of these other LCD sides will produce a different grid.

The triacon has been popular with geodesic dome builders because it minimizes the number of parts for any given frequency when compared to other class schemas. The triacon creates a very uniform distribution of vertices on the sphere, which results in the fewest number of different chords, triangles, areas, and shapes. A wide variety of face types—including triangles, diamonds, and hexagon-pentagons—is possible, because all Class II subdivisions have triangulation numbers, evenly divisible by three.

---

[24] (Clinton 1971, I-26–I-35)

8.7   Class III: Skew

Perhaps the most unusual spherical subdivisions are Class III subdivisions, or *skew* as they are sometimes called. Class III is quite different from the other classes. Tessellations are not parallel or perpendicular to the spherical polyhedron's edges; they cross at oblique angles, as shown in Figure 8.38. Because of this, their reflections are not symmetrical, either. Unlike Class I and II schemas, Class III schemas do not have convenient equal-chord and equal-arc reference points that can be used to start the subdivision process; thus, we will use a totally different technique to subdivide the PPT.

The method we will use here (there are others) is projection. We start by making a 2D triangular grid that meets the geodesic specification. This grid is transformed to three dimensions and is projected onto the surface of the PPT. This chapter describes how the initial 2D trigrid is defined, converted into a 3D grid, and then projected onto the surface of a PPT. Although triangular grids are developed here, the same technique can project a wide variety of nontriangular grids and designs onto the sphere. The projection technique can create Class I and II grids as well, though the results are not as optimal as the methods already presented for those classes. Optimization aside, projection is the most versatile technique we present.

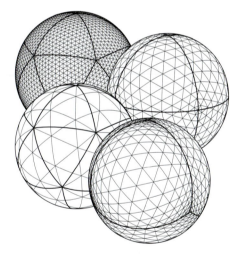

Figure 8.38. Examples of class III subdivision schemas.

8.7.1   What Is Class III?

Like the other class subdivisions, Class III subdivisions apply to spherical deltahedra—tetrahedra, octahedra, and icosahedra. Class III subdivisions are hybrids: their grids are oriented somewhere between Class I and II subdivisions. Tessellation frequencies can be even or odd, but their geodesic coefficients b,c are never equal to each other (Class II) or one of them equal to zero (Class I). Figure 8.38 shows some of the many Class III possibilities. From back to front they are icosahedron $\{3,5+\}_{12,8}$, octahedron $\{3,4+\}_{7,5}$, icosahedron $\{3,5+\}_{2,1}$, and tetrahedron $\{3,3+\}_{5,13}$. In the figure, the spherical polyhedron's edges are highlighted to show the tessellation orientation relative to the polyhedron's edges, but the polyhedron's edges are not part of the tessellation. Even without these accented edges, the polyhedral base can be identified by locating points where the grid valence changes. The change occurs at each polyhedral vertex. In triangular tessellations, the majority of grid points are *hexavalent*; that is, six grid members meet at a grid point. However, a different valence occurs when a tessellation point coincides with a polyhedron's vertex, and these grid points will be pentavalent (5), tetravalent (4), or trivalent (3) at the respective vertices of spherical icosahedra, octahedra, or tetrahedra. We can see the valence changes in Figure 8.38 if we ignore the polyhedral edges, which are not part of the Class III tessellation. Notice that most grid points on any sphere are hexavalent, but they change to a lower valence at a polyhedral vertex.

8.7.2   Snubbed Relatives

Class III subdivisions of the octahedron and icosahedron are related to the snub cube and snub dodecahedron.[25] Figure 8.39 shows them in symmetric positions where the octahedron's 6 vertices align with the snub cube's 6 square faces and the icosahedron's 12 vertices align with the 12 pentagonal faces of the snub dodecahedron. If the square and pentagonal snub faces were triangulated, the tessellations would have Class III geodesic notations $\{3,4+\}_{1,2}$ and $\{3,5+\}_{2,1}$. The snub cube's six square faces circumscribe and center over the octahedron's vertices in Figure 8.39(a). The snub dodecahedron's faces circumscribe the icosahedron in Figure 8.39(b), and its pentagonal faces are centered over the icosahedron's vertices.

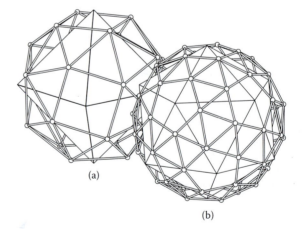

(a)

(b)

Figure 8.39. Snub cube and dodecahedron, octahedron, and icosahedron.

8.7.3   Enantiomorphs

One of the most distinguishing features of Class III subdivisions is their asymmetry. Enantiomorphs are geometric forms that are the mirror images of each other. Class III grids are rotationally symmetric, but their reflection symmetry is enantiomorphic or opposite-handed. Figure 8.40 compares the reflections of two icosahedron 8^v subdivisions. The top row

---

[25] Snub cube—24 vertices, 60 edges, 6 square faces, and 32 triangular faces. Snub dodecahedron—60 vertices, 150 edges, 80 triangular faces, and 12 pentagonal faces. Schläfli symbols $S\{3/4\}$ and $S\{3/5\}$, respectively. The tetrahedron has no snub relative. See Section 6.4.6.

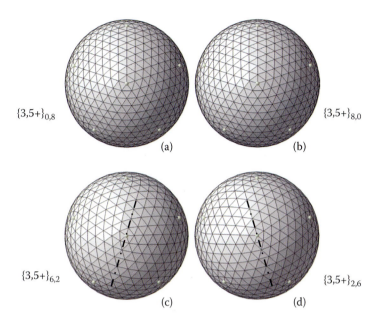

{3,5+}$_{0,8}$

{3,5+}$_{8,0}$

(a)

(b)

{3,5+}$_{6,2}$

{3,5+}$_{2,6}$

(c)

(d)

Figure 8.40. Reflection symmetry and enantiomers.

(Figure 8.40(a) and (b)) is Class I, and their left-right reflections are perfectly symmetric. However, left-right reflections for the Class III tessellation in the bottom row (Figure 4.40(c) and (d)) are opposite-handed. The { }$_{b,c}$ specifications, each Class I and Class III examples, are simply the reverse of each other. In the Class I example, reversing the { }$_{b,c}$ specification has no effect on the grid layout; however, this reverses the handedness of the Class III grid in Figure 8.40(b).

8.7.4 Harmonics

Just like the other class subdivisions, Class III subdivisions can contain harmonic frequencies. This can occur if b and c are not prime numbers. The number of evenly spaced points coincident with the PPT's edge chord is one less than the *greatest common factor (GCF)* of b,c.[26] The GCF of two whole numbers is the largest whole number that divides evenly into each of them. If the geodesic notation b,c coordinates of any coincident edge-point sum to a frequency that is a multiple of the overall subdivision frequency, it is a point on a harmonic grid.

A simple way to find the GCF of b and c is to list all the factors for both specifications. The GCF will be the largest factor common to both of them. Next, we show how this works. In Example 1, the overall PPT tessellation b,c = 4,16 or 20^v. The greatest common factor of 4 and 16 is 4; thus, three grid points (GCF-1) are coincident at each PPT edge. Their b,c coordinates and frequencies are listed in the figure for one PPT edge. A harmonic occurs at those points where $b + c$ frequencies evenly divide into the overall frequency. There are two harmonics, 5^v and 10^v, in Example 1. In Example 2, b,c = 12,18 or 30^v and five points are coincident at each PPT edge. Three of the edge points have frequencies that evenly divide into the original 30^v grid: 5^v, 10^v, and 15^v.

---

[26] Also called the greatest common denominator (GCD), or highest common factor (HCF).

Example 1

$b,c = 4,16$ or 20^v
4 factors 1,2,4
16 factors 1,2,4,8
GCF = 4
b,c points on PPT edge = GCF-1 = 4-1 = 3
1,4 5^v; 2,8 10^v; 3,12 15^v

Example 2

$b,c = 12,18$ or 30^v
12 factors 1,2,3,4,6
18 factors 1,2,3,6,9
GCF = 6
b,c points on PPT edge = GCF-1 = 6-1 = 5
2,3 5^v; 4,6 10^v; 6,9 15^v; 8,12 20^v; 10,15 25^v

Figure 8.41 shows the initial tessellation grid and its first harmonic for the two examples. In the figure, the original PPT tessellation is above and its first harmonic below. Grid points coincident to the PPT edge are highlighted, as are the edge points that are retained in their harmonic. A sample triangle within each tessellation grid is highlighted to make it easier to see how one tessellation relates to the other.

Class III tessellations tend to have fewer harmonics than the other classes because neither b nor c can be prime, even if their $b + c$ frequency is not prime. Harmonics also occur whenever b,c values double or are half, as long as the result is a whole number. Again, grid points coincident to the PPT's edges indicate a possible harmonic. Harmonic grids, like any other Class III grid, are rotationally symmetric but not reflection symmetric. All reflections are enantiomorphous or opposite-handed.

8.7.5   Developing Grids

Class III lacks the convenient 3D equal-chord and equal-arc references that were the starting points for Class I and II subdivisions. For Class III, we use a different approach and create a 2D grid, which we will eventually project onto the PPT's surface. Although we will develop a triangular grid in our example, it is possible to use this same technique to project any 2D geometry or pattern onto a sphere. Figure 8.42 graphically summarizes the entire sequence we will use; the remainder of this chapter details these steps.

The PPT's tessellation is first defined as a 2D trigrid with the required frequency and orientation relative to the PPT's edges. Figure 8.42(a) shows the grid and the triangular faces that are completely within the PPT. Trigrid point coordinates are converted to Cartesian values and positioned so that one of the grid's directions is perpendicular to the *xz*-plane and centered over the origin, where it is scaled to correspond to the final PPT's chord length, as shown in (b). The scaled grid is then translated up the +*z*-axis, a distance equal to the polyhedron's insphere radius. The center of the grid is tangent to the insphere and its apices are tangent to its circumsphere (c). Trigrid points are projected from the origin to the surface of the sphere by normalizing each grid point's coordinates. An arrow from the

Example 1: b,c = 4,16 20^v Example 2: b,c = 12,18 30^v

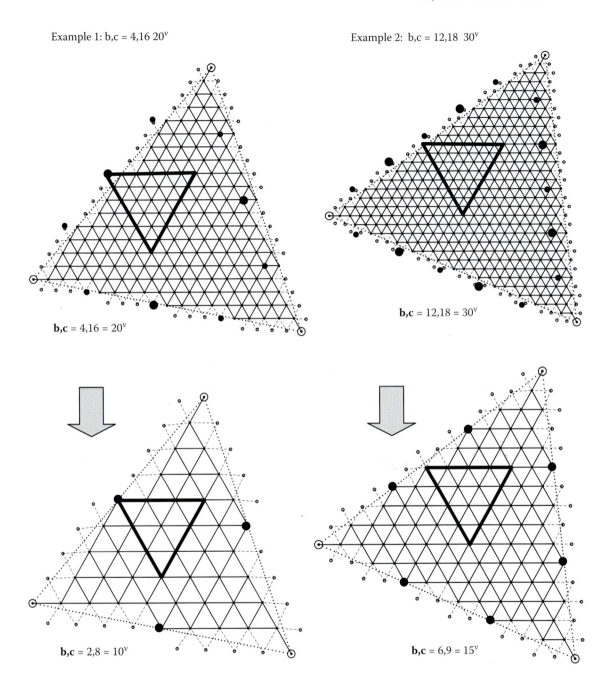

b,c = 4,16 = 20^v

b,c = 12,18 = 30^v

b,c = 2,8 = 10^v

b,c = 6,9 = 15^v

Figure 8.41. Examples of Class III tessellation harmonics.

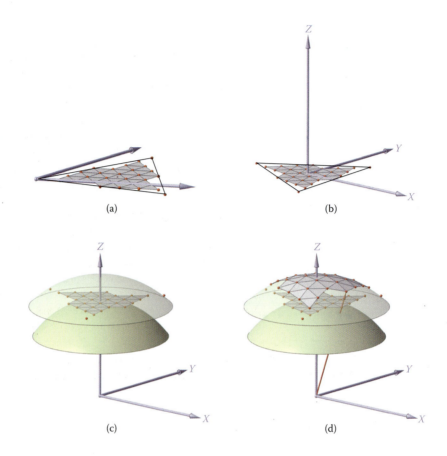

Figure 8.42. Class III grid development and PPT projection sequence.

origin through one grid point shows its projection to the sphere (d). The result is a single face-zenith tessellated PPT. Copies are then replicated to cover the rest of the sphere.

8.7.6   The *b,c* Trigrid

The first step is to define a 2D grid based on the geodesic specifications *b,c*. These two specifications determine the frequency and the grid's orientation relative to the PPT's face. Figure 8.43(a) shows the basic trigrid system, its origin, *b,c* axis, and reference grid oriented at 0°, 60°, and 120°. The standard grid unit is *d*, and the spacing between parallel grid lines is $d$sin(60°) or .866025 when $d = 1$. In Figure 8.43(b), a 2D 8^v PPT ($b,c = 2,6$) is mapped out on the trigrid. The PPT's trigrid apex coordinates, in clockwise order from the origin, are 0,0; 2,6; and 8,−2.

Trigrid points within the PPT or its edges are the points that will be projected onto the sphere. We are not concerned with any trigrid points that fall outside the trigrid PPT. Although this grid is not yet applied to a polyhedron's face, its geodesic specification is nearly complete. Lacking only the valence (v), the geodesic notation for the trigrid so far would be: $\{3,v+\}_{2,6}$.

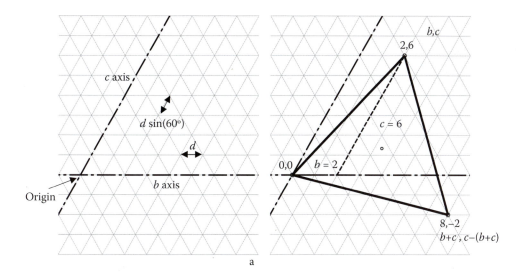

Figure 8.43. Trigrid definition for 8^v tessellation $\{3,v+\}_{2,6}$.

8.7.7   From Two to Three Dimensions

The b,c coordinate system, shown in Figure 8.43, is convenient for specifying the tessellation grid, but it is not convenient for projection onto the sphere. Instead, we need 3D Cartesian coordinates. Converting trigrid coordinates to Cartesian coordinates is simple. If d is the length of one grid unit and b,c are trigrid coordinates, 3D Cartesian values for any trigrid point can be found by

$$x = d\left(\cos 60^\circ c + b\right),$$

$$y = d\left(\sin 60^\circ c\right), \text{ and}$$

$$z = 0.$$

In the example shown in Figure 8.44, the Cartesian coordinates for the trigrid PPT's apices A, B, and C are $(0, 0, 0)$, $(5.0, 5.19615, 0)$, and $(7.0, -1.73205, 0)$. In each case, the z-coordinate is 0 because the trigrid system is still on the xy-plane.

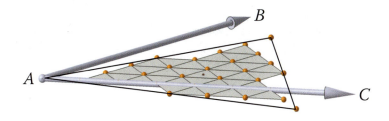

Figure 8.44. PPT b,c coordinates.

8.7.8   Scale and Translate

The trigrid PPT is not the right size for projection; we need to scale it so that the distance between each of its PPT apices is the same as the chord distance of the polyhedron's face apices. To do this, all PPT points are scaled by the ratio of the polyhedron's chord length (inscribed in a unit radius circumsphere) and the chord of the PPT grid. The icosahedron's chord length is 1.05146[27] and the grid is 7.21110. The scale factor is the ratio of the polyhedron's chord length and the PPT's chord lengths. This is shown in the following formulas:

$$\text{scale factor} = \frac{\text{polyhedron's chord length}}{\text{PPT's chord length}},$$

$$\text{scale factor} = \frac{1.05146}{7.21110},$$

$$\text{scale factor} = 0.14581.$$

By multiplying the PPT's point coordinates by the above scale factor, the vertex distance of the PPT grid becomes exactly the same as the vertex distance of the spherical icosahedron with a *unit circle* circumsphere.

8.7.9   PPT Standard Position

For Class III projections, the trigrid PPT must be repositioned with one of its edges perpendicular to the *xz*-plane and centered at the origin in a face-zenith orientation. The simplest way to reposition the edge is to rotate the PPT point set around the *z*-axis until the edge opposite the origin is perpendicular to the *x*-axis. One of the PPT's apices is already at the origin, so we simply rotate the PPT around this point. Earlier, we found the Cartesian value of apex $b,c = 2,6$ apex to be (5.0, 5.19615, 0). This point is 46.10211° with respect to the *x*-axis. If it were 30°, the grid's edge opposite the origin would be in the desired position. To reorient the PPTs, we simply rotate the grid points counterclockwise around the *z*-axis.

Figure 8.45. PPT centered, scaled, and edge normal position.

---

[27] Chord lengths for tetrahedra and octahedra with a unit radius circumsphere are approximately 1.63299... and 1.141421..., respectively.

The amount of rotation is 46.10211° – 30° = 16.10211°. To orient the PPT face-zenith, we only need to translate (slide it over) so its center is over the Cartesian origin. It is easy to do this. The PPT's centroid is simply the average of the sum of the x-, y-, and z-components of the PPT's apex points. Subtracting the centroid coordinates from the coordinates in the PPT's point set centers it perfectly. Figure 8.45 shows the PPT grid after the translation and rotation.

To review: a 2D trigrid on the xy-plane defines the vertices of the PPT based on the geodesic notation b,c specification. The trigrid ensures the correct frequency and grid orientation with respect to the PPT's edges. Only grid points, within or on, a PPT's edge are of interest. Grid points are converted from the 2D trigrid b,c system to 3D Cartesian coordinates. All points are still on the xy-plane; their z values are 0. The PPT is rotated around the z-axis to position the edge opposite the origin perpendicular to the positive x-axis and then centered over the origin of the Cartesian axis. The PPT's grid points are scaled to the size of the spherical polyhedron's face.

8.7.10   Projection

In Chapter 6 we saw that Platonic vertices are tangent to their circumsphere and their mid-face points are tangent to their insphere. We now translate the trigrid PPT up the $+z$-axis the distance of the insphere radius of the icosahedron by simply adding the radius value (.7946544) to the z-component of every point (which was zero prior to this addition). The effect is to move the grid up the positive z-axis where the grid's center is now tangent to the icosahedron's insphere and its three apices are tangent to the circumsphere, as shown in Figure 8.46. The grid is now in the proper position for projection.

Projecting the grid is easy; we normalize each grid coordinate. This makes their vector length equal to one, the same as the unit radius sphere. In effect, normalizing pushes each point onto the sphere's surface, projecting it along a line from the origin to the sphere's surface. Figure 8.47 shows the grid before and after projection. An arrow shows one point being projected from the origin to the surface of the circumscribing sphere. Normalizing the grid will not affect its three apex points; they are already on the surface of the unit

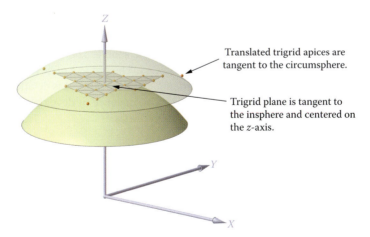

Figure 8.46. Tessellation grid preparation for projection.

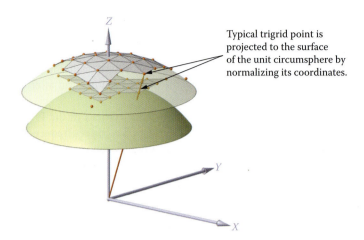

Typical trigrid point is projected to the surface of the unit circumsphere by normalizing its coordinates.

Figure 8.47. Projecting a grid to the surface of the sphere.

sphere, a result of the translation in the previous step. This last step completes the tessellation of the PPT. The result of projection is a tessellated, face-zenith PPT with one edge "parallel" to the *yz*-plane.

8.7.11   Class III PPT

Figure 8.48 illustrates the PPT we just created using the same graphic standards as before. The schematic in Figure 8.48(a) shows the PPT's projected points as well as the points in its neighboring PPTs, so the triangle faces that straddle the edge are easier to read. In Figure 8.48(b), the resulting great circle arcs between vertices are shown along with the PPT's edge (red), which is not part of the final tessellation. Some tessellation surface treatments are suggested in Figure 8.48(c) and (d). PPT apices are marked with small black dots.

Note that in Class III subdivisions, tessellation triangles that straddle the edges of the polyhedron can only be created after the initial PPT is replicated and rotated because these triangles are created from points in PPTs on both sides of the edge. These edge triangles are included in (c) and (d), however. Figure 8.49 shows the complete spherical subdivision, in which the PPT has been replicated to cover the entire sphere. The total number of faces, edges, and vertices for the entire sphere are also listed. The process of covering the whole sphere is exactly the same process used in earlier Class I schemas. It is detailed in Appendix E. The face-zenith orientation of the PPT is a consequence of developing the original trigrid on the *xy*-plane before translating the positive *z*-axis.

8.7.12   Other Polyhedra and Classes

The projection technique just described for Class III subdivisions applies equally well to the other two deltahedra—spherical tetrahedra and octahedra. Only three changes are required. First, the PPT trigrid must meet Class III *b,c* specifications. Second, the PPT grid must be scaled so that its vertices are the same distance apart as the polyhedron's face vertices, and third, the PPT is translated up the +*z*-axis a distance equal to the polyhedron's insphere radius. The scale factor and insphere radius are unique to each polyhedron. These are the only changes needed.

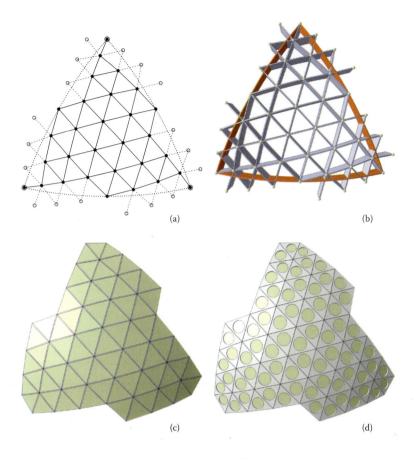

(a)   (b)

(c)   (d)

Figure 8.48. Class III tessellation schema.

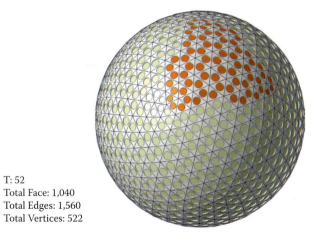

T: 52
Total Face: 1,040
Total Edges: 1,560
Total Vertices: 522

Figure 8.49. Class III tessellated sphere.

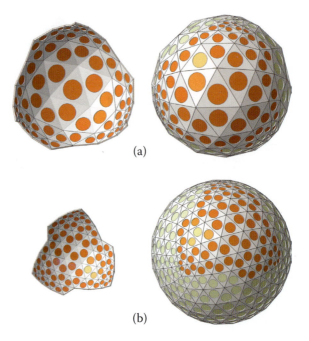

(a)

(b)

Figure 8.50. Class III 8^v tetrahedron and octahedron.

Figure 8.50 shows an inside view of the spherical PPT and a fully subdivided tetrahedron (a) and octahedron (b). The radius of each incenter circle is 80 percent of the distance from the incenter to the PPT's edges. You immediately notice the significant variation in face sizes, particularly with the tetrahedron.

Class I and II tessellations can be made by using the projection technique just used here for Class III.[28] Figure 8.51 shows Class I and II 8^v trigrids in (a) and (b) and their resulting PPTs in (c) and (d) obtained by projecting the trigrids. The question arises, why not just use projection for all spherical subdivisions, since it can handle any class and any frequency?

The answer is the results are not as good as some of the techniques we have presented. For Class I Equal-chords, trigrid projection produces exactly the same result because Equal-chords is based on an equal subdivision of the PPT's planar edge and this is essentially the same as the planar trigrid before projection. Both subdivision surfaces are superimposed in Figure 8.51(e), and they are perfectly aligned and there is no difference. Trigrid projections for Class II grids (b) and (d) have similar results, as the triacon technique described earlier; however, they are not the same and give up many of the triacon's advantages.

In Figure 8.51 the trigrid projection results for Class II (d) are superimposed on a regular triacon subdivision for the same PPT and the same frequency. Unlike Figure 8.51(e), where Class I projection and Equal-chords methods produced identical results, it is obvious from (f) that the two Class II methods produce different results. The triacon's vertices and surface (in red) do not align with the trigrid's projection (in gold). Common vertices are

---

[28] In 1990, Christopher Kitrick developed a unified approach to generating the geometry of the Class I, II, and III grids by defining a set of mathematical algorithms for ten different geometric methods of generating geodesic triangulated grids. Several of these had not been previously found in the literature (Clinton 2002, 438).

both red and gold. Note that only the PPT's mid-face and mid-edge vertices are the same in both schemas; they are circled in the figure. The trigrid projection results in many more different chords as well as more triangle area, as the two surface colors show. Thus, the big advantages of the tri-acon method are lost if projection is used. This result is not a total surprise. The triacon equally subdivides the spherical polyhedron's arc edge with a series of spherical right triangles. Equal-arc references along the polyhedron's edge, drive the grid development In contrast, projecting a b,c grid is more like a projection of an equal-chord grid. It produces a Class II grid that is perpendicular to the edges of the polyhedral face, but it does not offer any of the benefits of the triacon, like a minimum number of different chords or face areas.

8.7.13 Summary

Class III skew applies to spherical deltahedra—tetrahedra, octahedra, and icosahedra. Tessellations grids are not parallel or perpendicular to the spherical polyhedron's edges, but cross at oblique angles. Their geodesic b,c must be positive, not equal to zero or equal to each other. Subdivisions of the octahedron and icosahedron are related to the snub cube and snub dodecahedron. Tessellations are rotationally symmetric, but their reflections are enantiomorphic, or opposite-handed. Like Class I and II tessellations, Class III tessellations may contain harmonics depending on b,c's common factors.

Class III tessellations lack convenient equal-chord or equal-arc references on the polyhedron's spherical polyhedron edge; thus, projection is used (other methods are possible too). Subdivision starts with a 2D trigrid on the xy-plane,

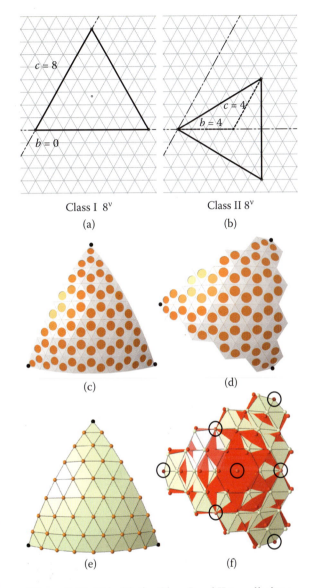

Class I 8^v
(a)

Class II 8^v
(b)

(c)

(d)

(e)

(f)

Figure 8.51. Trigrids for Class I and II tessellations.

where we define the final frequency, grid orientation, and included grid points. We convert points within the trigrid PPT or on its edge to Cartesian values and the points are then translated to the origin, scaled, and translated up the positive z-axis. The center of the grid (still planar) is tangent to the insphere, while the trigrid PPT's apices are tangent to the circumsphere. We then project grid points to the surface of the unit sphere by normalizing their coordinates.

We can also apply projection methods to other classes. When we apply these methods to Class I grids, the results are exactly the same as they are for the Equal-chords schema, but not the same as they are for the other Class I schemas we developed. For Class II schemas, the results are similar to but not the same as the triacon because Class III does not divide the polyhedron's edge into equal arc segments like the triacon does.

Class III projections involve a series of coordinate transformations: scaling, rotating, and translating. They can all be combined into a single transformation by combining their specifications. We give examples in Appendix C.

8.8   Covering the Whole Sphere

Each of the schemas developed in this chapter resulted in one tessellated PPT or diamond. But how does this describe the whole sphere? The answer is simple—rotation! Essentially, we copy the tessellated PPT or diamond geometry and use rotation to position the copy wherever we want it. We rotate copies of one PPT or diamond at a time or we can combine small groupings of PPTs or diamonds into a larger grouping and rotate the larger grouping to cover the whole sphere. We show several examples here and others in Chapter 10.

Rotation recomputes new coordinate values, in effect finding new values for coordinates, as they revolve around the x-, y-, or z-axis to their new position. Rotation does not alter the relationships of any grid point to one another within the PPT or diamond being rotated. All angles and distances between points and the origin are preserved during the rotation.

When making a complete sphere from a single PPT or diamond, it is often advantageous to create subassemblies and then copy-rotate them rather than handle each PPT or diamond separately. Figure 8.52 shows the exterior and interior of a Class I subassembly of four PPTs. In (a), the subassembly resembles a spherical lune. It is created by rotating three instances of the original principal triangle, the topmost triangle in (a), and positioning them back-to-back, creating a lune-like segment that is one-fifth of the sphere. The assembly in (a) is not yet fully transformed in order to clearly show each PPT copy in the subassembly. In Figure 8.52(b) the subassembly in (a) has been copy-rotated four more times around the z-axis to complete the entire sphere.

Figure 8.53 shows how subassemblies might be used in Class II subdivisions. An upper pentagonal cap subassembly is created from the original subdivided diamond by copy-rotating it nine times. The cap, when completed, has ten diamonds. A lower cap is created by copy-rotating the entire upper cap subassembly. Assume the vertical axis of the sphere is the z-axis; the upper cap is copy-rotated 180° around either the x- or y-axis and then again 36° around the z-axis because the icosahedron's upper-lower pentagonal cap is offset by this much. The central ring of ten diamonds is accomplished by copy-rotating two

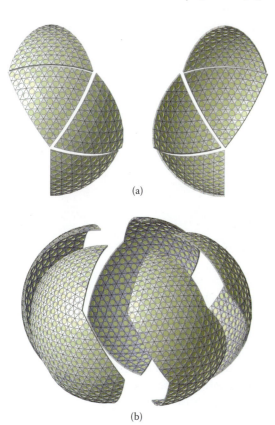

(a)

(b)

Figure 8.52. Class I Equal-chords lune assembly.

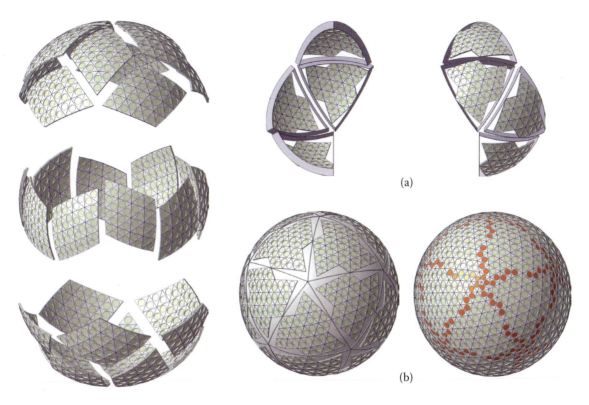

Figure 8.53. Class II pentagonal cap and ring assembly.

Figure 8.54. Class III lune subassembly.

diamonds from the upper (or lower) cap and positioning two adjacent to one another with their smaller diamond sides tangent. This subassembly of two diamonds is then copy-rotated four more times around the z-axis, each instance offset from the previous one by 72°. When all of the rotation transformations are fully applied, a completed sphere is the result.

Class III rotations can use the same strategy as Class I rotations. Figure 8.54(a) shows a lune subassembly. The triangles that straddle the PPT's edges can be completed from tessellation vertices of PPTs that share the edge. The spherical icosahedron edges, shown in Figure 8.54(b), are not part of the final tessellation, but show where straddling triangles fill in to complete the sphere.

Additional Resources

Clinton, Joseph D. *Advanced Structural Geometry Studies Part I—Polyhedral Subdivision Concepts for Structural Applications.* National Aeronautics and Space Administration contractor report NASA CR-1734, Carbondale, IL: Southern Illinois University, 1971.

——. "Geodesic Math." *Domebook 2* ed. Lloyd Kahn et al., Bolinas, CA: Pacific Domes (1971): 106–113.

———. "A Group of Spherical Tessellations Having Edges of Equal Length." West Chester, PA: Clinton International Design Consultants, undated.

———. "A Limited and Biased View of Historical Insights for Tessellating a Sphere." *Space Structures 5* ed. G. A. R. Parke and P. Disney. London: Thomas Telford (2002): 423–431.

Gailiunas, Paul. "Twisted Domes." Newcastle, UK: Unpublished paper, http://web.ukonline.co.uk/polyhedra/Twist.pdf, 2008.

Kenner, Hugh. *Geodesic Math and How to Use It.* Berkeley: University of California Press, 1976.

Stuart, Duncan R. "The Orderly Subdivision of Spheres." *The Student Publications of the School of Design,* North Carolina State University, monograph, 1963.

Wenninger, Magnus J. *Spherical Models.* Cambridge [Eng.], New York: Cambridge University Press, 1979 (new appendix, 1999).

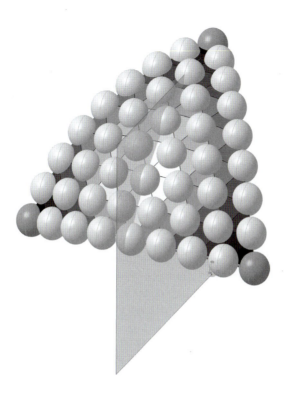

9 Comparing Results

We have developed six subdivision schemas, and it is natural to wonder which is best. The answer depends on the specific geodesic application and its design goals. No single schema fits every need. In this chapter, we offer a series of visual and statistical metrics to compare schemas. Some compare distances between vertices, triangle area, or volume of the subdivided sphere; others look at face shapes and their orientation. Although each metric highlights one characteristic and excludes others, there are common threads throughout and each provides a partial answer to the question of which is better.

What our analysis will show is that Class I Equal-arcs (three great circles) and Class II triacon produce the most uniform distribution of points on a sphere, and the most uniform face orientations of the schemas we developed. They also offer additional advantages. Triacon has the fewest different chord lengths, the fewest different face angles, and the largest number of equal-sized and equal-shaped faces by far. It is a superior performer. Equal-arcs (three great circles) is also a superior performer, with more variation than the triacon, but very uniform overall. It offers superior face orientations and a high percentage of symmetrical parts. However, these two schemas are not the last word. Mid-arcs and Equal-chords both offer unique features that make them preferred choices in certain applications.

So once again, the answer to the question of "which is best?" depends on what you are trying to accomplish and your design requirements. We begin our analysis by discussing an intimate subject: kissing-touching.

9.1 Kissing-Touching

Chapter 4 introduced the concept of balls "kissing" and "touching." These terms, borrowed from the game of billiards, offer us a way to see how closely things pack together. Closely packed balls on a sphere measure the compactness and distribution of tessellation vertices. When the largest balls possible, all with the same diameter, are placed at each tessellation vertex, they give us a 3D way to visualize the symmetry, distribution, and maximum packing a schema affords. Their pattern also makes it easy to see where the smallest and largest tessellation triangles can be found.

Some vertex balls will kiss and touch, but none will penetrate or overlap another. The spacing between balls and the maximum diameter possible for a schema indicate how uniformly the subdivision distributes points on the surface of the sphere. The best subdivision schema, in terms of point distribution, will have the closest packing of the largest diameter balls when compared to other schemas.

Kissing-touching does not measure the same parameter as spherical caps (discussed in Chapter 4). Both techniques are indicators of packing, but spherical caps measure how much of the sphere's surface area is covered. It is a density measurement, which will be used shortly. In the spherical cap technique, the cap's radius is based on the smallest *arc* distance between tessellation points. In contrast, kiss-touch ball diameters are based on the smallest chord length between them. Table 9.1 lists the kissing-touching distance between the closest balls in our schemas on the unit sphere.

Figure 9.1 shows kiss-touch vertex balls for the schemas developed earlier in the chapter for unit spheres. The viewpoint for each figure is from *inside* the sphere, where we get the least-obstructed view of their PPT or diamond packing. Contacts between balls, especially near the PPT's apices (red ball), are somewhat obstructed in the figure, but views from inside the sphere are still better than from outside.

Kissing-Touching Ball Diameters	
Schema	Smallest
Class I	
Equal Chords	0.119459
Equal Arcs (2 GCs)	0.132679
Equal Arcs (3 GCs)	0.138283
Mid Arcs	0.138283
Class II	
Triacon	0.153316
Class III	
Full Face	0.128646

Table 9.1. Kissing-touching distances.

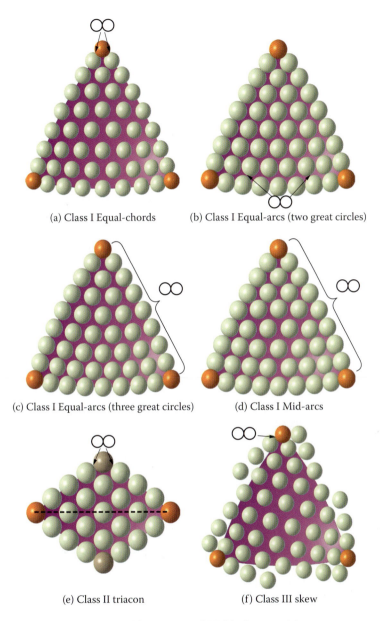

(a) Class I Equal-chords

(b) Class I Equal-arcs (two great circles)

(c) Class I Equal-arcs (three great circles)

(d) Class I Mid-arcs

(e) Class II triacon

(f) Class III skew

Figure 9.1. Class I, II, and III kissing-touching.

Figure 9.1(a), (b), (c), and (d) are Class I Equal-chords, Equal-arcs (two great circles), Equal-arcs (three great circles), and Mid-arcs schemas, respectively. Class II triacon and a Class III skew are shown in (e) and (f). Each schema is represented by a PPT or diamond surface (pink) and its kiss-touch balls. PPT and diamond apex balls are indicated (red) and correspond to the vertices of the foundation polyhedra, the icosahedron in our examples. The figure points out typical kiss-touch pairs.

In Equal-chords, Figure 9.1(a), kissing-touching starts at each PPT apex (red ball) and the balls immediately around it. All other chords within the PPT are longer, and space appears between the balls. For Class I subdivisions, the general pattern is that chord lengths increase from the PPT apices to the center of the PPT edge, and from the center of the edge towards the center of the PPT. This pattern is quite visible in the figure. The balls show that the size of the triangles increases as they progress from the PPT vertices and edges towards the center of the PPT face. The biggest gaps are in the middle of the PPT. Kissing-touching makes these relationships quite visible.

Figure 9.1(b) shows Equal-arcs (two great circles). Its kiss-touch ball diameter is larger than the diameter of Equal-chords, and the smallest chord length does not occur at a PPT apex, as might be expected. Instead, it occurs on the bottom PPT edge between the two tessellation vertices immediately inside to the left and right of the center edge. They are pointed out in the figure. Since these two balls set the maximum diameter possible for the others, none of the balls along the PPT edge kiss-touch, even though the schema is based on an Equal-arcs PPT reference point. All the chords along the PPT's edges are larger. Thus, there is some space between balls. Notice, also, all the interior balls are displaced towards the bottom PPT edge, a consequence of defining tessellation points with just two intersecting great circles. As a result, the PPT's tessellation points are not rotationally symmetrical.

Equal-arcs (three great circles), shown in Figure 9.1(c), demonstrates a highly uniform distribution throughout. All PPT edge balls kiss and touch, because the edge reference points are evenly spaced and all have the same shortest chord lengths. The distribution of tessellation vertices is symmetrical in terms of face, edge, and PPT apices. The uniform distribution across the PPT's face is a direct result of intersecting three great circles and handling the resulting window artifact. Earlier, we showed that the intersection of three great circle planes in this schema create three points, not one, on the surface of the PPT (we ignore the three on the opposite side of the sphere). The window's projected centroid is used as the tessellation grid point. The combined effect of using three great circle intersections and the extra step of processing the window points is well worth the extra computation to eliminate windows. Its close packing, shown in (c), demonstrates that it is the most uniform of the Class I schemas.

Figure 9.1(d) shows Mid-arcs. The nesting pattern of triangles within triangles, the basic subdivision method used, is evident in the ball pattern. Kissing-touching occurs along all three PPT edges because this schema recursively divides every arc at its mid-point and this includes the entire length of all three PPT edges. In this respect, it is similar to Equal-arcs (three great circles). All interior chords are larger than the edge chords; thus, all balls along the PPT's edges kiss and touch. Notice, also, the spacing between balls increases as they get closer to the center of the PPT. The three balls at the middle of the PPT mark the vertices of the largest tessellation triangle within the PPT.

Figure 9.1(e) shows the triacon diamond, which is symmetric to every edge of the foundation polyhedra. The method divides the polyhedral edge into equal arc segments, which is reflected in the even spacing of kissing-touching along the icosahedral edge (dashed diamond axis between the red balls). Unlike the previous schemas, kissing-touching in a triacon diamond first occurs at mid-PPT vertices (brown) instead of at the PPT apices (red). The triacon is the only schema that consistently defines a vertex in the middle of the PPT's face.

Figure 9.1(f)[1] shows a Class III skew. Of all the schemas developed, skew has the greatest difference in chord lengths, as evidenced by how quickly the kiss-touch balls spread out towards the center of the PPT. We will see just how much they differ when we compare chords in the next section.

In summary, kissing-touching is an effective metric for evaluating the symmetry, distribution, and compactness of subdivision schemas. The larger the balls and the more closely they are packed, the better the points distribution. From the point of view of kissing-touching, Class I Equal-arcs (three great circles) and triacon perform best. Mid-arcs performs well, though not as well as the other two. Class III subdivisions perform the worst. Equal-arcs (two great circles) has good packing, but its asymmetry may preclude it from many applications.

9.2   Sameness or Nearly So

Statistics show how similar (or dissimilar) objects are, and we can use these statistics to quantify the variation of chord lengths and triangle areas within a schema. Schemas with low statistical variation tend to have more uniform grids, but there are important exceptions to this rule.

In statistics, *variance* measures how much a set of values spread out. These values could be the chord or arc lengths, triangle area, or other geometric measurements. Variance is the average squared deviation of these values from their mean. The *standard deviation*, another important statistic, is simply the square root of variance. The lower the standard deviation, the closer the values are to being the same or alike.

Is there a spherical baseline with no variation where every chord and face area is the same? Yes, the foundation polyhedron itself is a baseline. All chords are equal, all faces are congruent regular polygons and all have equal area, all central and surface angles are the same, and all vertices are evenly distributed on the sphere. For example, the icosahedron's 12 vertices are evenly distributed on its circumsphere, all faces are equilateral and congruent, and all 30 chords and central angles are the same. This means that a basic

Tessellation Vertices			
Schema	T	PPT/Dia	Sphere
Class I			
Equal Chords	64	45	642
Equal Arcs (2 GCs)	64	45	642
Equal Arcs (3 GCs)	64	45	642
Mid Arcs	64	45	642
Class II			
Triacon	48	34	482
Class III			
Mid Face	52	30	522
Straddling Edge			

Table 9.2. Schema vertex comparisons.

---

[1] The Class III example is an icosahedron $\{3,5+\}_{2,6}$. The view in the figure is from the *inside* of the sphere, and this creates the impression that the subdivision is $\{3,5+\}_{6,2}$.

Tessellation Chords									
Schema	PPT/Dia	Sphere	Unique	Shortest (S)	Longest (L)	S/L	Variance	Std.Dev.	Rank
Class I									
Equal Chords	108	1920	20	0.119459	0.164647	72.6%	1.130E-04	1.063E-02	3
Equal Arcs (2 GCs)	108	1920	44	0.132679	0.176238	75.3%	1.292E-04	1.136E-02	4
Equal Arcs (3 GCs)	108	1920	17	0.138283	0.162173	85.3%	5.369E-05	7.327E-03	1
Mid Arcs	108	1920	15	0.138283	0.164647	84.0%	9.604E-05	9.800E-03	2
Class II									
Triacon.	56	1440	8	0.011174	0.013765	81.2%	1.767E-04	1.329E-02	6
Class III									
Mid Face Straddling Edge	60	1560	20	0.128646	0.182470	70.5%	1.490E-04	1.220E-02	5

Table 9.3. Schema chord comparisons.

Tessellation Triangle Areas									
Schema	PPT/Dia	Sphere	Unique	Smallest (S)	Largest (L)	S/L	Variance	Std.Dev.	Rank
Class I									
Equal Chords	64	1280	15	0.006780	0.011738	57.8%	1.602E-06	1.266E-03	5
Equal Arcs (2 GCs)	64	1280	35	0.008070	0.011102	72.7%	6.943E-07	8.333E-04	2
Equal Arcs (3 GCs)	64	1280	15	0.009083	0.010308	88.1%	9.657E-08	3.108E-04	1
Mid Arcs	64	1280	15	0.009083	0.011738	77.4%	7.062E-07	8.403E-04	3
Class II									
Triacon	32	960	4	0.011174	0.013765	81.2%	8.550E-07	9.247E-04	4
Class III									
Mid Face Straddling Edge	70	1040	16	0.007862	0.014417	54.5%	2.773E-06	1.665E-03	6

Table 9.4. Schema triangle area comparisons.

Platonic solid, planar or spherical, has zero variation and zero standard deviation in each of these characteristics. However, when we subdivide a polyhedron, everything changes. Many chord lengths and triangle areas result; their variance tells us how different they are.

Intuitively, one would expect the best-performing schemas to have the lowest standard deviations and variances, but this is not always the case. We will find excellent schemas with somewhat high deviations, which eliminate trivial differences in chord length by making more pronounced differences in the few chords they create.

Tables 9.2, 9.3, and 9.4 summarize the vertex counts, chord lengths, and triangle areas that result from the subdivision schemas developed. Table 9.2 shows the triangulation number, T, and the total number of vertices in a PPT/diamond and on full spheres. There is quite a range in total points distributed, even though the foundation polyhedra and subdivision frequency are the same for every schema developed. The Class I schemas all distribute the same number and the most points, while triacon distributes the fewest. The table lists the number of vertices falling within the PPT, diamond, or on any edge.

Table 9.3 compares the variation in chord lengths for each schema. The table tallies the number of chords for each PPT/diamond and for the entire sphere. The number of unique chords, the shortest and longest, and percentage difference is also listed. The table calculates the variance and standard deviation for the entire set of chords on the sphere, and ranks schemas from lowest to highest variation. It is remarkable that an 8^v triacon requires only 8 unique chords, while Equal-arcs (two great circles) requires 44.

It is no surprise that Equal-arcs (three great circles) is the best performer. We have seen the pattern repeatedly in our previous analysis. It has few unique chords, and the difference between the largest and smallest length is small. The Equal-chords schema performs reasonably well with 20 unique chords and low variance in their lengths, but Equal-arcs (three great circles) and Mid-arcs perform far better, with much less variation in length and in the number of unique chords. Equal-arcs (two great circles) performs poorly because, compared to other schemas, it has more than twice the number of unique chords. The relatively low variance in its lengths tells us the difference between the members of this large set of unique chords are very small, which is not a desirable characteristic.

Once again, the triacon stands alone. Its tessellation grid has only eight unique chords. The fact it develops a diamond symmetric to the polyhedral edge also helps. Unlike other schemas, in which small differences in length resulted from lots of unique chords, triacon eliminates trivial differences in length by increasing the differences between the ones it creates. Thus, its high variance is not a surprise, and the variance serves a useful purpose. Although we have not yet focused on geodesic applications, it is worth pointing out that from a manufacturing point of view, it is usually desirable to have fewer components used many times, rather than many different components used just a few times. The triacon offers many advantages, when manufacturing is a consideration. The Class III performance is quite poor from a chord length perspective. Its chords have high variation in length. However, unlike the triacon, where variation led to a dramatic reduction in total unique lengths, variation here only led to lots of small differences in a high number of unique chords.

9.3   Triangle Area

Chord lengths directly affect a triangle's shape and area. Here we compare just the area of tessellation triangles. We will consider their shape in a moment. Our analysis is both statistical and graphic. Table 9.4 compares the variance in triangle area of each schema developed, and clearly they are correlated with chord length. All triangles on the sphere are considered. This chord table, like the previous one, lists such information as the number of unique areas, the smallest areas, and the largest areas. Once again, Equal-arcs (three great circles) is a superior performer with the least variation in face areas. Class III subdivisions have the most variation. The contrast between triacon and the other schemas is dramatic. Triacon has only four unique triangles, while the other schemas have 15 or 35. This is an amazing reduction. triacon achieves this by increasing the differences in the face areas it does create. Equal-arcs (two great circles) has the highest number of unique triangles, a consequence of having so many different chord lengths.

Figure 9.2 graphically shows the differences indicated in Table 9.4. The figure gives the triangles in each schema a gray value from black (smallest triangle) to white (largest triangle) to indicate their increasing area. The scale is relative to each figure and provides a quick way to see where similar faces are located and where symmetry occurs. In Figure 9.2(e) triacon is the easiest scale to read, with only four different-sized faces. The largest faces are along the diamond's long axes. Although they appear equilateral, they are slightly isosceles. We examine their shape more closely in a moment. The smallest faces are those at the diamond's short axis and fall near the midpoint of the principal triangle.

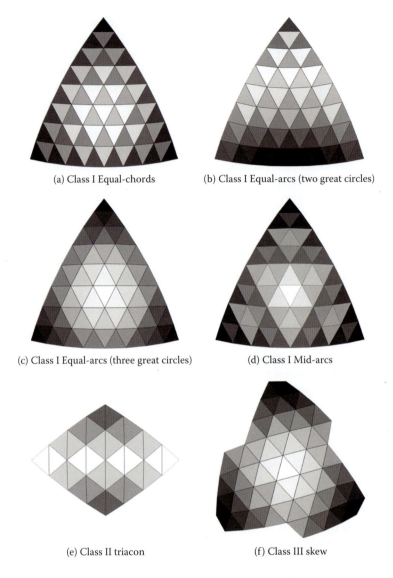

(a) Class I Equal-chords

(b) Class I Equal-arcs (two great circles)

(c) Class I Equal-arcs (three great circles)

(d) Class I Mid-arcs

(e) Class II triacon

(f) Class III skew

Figure 9.2. Tessellation face area comparisons.

Triacon is unique in another way. Its largest triangles are adjacent to the PPT's apices and the smallest ones surround the PPT's center, just the opposite of Class I or III schemas.

All four Class I schemas have different patterns, but there are some common features. The smallest faces are usually located next to the PPT's apices and the largest in the middle of the PPT. Triangles along the PPT's edge increase in size as they approach mid-edge and as they move closer to the center of the PPT. However, Equal-arcs (two great circles) does not follow this pattern exactly. Its smallest triangles cluster along the bottom edge of the PPT, instead of around the PPT's apices; they are quite apparent in (c). There is also another exception to the pattern; its largest triangle is not in the middle of the PPT, it is above the PPT's center.

The triangle areas in Class I Mid-arcs, Figure 9.2(d), clearly reflect the recursive triangle-in-triangle subdivision method. The triangles within the PPT's central equilateral area (outlined in the figure) are all larger than the triangles in the three outlying groups next to the PPT's apices. All triangles within the large central equilateral area are larger than those in the three surrounding groups next to the PPT's apices.

Equal-arcs (three great circles) has the least deviation in the area in 15 unique triangles, while triacon is ranked second, with four unique triangles. Class III skew shows the highest variance and ranks last of the schemas.

9.4   Face Acuteness

The schemas developed in this chapter show the influence the foundation polyhedra and subdivision schema have on the final tessellation grid and shape of the resulting triangles. Figure 9.3 shows just how much influence they have by comparing three deltahedra 8^v subdivisions. From left to right, the figure shows a single tetrahedron, octahedron, and icosahedron PPT. What we see repeatedly is the icosahedron producing more uniform results than either the tetrahedron or octahedron. The basic reason is that the icosahedron's spherical excess (its area) is smaller than the spherical excess of the others. Thus, for the same frequency, the tessellation faces are smaller and have less variation in shape. Within a given polyhedron, and for the same frequency subdivision, there is still variation in the shape of triangles that result from the subdivision schema itself. A direct approach would compare a triangle's face angles. However, it is still hard to see overall patterns this way. Instead, we will use a simple graphic convention, which allows us to see the shape of a triangle very quickly—we will use the Euler line.

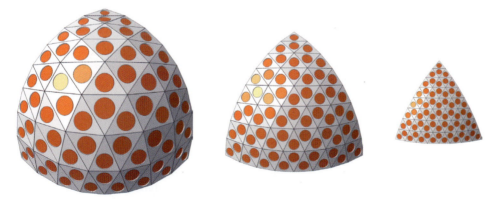

Figure 9.3. Face acuteness in subdivided deltahedra.

9.5   Euler Lines

In Chapter 4, we presented three important centers of a planar triangle—orthocenter, incenter, and circumcenter. Recall that in nonequilateral planar (not spherical) triangles, these points lie on a line called the Euler line. The length of the line segment between orthocenter and circumcenter is very sensitive to the ratio of the triangle's sides. A slight change in any

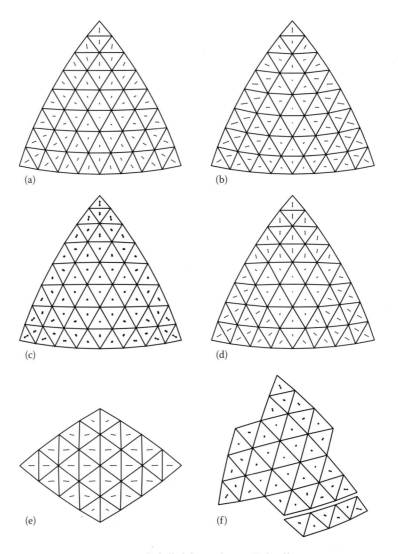

Figure 9.4. Subdivision schema Euler lines.

of the triangle's chords causes a large change in the length of this segment and the angle it takes on the triangle's surface—herein lies the Euler line's analytic value. We can use this sensitivity and the Euler line segment to see how close or far away triangles are from being equilateral, and to analyze shape. We can also display the line segment itself and graphically tell if the triangle is equilateral (no segment, just a point), isosceles, or acute. Figure 9.4 displays the Euler line segments for all of the schemas developed. Illustrations (a) through (d) are Class I, while (e) and (f) are Class II and III, respectively. The orthocenter and circumcenter of each subdivision triangle is found, and the Euler line segment between them is shown. If the face is equilateral, both centers are coincident and there is no Euler line, only a point. Class I and III figures (a), (c), (d), and (f) show this. Their PPT's center

Euler Line Segments					
	PPT / Diamond		Full Sphere		
	Total	Euler Line	Faces or	Length	
Schema	Triangles	Length	Edges	Sub-Total	Total Length
Class I					
Equal Chords	64	0.970213	20	-	19.40
Equal Arcs (2 GCs)	64	2.242257	20	-	44.85
Equal Arcs (3 GCs)	64	1.817358	20	-	36.35
Mid Arcs	64	1.235249	20	-	24.70
Class II					
Triacon	32	1.738888	30	-	52.17
Class III					
Mid-face	34	0.237650	20	4.75	
Straddling Edges	6 (per edge)	0.067179	30	4.03	8.7838

Table 9.5. Euler line segment comparisons.

triangle is equilateral, and there is no Euler line, only a point in the middle of this triangle. All other triangles in these schemas are either isosceles or acute, and their Euler line segments have measurable length and direction. There are no right triangles in these schemas, but if there were, the Euler line segment would touch the hypotenuse, or the apex opposite it, and be perpendicular to the hypotenuse.

Table 9.5 tabulates these lengths. For each schema, the total Euler line segment length is listed for the basic PPT or diamond and then extrapolated to the total length of all segments on a full sphere. Classes I and II are straightforward, but the Class III schema require subtotals because some subdivision triangles lie totally within the PPT, while others straddle the edge. The last column lists the total length of the Euler line segments.

It is evident from the table that there is considerable variation in grid triangle shape. Clearly, Class I Equal-chords and Class III subdivisions (which are also based on a projection of Equal-chords) have the most uniform (near equilateral) faces. This result is consistent with what we saw in kissing-touching. The greatest deviation from equilateral faces is found in triacon. There is not a single equilateral face in the diamond, and those that straddle the polyhedra edge (long diamond axis for icosahedra) have the longest Euler line segments, as shown in Figure 9.4(e).

We have compared kissing-touching distances, chord lengths, triangle areas, and triangle shapes. We now need to see how subdivision grid components are affected by changes in frequency. We will find the triangulation number, T, very handy for this kind of analysis.

9.6   Parts and T

The total number of parts in a fully tessellated spherical polyhedron depends on the polyhedron's basic characteristics: faces, edges, and vertex valence. It also depends on the schema's b,c frequency specification. We can use the triangulation number, T, and these characteristics to count the parts on the full sphere. Table 9.6 restates the faces, edges, and vertex valence for each deltahedron. The basic formulas for finding the total faces, edges,

Deltahedra Valences			
Deltahedron	**Faces**	**Edges**	**Valence**
Tetrahedron	4	6	3
Octahedron	8	12	4
Icosahedron	20	30	5

Table 9.6. Deltahedra valences.

v	Class	b	c	T	Tetrahedron			Octahedron			Icosahedron		
					Faces	**Edges**	**Vertices**	**Faces**	**Edges**	**Vertices**	**Faces**	**Edges**	**Vertices**
2	I	0	2	4	16	24	10	32	48	18	80	120	42
	II	1	1	3	12	18	8	24	36	14	60	90	32
	I	0	4	16	64	96	34	128	192	66	320	480	162
4	III	1	3	13	52	78	28	104	156	54	260	390	132
	II	2	2	12	48	72	26	96	144	50	240	360	122
	I	0	8	64	256	384	130	512	768	258	1,280	1,920	642
8	III	1	7	57	228	342	116	456	684	230	1,140	1,710	572
	III	2	6	52	208	312	106	416	624	210	1,040	1,560	522
	III	3	5	49	196	294	100	392	588	198	980	1,470	492
	II	4	4	48	192	288	98	384	576	194	960	1,440	482
	I	0	16	256	1,024	1,536	514	2,048	3,072	1,026	5,120	7,680	2,562
	III	1	15	241	964	1,446	484	1,928	2,892	966	4,820	7,230	2,412
	III	2	14	228	912	1,368	458	1,824	2,736	914	4,560	6,840	2,282
16	III	3	13	217	868	1,302	436	1,736	2,604	870	4,340	6,510	2,172
	III	4	12	208	832	1,248	418	1,664	2,496	834	4,160	6,240	2,082
	III	5	11	201	804	1,206	404	1,608	2,412	806	4,020	6,030	2,012
	III	6	10	196	784	1,176	394	1,568	2,352	786	3,920	5,880	1,962
	III	7	9	193	772	1,158	388	1,544	2,316	774	3,860	5,790	1,932
	II	8	8	192	768	1,152	386	1,536	2,304	770	3,840	5,760	1,922
	I	0	32	1,024	4,096	6,144	2,050	8,192	12,288	4,098	20,480	30,720	10,242
	III	1	31	993	3,972	5,958	1,988	7,944	11,916	3,974	19,860	29,790	9,932
	III	2	30	964	3,856	5,784	1,930	7,712	11,568	3,858	19,280	28,920	9,642
	III	3	29	937	3,748	5,622	1,876	7,496	11,244	3,750	18,740	28,110	9,372
	III	4	28	912	3,648	5,472	1,826	7,296	10,944	3,650	18,240	27,360	9,122
32	III	5	27	889	3,556	5,334	1,780	7,112	10,668	3,558	17,780	26,670	8,892
	III	6	26	868	3,472	5,208	1,738	6,944	10,416	3,474	17,360	26,040	8,682
	III	7	25	849	3,396	5,094	1,700	6,792	10,188	3,398	16,980	25,470	8,492
	III	8	24	832	3,328	4,992	1,666	6,656	9,984	3,330	16,640	24,960	8,322
	III	9	23	817	3,268	4,902	1,636	6,536	9,804	3,270	16,340	24,510	8,172
	III	10	22	804	3,216	4,824	1,610	6,432	9,648	3,218	16,080	24,120	8,042
	III	11	21	793	3,172	4,758	1,588	6,344	9,516	3,174	15,860	23,790	7,932
	III	12	20	784	3,136	4,704	1,570	6,272	9,408	3,138	15,680	23,520	7,842
	III	13	19	777	3,108	4,662	1,556	6,216	9,324	3,110	15,540	23,310	7,772
	III	14	18	772	3,088	4,632	1,546	6,176	9,264	3,090	15,440	23,160	7,722
	III	15	17	769	3,076	4,614	1,540	6,152	9,228	3,078	15,380	23,070	7,692
	II	16	16	768	3,072	4,608	1,538	6,144	9,216	3,074	15,360	23,040	7,682

Table 9.7. Faces, edges, and vertices for 2^v, 4^v, 8^v, 16^v, and 32^v, Class I, II, and III tessellations.

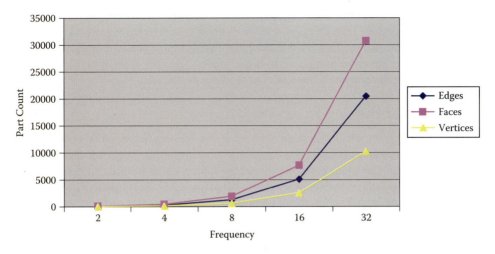

Figure 9.5. Icosahedron Class I part counts for 2^v, 4^v, 8^v, 16^v, and 32^v.

and vertices on the sphere are shown below:

$$T = b^2 + bc + c^2$$

Total subdivision faces $= T \times$ faces in base polyhedron,

Total subdivision edges $= T \times$ edges in base polyhedron,

$$\text{Total subdivision vertices} = \left[T \frac{(2 \times \text{valence})}{(6 - \text{valence})} \right] + 2.$$

The frequency of subdivision has a huge effect on part counts. Based on b,c and T values, Table 9.7 lists the face, edge, and vertex counts for 2^v, 4^v, 8^v, 16^v, and 32^v tessellations for the deltahedra. Shaded areas indicate the 8^v icosahedron schemas developed earlier in this chapter. Although it is not apparent from the table, the subdivision method within a class does not affect these counts. For example, Equal-arcs and Equal-chords schemas produce the same part counts; only their face areas, angles, and chords differ. Reversing coefficients b,c will not affect the counts. A 2,6 tessellation will have the same part counts as 6,2 for the same polyhedron.

For a given frequency and polyhedron, Class I tessellations create the highest number of parts, while Class II tessellations create about 25 percent fewer. This is a significant difference. Class III counts lie somewhere in between Classes I and II. For the same polyhedron and frequency, the greater the difference between Class III b,c values, the closer their part counts are to Class I, and vice versa. Likewise, the smaller the difference, the closer their counts are to Class II.

While part counts are critical factors in most geodesic applications, the way b,c combinations affect these totals is also critical. That's because the choice of polyhedra and frequency are made early in the design process, and the choice significantly impacts all subsequent work and development. Figure 9.5 graphs part counts for the Class I icosahedral tessellations for various frequencies. The graphs for Class II tessellations are very similar; only the part counts are less. The curves show that part counts increase as a power, usually squared, of the frequency. This rapid buildup implies that most geodesic applications will tend to use lower frequencies rather than higher ones, especially if the spherical object is manufactured from separate components (as geodesic domes are).

9.7   Convex Hull

The convex hull is the smallest convex polyhedron that surrounds the points on the sphere. The convex hull of a subdivided sphere measures how uniformly points are distributed by "shrink-wrapping" the whole envelope around the sphere with a surface that defines a volume. In some cases, the convex hull resembles the planar surfaces of tessellated spheres. In Chapter 4 we compared the convex hull volumes of the five Platonic polyhedra, each with a unit radius circumsphere, and saw that as the number of vertices increased, the convex hull volume increased. The dodecahedron distributes the most vertices and has the largest convex hull volume of the five. As more and more points are distributed, the convex hull faces become smaller and the overall shrink-wrapping looks more "spherical." Its volume also approaches, but never quite reaches, the unit sphere's volume. We also saw that when two spheres have the same radius and number of points on their surfaces, the one with the most uniform distribution or points will have the largest volume. Herein lies the use of convex hulls for comparing subdivisions. We can use convex hull volumes to compare subdivided spheres. The one with the largest volume has the most uniform distribution of points.

For the icosahedral schemas we developed earlier, we can use the surface of their triangular faces to define their convex hulls because all vertex chords connect the closest neighboring pairs of points. However, this is not always the case, as the 8^v triacon diamond in Figure 9.6 shows. For the same set of tessellation vertices, there is quite a difference between the schema faces for this octahedron (a) and its convex hull surface (b). The vertex points in both figures are the same.

Convex hull analysis demands precise surface definitions and volume calculations. The differences in our schemas are very small, and there is a computational trade-off here. We can find the convex hull of the whole sphere at once, or find the volume of a smaller representative part and then extrapolate it to the whole sphere. We choose the latter approach. Figure 9.7 shows examples of representative convex hulls for each class we developed. We

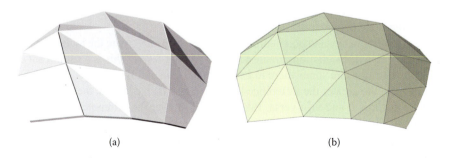

(a)   (b)

Figure 9.6. Tessellation triangles and their convex hull surface.

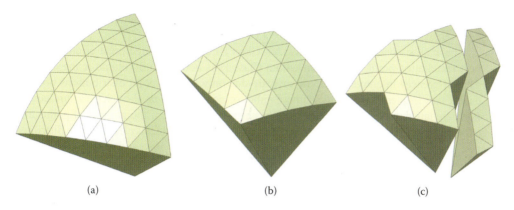

(a)   (b)   (c)

Figure 9.7. Example Class I, II, and III convex hull models.

calculate the volume of these sections and then extrapolate (multiply) it by the number of times required to define the total sphere's convex hull. The convex hull surface for each schema covers the faces of their subdivision triangles and their sides (to the origin of the sphere) to enclose a complete volume. The volume sides are created from a series of triangular surfaces defined by the end points of chords that run along the PPT's or diamond's edge and the origin of the sphere. For Class I and II convex hulls, a single 3D volume model is sufficient. For Class III convex hulls, two volume models are needed, one to define the volume for the triangles totally within the PPT and another to define the volume of triangles that straddle the PPT's edge. Each one has its own volume, and a particular number of times its volume is multiplied to describe the whole sphere.

Table 9.8 tabulates the convex hull volumes for each unit radius spherical schema we developed. The table lists the basic section, volume, and the number of times this volume is replicated to define the entire sphere's convex hull volume. Class III schemas require two volumes and the total number of times each appears in the total volume of the sphere; their

Subdivision Schema Convex Hulls						
Schema	Volume	Instances	Sub-Total	Total	% Sphere	Rank
Class I						
Eaual Chords	0.207634	20	-	4.152685	99.138%	3
Equal Arcs (2 GCs)	0.207629	20	-	4.152571	99.135%	4
Equal Arcs (3 GCs)	0.207652	20	-	4.153034	99.146%	1
Mid Arcs	0.207637	20	-	4.152740	99.139%	2
Class II						
Triacon	0.138018	30	-	4.140535	98.848%	6
Class III						
Mid Face	0.144326	20	2.886529			
Straddling Edge	0.041928	30	1.257845	4.144374	98.940%	5

Table 9.8. Subdivision schema convex hulls.

subtotals are listed separately. The total convex hull volumes are then listed and ranked as a percentage of an undivided unit sphere's volume.

The analysis once again indicates that Class I Equal-arcs (three great circles) has the best tessellation point distribution within Class I; its convex hull volume is the greatest. The differences in convex hull volumes are due only to their point distributions because each Class I schema placed the same number of points (642) on the sphere's surface and none are tangent. Convex hull comparisons can still be made if the number of points distributed by several schemas are not all the same. Class II and Class III schemas distribute 482 and 522 points, respectively. We are comparing volumes only, and neither of these two schemas are as good as Class I schemas.

Convex hull volumes are excellent ways to compare two or more tessellation point distributions. For comparisons to be meaningful, each sphere must have the same radius and it is best to have the same number of points on their surfaces. In all comparisons, high-precision surface definitions are required because volume differences can be small.

9.8   Spherical Caps

A spherical cap is a slice of a sphere created by passing a plane completely through the sphere, without intersecting the origin. A cap looks somewhat like a contact lens and each one, centered over a tessellation vertex, has a spherical surface area and volume. Figure 9.8(a) shows caps for a Class II triacon icosahedron 10^v. All have the same radius, and they are as large as they can be without causing overlaps. We can use their total surface area to measure the efficiency of the subdivision schema. The ratio of the total cap area to the sphere's surface area is a density factor that tells us the percentage of the sphere's surface they cover. The subdivision schema with the highest density factor for the same number of points distributed on the sphere has the better distribution of those schemas.

Caps kiss and touch, just as kissing-touching balls do, and they appear to be measuring the same thing. However, the cap's lesser circle diameter is based on the arc distance between the two closest tessellation vertices, as shown in Figure 9.8(b). The arc distance is measured along the surface of the sphere (accented arcs). The ball's diameter is also based on the two closest tessellation vertices, but it uses the chord distance between them instead

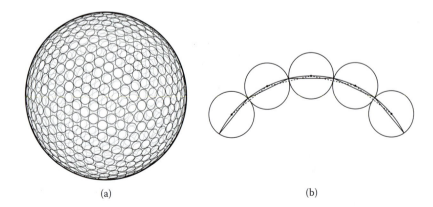

(a)   (b)

Figure 9.8. Spherical caps and kissing-touching balls.

(dashed lines). The result is the spherical cap's lesser circle diameter is a little larger than the kissing-touching ball radius for the same subdivided sphere.

The size and distribution of spherical caps is an important measure of how evenly distributed grid points are on the sphere's surface. We can compare one schema to another by comparing how much of the sphere's surface is covered by their caps. We can also make stereograms and visually see how spherical caps are distributed.

9.9   Stereograms

A *stereogram* is a 2D projection of the surface of a sphere onto a plane called the *primitive*. Every point on the surface of the sphere can be projected onto the primitive except one point: the projection point itself. In a sense, a stereogram is a 2D map, with some distortion, of the sphere's 3D surface (Appendix B describes the technique in detail).

Figure 9.9 shows the dimensions and formulas we use to find a cap's volume and area. Table 9.9 uses these formulas to find the area and density of the caps for the schemas we

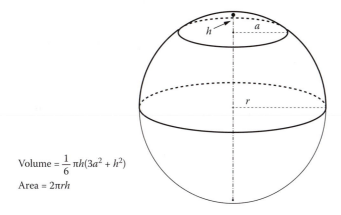

$$\text{Volume} = \frac{1}{6}\pi h(3a^2 + h^2)$$

$$\text{Area} = 2\pi rh$$

Figure 9.9. Spherical cap dimensions.

Spherical Cap Comparisons						
	Cap's					
Schema	Radius (a)	Height (h)	Area	Total Caps	Density	Rank
Class I						
Equal Chords	0.059729	0.001785	0.011218	642	57.31%	5
Equal Arcs (2 GCs)	0.066339	0.002203	0.013841	642	70.71%	4
Equal Arcs (3 GCs)	0.069142	0.002393	0.015037	642	76.82%	1/2
Mid Arcs	0.069142	0.002393	0.015037	642	76.82%	1/2
Class II						
Triacon	0.076658	0.002943	0.018489	482	70.92%	3
Class III						
Skew	0.064323	0.002071	0.013012	522	54.05%	6

Table 9.9. Spherical cap comparisons for unit radius sphere.

developed in the last chapter. Figure 9.10 through Figure 9.12 are the stereograms for each of the schemas. Each stereogram shows tessellation vertices and their caps in the northern hemisphere of the icosahedron. The ones in the southern hemisphere produce the same stereogram (but are rotated 32° around the center of the stereogram).[2] They do not add new information and, if superimposed, make stereograms difficult to read, so we omit them.

To facilitate comparing one schema with another and to highlight how caps relate to the base spherical polyhedra, the icosahedron is kept in the same vertex-zenith position in each figure. Our view is from the positive z-axis looking directly at the primitive on the xy-plane. The icosahedron's great circle edges (solid) and face medians (dashed) are projected as well to make it easier to interpret the pattern and see how points and their caps relate to the icosahedron's faces.

In each stereogram, one point (dot) surrounded by a circle appears on the primitive for every vertex-cap pair in the northern hemisphere or along the sphere's equator. Although the caps on the actual sphere all have the same radius, their projection on the primitive does not; their radii increase as they get closer to the circumference of the primitive. Projected circles are not centered on their corresponding vertex points, either. These subtleties are difficult to read at the scale of these stereograms, but the reasons are explained in Appendix B.

Figure 9.10(a) and Figure 9.10(b) show spherical cap stereograms for Class I Equal-chords and Equal-arcs (two great circles), respectively. Caps are evenly distributed in Equal-chords, but few touch. The overall density (percentage of sphere's area covered) is relatively low—only 57.31 percent of the sphere is covered, as shown in Table 9.9. Caps appear around the circumference of the primitive, one of the icosahedron's six mid-edge great circles. The pattern of their spacing is regular, but the distances between points are not all the same. The cap size and vertex distribution of Equal-arcs (two great circles) are better than the cap size and vertex distribution of Equal-chords, despite the asymmetry of the points within each PPT. The largest cap possible is determined by the arc distance between points near one edge rather than at the PPT's apices. They are the same points that were highlighted in kiss-touch balls in Figure 9.8(b). If you look closely at the caps along the PPT's edges, you will notice they are not all the same distance from the edge. Compare the caps along the two edges circled. The ones along the edge labeled A in Figure 9.10(b) are closer to the edge than the ones along the edge labeled B. This is because this Equal-arcs schema defined tessellation points by intersecting only two great circles, rather than three.

Figure 9.11(a) and (b) show the spherical cap distributions of Equal-arcs (three great circles) and Mid-arcs, respectively. Equal-arcs' caps, along every PPT edge, are evenly spaced and touch each other. Caps also follow every PPT median, though they do not contact each other. From prior analysis, we expect this Equal-arcs schema to be a good performer and its stereogram confirms this. Caps are large, evenly distributed, and cover 76.82 percent of the sphere's surface, the highest percentage thus far.

Figure 9.11(b) shows Mid-arcs' caps, and they cover 76.82 percent of the sphere's surface. Surprisingly, their density is identical to Equal-arcs (three great circles); see Table 9.9. Like Equal-arcs (three great circles), its cap radius is determined by the closest two points in the grid and this occurs along the PPT's edge, where all of these caps touch. These two schemes perform identically and better than the others. They are tied for the first two rankings.

---

[2] For the vertices of an icosahedron, there are six axes with five-fold rotation symmetry. The icosahedron can be rotated around a vertex axis any multiple of 72°, and the appearance of the icosahedron is not changed. Faces in the southern hemisphere are rotated 32° with respect to those in the northern hemisphere. See Chapter 6 for details.

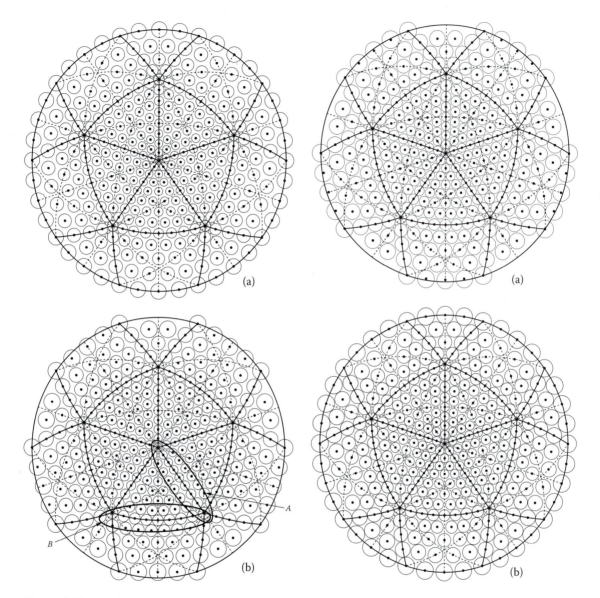

Figure 9.10. Equal-chords and Equal-arcs (two great circles) spherical caps.

Figure 9.11. Equal-arcs (three great circles) and Mid-arcs spherical caps.

Mid-arcs' stereogram reveals one of its symmetry features. Recall from Section 2.1 that one of the icosahedron's many sets of great circles are six that span from mid-edge to mid-edge across each of its faces (see Section 2.5). These circles define equilateral triangles within every face of the icosahedron, exactly like Mid-arcs' initial subdivision does. As Mid-arcs continues to subdivide the PPT, the edges of this inner equilateral become increasingly subdivided. Its frequency is half the overall frequency. In our 8^v example, its edges will be 4^v. The

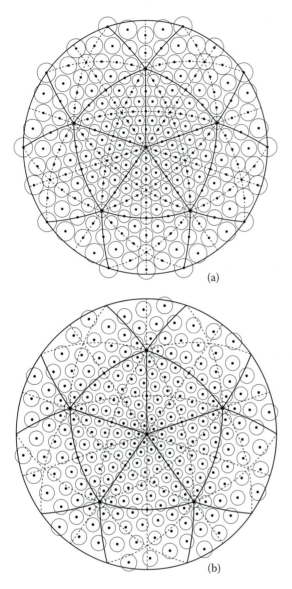

(a)

(b)

Figure 9.12. Class I triacon and Class III spherical caps.

equally spaced points along these inner-PPT edges project as equally spaced caps around the circumference of the primitive. We can verify that these caps belong to one of the six great circles. Each edge of the inner-PPT equilateral consists of four segments with arc lengths of 9° for a total of 36° per edge. Each of these six great circles passes through ten such edges as they complete their 360° tour of the whole icosahedral sphere. Unlike the caps along the PPT's edges, these caps do not touch because their tessellation vertex points are farther apart than those along the PPT's edges. Caps aside, the fact Mid-arcs subdivisions result in equally spaced points on these six great circles may be exploited in some design situations.

Triacon schemas result in a diamond pattern (four LCDs back to back) symmetrical to each edge of the spherical polyhedra. Figure 9.12(a) highlights a typical diamond. The pattern of caps within a diamond is also distinct. They are evenly spaced, but not tangent, along both of the principal triangle edges. The triacon is limited to even frequencies. Thus, they always create a vertex and a cap centered at the intersection medians in the middle of every PPT face. None of the Class I schemas do this. Unlike the other schemas, the smallest arcs are found in the middle of the PPT's face and not at its apices. The triacon caps are the largest of the schemas, but its grid distributes only 482 vertices over the sphere, the lowest number, when compared with the other schemas. Even though its cap area is the largest, they cover only 70.92 percent of the sphere.

Figure 9.12(b) shows spherical caps for a Class III schema. Inspection of the spherical caps shows the vertex distribution is not very good. For starters, this 8^v subdivision distributes only 522 vertices on the full sphere, compared to the 642 distributed by each of the Class I schemas. The smallest arc distance between grid points is one of the smallest in the six schemas we review, and this determines the cap's radius. This schema also has the largest arc distances between vertices, which means spherical caps near the middle of each polyhedral face are quite separate from their neighbors' caps. The stereogram clearly shows this. Caps in this schema cover only 54 percent of the sphere, the lowest density of any of the schemas we are comparing.

To summarize, spherical caps are slices of the sphere where the plane of the slide does not pass through the origin of the sphere. The best schemas distribute the most vertices with the shortest and least variation in arc lengths between vertices. The radius of the cap's lesser circle is determined by the arc between the two closest vertices within a subdivision schema. A spherical cap surrounds each tessellation vertex within the schema. All have the same radius and, while some caps kiss-touch, none will overlap. The ratio of the total cap surface area to the sphere's surface area is called the density factor and it tells us how much of the sphere is covered. The more area covered, the more uniform the tessellation vertices are distributed.

The larger the cap area, as a percentage of the sphere's area, the more uniformly the tessellation vertices are distributed. Stereograms provide a convenient way to see the cap distribution. Various symmetries within the schema are apparent, and they can be compared side by side or overlaid. Of the schemas reviewed, Equal-arcs (two and three great circles) and Mid-arcs schemas perform the best. Class II triacon was also a strong performer, but not as good as the other two. Triacon has larger caps than the others, but it also distributes the fewest tessellation vertices on the sphere of any schema.

9.10   Face Orientation

Thus far, we have compared kissing-touching distances, chord lengths, triangle areas, and shapes, convex hulls, and spherical caps. These metrics are very helpful when deciding which schema is best for an application. However, none of them tells us about the orientation of tessellation faces. Different shaped triangles can have the same kissing-touching distance, but very different orientations. We saw this earlier when we created triacon subdivisions for each of the Platonic solids. Here we measure the direction of faces. In some applications, face orientation may be the deciding factor when choosing one subdivision schema over another.

Every planar face we create in a subdivision is part of an infinite plane in 3D space, and each one has a normal. In Chapter 4 we described a normal as a mathematical entity that indicates both direction and magnitude. The direction of a normal to a plane is always perpendicular to the plane. Depending how the plane (or triangle) is defined, the normal's direction points towards the origin of the sphere, or away from it. In either case, however, the normal always points perpendicular to the plane of the triangle and everything on it.

Another way to visualize a normal is to think of its direction as a line that radiates from or towards the origin of the sphere and passes perpendicular to the plane of the triangle. The sphere's surface has an infinite number of planes tangent to its surface; thus, it also has an infinite number of normals. However, when a sphere is subdivided, triangular faces result and their planar versions each have a normal. Normal vectors have length, in addition to direction, and their length depends on the perpendicular distance between the triangle's plane and the origin of the sphere. We can use *face normals* as pointers to see which direction a face point is relative to the center of the sphere. Since the line of the normals pierces the sphere's surface at different points, we can use the normal to derive a Cartesian point that is on the surface of the sphere. Stereograms of derived points from the normals of all the subdivision faces tell us how uniformly subdivision faces are oriented. It is easy to define these points. All we have to do is "normalize" the normal, which means make its length equal to one, the radius of the sphere we are working with.

Figure 9.13 illustrates the relationship between subdivision faces, their face normals, and the projected point on the sphere derived from them. The figure shows a single Class II 8^v diamond subdivision for a spherical tetrahedron. Each arrow is perpendicular to the face at its base. The mixed orientation of triangles is evident. The arrows (blue) indicate the normals

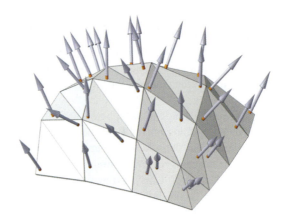

Figure 9.13. Projected face normals.

for their corresponding face. The arrow's base is tangent to the sphere's surface, indicating where the normalized (unit) normal is on the surface of the sphere.

In Figure 9.13, the normal arrow's blue part is totally outside the sphere, and all arrows are the same length to make it easier to compare their directions. The base of each arrow (red) is extended inside the sphere's surface to be tangent to the point where the normal direction line intersects its triangle. The extended arrow lengths (red) indicate how far above the plane of the triangle the projected normal point is above the sphere. Clearly, they are not all the same length. Their lengths decrease to zero as their intersection points get closer to one of the triangle's vertices. Although there are no equilateral triangles in this example, the triangles that are nearly equilateral have normal arrows located closest to their centers.

In Figure 9.14 we see the normals for fully subdivided triacon spheres. Both are Class II 8^v subdivisions, and each triacon diamond in both spheres has 32 faces. Figure 9.14(a) shows a subdivided tetrahedron, while (b) shows an icosahedron. It is obvi-

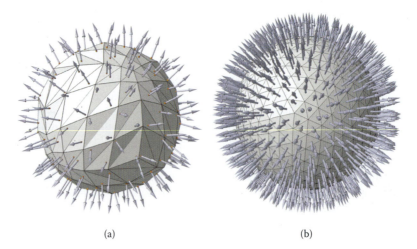

(a)   (b)

Figure 9.14. 8^v Class II spherical tetrahedron and icosahedron face normals.

ous that the icosahedron's face orientations are more uniform. The perpendicular face arrows point outward more consistently than the tetrahedron's. The main reason is that the icosahedron's triacon diamond is smaller than the tetrahedron's, which means more of them are needed to cover the sphere. More and smaller diamonds mean more vertices are distributed on the sphere and they are closer together, forming smaller triangles that are nearer to being equilateral than the tetrahedron's. The icosahedron has 30 characteristic diamonds (one diamond for each polyhedral edge), and this creates 960 triangles in the overall sphere, while the tetrahedron has only 6 characteristic diamonds, which create 192 overall faces. The improvement in face orientation is easy to understand. More diamonds and more points mean smaller faces with near-equilateral shapes. It all adds up to a better distribution of face orientations and its normals (arrows) show it.

To better understand face orientations, we use stereograms to compare the subdivision face normals from one schema to another. Normals to the planes of subdivision triangles cannot be used directly to make stereograms; however, we can derive a Cartesian point from the normal that is on the surface of the sphere and can be stereoprojected. The derived point will be located where the normal's ray pierces the surface of the sphere and it is this derived point that we stereoproject (see Appendix B for a complete explanation of how this is done.)

Figure 9.15(a) and (b) show the projected derived points for the face normals of the Class II tetrahedron and icosahedron subdivision just discussed. The fact that the projected points are more spread out in (a) than in (b) indicates that the tetrahedron's faces have a greater dihedral angle between them than the icosahedron's. The stereogram clearly shows this. In Figure 9.15(a), the tetrahedron's projected derived points are not only spread out more but they pair up in rows parallel to the diamond's long axis. This also happens in the case of the icosahedron (b), but to a far lesser extent. These stereograms confirm what we have already observed in the 3D models. For the same subdivision schema and frequency, the icosahedron has more uniform face orientations.

The differences, shown in Figure 9.15, are quite dramatic, largely due to the choice of foundation polyhedra (tetrahedron versus icosahedron). Here, stereographics merely verified

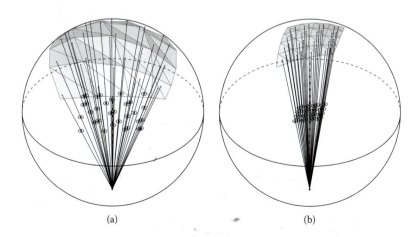

(a)   (b)

Figure 9.15. Class II 8^v tetrahedron and icosahedron stereograms.

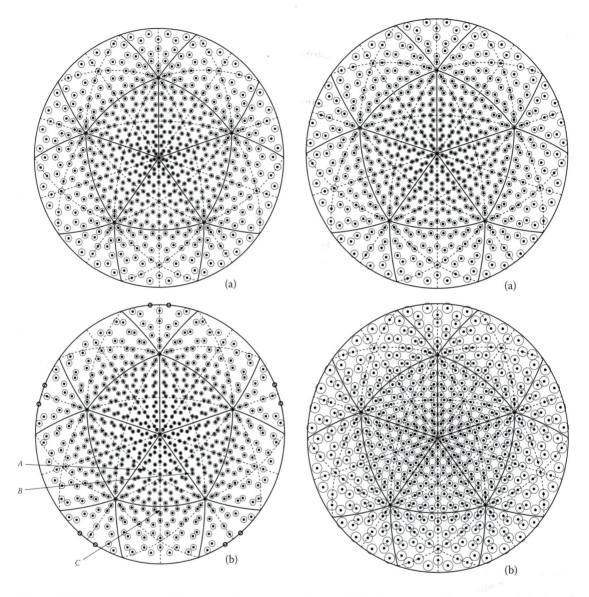

Figure 9.16. Equal-chords and Equal-arcs (two great circles) face orientations.

Figure 9.17. Equal-arcs (three great circles) and Mid-arcs face orientations.

something that was already obvious. However, when the foundation polyhedra are the same, the differences in face orientations between one schema and another for the same frequency can be subtle, and here stereographics prove to be an invaluable tool.

Figure 9.16(a) and (b) show stereograms for Equal-chords' and Equal-arcs' (two great circles) points derived from each subdivision face normal. Their stereograms are similar, but there are key differences. For example, it is apparent that derived points tend to pair up, more in Equal-arcs schema (b) along the icosahedron's edges marked A and B, whereas the points for Equal-chords (a) tend to be more uniformly distributed. Neither schema projects derived points on the icosahedron's edge, but both distribute points along face medians. Only trivial differences occur in the derived points within a principal triangle face.

Overall, face orientations for the Equal-arcs (two great circles) are not as uniform as Equal-chords. Equal-arcs' faces are only symmetric to one median per principal triangle, and the tessellation grid (and the faces it creates) is not rotationally symmetric around the center of the icosahedron's face. This can be seen in the projected derived points along PPT edge C. They are not the same as those along edges A and B, and this is a serious drawback for some applications. The stereogram clearly shows the problem.

Figure 9.17 shows the projected derived points from the face normals of both Equal-arcs (three great circles) and Mid-arcs. Both look similar and the differences are subtle. Mid-arcs' points are more closely paired up than Equal-arcs'. Of the two, Equal-arcs' faces are more uniformly oriented.

Triacon schemas are based on a diamond layout that is symmetrical to the polyhedron's edges. Figure 9.18(a) highlights a typical diamond. The most dominant features of the stereogram are the pairs of projected points in rows parallel to the icosahedron's edges. These rows are somewhat aligned to polyhedral edges on which the subdivision schema is based. The triacon schema also creates isosceles faces symmetric to every polyhedral edge; thus, it is no surprise to see projected derived points fall on the icosahedron's edge arcs. None of the other schemas routinely have derived points along the edges, though some frequencies of Class III will produce a few points there. Since the triacon does not produce a triangle in the middle of the PPT's face, there is no point at or near that point like there are in Equal-

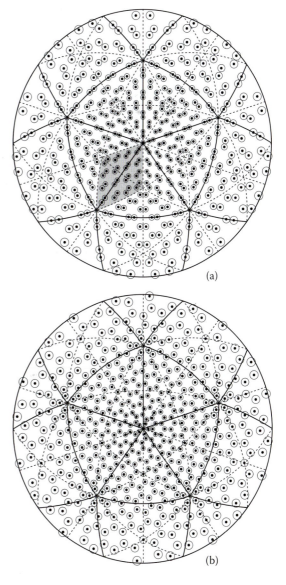

Figure 9.18. Class II triacon and Class III face orientations.

chords, Equal-arcs (three great circles), and Mid-arcs schemas. No triangle straddles the principal triangle's three medians. Thus, we will never find a projected derived point in the middle of the principal triangle, as we would in most Class I and some Class III schemas.

At first glance, Class III schemas seem almost impossible to compare to the others. The highly skewed nature of the overall tessellation grid, with respect to the icosahedron foundation, is immediately apparent in Figure 9.18(b). Unlike previous schemas, only one derived point per principal triangle falls on a face median. It is located at the midface point of the icosahedron. None of the other points fall on an icosahedra edge, median, or vertex. Despite all of this, the stereogram indicates a surprisingly uniform distribution of face orientations.

Uniform face orientations are a direct result of two factors: the number of points distributed on the sphere, and the degree to which the chord distances between them are equal. When the subdivision frequency is the same, subdivisions of spherical icosahedra produce better results than can be achieved with octahedra or tetrahedra because they start with more subdivided principal triangles and smaller chords. However, within the same polyhedral base, Class I Equal-arcs (three great circles) and Mid-arcs have the most uniform face orientations.

9.11 King Icosa

By almost any measure, the icosahedron is the best polyhedral foundation to use, when the goal is to create the most uniform point distribution possible, regardless of which subdivision schema is used. The main reason is its small face size: they have the least area (or least spherical excess) of any regular polyhedron for a unit sphere. The tessellation grid for a single icosahedral PPT defines only 5 percent of the total sphere's surface area, while the same frequency subdivision for an octahedron or tetrahedron must cover 12.5 percent and 25 percent of the sphere's area, respectively. We can illustrate this point graphically by subdividing all three polyhedra with the same frequency grid and simply noting the spac-

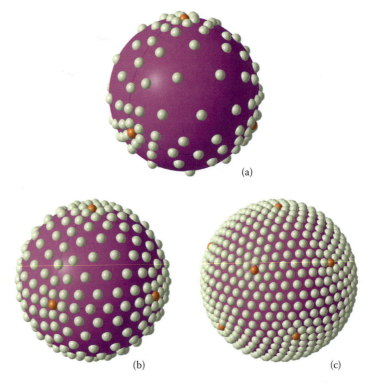

(a)

(b)   (c)

Figure 9.19. Polyhedra foundations compared.

ing between grid vertices. Figure 9.19 marks the vertices of an Equal-chords 8^v grid on the tetrahedron (a), octahedron (b), and icosahedron (c). In each case, kissing-touching occurs around the polyhedron's vertices, which are designated by red balls. The uniform distributions achieved with an icosahedral foundation (c) cannot be disputed. Its smaller and more numerous faces lead to a significantly more uniform distribution than the other two. The icosahedron is king!

9.12   Summary

In the previous chapter, we introduced three basic classes of spherical subdivision and a few of their variations. In Class I, the polyhedron's principal triangle was subdivided by defining edge-reference points (Equal-chords or Equal-arcs). These reference points were used to position great circle arcs intersecting where tessellation vertices occur. Class I grids run approximately parallel to the edges of the PPT, and the edge remains part of the final tessellation grid. Subtle differences in the way great circles are intersected lead to subtle differences in the way points on the PPT are created. These differences influence chord lengths and triangle area, shape, and orientation, and affect rotational and reflection symmetry.

Class II subdivisions produce grids that are parallel to the altitudes of the principal triangle. Instead of subdividing the entire principal triangle, they subdivide a diamond-shaped area that is symmetrical to the polyhedron's spherical edge. This diamond has a major and minor axis, which further breakdown the diamond into four back-to-back right triangles. Subdividing just one of them defines the grid for the other three. Since Class II schemas are based on symmetrical areas around the polyhedron's edges (not faces), they can be applied to all five Platonic solids.

Class III is a variation of Class I, but its grid is skewed relative to the principal triangle's edges. Like Class I, Class III can be applied to tetrahedra, octahedra, and icosahedra, but not to spherical cubes and dodecahedra.

In this chapter, we analyzed the differences between each of the schemas for these three classes. The differences in the tessellation grids created from the various schemas are subtle, but significant. What we found is that there is no single "perfect" subdivision, each has characteristics that make one or two of them more appropriate for an application than others. Although we did not subdivide other polyhedra, it should be clear, the icosahedron generally provides the best spherical framework if uniform grids are the objective. Our hope is that the metrics developed here will provide a way of deciding which is best for your needs.

Additional Resources

Subdivision Schemas

Clinton, Joseph D. *Advanced Structural Geometry Studies Part I—Polyhedral Subdivision Concepts for Structural Applications.* National Aeronautics and Space Administration contractor report NASA CR-1734, Carbondale, IL: Southern Illinois University, 1971.

———. "A Group of Spherical Tessellations Having Edges of Equal Length." West Chester, PA: Clinton International Design Consultants, undated.

Wenninger, Magnus J. *Spherical Models.* Cambridge [Eng.], New York: Cambridge University Press, 1979 (new appendix, 1999).

Geodesic Math

Clinton, Joseph D. *Advanced Structural Geometry Studies Part I—Polyhedral Subdivision Concepts for Structural Applications.* National Aeronautics and Space Administration contractor report NASA CR-1734, Carbondale, IL: Southern Illinois University, 1971.

——. "Geodesic Math." *Domebook 2*, ed. Lloyd Kahn et al., Bolinas, CA: Pacific Domes (1971): 106–113.

Kenner, Hugh. *Geodesic Math and How to Use It.* Berkeley: University of California Press, 1976.

Kitrick, Christopher J. "Geodesic Domes." *Structural Topology* 11 (1985).

——. "A Unified Approach to Class I, II and III Geodesic Domes." *International Journal of Space Structures* 5.3–4 (1990): 223–246.

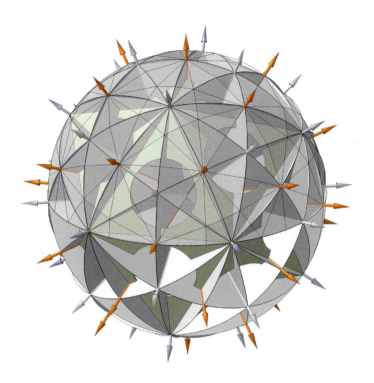

⑩ Computer-Aided Design

T he goal of this chapter is to introduce you to the basic concepts of computer-aided design (CAD) for spherical work and the many possibilities it offers and to provide examples of how CAD can be applied to spherical design. Very few spherical examples are provided in CAD literature; the focus is mostly on mechanical parts or sculptured surfaces for products such as cars, consumer electronics, industrial equipment, or aircraft. We assume the reader already has some basic understanding of CAD. Therefore, we will focus on approaches to spherical design and on practices that maximize the benefits of using CAD.

Most CAD systems offer rich and precise geometric modelers. One way to use these modelers is to rely totally on their geometric tools to subdivide spheres and use the resulting subdivision grid as reference geometry for subsequent designs. Another way to use CAD is to develop separate computer programs to generate the reference geometry, such as sets of grid points, and then import this geometry into the CAD program for further design refinement and visualization. In this chapter, we will show both approaches.

CAD is more than an "electric pencil"—it is one of the most significant resources available to the spherist. CAD programs offer superb geometric modeling, precise analysis, and outstanding visualization. Users benefit from many add-on applications, which use and enrich the initial geometry. Once a spherical subdivision or layout is created, many reports, visualizations, analysis, and data extractions can be made. CAD opens the door to many design possibilities, including automatic drafting, interference analysis, kinematics, and numerical control output that can automatically fabricate and assemble spherical parts.

Why use CAD at all? The short answer is you don't have to. However, when spherical subdivision frequencies increase, using calculators (even programmable ones) or writing your own display programs can be time-consuming, error-prone endeavors. In the author's experience, it is more productive to write a few simple geodesic programs to create starter geometry, and then import that geometry into the CAD program for further development, visualization, and analysis.

The examples developed in this chapter demonstrate the wide spectrum of CAD capabilities. The CAD program used is CATIA (Computer-Aided Three-Dimensional Interactive Application), a leading-edge geometric modeler developed by Dassault Systèmes. There are many other CAD programs with similar functionality. We prefer CATIA because of its flexibility, precision, and ease of use. In the next chapter we discuss techniques that will increase your productivity and the quality of your results. Some are more useful when there is a collaborative project and designs need to be exchanged or integrated, or when work needs to be coordinated within a team. We also discuss how to develop useful spherical CAD tools that might be missing from the CAD program you are using. However, before we start working with spherical designs, let's look at the short and dramatic history of CAD technology itself.

10.1   A Short History

CAD systems have existed since the early 1960s. The first systems were primarily used for 2D drafting in the automotive and aerospace industries. Until the advent of minicomputers, large mainframes were needed to power these applications, and peripheral devices, such as drum and flatbed plotters or cathode-ray tube displays, were specialty instruments found only at research centers or universities. Today, affordable CAD programs available for personal computers offer enormous geometric and visualization power for spherical research and development. Today's leading systems are 3D, and they are used in virtually every field, including architecture, plant design, consumer goods, microelectronics, industrial equipment, packaging, automotive, aerospace, and shipbuilding.

CAD evolved in stages. Early systems substituted light pens and computer monitors for T-squares and tracing paper. The emphasis was on high-quality drafting and managing changes and updates to drawings. In effect, CAD was a glorified electric pencil with some automation, such as dimensioning, symbols, and text generation. Two-dimensional drafting systems were followed by 3D modeling programs that represented objects as "wireframes" or a series of simple points and lines. Research and development, largely driven by the automotive and aerospace industries, eventually led to highly accurate solid modeling and surface-modeling techniques that could be further refined with features, such as holes, fillets, and chamfers in planar or curved forms. These geometric models could be analyzed structurally for weight, balance, surface area, and many other characteristics. It was now possible to automate drawing production, a huge cost savings, and to pass the product model information onto production equipment for milling, cutting, molding, or welding. At this point the need for a company to make costly prototype mockups from paper, clay, or wood was largely eliminated. Companies could go directly from a computer representation to a manufactured item.

One of the most impressive advancements in CAD has been visualization. Early CAD displays were limited to simple green glowing lines on cathode-ray tubes. But even modest CAD systems today offer full-color, solid, and transparent rendering; textured surfaces; hidden line removal; and photo-realistic scenes, shades, and shadows. Displays have improved, as well, and most offer high-definition digital images. Home entertainment systems, game stations, and the movie industry have accelerated the development of sophisticated graphics hardware, which offloads much of the computational work that was once done by software. Users can interact with CAD systems with electronic gloves, body sensors, and special tracking devices. Head-mount displays and 3D stereo displays add amazing depth perception as well. For the spherist, all of these developments make it easier to understand and work with complex spatial designs.

As users began to model more complex product designs, design rules and *knowledge-ware* applications appeared. These applications are part of the CAD system and guide the user's design process, informing the user when a particular arrangement of parts would be better than another, or when design features used in a previous project could be reused elsewhere. Of course, the user must define these rules ahead of time, but once they are established, anyone using the CAD system benefits. These applications greatly enhanced the business value and productivity of systems. Best practices and design processes unique to the way a company does business could be preserved and automated for others to use.

The Internet has greatly extended the reach of CAD systems. Not long ago, access to CAD was limited to specially trained operators. But today, many systems offer viewing or display programs compatible with standard web browsers, so virtually anyone with an Internet connection can view and extract reports from CAD models, without directly using the CAD system itself. The Internet has broadened the appeal of CAD and lowered its cost to companies by increasing the number of people who benefit from the data. As CAD was applied to larger and larger problems, CAD users soon became part of a larger collaborative effort, in which data-exchange standards, change control, and configuration management became important. Today, CAD systems are so prolific that the management of their data has spawned an entirely new generation of software, just to manage all the 3D models, their changes, and the documents created from them. This broad category of software is called *product data management (PDM)*. Most large CAD implementations use CAD and PDM systems together.

Although PDM is not CAD per se, it is a critical component whenever a project is complex, involves teams and design disciplines, or is subject to frequent revision and change. PDM technology is a sophisticated database mechanism for managing all the 3D models CAD produces and all their ancillary products, such as drawings, analysis, catalogs, and so forth. PDM also manages work between groups of CAD users, such as informing one group that parts of their design interfere with another group's design or some design change by one group requires changes from another group. In addition to providing change and interference management, PDM helps users organize and retrieve the models that are relevant to the work they need to do. For example, if a company makes several versions of a product, a designer might use PDM to retrieve only the 3D models that relate to the one on which he or she is working. Thus CAD and PDM go hand in hand when projects are complex, designs change a great deal, or work requires extensive collaboration.

The future of CAD is very bright and the 3D world it created is taking on new forms in gaming, social networking data mining, online shopping, 3D catalogs, electronic books, and virtual reality. The spherist has a lot to look forward to.

10.2 CATIA

Today's market offers a number of excellent 3D CAD systems. We use CATIA because it is highly precise and one of the most advanced geometric modelers available today. CATIA is the leader in geometric modeling, especially in high-precision surface definition. It also offers an extensive portfolio of applications, including kinematics, numerical control, automatic drafting, analysis, and knowledgeware. Each of these applications builds on and enriches the digital product model it creates. This chapter demonstrates many of these features.

CATIA, developed and supported by the French multinational company, Dassault Systèmes, is marketed and supported worldwide by Dassault, its subsidiaries, and business partners. CATIA is widely used in industry for the design and manufacture of everything from the Hubble telescope and Formula 1 racing cars to nuclear submarines and soda bottles. In addition to providing excellent geometric modeling, visualization, and analysis capabilities, CATIA's open architecture supports many industry-standard data exchanges. It also provides extensive customization and user-programming options. This chapter makes use of CATIA's version 5 visual basic application programming interfaces (APIs) to automate many common spherical design tasks.

We begin our discussion of CAD by recreating Synergetics, Inc.'s famous 1950s octet truss construction system connector developed for the Museum of Modern Art exhibition.[1] It introduces the essential CAD techniques, terms, and methods that we will use throughout the rest of the chapter.

10.3 Octet Truss Connector

The octet truss structural system is a revolutionary design offering high strength-to-weight ratios, which can be exploited in totally new architectural solutions to long span architec-

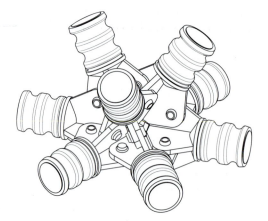

Figure 10.1. Cluster of octet truss structural connectors.

---

[1] Fuller applied for patent in February 1956 but was not awarded his patent until May 1961.

tural structural enclosures. One of its key innovations is the connector used to join structural members.

The octet truss connector joins truss members made of thin-walled metal pipes. Figure 10.1 shows a configuration of nine octet truss connectors (without pipes) and how, when they are joined together, they form a single structural joint. The geometry of both the connector and the overall joint, where multiple connectors come together, is based on the great circle planes of the spherical cuboctahedron. This connector, because of its simplicity, is an ideal way to show spherical design principles and introduce CAD concepts. We will reconstruct this famous connector using modern CAD technology. You may be surprised by how little geometric data is needed to fully describe this intricate connector.

10.3.1 Connector History

In 1959, the Museum of Modern Art (MoMA) in New York City featured work from designers and architects at various Fuller offices in an exhibition entitled "Three Structures by

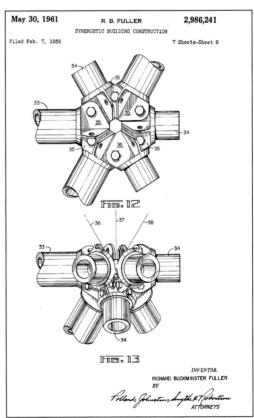

Figure 10.2. Buckminster Fuller's octet truss patent.

Buckminster Fuller."[2] The three structures were a tensegrity mast (made of pipes suspended in a maze of cables designed by Shoji Sadao), a geodesic dome (similar to the DEW Line radomes, shown in Chapter 2 developed by Bernard Kirschenbaum and William Wainwright), and an octet truss space frame designed by T. C. Howard.[3] The space frame, weighing approximately 8,000 pounds, was 100 feet long, 35 feet wide, and 60 feet high. Composed of pipe struts arranged in a tetrahedron-octahedron design, the pipes that made up the structural frame were joined by cast aluminum connectors, the ones shown in Figure 10.1. The tremendous strength of the structure was due to the triangulation of members, plus a connector design that kept all forces directed along the centers of the pipe struts. It was reported that a typical octet truss made of pipes and cast aluminum connectors and weighing only 65 pounds could support a load of 12,000 pounds.

In 1956, Fuller patented the octet truss and called it the "Synergetic Building Construction" (patent 2,986,241).[4] The patent illustrated two prototypes. One design employs extruded aluminum struts, such as the system used on the early 1950s Ford Rotunda dome. The other design uses tube struts joined together with cast-aluminum end connectors. It is the one shown in Figure 10.2.[5] The CAD example we will build is based on the second design, but we will use dimensions taken from the actual octet

---

[2] (Museum of Modern Art 1959)

[3] (Chu 2009, 139)

[4] (Fuller 1961, sheet 6/7, figs. 12 and 13)

[5] (Fuller 1956, sheet 6/6, figs. 12 and 13)

truss originally designed by T. C. Howard for the MoMA exhibition rather than the version Fuller patented. Howard's MoMA connector is highly refined with many subtle and creative features.

10.3.2 Connector Concept

Octahedra and tetrahedra can be maximum-packed. In other words, they can be arranged to totally occupy a volume without leaving voids. Figure 10.3, an illustration from Fuller's patent,[6] shows the basic octahedron-tetrahedron grid. Note that the tetrahedron to the left packs perfectly into the octahedron on the right, resulting in a triangular grid along the top and bottom.

Figure 10.4 replicates the octahedron-tetrahedron grid to cover more area and uses octet truss connectors to join the structural pipes. Several features are evident. For example, every octet truss connector has identical geometry. Each will join between four and nine tube struts. (There are exceptions for double-layer grids, or for cases where columns mate to the truss and 12 tubes can be accommodated by a single joint.) Notice that the trussing grid defines a series of planes. The dihedral angle of a tetrahedron is 70.528779366° or 70° 31' 43.605718", and its complement angle for the octahedron is the key to understanding the entire connector. We will use these angles to create our CAD model of the connector.

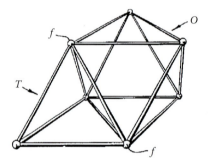

Figure 10.3. Octahedron-tetrahedron grid.

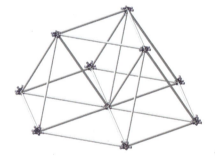

Figure 10.4. Octet truss general arrangement.

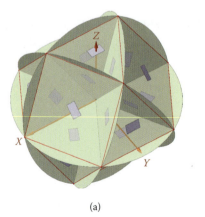

(a)

(b)

Figure 10.5. Octet truss connector with spherical cuboctahedron reference.

---

[6] (Fuller 1956, sheet 3/7, fig. 5)

The octet truss connector can be visualized as a series of great circle planes passing through the middle octahedron-tetrahedron grid vertices. Figure 10.5 shows both the spherical cuboctahedron (a) and the relationship of its great circle planes to a nine-way joint (b). The reference great circle planes (not an actual part of the connector) are shown in transparency in (b). A key concept behind the connector is the fact that one of the faces of every mounting "ear" is tangent to the plane of one of these great circles, and that the ears alternate sides of the plane, or a great circle, as they radiate out from the center of the sphere. We will show detailed views and exact angular relationships later. Notice, too, if you connect the ends of the protruding pipe hubs, they create the triangles and square patterns of the cuboctahedron.

In the next section we will show the basic steps involved in making a 3D representation of this connector. Then we will show a variety of visualizations and documents that can be extracted from the 3D model, such as orthographic drawings, sections, bills of material, weights, areas, and volumes. If this CAD model were to be used to actually manufacture connectors, it would likely be input into a computer numerical controlled (CNC) machine that would then make a female mold of it capable of mass-producing the connector in metal, plastic, or other moldable material. We do not describe the CNC or manufacturing processes.

10.3.3   Octet Truss Design Unit

Three-dimensional modeling of the basic octet truss connector is surprisingly simple. Keep in mind, the final model we will make will be used to make a female mold with a computer numerical controlled machine. The mold, in turn, will make connectors in metal, plastic, or other moldable material. Thus, an actual octet truss connector is not manufactured by welding small parts together; it is one, solid molded unit in aluminum.

To make a geometric model of the connector, only CATIA's basic part design and assembly applications are needed. We will model the ear first. A simple 2D sketch, also called a profile, is made of just the ear's perimeter and hole. Next, the 2D profile is then extruded to make a 3D form (CATIA calls it a pad) to the required thickness. A completed connector ear is shown in Figure 10.6(a). The height of the extrusion, in this case, is the required thickness of the octet truss ear.

Next, we model the pipe hub. It is even simpler to define than the connector ear. Once again, the starting point is a 2D sketch. The profile traces the required longitudinal cross

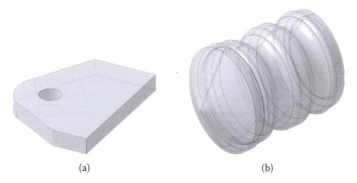

(a)   (b)

Figure 10.6. Sketcher profiles for octet truss pad and shaft.

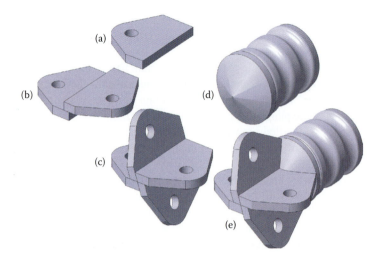

Figure 10.7. Octet truss connector modeling sequence.

section of the hub, which will define the crenellations that will attach the pipes when they are knurled onto the hub. The profile is shown in Figure 10.6(b). We rotate this profile completely around the center line of the hub. The resulting swept-surface defines the surface of the whole pipe hub. We convert the surface, essentially a hollow envelope, to a solid body. With the ear and hub defined, we are ready to make a full connector. Figure 10.7 shows the steps.

Referring to Figure 10.7(a), we build the multi-ear assembly first. The original ear is copied and the copy is rotated 180° around one of its edges, creating the double-ear shown in (b). Notice that both ears are perfectly symmetrical, share a common edge, and have faces that are 180° apart and aligned so that one face in each ear is coincident with the same plane. To define the other two ears, a copy of the double-ear in (b) is rotated, but instead of rotating the pattern 180° as before, we will rotate the pair only 70.52877° relative to the plane of the first pair. This is the dihedral angle, shown in (c). At this point, all four ears have been defined and positioned with respect to each other. Keep in mind that the four-ear assembly shown in (c) is nothing more than three copies of the original ear created in (a), the one made by extruding a simple profile and a circle. Already we have made the major part of our connector and we have done so with only a few CAD operations. In a small way, this demonstrates the amazing power of CAD.

All that remains is to geometrically combine the four-ear assembly (c) and the pipe shaft (d) together into a single logical and geometric unit. In CATIA, two or more individual 3D parts are joined together with solid modeling functions to make a single 3D part. Figure (e) shows the result of combining (c) with (d). Different views of the completed connector are shown in Figure 10.8.

As CATIA builds its 3D models, it captures all geometric information and logical relationships between geometric entities as you work. This information is maintained and displayed in a hierarchical list called a feature tree. In a sense, the feature tree is the life

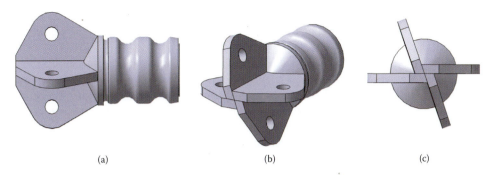

(a)   (b)   (c)

Figure 10.8. Octet truss pipe connector.

history of your design. Should you need to change the design, perhaps increasing the thickness of the mounting ears, or adding a different dihedral angle between the ears or a larger diameter pipe hub, you simply have to change only the feature affected in the tree and the entire connector model will automatically update. The productivity and data integrity value of this capability cannot be overstated.

Rivets. We need a way to join pipe connectors. The MoMA exposition's octet truss used two-part rivets and a special crimping tool to join them. A cap rivet was placed through the aligned holes of two adjacent connectors from the tetrahedral side. A collar was slipped over the protruding end and crimped from the octahedral side. To assist with visualization, structural analysis, extracting drawings, and listing bills of material, it is desirable to model and include the rivets in the final assembly. We use the same modeling technique for the rivet that we used for modeling the pipe hub. We make a 2D profile of the longitudinal cross section of the whole rivet and revolve the profile around its centerline axis using CATIA's shaft tool. It could not be simpler. The final rivet is shown in Figure 10.9.

We could have modeled both parts of the rivet separately; that is, one model for the cap rivet and another for its collar. This extra step in modeling both parts of the rivet, instead of modeling it as a single unit, has important benefits. CAD applications can innumerate every part and in some design situations, it is desirable to have a bill-of-material or parts lists that count every distinct part. Having separate part counts might be important. For example, there may be several different types of rivet caps or they may be sourced from different vendors, and knowing how many is important. In our example, however, separate part counts are not important; thus, we model the entire rivet as a single unit.

Figure 10.9. Octet truss rivet fastener.

10.3.4 Octet Truss Assembly

Thus far, we have built highly accurate 3D models of a single octet truss pipe connector and a single rivet. Let's assume these two models are not models for a casting mold but that they are final 3D models of the connector and rivet, as if they were already made in metal. In an octet truss structure, between four and nine struts typically form a joint in the tetrahedron-octahedron grid. Twelve-connector joints are possible, but not common. Here we review the steps for assembling a nine-way joint from a 3D model of a single connector and rivet.

Before we begin assembling a joint, it is important to understand the concept *degrees of freedom*. Any rigid body—a single connector or rivet, for example—has six degrees of freedom to move in 3D space. Any body can move or translate in the x, y, or z direction, and it can also rotate around the x-, y- or z-axis. Thus, three of these movements are translational and three are rotational. The same is true when two or more connectors or rivets need to be positioned or assembled together at a joint. Almost always, all six degrees of freedom have to be eliminated between pairs of bodies as the assembly is built up from individual parts (or other subassemblies). The challenge in assembling octet truss connectors is to arrange them in their proper orientation relative to each other and to remove all degrees of freedom so that there is no ambiguity about their orientation and position in the final joint. To do this, we will use *constraints*.

10.3.5 Constraints

In CATIA, constraints are mathematical entities added to the assembly's feature tree. There are several types of constraints, but they all define how two or more part models are positioned with respect to one another. A quick example illustrates the concept. Let's assume we are modeling a bolt and a plate with a hole in it to receive the bolt. We create two separate models, one for the bolt and another for the plate with a hole. To place the bolt in the plate's hole and unambiguously define its position there requires at least two constraints. The first constraint, a *coincident constraint*, is applied between the bolt's centerline axis and the hole's centerline axis. The bolt properly aligns along the hole's axis when the constraint is applied (added to the feature tree). Although the bolt is constrained to remain aligned with the hole, this one constraint is not enough to position the bolt *in* the hole; we need one more constraint. To insert the bolt *into* the hole and secure the bolt's head snug against the surface of the plate requires a *contact constraint* between the plate's surface and the underside of the bolt's head. When this second constraint is applied (added to the feature tree), the bolt is fully inserted into the hole. To review, the coincident constraint aligns the bolt with the hole; the contact constraint brings the underside of the bolt's head into contact with the plate's surface. With these two constraints, the bolt's placement in the hole is unambiguous with respect to the plate. If the plate is moved around in three dimensions, the bolt follows along and always maintains its snug placement.

We should point out that one degree of freedom remains in our example; the bolt is free to rotate or spin around in the hole. This is because the coincidence and contact constraints together did not remove the rotational degree of freedom between the two bodies. In this particular application, the remaining degree of freedom is not a concern. In fact, in applications where relative motion between parts is simulated, some degrees of freedom are intentionally left unconstrained; otherwise, no relative motion between parts would be possible.[7]

With this brief introduction to degrees of freedom and constraints, let's take a look at the other types of constraints. We will use many of them to assemble the octet truss and spherical designs later in the chapter.

---

[7] When parts are assembled in a mechanism where some parts must move relative to one another, such as a wheel spinning around an axis, some degrees of freedom are necessary. However, the omitted constraints are specifically limited to degrees of freedom matching the desired motion in the mechanism.

A part's location in 3D space can be completely fixed by assigning an *anchor constraint* to it. Once applied, the entire part has no degrees of freedom. It is completely "anchored" and cannot translate or rotate in any direction in 3D space. This also means that the axis system, used to define the part, becomes the axis system for any other parts constrained to it. When assembling parts, most CATIA users anchor the initial part and then proceed to include other parts, some or all of which will be constrained to the anchored part. To assemble our octet truss joint, we will anchor one connector in the assembly and then constrain the other connector's position to this single anchored one.

An *angle constraint* fixes the angular relationship between planes, edges, axes, or some other geometric feature between two parts. We use angle constraints to fix the center line angles between adjacent pipe hubs. Many hubs form a hexagonal pattern; therefore, the constraint angle will be 60° between adjacent hubs. Not every pair of connector hubs need angle constraints, however. It turns out that once the angle between one or two pairs of connector hubs is set, the remaining hubs will line up properly, when overlapping holes in their ears line up (with coincidence constraints). Four degrees of freedom are removed from each hub by the angle constraints, but each connector can still rotate around and translate along their centerline axis.

Offset constraints position one part at a fixed offset distance from another. The constraint can be applied to surfaces, planes, edges, and so on. It is not used in this octet truss assembly, though it is quite useful in general. An offset constraint can remove up to three degrees of freedom. The parts are still free to rotate around the axis of the offset constraint and translate on the plane of the offset, while maintaining the offset distance.

Contact constraints are just what you would imagine them to be. When this constraint is applied between two surfaces, edges, or faces on two different parts, the two parts remain in contact and three degrees of freedom are removed between them. The parts are still free to slide and spin across each other. Contact constraints are key components of our octet truss assembly. Each octet truss connector has four ears, and the surface of each of those

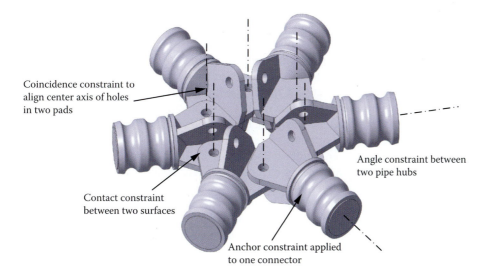

Coincidence constraint to align center axis of holes in two pads

Angle constraint between two pipe hubs

Contact constraint between two surfaces

Anchor constraint applied to one connector

Figure 10.10. Nine-way octet truss partial assembly.

Figure 10.11. Disassembled nine-way connector.

ears is in contact with the surface of the ears belonging to adjacent connectors. We use contact constraints a lot when we assemble complete spheres from part models of subdivided PPTs or LCDs. Figure 10.10 shows six of the nine connectors in a pinwheel layout around a center point. Four types of constraints are used to position each connector with respect to its neighbors and the original anchored one. An anchor fixes the first connector and an angle constraint fixes the angle between one pair of pipe hubs. Coincident constraints align the holes between overlapping connector ears and contact constraints keep the ears of one connector in contact with the ears of its adjacent connectors. Contact constraints insure that surfaces of connector ears stay in contact.

Figure 10.10 shows these constraints. When all nine connectors are assembled into a single joint, and the constraints applied, no translation or rotation of the joint can occur. All parts are unambiguously positioned in three dimensions. Each connector, in effect, locks-in (constrains) the positions of its neighbors.

We are not quite finished assembling the joint. We need to add the rivets. To assemble rivets, we can choose from at least two methods. The first includes the rivet model in the assembly and references the rivet wherever one is needed. We would apply coincidence constraints to align each rivet to the centerlines of the two holes from the two ears to be joined. A contact constraint between the bottom of each rivet's head and the top of the connector's mounting ear will seat the rivet. The technique is the same as the bolt-plate example given earlier. For a nine-way joint, this method would be repeated for all 15 rivets in the joint and require 30 constraints.

Another method, however, significantly reduces the effort, eliminates repetitive entries in the assembly's feature tree, and produces the same end result. Two rivets can be preassembled into a single connector, one each on two opposing mounting ears. When references to the connector are made in the assembly, the reference includes the rivets as well, since they are now part of the preassembled connector. When the preassembled connectors are constrained together, rivets from one connector fill empty holes in adjacent connectors. Thus, by preassembling just two rivets in our original connector model, the octet truss joint assembly is automatically riveted together, as each new connector is added to the joint. You can convince yourself that this simple technique works by studying the partially disassembled nine-way connection, shown in Figure 10.11. Notice that there are two preas-

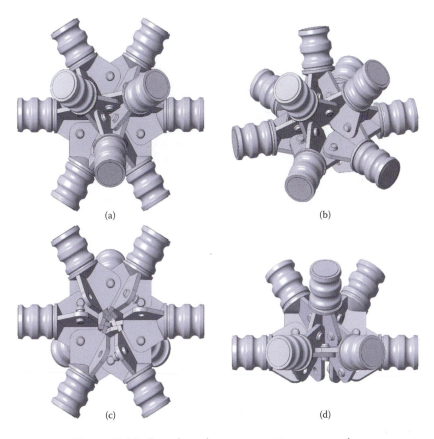

(a)

(b)

(c)

(d)

Figure 10.12. Complete nine-way octet truss connection.

sembled rivets per connector. This method, however, places unnecessary rivets whenever a connector does not have an adjacent counterpart. You can see unused rivets in the bottom of the image in Figure 10.12(c) and (d); however, they can be easily deleted.

10.3.6 Photo Realism

We have already presented many images of the octet truss, but here we show a specialized example of highly realistic views with texture, shading, and shadows. There are many uses for realistic views of a design. Often, a design must be communicated to nontechnical people and such visualization is a help. In some cases, photo realism is an essential CAD requirement. Consider the styling of automobiles or the design of a smartphone. How consumers react to its design, color, texture, and so on is critical to the product's success. The images in Figure 10.13 are simple examples of what can be done, even with an unadorned cast aluminum connector.

10.3.7 Drafting

Perhaps no area has benefited more from CAD than drafting. Previous generations of CAD systems focused entirely on producing 2D drawings; CAD functioned like an electric pencil.

Figure 10.13. Photo realistic octet truss connector and joint.

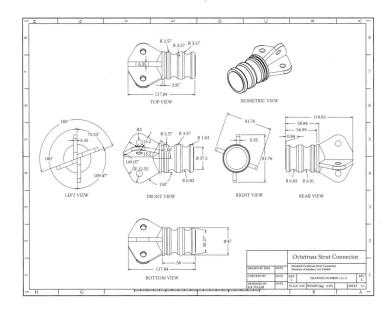

Figure 10.14. Octet truss connector automated drawing.

Today's systems automatically produce drawings directly from the 3D model. Drafting is no longer the objective of CAD; it is now a by-product of design.

CATIA's drafting application automatically produces standard views directly from its 3D models. It does so by maintaining a logical link between the 3D model and geometric elements, shown on the 2D drawing; if the model changes, the drawing changes. This way, drawings are always synchronized with the model.

To appreciate the value of this feature, consider how design changes were administered before the use of 3D CAD. In the past, whenever a change occurred, all affected drawings had to be located. This in itself could be a big task. Industrial products, plants, cars, airplanes, or ships involve thousands of drawings, and not all are kept in a single repository. The design change may impact any number of drawings. Moreover, a single drawing sheet may have several views of the affected area and each view may look quite different, depending on the type of projection, scale, or graphic standard used.

No wonder change management and drawing synchronization was (and still is) a major effort. The opportunities for error are considerable. Today, however, systems such as CATIA work seamlessly with PDM systems, such as ENOVIA, to keep track of which drawings include views of each model. This integration synchronizes 3D models and 2D drawings. The benefits are increased CAD-user productivity and reduced errors. These are important reasons why so many companies implement CAD and PDM systems together.

Figure 10.14 shows a fully automated drawing with standard American National Standards Institute (ANSI) views (top, front, side, isometric, etc.). The dimension standards are somewhat arbitrary, but enough cases are shown to suggest the power of using CAD and automatic drafting.

10.3.8 Mass Properties

A great deal of analysis can be performed on individual parts or assemblies of parts. Mass property analysis, for example, can include an object's volume, surface area, principal moments, and center of gravity. This data is very useful in cost estimation, production planning, and manufacturing. Even a simple thing, such as the amount of paint needed to paint a truss structure, can be calculated. Table 10.1 summarizes key properties for a single octet truss connector. For example, the weight of the connector, shown in Figure 10.13, is 7 pounds, 13 ounces (3.543.7 kg). The predicted weight for die-cast aluminum was within 2 percent of actual. This information is invaluable for estimating the materials needed for casting the joint, preparing shipping containers, or finding the total weight of a structure, if it is to be lifted by a crane. Structural calculations, where bending moments are considered, also depend on mass properties. Mass properties are automatic extractions of well-developed 3D geometric models, just as drawings are.

Product : Octetruss Joint		
Component		Area[m2]
		0.438
Volume[m3]	Density[kg_m3]	Mass[kg]
0.00139559	1000.000	4.096
Gx[mm]	Gy[mm]	Gz[mm]
95.577	0.027|	0.015
M1[kgxm2]	M2[kgxm2]	M3[kgxm2]
4.14187e-005	0.000157924	0.000163808
IoxG[kgxm2]	IoyG[kgxm2]	IozG[kgxm2]
4.14189e-005	0.000159194	0.000162538
IxyG[kgxm2]	IxzG[kgxm2]	IyzG[kgxm2]
1.33505e-007	7.61432e-008	-2.42019e-006

Table 10.1. Octet truss connector mass property report.

10.3.9 Sectioning and Space Analysis

CAD offers many types of space analysis. They can be as simple as measurements between points, edges, or faces or as complex as interferences and sectioning. Figure 10.15 shows a dynamic section through the horizontal plane of the octet truss connector. Part (a) is the profile of a section cut through the connector. Part (b) is a view from underneath the connector at the plane of the section cut. The connectors are shown as shells or thin surfaces to make the relationship of the individual connectors and their features easier to analyze.

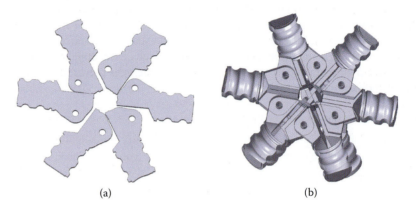

(a)   (b)

Figure 10.15. Octet truss connector sections.

10.3.10 Summary

The octet truss connector example only scratches the surface of what CAD techniques can do. From a modest start with two simple 2D profiles (three if you include the rivet), an accurate 3D model of an intricate connector is built and assembled. Based on this product model, we are able to visualize and analyze the whole design. We are also able to extract drawings from the product model. The reminder of this chapter applies these CAD techniques to designing spherical parts and assembling them to cover an entire sphere.

10.4 Spherical Design

In the preceding section, we demonstrated basic CAD capabilities. In this section, we develop a series of spherical designs based on subdivision schemas presented in earlier chapters. To make it easy to compare techniques, each example will follow the same format:

- *Design preview*. Gives a quick overview of the design to be developed.

- *Reference geometry*. Provides the starting point for 3D modeling, including the first points, lines, and edges.

- *Basic unit design*. Explains the steps taken to model the essential parts of the design.

- *Design assembly*. Puts the basic units together to make a full sphere.

- *Analysis, visualization, and data extractions*. Demonstrates what can be done with the completed sphere.

- *Summary*. Recaps the highlights of the design example.

Each of our design examples start with a reference model that is used to define the spherical design's theoretical framework (points, lines, planes, and surfaces). The reference model does not contain any geometric information regarding a real connector, panel, or strut. For example, the reference model might contain a line indicating where the center line of a pipe will be, but in the final design, it does not contain the geometry of the pipe itself. The reference model might also contain planes. They might indicate the limits of a surface or pad in the final design.

In the next step, the reference model is used to build up the actual spherical design parts, such as struts, connectors, surfaces, or panels. For example, in this step, a pipe of a particular diameter, length, and wall thickness is created using the center line in the reference model. Most of the time, spherical designers develop parts for only a single LCD or PPT. When the final design of this small area is complete, it is instantiated many times. When a part is instantiated, it is not copied. Instead, a reference to the part's original geometry is created and new location information is added, which describes how the original part (the one referred to) is to be transformed (moved, rotated, scaled, etc.) to position it somewhere else. Instancing allows us to cover the entire sphere with a design based on just one LCD or PPT. Each instance of the small part is positioned side by side and aligned properly on the sphere by applying constraints between points and edges of adjacent parts. A group of constrained parts defines an assembly (of LCD or PPT parts). Due to the symmetry in most spherical designs, the assembly itself can be instanced many times and constraints applied to make still larger assemblies. For instance, there are 120 LCDs in a spherical icosahedron. A single spherical design for one LCD could be instanced six times to make a single assembly for one icosahedral PPT. This assembly, in turn, could be instanced 20 more times. Constraints between each instance positions them to cover the entire sphere. This scenario is only one possibility. We will demonstrate some designs where there are several levels of assemblies and subassemblies.

In a sense, what we have just described is using CAD such as a digital brick layer. We design one brick—a very precise and intricate one—and then lay instances of this brick (references to the original brick, plus geometric transformation information describing its new location) side by side to make a wall. Our bricks and wall, however, are spherical triangles, hexagons, and hyperbolic surfaces. Mathematical constraints between assemblies act as mortar. The reference methodology we use here is applicable to any spherical polyhedron, subdivision method, design layout, or frequency. It exploits the high degree of symmetry and repetition found in spherical designs. Our method begins with reference models.

10.4.1 Start with a Few Good Points

Reference models greatly facilitate spherical design. Every example in this chapter uses one. The first three examples reuse the same reference model, while the last two use a different one. A reference model is made with CATIA and it contains only points, lines, surfaces, and planes. There are no physical design elements in the reference model; that is, it does not contain elements such as connectors, panels, or struts.

The reference model is highly precise, and its layout is carefully chosen to include only the minimum number of elements that will define the reference geometry needed

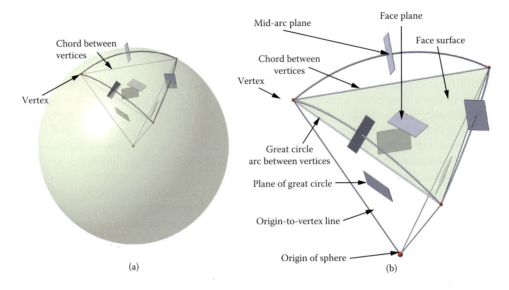

Chord between vertices

Vertex

Mid-arc plane

Face plane

Face surface

Chord between vertices

Vertex

Great circle arc between vertices

Plane of great circle

Origin-to-vertex line

Origin of sphere

(a)

(b)

Figure 10.16. Simple spherical reference model.

to build our parts. Reference models might seem an unnecessary step, but they are well worth the effort. If they are accurate, every spherical design, based on them, will be accurate as well.

Reference models can be reused repeatedly to make a surprising variety of spherical designs. Our first three examples all use the same reference model, but their physical designs could not be more different. Figure 10.16 shows an overall (a) and detailed (b) view of a very simple spherical triangle reference model. None of this reference geometry is visible, but we have used small spheres, cylinders, pads, and coloring to indicate where points, lines, planes, and surfaces are located.

To build this model, we calculated four precise points. Three points lay on the surface of the sphere and define the triangle's vertices. They have the exact angular relationship to one another we want, and they are perfectly tangent to the surface of the sphere. The forth point defines the sphere's origin. The points used in this figure are the vertices of a standard spherical icosahedron face; thus, this model also describes a spherical equilateral triangle.

Using just these four points, CATIA part design and surfacing tools will define everything else. A line tool defines vertex-to-vertex lines (the triangle's edges or chords) and vertex-to-origin reference lines. A plane tool defines planes at the triangle's face and at the three great circles, which make the sides of the spherical triangle. Three more planes are defined normal to both the triangle's mid-chord and mid-arc sides. A circle tool defines great circle arcs on the planes between vertex points.

The reference model is now complete and includes one spherical surface (only shown in Figure 10.16(a)), three points on the spherical surface (which define the icosahedron's face triangle), one origin point, three vertex-to-vertex chord lines, three vertex-to-origin lines, one triangle face plane, one triangular surface, three great circle planes, and three mid-chord/arc normal planes. CATIA defined all of these elements from the initial four input points.

Most of the time, the number of starter points is quite modest, and you can calculate them with a scientific calculator, using the equations provided in previous chapters. In other cases, you may find it easier to use the CAD program itself to calculate the points. We use both techniques in this chapter. Regardless of the input method you use, you may be surprised at how few points are actually needed to build the reference model of a complex spherical design.

The spherical designs in this chapter are based on equal-arc (thee great circles),triacon, and skew subdivisions. A triacon reference model of a single-edge diamond, or even half-diamond, is usually enough to describe a design covering the entire sphere. For equal-arc subdivisions, a single LCD or PPT is used.

Most CAD systems have reflection or mirroring tools. Any geometry can be copied and reflected around a specified plane. Thus, the reference geometry for a left-handed LCD can be reflected to make a right-handed LCD with minimum work. If the design was completely symmetrical around one great circle, the design model for one hemisphere can be reflected to complete the other hemisphere. Reflection is just one of many labor-saving CAD tools.

Some subdivisions result in enantiomorphic shapes. For example, triangular, pentagonal, or hexagonal shapes that do not have equal-length sides often reoccur in other parts of the subdivision as left- or right-handed versions. Copy-reflecting can solve most of these cases. The Ford hex-triangle example, developed later in this chapter, demonstrates enantiomorphic conditions, which are easily solved by mirroring the proper reference model geometry.

10.4.2   Spherical Assembly

The octet truss example introduced constraints and their role in assembling 3D models. In that example, individual octet truss connectors were assembled, along with their rivet fasteners, into a large, nine-way connector.

Spherical assembly makes use of the same techniques, but because the number of parts is so much bigger, even greater productivity gains are possible with careful assembly techniques. A complex and detailed section of the sphere can automatically define every other location on the entire sphere. This is usually accomplished by assembling basic design units into a larger unit and then assembling those units into an even larger assembly.

There are many benefits to using assemblies. Any change to the basic design unit instantly propagates wherever the unit is instanced. This benefit alone makes spherical assembly worth using. There are no restrictions. Automatic drawings can be made, analysis performed, and all the usual visualizations and space analysis can be performed.

Spherical assemblies result in surprisingly small CAD models. This is because they contain just the geometry of the basic design unit, its reference model (which could be deleted after the design unit is created), and the mathematical constraints, which define how the design units are instanced and constrained in larger assemblies. In other words, the design model is not copied repeatedly; it is instanced. This is a key concept, because without it, CAD geometric models would be huge and keeping them consistent would be a major task. There is only one copy of the basic design part, plus some administrative overhead (mostly constraints and visualization options, such as colors or materials), which apply to each instance. The CAD user will certainly notice when spherical assemblies are used; system response does not degrade much, even as he or she works with increasingly large frequencies and more complex part detail. Any spherist anticipating using CAD would be well advised to understand spherical assemblies and to use them wherever there is a high degree of symmetry in the application.

10.5 Three Class II Triacon Designs

Next we will develop three Class II triacon designs. All of them are based on icosahedral 4^v triacon breakdowns and are easily visualized. We will make one comprehensive reference model and reuse it to make very different spherical designs.

The standard unit of symmetry in triacon breakdowns is the polyhedral edge-symmetric diamond. Two of the diamond's vertices are tangent to adjacent polyhedra vertices. The other two diamond vertices are tangent to the polyhedron's face midpoint (see Section 8.6). Figure 10.17 (a) shows a 4^v triacon edge-zenith diamond in the context of a complete spherical icosahedron reference model. Again, the small planes, arcs, and colored surfaces, shown in the figure, simply serve to make these reference elements visible. Only one diamond reference diamond model is needed because the 29 others around the sphere (one diamond straddles each of the icosahedron's 30 edges) are identical except for their absolute vertex coordinate values. Thus, we can create a single design unit based on this one diamond and then copy it to all the other positions on the sphere. Figure 10.17 (b) is a close-up view of just the edge-zenith triacon reference model. The surface of the sphere is shown in gold, spotlighting the diamond's key elements. The *xyz*-axis is shown again to help associate illustrations (a) and (b).

This model contains the same reference elements as the spherical triangle example, shown in Figure 10.16, except that there are more of them: 9 spherical surface points, 1 origin point, 8 triangular surfaces, 16 vertex-to-vertex edges and arcs, and 16 great circle planes (one of every vertex-to-vertex arc). Small red spheres indicate the two diamond points that are tangent to the icosahedron's vertices. Notice that the triacon diamond is symmetrical around both the *zx*- and *zy*-planes. This means you could start building this reference model with as few as five or six points and use the CAD system to complete the

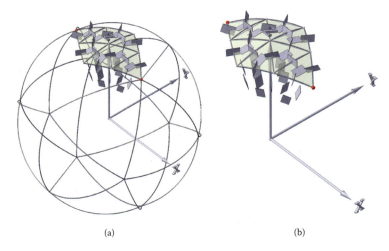

(a)   (b)

Figure 10.17. Icosahedron 4^v triacon reference model.

other half by mirroring them to the other side. Note also that the midpoint of the diamond is at the zenith of the +z-axis. On a unit sphere, this point has coordinates (0, 0, 1).

10.5.1 Summary

The triacon 4^v reference is an accurate 3D model of a single edge-zenith triacon 4^v icosahedral diamond. It contains only points, lines, arcs, planes, and surfaces. Except for the fact that the initial points input are for the edge-zenith diamond, the distances of all points to the origin and all angular relationships between them are the same as the other 29 triacon diamonds on the icosahedral sphere.

Said another way, we can rotate this edge-zenith diamond to any of the other 29 positions, and the reference model would describe every other position. This is exactly what we will do later when we develop our spherical designs. We will reference and assemble one detailed design unit using this single reference model and then rotate and assemble it to cover the entire sphere with the design.

The examples of spherical design that follow take advantage of different combinations of reference elements. None uses them all. The first example, a panel sphere, relies heavily on the triacon grid points and triangle face planes. The second example, a strut sphere, uses the vertex-to-vertex arcs, and their associated great circle planes. The last example reuses the grid points and triangle face planes, but adds points and lines to create a parabolic surface design.

10.6 Panel Sphere

Now that we have a triacon reference model, we can use it to build a 3D model of an actual sphere. In our first example, we will make a design based on panels with holes. The edges of adjacent panels form beam-like struts or chords along the spherical grid. Figure 10.18 shows the completed design we are going to build. For this design, we only need to use half of the triacon diamond reference model because the panel geometry we create from it can be reflected to complete the other half of the diamond. The full diamond, when replicated properly, will cover the entire sphere without overlaps or gaps.

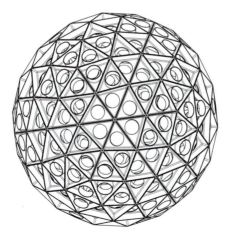

Figure 10.18. Panel sphere.

10.6.1   Panel Reference Model

Figure 10.19 shows the design of the panel unit superimposed on the triacon reference model. This design does not use the great circle arcs or the mid-arc and mid-chord planes to make the model easier to understand.

The steps required to build all four panels are basically the same. A triacon triangle face plane is selected from the reference model, and a part sketch is created on it connecting the three points defining its face triangle. When the triangular profile is complete, these same three points are projected towards the origin of the sphere a certain distance to define three more points, which become the corners of the interior face of the panel. The projected points lay on a plane parallel to the original face plane. One is illustrated in Figure 10.19. The projection distance is not important; we are only interested in establishing a thickness for the triangular panel.

At this point, we have two triangle sketch profiles, one for the outer face of the triangular panel and another for the inner one. To create a solid panel with tapered sides, we select the two profiles and use CATIA's lofting tool to generate a pad with tapered sides. Unlike the rectangular pads we have seen earlier, these pads have tapered sides because the panel's inner triangular face is smaller than the other one.

Our design includes a hole in the middle of each panel. The hole is centered at the outer triangle's incenter. The incenter is the point in the triangle where the perpendicular distance to each side is the same.

By using the incenter, we ensure any circle centered on it will have an equal amount of clearance to each edge. This fact is particularly important when triangles are isosceles or scalene. The radius of the circle was chosen to provide enough clearance for the hole collars and to allow some fillet treatments inside the panel. Once a circle is sketched, CATIA's pocket tool puts a hole through the entire triangular panel.

There is a slight difference in modeling techniques used for the two icosa-edge panels and the two remaining panels. All panels are shelled (another CATIA term, which means "to hollow out a solid") and preserve a given thickness at the edges and bottom. However, in the case of the icosa-edge panels, a collar effect is created. This is useful because it helps distinguish which panels are which when we see the final spherical assembly. Since the collared holes follow the icosa-edge, the final spherical assembly will highlight the basic

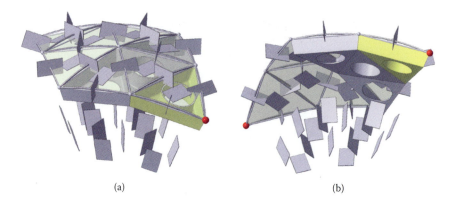

(a)   (b)

Figure 10.19. Panel reference model.

icosahedron, which is at the core of this subdivision. Only one more dress-up feature is applied to the half-diamond model. Inside the shelled area, all interior surfaces are filleted.

These details help demonstrate the value of spherical assemblies. Any detailing performed on the half-diamond model will propagate throughout the entire sphere. This is a huge productivity gain. In actual spherical applications, dress-up features might include openings, local panel reinforcing, covers, gaskets, fasteners, and so forth. Figure 10.20 shows the final result of the half-diamond panel modeling. The underlying reference model is not displayed.

10.6.2   Panel Unit

The panel sphere's basic design unit is four triangular, wedge-shaped hollow prisms. Each has a hole, and two of them have collars around their holes. All four prisms were made the same way.

The plane of a triangle is selected from the reference model. A 2D sketch is made on that plane by projecting the three corners of the triangle onto the sketch plane and then drawing lines between them. This triangle defines the upper face of the prism. The incenter of the triangle is geometrically defined using CATIA's sketching constraints. One of the circle tools will draw a circle tangent to three sampled points or lines. By choosing the three sides of the triangle, the incenter is found and a circle centered on it. The radius is adjusted to provide clearance between the circumference and the three sides of the triangle.

Next, the three corners of the triangle are projected again, but this time towards the origin of the sphere. The amount of projection is arbitrary, but it results in the thickness of the prism.

A plane is defined for the bottom of the prism based on the three projected points. Another profile sketch is created for the triangle, which makes the bottom of the prism. At this point, we have two sketches, each on separate but parallel planes. The upper sketch includes a circle.

CATIA's lofting tool creates a prismatic solid between the two triangular profiles and extrudes the circle in the upper profile into a hole.

Next, we apply CATIA's shell tool. We select the upper face for removal and specify a minimum thickness to be maintained. The result is a hollowed-out triangular prism with a hole. Shell options specify if the hole is to be trimmed flush, or if a collar is to be left around it.

One last dress-up feature is applied to the hollow prism. Edge-to-edge and corner fillets are applied to the inside of the prism. The result is to create smooth transitions from the interior sides to the bottom of the prism. The final part is shown in Figure 10.20. Notice each of the four subpanel units is visually different: no holes, a simple hole, a hole with a collar, or a hole with a collar in yellow. This is intentional. These visual cues make it easier to follow the assembly process and to see how assemblies form the complete sphere.

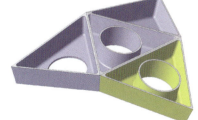

Figure 10.20. Panel unit details.

10.6.3   Assembly

The assembly technique used for the panel sphere (and the others in this chapter) is very simple. Coincidence constraints are applied between pairs of symmetrical points on two-part models. The constraints cause the unanchored part to be joined to the anchored one,

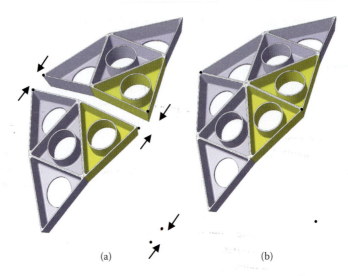

(a)   (b)

Figure 10.21. Assembly with constraints.

and almost every assembly case can be solved with just three pairs of constraints. When the models are updated and the constraints take effect, separate models are instantly repositioned side by side, as needed.

The panel unit, detailed in Figure 10.20, was built from points, lines, and planes within the reference model. Four reference points are particularly important in this assembly: the origin, the icosahedron vertex, and two icosahedron midface points. Figure 10.21(a) shows an instance of the panel unit alongside the original panel unit. It does not matter which is which. They are positioned near each other, but not constrained to form an assembly. Small dots indicate points in each panel unit's geometry we want to constrain. The dots at the corner of the two yellow panels indicate icosahedra vertices, and the one on the blue panel is the icosahedral midface point. Two more dots appear at the origin of each panel unit.

To assemble these two units, an anchor constraint is first applied to one of the units. This makes its *x*-, *y*-, and *z*-axis the one all the other assembled units will adapt to. In this example, the upper unit was anchored, but the choice was somewhat arbitrary. However, in practical applications this detail is not arbitrary because the anchored unit's *xyz*-axis system subsequently defines the meaning of orthographic views, such as top, front, and side. This is an important consideration if you are making drawings and expect your drawing views to correspond to standard conventions.

After the anchor constraint is applied, three pairs of coincidence constraints are applied between the corresponding design features of the two design units. One pair of constraints is applied to both origin points. Likewise, a pair of coincidence constraints is applied between the two icosahedral vertices and another pair between the two midface points. The arrows in Figure 10.21(a) show all three pairs of constraints and the direction the upper panel unit travels, when it is updated and the constraints take effect. The upper unit will automatically translate and/or rotate to satisfy the constraint condition. It moves towards the upper unit because that one is anchored. Figure 10.21(b) shows the result after the update. Notice, also, only one origin dot appears. In actuality, the two-part model origins are now superimposed.

10.6.4   Pent-Cap Assemblies

The basic half-diamond panel set and the assembly method just described are all that is needed to build a full sphere. The assembly process starts by clustering five half-diamond panel units around their icosahedral vertex point. The result is a *pent-cap* assembly. Figure 10.22 shows the assembly sequence.

Figure 10.22(a) shows how the assembly starts by joining two half-diamond panel models together. Constraints are applied to the pairs of origin points (shown as black dots) and to symmetrical points along their shared edges. The result of applying these constraints is shown in (b).

Three pairs of coincidence constraints are applied between any two instances. One pair joins their local origin points. The other two pairs of constraints join symmetrical points together, one from each of the two panels. When the assembly is updated and the constraints take effect, the half-diamond panels join together. The result is shown in Figure 10.22(b).

Three more half-diamond panels are instanced and added between them are constraints. Figure 10.22(c) and (d) show the progression as we repeat the use of constraints on other half-diamonds instances. When the last set of constraints are applied in (d), the pent-cap assembly is complete.

It is worth repeating what an instance of a geometric object is. An *instance* is a reference to a geometric model we have made; it is *not* a copy of it, it is only a pointer to it. To

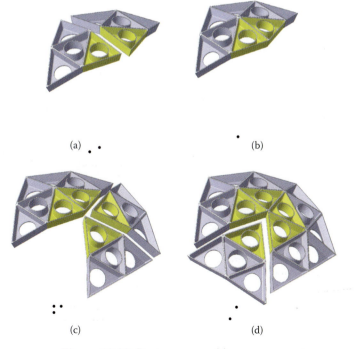

(a)   (b)

(c)   (d)

Figure 10.22. Pent-cap assembly sequence.

replicate the geometry elsewhere, we take the reference and add to it the location information that will place the original geometric model in the new location. In a sense, the new geometric model we have created is a virtual model of the original one. In other words, instance models are simply references to another model and some administrative information saying where the referenced model is being placed. There is no need to make copies of the original geometric model. This is an important point. Many of the advantages of assembling large models from instances of smaller ones would be lost if repetitive parts were to be copied repeatedly. The effects on inflating the model size and slowing computer performance are obvious, not to mention the fact that any changes would have to be repeated for every copy.

10.6.5   Full Sphere

A spherical icosahedron is made of 12 pent-caps, so a full sphere can be assembled from 12 pent-caps subassemblies. Figure 10.23 shows the clockwise sequence for achieving this. The exact same constraint procedure is used between instances of subassemblies. Coincidence constraints are applied to key points on adjacent pent-caps and their assembled model origins. This process is repeated 11 times to define the full sphere.

Pent-cap assemblies offer an advantage that other assemblies do not. There is only one way they can fit together. This makes working with high-frequency subdivisions much easier, and eliminates one source of error—choosing the wrong points when constraining two instances of a model.

Figure 10.24 shows the completed panel sphere. The panels on the far side are visible through the holes in the front side panels. The complete model size is equivalent to ap-

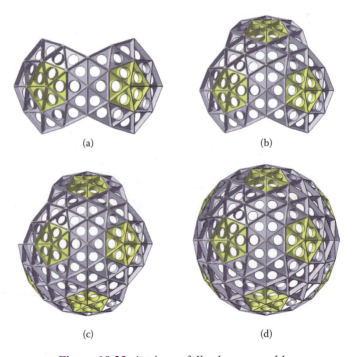

(a)

(b)

(c)

(d)

Figure 10.23. 4^v triacon full sphere assembly.

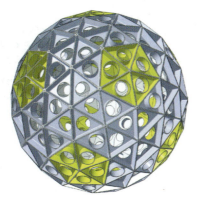

Figure 10.24. Panel sphere.

proximately twice the size of the original part model, demonstrating once again the major advantage of using spherical assembly techniques.

10.6.6   Analysis, Visualizations, and Data Extractions

Spherical assembly is relatively straightforward. The fact the panel model creates a totally symmetrical pent-cap assembly helps a great deal in eliminating errors. An error in assembly would be quite noticeable. However, some subdivisions are subtler than others, and great care must be taken to ensure that the proper points are selected for constraints. One convenient validation method is to "explode" the assembly at various levels. Figure 10.25

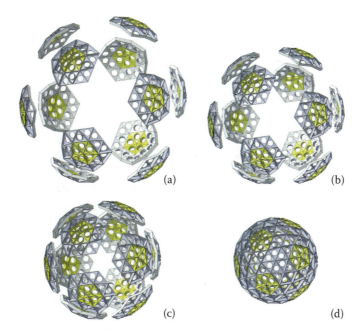

Figure 10.25. Assembly validation.

shows a typical sequence. Within CATIA's assembly workbench, the explode tool can take apart or put back together each assembly sublevel and replay it step by step to verify that the order and sequence are correct. This is an extremely useful tool.

10.6.7   Summary

The triacon 4^v half-diamond reference model provides the basis for the panel sphere. The basic design unit consists of four wedge-shaped panels. The two panels, along the icosahedral edge, feature collared holes and are identical except for their color. The other two triangular wedges have holes but no collars, and are enantiomorphic (mirrors of each other).

We create each individual panel by using three triacon grid points to define one surface of a triangular wedge. We project these same points a short distance towards the origin, creating three more points. These last three points define the inner surface of the triangular wedge.

Next, a loft or tapered pad is created. The pad's edges tapers towards the origin of the sphere, somewhat like a shallow wedge. A hole is made at the incenter of the upper triangular surface. For some triangular wedges, a hole is created before the wedge is "shelled," which results in a collar around the hole. For others, the hole feature is applied after the wedge is shelled and no collar results.

We perform two stages of assembly. The first stage groups five half-diamond panels into a single pent-cap assembly. The second stage completes the sphere by assembling 12 pent-cap subassemblies. In all cases, three pairs of coincidence constraints are applied to join elements—one pair joins their local origins, the other two pairs join points on each element on a shared edge. We check the final assembly by "exploding" the subassemblies and verifying that each subassembly is properly positioned in the full sphere.

10.7   Class II Strut Sphere

In our next example, we use the same Class II triacon reference model from the previous example, but instead of developing a panel design, we make a different design in which each subdivision grid becomes a strut. Figure 10.26 shows a perspective drawing of the design we are planning to build. The design is based on the same 4^v triacon icosahedral

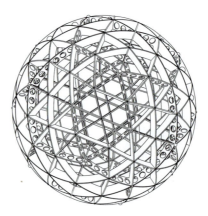

Figure 10.26. Strut sphere.

reference model as the panel sphere. Unlike the panel sphere, this design takes advantage of the vertex-to-vertex great circle arcs and their reference planes.

We will use a similar assembly strategy for the panel sphere. Because the standard design unit is so different, the final result will also be quite different, even though the underlying reference models are the same. To demonstrate one more benefit to using spherical assemblies, we will include a structural connector to join certain struts. Although the connector is simply a stylized pin, it is enough to demonstrate how the connector is designed and how it is introduced at a high level in the assembly process. Just as we did before, we start with the reference model and build the strut members from it.

10.7.1   Strut Reference Model

Figure 10.27 shows top (a) and bottom (b) views of the familiar 4^v triacon reference model (edge-zenith icosahedral diamond) that we will use to develop our strut unit from. The unit consists of six strut members joined in a zigzag pattern. They are best seen in view (b).

The outermost arc of each strut is tangent to the great circle arc between the two vertices spanned by the strut. The arc's center is the sphere's origin. The inside arc is concentric to the outer one, lies on the same great circle plane, and is centered on the origin as well. Two struts (yellow) are *coplanar* to the triacon diamond's edge. They also have holes to visually distinguish them from the others and to create a repetitive pattern over the completed sphere, which will highlight the triacon grid. The other strut members follow the same pattern, only they do not have distinguishing holes.

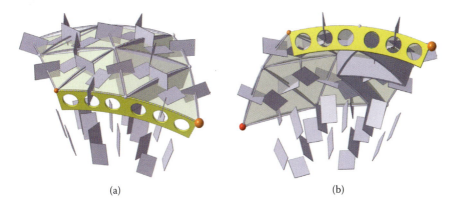

(a)   (b)

Figure 10.27. Strut reference model.

10.7.2   Detailing Struts

The first step in creating a strut is to make a 2D sketch of its outline. The sketch is made on the plane of the great circle arc that spans the two triacon vertices receiving the strut. Although they appear as one in Figure 10.28, there are actually two of them positioned end to end, spanning the three triacon grid vertices indicated with small red spheres. The triacon vertices, mid-arc planes, great circle planes, and their surfaces are also shown. These two struts are tangent to the triacon diamond edge, and both of their great circle planes are identical because both chords are defined by the same overall great circle. Although their

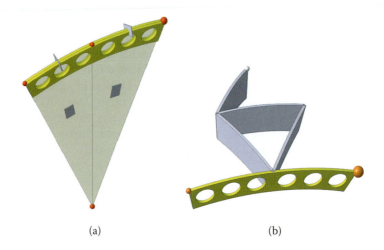

(a)   (b)

Figure 10.28. Strut unit details.

vertex-to-vertex arc lengths look equal, they are slightly different, which is a characteristic of the triacon subdivision.

Strut detailing begins with a 2D profile sketch created on the plane of the strut's great circle. Since the diamond edge struts in Figure 10.28(a) share the same great circle arc, and the two reference planes describe the same surface, we can sketch both (yellow) struts at the same time and consider them a single unit. The sketch consists of two concentric arcs centered at the origin of the sphere.

One arc defines the upper chord, the other the lower chord of the struts. The outer arc spans the diamond edge end points. The inner arc spans a projection of these end points. The projection is towards the center of the sphere. A third arc is added midway between the other two and is used only as a reference to align the centers of the six circles, all of which have the same radius. These circles become holes when the 2D profile is extruded to make a 3D strut.

The strut's 2D profile is closed when two short line segments join the ends of the concentric arcs. These line segments are taken from the reference model's vertex-to-origin lines. To review, the 2D strut sketch consists of two concentric arc segments (at the top and bottom of the strut), and two end-line segments. The outline of the strut is complete.

An additional arc was added indicating the profile's center line. Six circles are added; their centers follow this center line. The sketch is complete and, when extruded, it makes a thin pad (a CATIA term) with six holes in it. In Figure 10.28(a), you can see that the 2D sketch was extruded symmetrically around the plane of the strut's great circle because half of its thickness is on either side of the great circle.

The strut detailing method just described is repeated—without holes—for all the other struts. When complete, the struts form the standard design unit, shown in Figure 10.28(b). The largest red sphere in (b) indicates a triacon diamond vertex. This point is also coincident with the spherical icosahedron's vertex. The red spheres are not part of the final design; it is shown only to indicate how the strut unit fits the triacon diamond. Figure 10.27 uses the same graphic conventions.

10.7.3 Assembly

The strut sphere employs the same assembly strategy used in the previous panel sphere example. In fact, the same points are constrained, even though the designs look quite different. The process starts with inputting the first half-diamond strut unit and applying an anchor constraint to it. This makes its *xyz*-axis and orientation the one used by the overall assembly.

Next we create a second instance—actually a reference to the original half-diamond strut unit. Coincidence constraint pairs are applied between their local axis origins (black dots in the figure) and two points, one on each strut unit, where they are joined together. When the model is updated, the constraints cause the second instance to move adjacent to the first one, as required. This process is repeated until five instances of the half-diamond strut unit are assembled.

Figure 10.29 shows the entire progression starting with (a). The resulting assembly looks somewhat like a pinwheel. The last image, (d), shows the nearly completed pinwheel just before the last update is applied to join them together. The full strut sphere is made by assembling 12 pinwheel subassemblies.

Figure 10.30 shows the sequence. Each pinwheel subassembly is joined to the next one by applying coincidence constraints between their local origins. Coincidence constraints are applied to pairs of points at common locations where two pinwheels will mate.

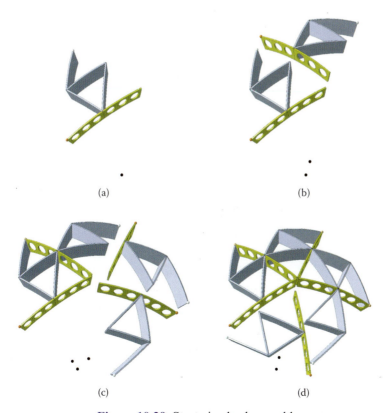

(a)

(b)

(c)

(d)

Figure 10.29. Strut pinwheel assembly.

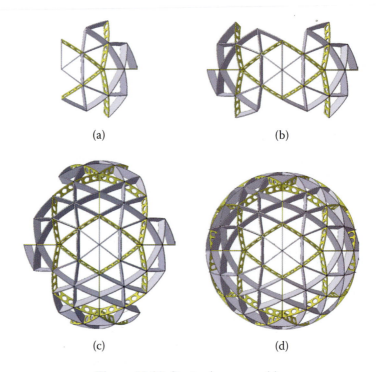

(a)   (b)

(c)   (d)

Figure 10.30. Strut sphere assembly.

The final image, Figure 10.30(d), shows the completed strut sphere. In this orthographic view, the struts on the back side of the sphere are completely hidden by the struts on the near and visible side. In this view, the sphere is perfectly symmetrical. Figure 10.31 also shows the completed strut sphere, but in perspective and from a different viewpoint. In this view, almost all the struts on the sphere are visible.

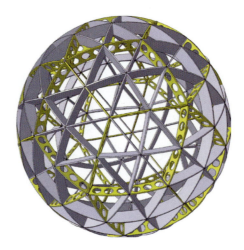

Figure 10.31. Strut sphere.

10.7.4 Summary

We base the strut sphere design on the great circle arcs between triacon grid points. We project the two points, which define an arc towards the origin an arbitrary distance and use the points to create a second arc on the same plane as the first one. Both arcs use the sphere's origin and are, therefore, concentric. Additional lines are added to define the complete profile of a single strut. The profile is extruded to make a pad, which looks like a curved plate. All six use the same design strategy.

For two struts, we add circles. They become holes when a pad is made from the profile. We do this to differentiate one strut from another and to see symmetry patterns in the completed sphere. These two struts are tangent to the edges of the triacon diamond. The final sphere highlights them with their yellow color and holes.

Again, we assemble the full sphere in two stages. The first stage assembles five strut units into a pinwheel assembly. The second stage joins 12 instances of the pinwheel subassembly from stage one into a full strut sphere. At each stage constraints are applied to assemble smaller geometric units into larger ones.

10.8 Class II Parabolic Stellations

Our third spherical example develops a "stellated" sphere with an outer hex-pent truss. Once again, we will reuse the Class II triacon reference model we used in the previous two designs. By now, you can see just how useful these reference models are and the variety of designs you can create from them. We characterize this design as "stellated," though, strictly speaking, it isn't a true stellation in the polyhedral sense of the word. Polyhedral *stellations* are extensions of the faces of a polyhedron, past the polyhedron's edges, until they intersect and create another set of faces. We use the term loosely; our raised or protruding constructions give the sphere a stellated appearance.

A single stellation will consist of three parabolic surfaces arranged in pinwheel fashion around the center of every triangle. For the truss system, we will add a small spherical connector to the apex of each stellation and then add struts between selected connectors. Figure 10.32 shows the overall design. This schema is similar to geodesic domes, such as the one at Epcot Center at Disney World in Orlando, Florida.

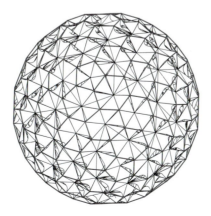

Figure 10.32. Stellated sphere.

Parabolic stellations are excellent demonstrations of the power of CAD in spherical design. First, surfacing is one of the prime tools of 3D CAD. Accurate descriptions and visualization are nearly impossible without three dimensions, and CATIA makes quick work of defining surfaces, as well as joining, cutting, and blending them. Analyzing surfaces (areas, intersections, etc.), determining mass properties, identifying interferences, and performing truncations are just as easily accomplished. Second, this example shows how to extend the standard single-diamond reference model to solve design challenges in which the design elements are related to two or more neighboring diamonds. Neither of the two previous examples demonstrated how parts, which span between two adjacent design units, could be done. Their design elements were planar and totally defined by reference geometry within the single edge-zenith diamond.

Modeling parabolic stellations is a little more complex than the process used in the previous two examples. For this reason, we suggest you look at Figure 10.32 and ahead to Figure 10.42 to see the completed sphere. Follow the pattern of the struts. They are mostly hexagons, but a pentagonal arrangement surrounds each of the icosahedron vertices. Notice that struts meet at small spherical connectors at the apex of each stellation. The connector is also directly over the incenter of each subdivision triangle in order for the layout of the 4^v triacon subdivision to become apparent. The parabolic surfaces are the most complex features, and three back-to-back surfaces share a common edge in the middle of each face.

CATIA is a superb surface modeler and offers a wide variety of techniques. Here, we employ the *fill surface* method (another CATIA term). A fill surface is simply a surface that spans a closed 3D boundary. If the boundary lies on a plane, the resulting fill surface will also be a flat plane. If it is not planar, a complex nonplanar surface results. In a sense, a fill surface is the mathematical equivalent of a child's pastime of making soap films to create bubbles. A wire frame loop is immersed into a soap solution, and when it is withdrawn a soap film fills in the wire loop. We will use CATIA's fill surface method to do the same. Our surfacing "wire frame" boundaries will be four straight lines, which do not all lay on a plane. A curved or warped surface can be generated by ruling lines between points along two opposite lines.

Figure 10.33 shows an example of a ruled surface. CATIA performs this operation with an advanced form of mathematics called *nonuniform rational b-splines*, or NURBS for short.

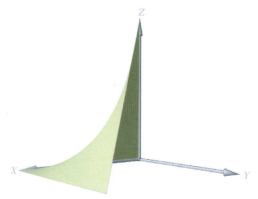

Figure 10.33. Ruled surface example.

The basic parabolic stellation design unit covers half of the edge-zenith diamond. We have used this approach in the two previous CAD examples. We start, as always, with the reference model.

10.8.1   Stellated Sphere Reference Model

Just as our previous examples, our starting point is the same 4^v triacon reference model we have been using all along. However, this time we will add an adjacent diamond. We need this extra diamond to compute some points for elements (struts), which span across two diamonds. The dual-diamond reference model is shown in Figure 10.34. Figure 10.34(a) locates these two diamonds in the overall edge-zenith spherical or icosahedron. Figure 10.34(b) shows a closer view without all the distracting diamond edges shown in (a).

The adjacent diamond geometry can be created in two ways. Points can be imported into CATIA, or a copy of the edge-zenith diamond can be made and then rotated to a position alongside the edge-zenith diamond. In this design example, we will only use the lines and

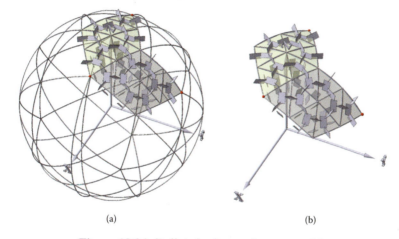

(a)   (b)

Figure 10.34. Stellated sphere reference model.

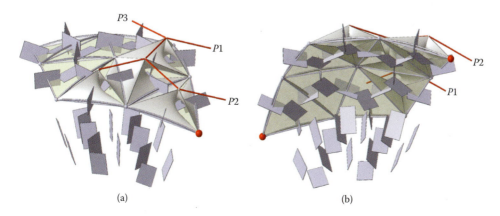

(a)   (b)

Figure 10.35. Single stellation reference model.

points for seven triangles: five from the edge-zenith diamond (yellow shading) and two from the adjacent one (gray shaded) in Figure 10.34.

Figure 10.35 shows the completed half-diamond stellation model from the same viewpoint as the panel and strut examples, shown in the previous two triacon examples. However, unlike those two examples in which all geometry was within the limits of the edge-zenith diamond, this design requires that elements (struts) span to an adjacent diamond. We needed to include the additional diamond in order to compute the locations for the end points of two struts, *P*1 and *P*2, in Figure 10.35. Strut *P*3 uses a projected point as well, but the triangle it needs for computation is already within the standard edge-zenith diamond.

10.8.2   Stellated Design Unit

The stellation design, shown in Figure 10.35, is made by clustering three parabolic surfaces around a central point on each triacon face triangle. In Section 4.17.2 we found that triangles have many types of center points: incenter, centroid, circumcenter, and orthocenter, to name a few. The one we want to use for this application is the triangle's incenter. The incenter is always inside the triangle and is an equal distance from each side. Within CATIA, a triangle's incenter can be found by geometric construction or by an equation in which the independent variables are the coordinates of the triangle's vertices. This example uses equations.

With the incenter defined, another point is created from it by projecting (or scaling) the incenter point beyond the surface of the sphere. This example scales the incenter by a factor of 1.10. Since we are working with a unit sphere, this means the projection is outside the sphere, 10 percent of the sphere's radius. One effect of scaling is that the new point is guaranteed to lie on a line projecting from the sphere's origin through the incenter.

Figure 10.36 shows a typical triacon triangle, its incenter point (labeled *I* for incenter), and its projection to point, *P*. It might appear if the incenters of a number of triangles were projected all to the same scale factor. They would be tangent to the surface of another sphere circumscribing the sphere defined by the triangle's apices. However, this is not quite the case. Not all triacon triangles are the same; thus, their incenters are not all the same distance from the sphere's origin. Therefore, when their incenters are projected by a constant scale factor, they would not all be on the same circumscribing spherical surface.

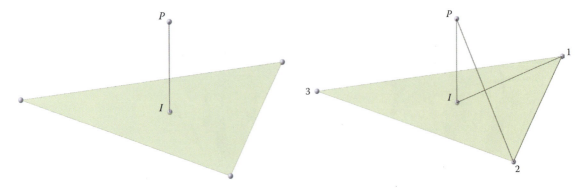

Figure 10.36. Incenter and projected point.   **Figure 10.37.** Parabolic surface boundary definition.

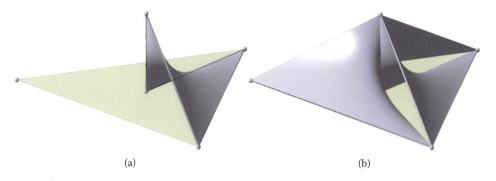

(a)   (b)

Figure 10.38. Stellation parabolic surfaces.

The stellated design unit requires three clusters of parabolic surfaces to be placed on four of the triacon triangles. A single stellation is made of three back-to-back parabolic surfaces, one for each sector of the triangle. To define each parabolic surface, a closed boundary of four lines is created. A typical surface boundary is shown in Figure 10.37 where the triangle's vertices are numbered 1, 2, and 3 along with the incenter I and its projected point P. The first parabolic surface boundary is made by connecting points 1-I-P-2-1. The second and third boundaries connect 2-I-P-3-2 and 3-I-P-1-3, respectively. For clarity, only the first boundary is shown in the figure. Keep in mind, all four boundaries are needed to completely define the stellation surfaces for this one triangle.

At this point, the reference model is complete. Using CATIA's fill surface tool, we create a parabolic surface for each of the boundaries defined. Figure 10.38(a) shows the parabolic surface for the boundary defined in Figure 10.37. Image (b) shows a completed stellation with three back-to-back parabolic surfaces (one is somewhat obscured by the other two). We repeat this process for the four triangles in our half-diamond model; thus, there are 12 parabolic surfaces in the final model.

Surfaces, such as points, lines, and planes, have no physical representation. In Figure 10.38, they are made visible by converting them to solids and giving them an arbitrary thickness. In actual design applications, this would also be the point where we would apply material definitions, texture, color, and visualization standards for rendering, if the application required it.

10.8.3   Hex-pent Details

The base stellation model is almost complete. To represent the hex-pent truss, we add a small spherical connector at each incenter projected point. Struts span selected connectors and are modeled as thin cylinders (see Figure 10.39). No attempt is made to further detail the connector or strut.

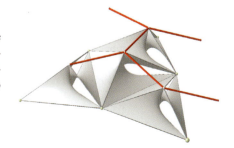

10.8.4   Assembly

The stellation assembly strategy is identical to the previous two triacon panel and strut sphere examples. Figure 10.40 shows the familiar pent-cap sequence in clockwise order, starting with the

Figure 10.39. Stellation unit details.

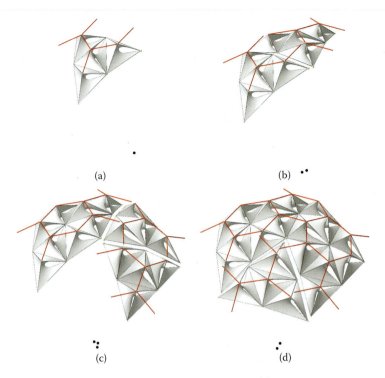

(a)

(b)

(c)

(d)

Figure 10.40. Stellated pent-cap assembly sequence.

upper-left image. The first half-diamond model (the first image in the figure) is constrained with an anchor. This ensures that its axis and orientation system are the ones used for the entire assembly. Any other parts constrained to it will adopt its axis system.

To position the next stellation model instance, we apply the now-familiar set of coincidence constraints. One pair is applied to the origin points of the two instances, and two more pairs are applied to the half-diamond end points (shown as small spheres in the upper right image). The remaining assembly is exactly the same as the panel and strut examples, but for the fact that the struts span from one half-diamond mate to their respective connectors on the adjacent half-diamond assembly. A perfect match is assured because we used a dual-diamond reference model and computed the exact projection point's mark where the stellation apex points will be located. The assembly strategy for the full stellation sphere is the same one used on the other designs (see Figure 10.41). Twelve instances of the pent-cap subassemblies are constrained together. The procedure is identical to the one used for the panel and strut examples, so we will not detail it further here. The full spherical assembly is shown in Figure 10.42.

10.8.5 Analysis, Visualizations, and Data Extractions

Accurate surface representation is a significant benefit to using 3D CAD systems, such as CATIA. Many spherical applications require sections and truncations. For example, geodesic domes are truncated at the ground, radomes are truncated at their support platforms, and in architectural applications, features like floors, walls, ceilings, and stairwells are often

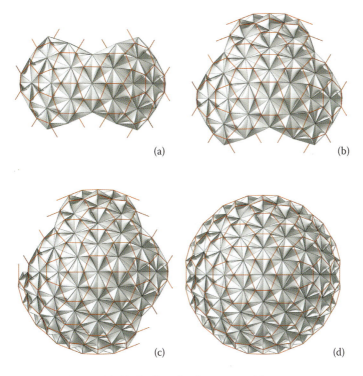

(a)   (b)

(c)   (d)

Figure 10.41. Stellated sphere assembly sequence.

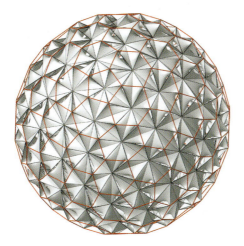

Figure 10.42. Stellated sphere.

based on planes that pass through the sphere. Without 3D CAD, it would be very difficult to determine the exact profile where the cut or sectioning plane passes through the sphere. And designs, such as this parabolic one, are even more complex.

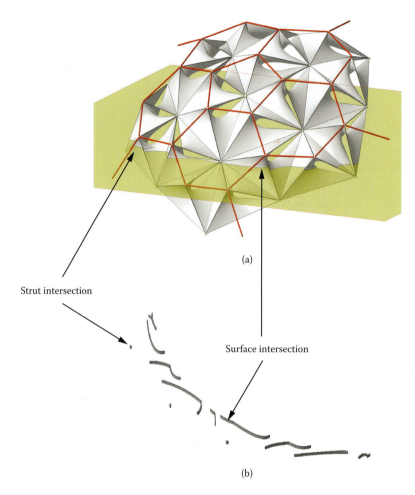

(a)

Strut intersection

Surface intersection

(b)

Figure 10.43. Stellated sphere truncation.

Figure 10.43 shows a case where an arbitrary plane is passed through a pent-cap assembly (a), and through the resulting surface and strut profiles at the plane (b). The intersections are quite complex, but CATIA makes short work of them.

When planes are passed through a solid body, CATIA captures the profile of all the curves defining the intersection. They are typically used in drafting or other modeling applications in which the boundary conditions are detailed. In a geodesic dome, they might be used to define the edge condition of a floor or wall that must contour right up against the surface. Both Figure 10.43(a) and (b) are shown from the same approximate viewpoint and are about the same scale. We have made small prismatic forms from the intersection profiles to make them easier to visualize. Typical strut and surface intersections are indicated.

10.8.6   Summary

Parabolic stellations entail complex geometric elements are easily solved with today's 3D CAD surfacing tools. In this example, a dual-diamond reference model provides the geometry to compute the incenter and projected this point to define stellation's surface boundaries. The projected points define the positions of the hex-pent connectors and strut ends, two of which span adjacent triacon diamonds.

This example is the third reuse of the 4^v triacon icosahedral reference model. The diversity of designs—panels, struts, and now, parabolic stellations—demonstrate a sampling of the design layouts that are possible from this reference model. All three examples also exploit the same assembly methodology. At this point, you may suspect you could create a single generic reference model and assembly definition for a family of spherical subdivisions and use it repeatedly for many design occasions. You would be right. In the next chapter we will revisit reference models and discuss how this is accomplished.

10.9   Class I Ford Shell

In our fourth spherical example, we will use a Class I subdivision method based on a 5^v Equal-chords subdivision of a spherical icosahedron. We described the Equal-chords schema earlier, in Chapter 8. We will call this design the Ford shell.

Class I subdivisions produce triangular grids, which run parallel to the edges of the spherical polyhedra. Depending on the subdivision frequency, you can select certain grid arcs or chords and make hexagonal and diamond patterns as well. This is precisely what we will do in our next example. The PPT reference model we will use here is a 5^v full spherical icosahedra face, and from it, we will create a design made of ten shell panels: three hexagonal shells, three triangular vertex shells, three triangular mid-edge shells, and one triangular midface shell. Figure 10.44 shows the design we will make.

This design is quite different from the three Class II triacon examples, shown previously. First, Class I Equal-chords grids run "parallel" to the edges of the icosahedron's face. Second, we will use an entire icosahedral face as our reference model. This means that when we complete our basic design, we will only need to instance it 20 times to cover the entire icosahedral sphere (the icosahedron has 20 faces), instead of 30 times, as in the previous three Class II triacon examples. The third difference in our design will rely on

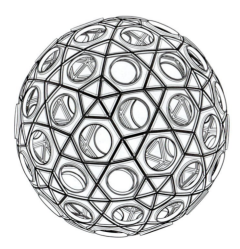

Figure 10.44. Shell sphere.

the sphere's surface definition in the reference model. The previous Class II examples used only the arc and planes in the reference model.

10.9.1   Shell Reference Model

Figure 10.45(a) shows the reference model we will use. The figure shows the edges of an entire spherical icosahedron with one of its faces subdivided. Planes for each of the arc's great circles and subfaces, resulting from the subdivision, are shown within the subdivided face. Arcs and vertices are represented by small arcs and spheres. In addition to the edges of the remaining icosahedral faces, the figure shows an *xyz*-reference axis, which shows the orientation of the whole reference model.

Figure 10.45(b) shows just the subdivided icosahedral face reference model and *xyz*-axis. Once again, the reference model uses small planes, arcs, and spheres to represent planes, arcs, and points in space. The planes, which appear to float below the subdivided face, are the planes of the great circle arcs that define the subdivision grid. Again, it is important to remember that the geometry in the reference model consists of simple abstract points in space and equations of planes and curves. We have made them visible by representing their locations with small 3D models of spheres, arcs, and pads. In Figure 10.45(c) we show our

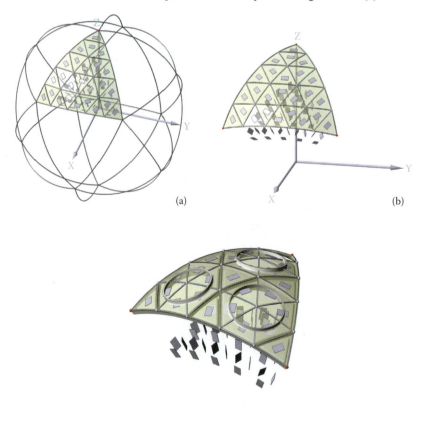

(a)

(b)

(c)

Figure 10.45. Ford shell reference model.

intended design parts superimposed on the reference model geometry. Our design parts are based on the points, lines, planes, and surfaces in the reference model. Let's see how this is done.

10.9.2   Shell Design Unit

Our Ford shell design consists of three hexagonal spherical panels with collared holes and seven triangular panels—one at each of the three apexes, three at mid-edge points, and one in the middle of the icosahedral face. The apex and mid-edge triangular panels are slightly isosceles, while the one in the middle of the icosahedron's face is equilateral. The hexagonal panels are all the same. We will only describe how to make the hexagonal unit. It is the most complex and demonstrates all the techniques needed to make the triangular panels, though the triangular panels do not include collared holes.

Figure 10.46 shows the sequence for defining a hexagonal panel. To define any of the hexagonal panels, you first define its limits on the spherical surface. This is easy to do because the reference model includes the plane of the great circle for every chord between two points. These planes can cut through the surface of the sphere and are used to define the hexagon's edges on the sphere's surface. Some are shown in Figure 10.46(a). Six planes are needed to define the limits of the whole spherical hexagon.

The next step is to convert the hexagonal spherical surface into a solid spherical element. This is shown in Figure 10.46(b). CATIA's thick surface tool easily converts the hexagonal surface piece into a solid element with thickness. Various options control the thickness on either side of the surface on which it is based. The next operation is to put a

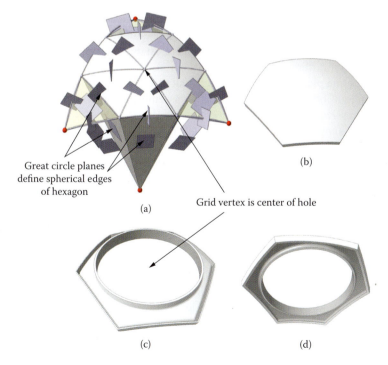

Great circle planes
define spherical edges
of hexagon

(a)

(b)

Grid vertex is center of hole

(c)

(d)

Figure 10.46. Shell unit details.

hole in the middle of the hexagonal element. We use the subdivision vertex, shown in (a), and create a hole in the middle of the element.

Our panel design calls for an edge lip and a collared hole treatment for both sides of the hexagonal panel. To achieve this, we use CATIA's shell tool again to remove material from the outer and inner surfaces of the hexagonal element. The shell tool removes a specified depth of material from the surface up to a certain distance from the edges of the hexagon and the hole. This simple action creates all the lips and the collar around the hole, shown in Figure 10.46(c). In effect, we have sculpted out material from the hexagonal shell, leaving the lips and collar we desire. We apply the same technique to the underside of the hexagonal panel, producing the results shown in Figure 10.46(d). As a final dress-up feature, fillets are applied between the hexagonal shell, and the lip and collars.

We will not repeat the detailed procedure to make the simpler apex, mid-edge, and central triangular shell panels. To summarize, they are made the same way the hexagonal panel was made, only they do not include the hole feature. Reference planes define the panel's edges on the surface of the sphere. We create a thick spherical shell element from this defined triangular surface. The outer and inner surface of the thick element is shelled to produce edge lips. We apply dress-up fillets between the lips and shell surface. Now that three triangular elements and one hexagonal element have been modeled, it is time to assemble them into a typical icosahedral face, and then into a full sphere.

10.9.3   Assembly

The shell sphere's standard face design is based on ten shell panels—three hexagonal and seven triangular ones. To achieve a full shell sphere, three subassembly levels are required. The first level assembles ten panels to make one standard icosahedral face. The next level assembles four icosahedral faces into a spherical gore—one-fifth of a sphere. The final level assembles five gores to make a complete sphere. At every level, constraints are applied between points on different models to join them all together, much like one might glue together pieces of a plastic model airplane. Small pieces join in making a larger assembly, which in turn joins others to make a still-larger assembly. In so doing, the entire spherical model is progressively built up.

Figure 10.47 shows the outside (a) and inside (b) views of the shell panels needed to assemble a face. The panels are in their relative positions and are separated from one another to make them easier to see. The panels required are three hexagonal shells, three triangular vertex shells, three triangular mid-edge shells, and one triangular midface shell.

Face assembly begins with the triangular apex panel, the uppermost one in both (a) and (b) views in Figure 10.47. When panel parts are inputted to CATIA's assembly program, each part retains its original local *xyz*-axis, the one used when it was made. The first assembly task is to apply an anchor constraint to the initial apex panel. This makes its *xyz*-axis system the one that defines the orientation of the remaining assembly. Since this panel was built using a vertex-zenith reference model, this means that when it is anchored in the assembly, the assembly will also be edge-zenith. All the other panels that are constrained to this one will have their local *xyz*-axes adjusted to keep the original anchored panel in place.

With the apex panel anchored, the next step is to assemble the topmost hexagonal panel. We want to position it adjacent to the triangular apex panel with their common edge tangent. To do this, coincidence constraints are applied to each of the two panels' local origin

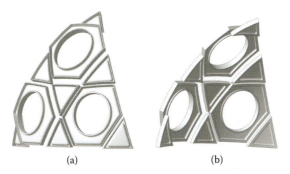

(a)   (b)

Figure 10.47. Face unit assembly.

points. Two more pairs of coincidence constraints are applied to the end points of their shared outer edge. When the assembly is updated and the constraints are resolved, the hexagonal panel repositions to mate alongside the apex triangle panel.

By now, positioning one panel next to another by applying coincidence constraints is our standard technique. We repeat the practice each time we join a new panel to the growing assembly. To assemble a complete face requires nine pairs of coincidence constraints between local origin points, and 24 pairs between the end points along various shared edges. We will use the same technique to assemble faces into a larger unit we call a gore. Since every panel in the face assembly is different, the assembly contains one reference to each of the original parts.

Both planar and spherical icosahedra can be made from five gore-like assemblies, each made of four spherical triangles. Figure 10.48 shows the outside and inside views of the gore subassembly. The face subassemblies have been separated to make their relationship easier to see.

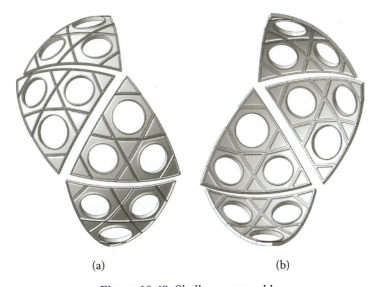

(a)   (b)

Figure 10.48. Shell gore assembly.

The same technique used to assemble the panels is used to make a gore. Gore assembly only requires three coincidence constraints between face subassembly origin points, and another six pairs between the end points of the shared edge between panels. The final gore assembly contains only one face subassembly and three references to it, along with their associated constraints. This is already quite a productivity gain. With just four face subassemblies, we have actually defined the position of 40 panels.

The final shell sphere is an assembly of five of the gore assemblies we just made. The same constraint technique is used, but this time they are applied between gores. Once again, the initial gore subassembly is input and anchored. This perpetuates the local axis orientation of the original vertex-zenith panel input at level one. Four more instances of the gore subassembly are created and constrained together. Ten pairs of constraints applied to the end points of shared edges, between gores, will bring them together.

Figure 10.49 shows various stages of gore assembly in a clockwise sequence starting with the upper left image. When the entire assembly is updated, all five gore subassemblies instantly assume their final position (as dictated by the applied constraints between them). Figure 10.50 shows the final shell sphere. It is the result of three levels of assembly. One level defines the full icosahedral face. The second level defines a gore made of four faces. The full sphere results at the third level, when five gores are assembled.

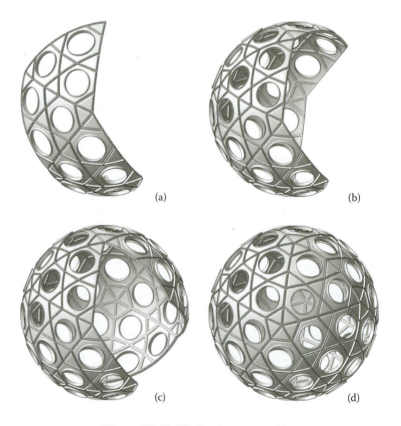

(a)   (b)

(c)   (d)

Figure 10.49. Shell sphere assembly.

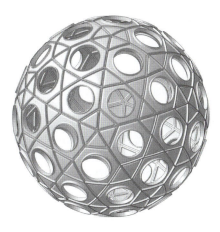

Figure 10.50. Shell sphere.

When you compare the Ford shell sphere assembly process with the one used for triacon spheres, it is clear the layout of the original design unit(s) is critical and has implications for the later assembly process. The basic unit can be a portion of a LCD, a whole LCD or a cluster of them in the form of a triacon diamond or a full PPT triangle. The size of the basic unit is related to the reference model you work with. We will revisit this topic in the next chapter.

10.9.4   Summary

The shell sphere uses the Equal-chords schema to subdivide vertex-zenith spherical icosahedra. A frequency, which is sufficiently high, will produce a grid in which selected vertex-to-vertex arcs form a triangular and hexagonal face pattern. The three apex triangles are tangent, as well as the three mid-edge triangles. The three hexagonal panels are also the same. The basic design unit is a full icosahedron face, and the design requires ten spherical shell elements to cover it (seven triangles and three hexagons).

First, we convert the spherical surface into a solid ball. Individual shell panels—triangular or hexagonal—are then "carved out" of the ball by passing great circle planes through it. The resulting spherical triangular and hexagonal wedges are further cut down by removing the faces toward the origin of the sphere, leaving a thin shell in the shape of a triangle or hexagon. We shell the inside and outside of the main face and produce a perimeter lip. The hexagonal shells receive midface collared holes. We apply fillets around the edge lips of each panel to demonstrate dress-up capabilities.

The first level of assembly joins ten shell panels into a single face unit. We assemble two panels at a time by applying coincidence constraints to their local origins and to points along their shared edge. We repeat this process until the face is complete.

The next level of assembly joins four face subassemblies into a spherical gore, which covers one-fifth of the final sphere. We use the same technique of applying coincidence constraints. The last assembly level instances five gore subassemblies to complete the entire sphere.

10.10   31 Great Circles

In 1949, Buckminster Fuller discovered the fundamental great circles of the spherical ico-sahedra. Documented in his day book, called *Noah's Ark II*,[8] this discovery quickly became the springboard for his passion for subdivision systems and techniques, which allowed symmetry to be exploited for architectural beauty, structural efficiency, and modular fab-rication. The 31 great circles sphere, shown in Figure 10.51, launched follow-up research, which lasted well into the 1970s (see Chapter 2).

We are including this spherical design, both to acknowledge this fundamental discovery, and to illustrate a case in which the reference model requires spherical vertex points that are not part of any of the standard systems covered in previous chapters. Many subdivi-sions will require the technique demonstrated here.

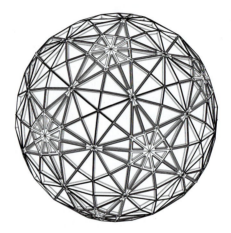

Figure 10.51. 31 great circles sphere.

10.10.1   31 Great Circles Reference Model

The 31 great circles sphere demonstrates another advantage to using CAD in geodesic work. Except for the usual spherical icosahedra features, such as vertices and edge arcs, most of the other points are not part of any standard subdivision method covered in earlier chapters. So, from where do the new reference model points originate? They are computed by the CAD system. Best of all, the simple technique demonstrated here is applicable to virtually any spherical subdivision you wish to make, though doing so can be tedious if the subdivision layout is not highly symmetrical.

Two great circles always intersect at two antipodal points, and these points can be added to the reference model. Figure 10.52 shows how an intersection point in the 31 great circles subdivision is found. Part (a) shows a typical spherical icosahedron face. Its edges, verti-ces, and origin are in red. Mid-edge points are marked with yellow dots, and the triacon diamond edges are marked in gray. In part (b) two great circle arcs are created between

---

[8] James Ward, editor of the four volume series entitled *The Artifacts of R. Buckminster Fuller—A Comprehensive Collection of His Designs and Drawings*, said, "The materialization of great circles was achieved in Fuller's sketchbook, *Noah's Ark II*, which alludes to the teleological role he expected his applied geometry eventually would play." See (Fuller 1985, 31).

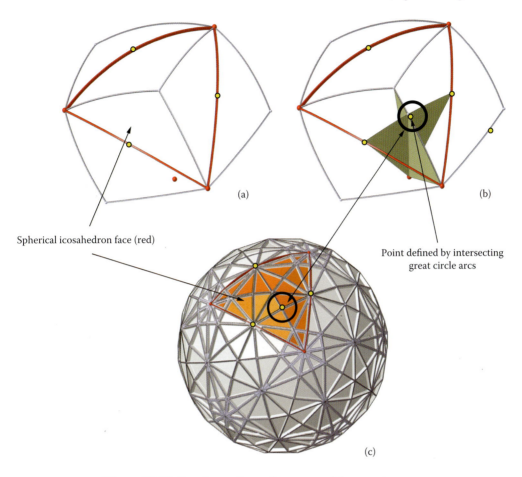

Spherical icosahedron face (red)

Point defined by intersecting
great circle arcs

Figure 10.52. Supplementing reference model geometry.

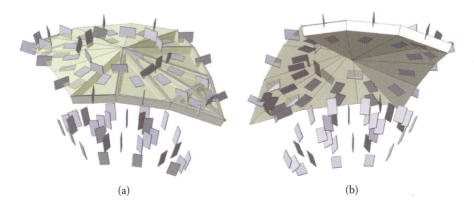

(a)   (b)

Figure 10.53. 31 great circles reference model.

selected mid-edge points. One of the two resulting intersection points is circled. This new point defines a needed vertex in the reference model. The same point is located in the completed sphere in part (c). The viewing angles of all three images in Figure 10.52 are the same to make it easier to compare them. The procedure just described is repeated three more times to get the necessary half-diamond reference points. Figure 10.53 shows the outside and inside of the completed half-diamond reference model. It was made by creating face planes for the triangles bounded by the points along the diamond's edge, as well as the point on the diamond's long axis near the icosahedral vertex.

10.10.2   Panel Unit

The standard panel unit in this design is a half-diamond, such as the one shown in Figure 10.54. Each of the eight panels is defined using the same technique as the panel sphere

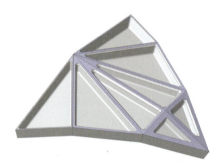

shown earlier. It should be noted, however, that four of the panels are mirror images of the other four. In this example sphere, each is modeled separately because there are only a few cases, and they are easy to make. In the next chapter we will describe a method for mirroring geometry. With this technique, therefore, only four panels in the 31 great circles plane unit would need to be specified. Figure 10.54 shows that a number of dress-up features have been added. Each panel is shelled, and the depth of the shelling varies. Panels nearest the icosahedral vertex are thicker than those near the diamond's short axis. Interior fillets have been applied, and the exterior edges have been chamfered. The shelled interior of each panel are colored white, and the edges given a light blue-gray color to highlight the subdivision grid.

Figure 10.54. 31 great circles panel unit.

10.10.3   Assembly

The assembly sequence demonstrated here uses the same two-stage sequence employed on the previous spherical designs. Figure 10.55 shows the sequence of assembling a pent-cap. Five half-diamond design units have their local axis origins constrained together, and two coincidence constraints mate adjacent pairs of units.

Figure 10.56 shows the same assembly process, but this time, the pent-cap subassembly is the basic unit. The full sphere requires 12 instantiations of the pent-cap. Figure 10.57 shows the completed sphere.

This subdivision has a number of striking features, not the least of which is the high number of hemispherical great circles—31, to be precise. Among the examples in this book, only certain frequencies of the triacon subdivision offer complete great circles, and the maximum number, which could be derived from spherical icosahedra, would be 15. One could argue that the spherical cuboctahedron, used as a reference model for the octet truss connector example, also qualifies. It offers six great circles. However, it is clear none of our previous examples come close to the 31 great circles presented in this subdivision. No wonder Fuller used it as a springboard for so many of his synergetic geometry theories.

10.10.4   Summary

The 31 great circles sphere demonstrates several useful techniques. The first uses the CAD system to define new points on the sphere by intersecting great circles. The steps outlined

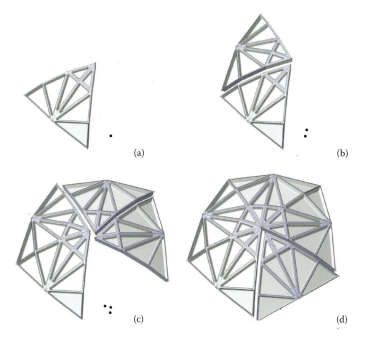

Figure 10.55. 31 great circles pent-cap subassembly.

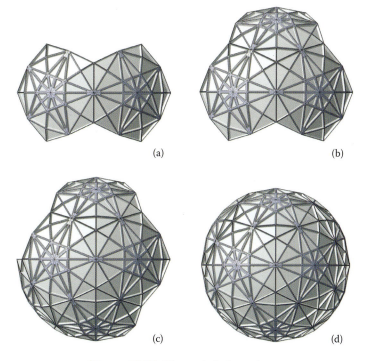

Figure 10.56. 31 great circles sphere.

Figure 10.57. 31 great circles sphere.

here are simple, though somewhat tedious, if a great number of points are needed. This point becomes relevant in the next chapter when we discuss user-written macros to automate a sequence of modeling steps. A macro program can perform the operation in seconds with little or no input from the user. The second important technique demonstrated by the 31 great circles sphere is the reuse of the assembly strategy, despite the differences in the overall layout. In the next chapter we discuss a generalized approach to reference modeling, which applies to the 31 great circles sphere as well as to the three triacon examples shown earlier.

10.11   Class III Skew

Class III subdivision grids are neither parallel nor perpendicular to the edges of the spherical polyhedra faces. These skewed grids introduce a new assembly challenge and lead to a useful technique, which can be widely applied to other classes as well. Figure 10.58 shows a complete icosahedral $\{3,5+\}_{2,1}$ 3^v sphere. Both the Class III grid and the spherical icosa edges are included. No design unit is developed, however. The icosa's vertices are evident because the pentavalent units, which surround them, clearly stand out. A solid sphere has been placed inside the model to make the subdivision elements easier to understand. Without it, we would see through the sphere to the members on the other side, and it would be difficult to follow the geometry and recognize which elements were on the front side and which were on the back side.

Despite its apparent complexity, very few unique 3D elements are needed to build this Class III sphere. Figure 10.59 shows the only parts needed; the icosahedron's face edges (red) are included to orient the grid components. The central Class III grid elements (gray) are centered over every icosahedral face. The Class $III_{2,1}$ subdivision also results in a face-centered equilateral triangle in which each vertex has a chord element to its respective icosahedron face vertex. These six (gray) elements are composed of just two different arc elements. The icosahedral edge elements (yellow) connect grid points over one icosa face with corresponding points on the adjacent face's grid. Like the center elements, only the end and center chords are unique.

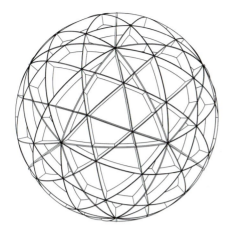

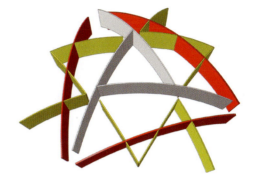

Figure 10.58. Skewed sphere.   **Figure 10.59.** Basic parts for a skewed sphere.

Many of the previous assembly examples used a pent-cap assembly strategy. This method could also be used here; however, we will demonstrate a different approach, which is somewhat related to the lune or the one-fifth sphere assembly method used in the Ford example. The strategy used here demonstrates a generalized way to handle grids, which span from one icosahedral face to another.

Figure 10.60 shows the assembly progression. A solid sphere is placed in the middle. It provides some spatial orientation and makes the geometry more obvious by hiding confusing elements on the back side of the sphere. Image (a) shows one icosa edge (red) and its cross chords (yellow). In image (b), three more edge units have been added, and the icosahedron's face edges are now apparent. As in the previous assembly examples, three pairs of coincidence constraints are needed wherever an edge unit is placed. One pair constrains the shared points at the origin of the sphere. Another pair constrains the icosa edge (red) ends together, and the third pair constrains matching cross chords (yellow). Figure 10.60(c) shows the edge assemblies and how they define four icosa faces. As a whole, the unit in image (c) covers one-fifth of the entire sphere. In this sense, it is the same approach used in the Ford example given earlier. What is interesting about the resulting assembly in (c) is only one of the icosa faces is completely defined with three of its edge units (red). The other icosa faces are missing one of their edge units. Figure 10.60(d) shows that the face-centered subassemblies (gray) have been added and the one-fifth "lune" is now complete. But why are some of the icosa face edge assemblies missing? This is not an oversight; it is deliberate. The missing edge assemblies in one lune will be supplied by its neighboring lune, when each is assembled side by side. Each lune completes its neighbor's missing edge assemblies. Figure 10.60(e) and (f) show this strategy does, in fact, work. The important lesson from this example is that it is possible to develop an assembly strategy that takes full advantage of the symmetry in the design and yet accommodates semisymmetrical cases, which result in some designs featuring elements that cross from one polyhedral face to another. Figure 10.60(a) and (b) review how one subassembly completes the next. In (a), two lune subassemblies are side by side, but not yet fully constrained. Note how the edge elements of one complement the other. Figure 10.60(b) updates and applies the constraints between the two lunes. The result is the fit we need.

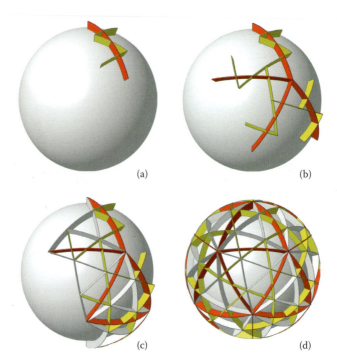

Figure 10.60. Class III assembly progression.

To summarize, Class III subdivision grids are not parallel or perpendicular to the polyhedron's edges. As a result, chord elements necessarily span the polyhedral edge connecting corresponding grid points on adjacent faces. An assembly strategy, which takes advantage of the fact that the edge elements are symmetrical about the center of the edge, makes easy work of what appears to be a difficult task.

There are numerous strategies for assembling the face-centered grid elements because there were so many symmetrically arranged identical pieces. Creating a master "lune" subassembly works particularly well in this case. Elements from one lune fill in missing elements in its neighboring lune when the lunes are finally assembled to complete the sphere.

10.11.1 Summary

In this chapter, we developed a series of design examples to show how a 3D reference model of points, lines, arcs, planes, and surfaces can be the basis of some very different designs. An octet truss model, panel, strut, and parabolic stellation designs were developed from reference models. And each design was further embellished with dress-up features, such as holes, chamfers, fillets, shelling, and slots. These embellishments just hint at the number of possible refinements.

All three Class II triacon designs (panel, strut, and stellated spheres) used the same two-stage assembly process to cover the entire spherical surface with just a small standard design unit. The first stage instanced five design units to make a pent-cap assembly. A second stage instances 12 pent-caps to cover the rest of the sphere. Each example introduced one or more analysis techniques such as sections, surface intersections, or exploded assemblies.

All the techniques were easily produced from the completed model; all would be typically performed by a spherist.

The flexibility of CAD was further illustrated by the Equal-arcs (three great circles) and 31 great circles spheres. Ford used a different reference model than the previous triacon designs and showed how mirroring techniques can solve left- and right-handed versions of the same design element. The 31 great circles sphere further demonstrated how new reference model points can be computed from existing geometry by intersecting great circles. This is a common requirement when nontraditional subdivision systems are used.

In the next chapter we look at how CAD systems can be customized to automate many of the spherical detailing we used in this chapter. The benefits of customization are increased productivity, consistent results, and the ability to consider many more design alternatives to arrive at the best design for a particular application.

Additional Resources

Hoschek, Josef and Dieter Lasser. *Fundamentals of Computer-Aided Geometric Design.* Wellesley, MA: A K Peters, Ltd., 1993.

Tickoo, Sham and Vivek Singh. *CATIA V5R18 for Designers.* Schererville, IN: CADCIM Technologies, 2008.

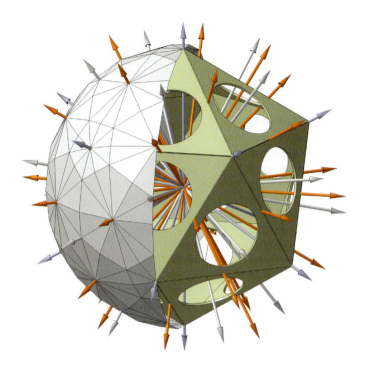

11 Advanced CAD Techniques

I n this chapter, we examine advanced CAD methods that lead to the best practices you may want to adopt when you use CAD for spherical work. Some methods are shortcuts and speed the subdivision process along. Other methods facilitate communication and coordination, when design projects involve large teams. Best practices have the highest benefit if implemented early in a project. Unfortunately, project teams are often confronted with a bewildering array of considerations when they start up; thus, we hope this chapter helps you to identify which decisions are the most important and your options. Early consideration of methods and best practices can save significant time and effort later, and maximize the benefits of using CAD. This is why these topics are covered here—to better prepare you for your own CAD work. The most important methods we will discuss here are reference models, design-in-context, mirroring, power copy, publishing, and user-written macros.

11.1 Reference Models

We have used reference models extensively in the previous chapter. Every CAD example we have developed so far started with one. However, more needs to be said because reference models are extremely useful. Sometimes called a *skeleton model*, a reference model is an auxiliary 3D model CAD users employ to make it easier to build new geometry. What makes reference models important isn't how they are made. In fact, the same geometric modeling workbenches we have already illustrated are used and, from that perspective, they are just another 3D model. The difference is in how they are used and the effect they have on other geometric models based on them.

A *reference model* is no more than a regular 3D model that remains visible in the background while new models are created. In collaborative projects, most general users would not be able to change a reference model, a privilege reserved for project administrators. Almost any of the geodesic subdivisions, which appeared in earlier chapters, could be thought of as reference models because they all have defined points, arcs, faces (planes), and so on. What a reference model is not, however, is a detailed final design. There are some key benefits of using a reference model:

- identifying key geometric elements (points, lines, planes, surfaces, axis systems, etc.) that locate where new modeling is to take place;

- facilitating collaboration among groups;

- promoting standards in methodology, report generation, naming, and so on by ensuring that work will be compatible and will be carried out to the same level of precision;

- centralizing project design units for dimensioning, weights, and measures or orientation;

- making team efforts cumulative; and

- maintaining data integrity.

11.2   An Architectural Example

Let's approach reference models with a nongeodesic example—the transition to geodesics will be easier. Let's suppose an architectural firm is going to design a multistory office building at a midtown site. The surrounding area is already urbanized. The project manager anticipates a design team of about 15 architects who will carry the project from its early conceptual design phase through to construction and eventual turnover to the owners. They will use 3D CAD throughout to develop the design and to prepare all necessary construction documents from the 3D models.

The project manager decides to build a CAD 3D reference model for everyone to use. The reference model will ensure everyone working on the project will be using the same base information. This base will be enriched and changed as the project evolves.

What might be included in the reference model? A number of categories of information would be helpful, including site characteristics, building parameters (or restrictions), project design standards, graphic conventions, data access, and exchange mechanisms.

When considering the site itself, most architects would expect the reference model to include an outline of the property and any key features already located there. If the site is not perfectly flat, elevation contour lines will be added or it might be decided to model the ground after it is excavated. Other geometry in the reference model would include survey points, such as benchmarks or latitude-longitude positions, and true and magnetic north direction indicators. Reference lines would define the center of the utilities (water, sewer, telephone) that pass near and through the site. The model would include any natural boundaries, such as rivers, creeks, high-water marks, or features in need of preservation, such as trees.

Building parameters could include horizontal planes defining the elevations of any floors built or the maximum height of the building permitted by building code. Vertical planes might define the property or building setback limits imposed by building codes. Site ingress and egress points might be defined in anticipation of vehicular or pedestrian access to the site.

Data standards would include the 3D model units (feet, meters, etc.) and the required level of precision—for example, dimensions in feet-inches–fractions are input and displayed to the nearest one-eighth inch. Although most architectural work would rely on Cartesian coordinate systems, spherical or cylindrical systems are also possible. Naming and labeling style for any 3D models or drawings would be part of the standards. The site's relative orientation to the model's *xyz*-axis to be consistent with other uses would be part of the reference model. In addition to standard 3D model orientations (top-bottom-side, etc.), the reference model might define additional views that are useful to the project. For example, the nonstandard view might be of the site from a vantage point across the street.

Visualization is a key benefit to using CAD, and the reference model is where the basic graphic standards would be defined for all 3D elements. Lines, points, planes, surfaces, solid bodies, and so on, as well as color schemas, would be defined.

In crowded building situations, reference volumes might define preexisting structures, or subsurface features, such as old foundations, basements, underground parking, or utility vaults. Some volumes might define reserved areas where no building can take place, or perhaps a future construction phase of the project.

A reference model, even for a modest project, can become quite complex. From a security and data integrity point of view, only selected team members are likely to be able to change this model once it is created and validated. If the reference model is wrong, everything linked to it will propagate the same error, and any change could set off an unanticipated ripple effect.

Key geometric features are likely to be named. CATIA refers to this as publishing, and it is covered later in this chapter. User setup preferences can limit user access to geometry with published names. Some features may only be visible to certain users as well. At first, these may seem like odd restrictions, but consider a congested geometric area where there are many overlapping reference points or planes or several architectural features that coincide such as the edge of the property, a building setback, or a utility line. Knowing which piece of geometry belongs to which architectural feature is important and published names and selective visualization can help users pick the right one as they work.

This example covers most of the essential elements of reference models. Notice that despite all the geometric details we just mentioned, we have not said one thing about the architectural design itself. That is work still to be done. However, that work will take place within the context of the reference model, and will build on references to it.

In this discussion of reference models, we have not said anything about the design the architects will create. We have only discussed the environment—the references—they will use. This is an important distinction because many different building designs could be developed using the same reference model.

11.3 Spherical Reference Models

Even modest geodesic subdivisions involve many points, lines, planes, surfaces, and geometric features. Figure 11.1 shows the minimum elements for a basic spherical icosahedron reference model. The sphere's surface is not shown, and the features are all exaggerated somewhat to make them visible. There are just 20 face planes, 30 edges, 30 edge-arcs, 30 edge-arc planes, and 12 vertices in this illustration. We include an *xyz*-axis reference to orient the model.

Now compare this model with the one in Figure 11.2, which shows the reference model for a relatively simple triacon 4^v icosahedral subdivision. It is evident that the number of reference features has dramatically increased and it is becoming hard to distinguish one geometric element from another. This underscores the value of methodology and of having a common reference model to organize data and facilitate sharing.

In the architectural example just given, the reference model was *not* a building design. It was only a reference framework to help architects work on their design. In the same way, reference models apply to spherical design as well. What might be included in this model? The specific application will determine the requirements, but most applications will find that the following elements are needed:

- the sphere's surface;

- face planes and edges;

- vertices and face normals;

- great circle planes;

- ground plane(s) if applied to a geodesic dome;

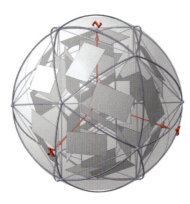

Figure 11.1. Spherical icosahedron reference model.

Figure 11.2. Icosahedron 4^v triacon reference model.

- the principal axis of each face, diamond, or other unit of symmetry;

- graphic standards—colors, line weights, and styles;

- axes of symmetry;

- initial view and orientation of the model when a new CAD session starts ;

- xy-, yz-, and zx-Cartesian reference planes with an origin point;

- spherical polyhedra features (vertices, faces, edges);

- mid-diamond or midface to origin reference lines; and

- incenter, centroid, circumcenter, and orthocenters for triangular faces.

11.4 Prepackaged Reference and Assembly Models

The triacon panel, strut, and hyperbolic surface spheres, presented in the previous chapter, all used the same 4^v half-diamond reference model and dual-stage assembly method. It seems natural to ask if there is a more general approach to building reference and assembly models that apply to any design. There is! Figures 11.3 and 11.4 show how this can be done.

In essence, a complete sphere is built using only a half-diamond reference model, the same one we have been using all along. The same two-stage assembly process is used to assemble them. The first stage assembles five half-diamonds into a single pent-cap, as shown in Figure 11.3. The second stage assembles 12 pent-caps (instances only) into the complete sphere, shown in Figure 11.4. What is different here is the resulting assembly sphere does not contain any physical design elements such as panels, struts, shells, or connectors. It contains only pure reference geometry—points, planes, lines, and surfaces.

The benefit of using a generalized reference and assembly model is now apparent. Any physical design created using the original half-diamond reference model is

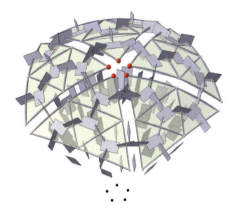

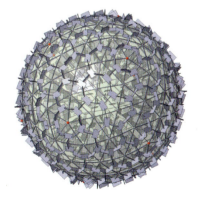

Figure 11.3. Triacon 4^v half-diamond pent-cap reference model.

Figure 11.4. Triacon 4^v full-sphere reference model.

instantly replicated all over the sphere with no additional effort. This is possible because the generalized full-sphere reference model simply consists of 60 instances of the original half-diamond reference model (five half-diamonds per pent-cap and 12 pent-caps for the entire sphere).

Each instance of the half-diamond references one version of it and adds only rotation and axis information to define where the new instance is located. For all its apparent complexity, the model in Figure 11.4 is quite simple. Computer-wise, the full reference model requires very little additional memory and storage in addition to what is needed to describe the original half-diamond reference model.

Generalized reference and assembly models can be used to create a wide variety of spherical designs. At first, it might seem you need a new model for every different frequency and subdivision method. While different subdivision methods will require their own models, it is not always true that a different frequency will require one. Due to frequency harmonics, it is possible to use some high-frequency reference models for lower-frequency design cases. In Chapter 8 we saw triacon 4^v vertices also appear in 8^v and 16^v breakdowns. Thus, a generalized reference model of a 16^v breakdown will cover these lower frequency cases as well. Reference models are extremely useful to the spherical designer. They should be used whenever a CAD application offers them.

11.5   Local Axis Systems

We are familiar with the basic *xyz*-Cartesian axis system, composed of an origin point and three orthogonal axes. Every spherical design example in the last chapter started with a reference model and its local axis system. Furthermore, the first step in every assembly process was to apply an anchor constraint to the first model reference to make its local axis system the one that applies to the overall assembly (see Section 10.3.4).

However, what if the original orientation of the reference model is not what is needed? Suppose each of the triacon spherical designs required a vertex-zenith orientation rather than edge-zenith, as used in last chapter's examples? Instead of recomputing (rotating) the entire reference model with all of its points, planes, surfaces, and so on, you could simply define a new local axis system.

To define a new axis system for the reference model, you simply define its origin and select edges, lines, or planes in the model that unambiguously define the new *xyz*-axis system you want. For a sphere, the old and new origin will be the same point. In addition, you can

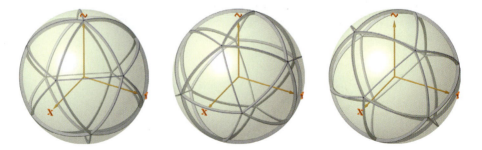

Figure 11.5. Vertex-, edge-, and face-zenith axis systems.

keep several axis systems for the same reference model and change them when they suit the application you need. Figure 11.5 shows one reference model with multiple local axis systems: vertex-zenith, edge-zenith, and face-zenith for the same spherical icosahedron.

Another possible use of local-axis systems is in situations where you want to do 3D modeling, using one of the planar features in the reference model as your new "ground plane." This is quite easy to do. It is often easier to build geometry in a local coordinate system than to create it in the absolute coordinate system of the overall reference model. You may have a situation in which you are developing a design or submodel relative to a polyhedral face or great circle plane. You can create a local axis to make the face or plane coincidental to a local xy-, yz-, or zx-plane.

How is a new local-axis system defined? Without getting into the details of particular CAD systems, we can say those that offer local-axis systems usually offer three ways to specify them. A standard definition for local-axis requires an origin point and three orthogonal directions, such as three mutually perpendicular lines or a plane and a perpendicular line. Actually, selecting any two will define the third automatically. A second definition specifies the local axis as a rotation. Here, the original model-axis system is used and an angle for the new one is computed from some user-selected geometric reference already in the model. For example, the y-axis from the standard-axis system is selected and a 15-degree angle is set in relation to an edge parallel to the x-axis. A third definition, especially important in aerospace applications of geodesics, is to use Euler angles to define local-axis. In this case, the user defines three angle values to establish the system.

11.6   Assembly Review

There are three key objectives in spherical assembly: (1) eliminate overlaps or gaps between parts, (2) use minimum constraints, and (3) ensure complete coverage. The examples in the last chapter showed how to achieve these objectives in a variety of ways, but one more example is shown here to reinforce the process and to illustrate a situation in which hemispherical subassemblies are possible. *Hemispherical assemblies* are convenient ways to package icosahedral pent-cap assemblies. They make the layout easier to understand and check, and they significantly reduce the number of constraints needed to complete a full sphere. Not all subdivision and design combinations offer hemispherical assemblies, but it is usually advantageous to use them when possible.

Figure 11.6 shows two half-diamond design units forming a standard triacon (Class II) diamond for a 4^v subdivision. The full icosahedral edge is indicated by the red elements in two back-to-back half-diamonds. A thin spherical shell is added under the curved pads to suggest the extent of the spherical surface defined by the design unit and to make it easier to visualize the assembly as it evolves. Our assembly will not use a diamond subassembly as this figure suggests. Instead, it will cluster five of the design units around an icosa vertex, resulting in a pent-cap subassembly similar to the ones shown in the last chapter. The difference in this example is how the pent-caps are assembled into hemispherical assemblies in the next-highest level of assembly.

Figure 11.7(a) and (b) show a pent-cap assembly just before all the constraints are applied. Five half-diamonds

Figure 11.6. Triacon diamond design unit.

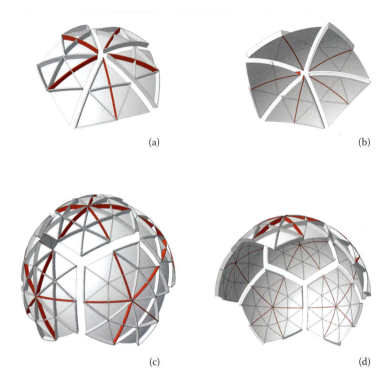

(a)   (b)

(c)   (d)

Figure 11.7. Class II triacon pent-cap assemblies.

design units are instanced, and constrained to each other to make the pent-cap. Like the previous examples, three coincidence constraints will position each pair of adjacent design units together; thus, the pent-cap will require 15 sets of coincidence constraints.

The next level of assembly is the hemisphere, shown in Figure 11.7(c) and (d). Not all subdivision classes and designs offer this, but this one does. The red elements in (c)

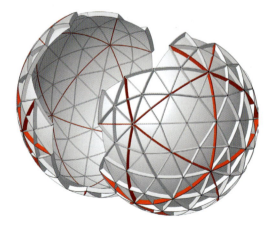

Figure 11.8. Class II triacon hemispheric assembly.

clearly show the evolving spherical icosahedron. Notice, too, that our pent-cap assembly resembles a subdivided face of the spherical dodecahedron; this is because we are actually assembling the icosahedron by assembling its dual.

The complete sphere is shown in Figure 11.8, which shows two hemispheres brought together. They are shown in their unassembled state to demonstrate how they will mate and to make it obvious that no minor parts or lower subassemblies are overlapped or duplicated. A significant advantage to using hemispheres is only three coincidence constraint sets are required to position the two halves together. Fifteen sets would be required if individual pent-cap subassemblies were used instead. Figure 11.8 shows just how easy it is to check for layout errors, duplicated parts, and coverage gaps.

11.7   Design-in-Context

Many 3D parts are designed with an axis system relative to the part. If an assembly is created from many independent parts, then constraints have to be applied to position each part relative to one other. Design-in-context offers an alternative approach. When a designer designs-in-context, he creates his new part geometry by using another piece's geometry or elements in a reference mode to pre-set the new part's position. For example, the designer can select a plane in the reference model and designate it to be a local reference system for his or her new geometry. If a dome requires special treatment for a triangular panel—say, the addition of a vent, access hatch, or window—the reference plane of that panel would be selected from the model and made to act as a new local xy-plane upon which to build the vent or hatch model. The new design work is being created "in the context" of this plane. This means that the CAD program derives all the orientation information for the new model from the reference model. Prior to the development of design-in-context methods, a designer would build the 3D element (vent, hatch, or windows in our example) relative to its own x-, y-, and z-axis and then translate, rotate, and perhaps scale it to properly position it as required on the panel.

Change management is another benefit to design-in-context methods. If the reference model changes—for example, if a great circle plane is changed—any geometry created with logical links to it will also automatically change. The value of reference models and the design-in-context methodology cannot be overstated. Later we will see another technique: publishing. When used with reference models, publishing increases the power of design-in-context methods.

11.8   Associative Geometry

With traditional CAD systems, users typically built subsections or individual parts, where each part has its own axis system. To assemble and correctly position the parts in the context of a larger model, the CAD user had to translate, rotate, and perhaps scale each one in order to correctly position it. Assembling these models was greatly facilitated by the use of constraints. Essentially, the user indicated which features in the part or subsection (edge, corner, face, etc.) were to be in contact with corresponding features in the overall assembly. Constraints are still widely used in modeling today.

Using *associative geometry* is an alternative to using constraints or to manually translating, rotating, and scaling individual parts. Here, the part is actually built using reference model elements as the starting point. Thus, the part's geometric elements are already aligned to where they will be used in the final assembly because they were initially built by way of the reference model. Part models built this way do not need constraints because during the building process they were positioned exactly as they will be in the final design.

There is yet another benefit to associative geometry. Geometry created within the context of a reference model can be instantly changed simply by modifying the reference. That is, geometry is associated by way of logical links to features in the reference model. Thus, if a single great circle plane is adjusted, any geometry with an associative link to it will automatically adjust as well. The same update is available if constraints are used to position the part, but constraints can become quite complex if a part requires many of them to define how parts are positioned relative to one another. Associative geometry is a powerful tool for maintaining data integrity when models are changed or the reference model updated.

11.9   Design-in-Context versus Constraints

Geometric constraints are essential when creating large assemblies. They can be used with or without reference models and are not limited to design-in-context projects. However, constrained elements are not the same as elements designed in context. The distinction is important, and the choice to use a reference model or not is a methodology decision the designer must make when starting a geodesic project.

Constraints were used extensively in each of the spherical designs in the last chapter. When two geometric models need to be assembled into a larger interim or final assembly, constraints were used to define relationships between them (for example, tangent, contact, coincidence, or anchor relationships). In the octet truss connector example, adjacent pipe connectors were grouped together in a nine-way joint by adding contact constraints and coincidence constraints. In the various spherical designs, coincidence constraints joined half-diamond models to make larger pent-cap subassemblies, and subsequently joined those subassemblies to make complete spheres.

At first glance, assembling models with constraints seems like the perfect solution, and for many instances, they are. However, there is a subtle difference in the result when compared to models built using design-in-context methods. Models designed in-context are already positioned where they will be used in the final assembly. The difference between the two methods is related to each model's local axis.

In typical design situations, a new part model is created around its own *xyz*-axis. Later, when assembling this model, users typically reposition parts in a larger assembly. This is accomplished by adding constraints between it and other models already in the assembly. If it is the first part model, then it is usually assigned an anchor constraint. The *xyz*-axis system of the anchored model establishes the axis system to which all other models are adjusted when they are brought into the assembly and constrained to take up their position. This means the local *xyz*-axis system for each assembled model is geometrically transformed by the constraint so that it assumes its proper position in the assembly. Thus the axis system used when they were created bears no relationship to where they will be used in the final assembly.

A part created with design-in-context methods undergoes an entirely different process. The part created around a *xyz*-axis system is derived from key positions on the assembly or on another model it is used with. Unlike a model constrained to be in the right position, a model designed in-context is already correctly positioned because its axis system was derived from other part models around it.

Let's assume a transparent cover is needed for one of the shell panels, shown in Figure 11.9. This is a unique case; not every shell with a hole-collar in this example will get a cover. One approach is to detail the cover as a separate model and then use constraints to position it over the hole-collar. To use the design-in-context methods to create this same design, the shell panel with the hole-collar is accessed and treated as a reference model. A new model for the cover is started, and its axis system is defined to be centered immediately over the hole-collar opening of the reference model. In this example, the xy-plane of the new axis system is tangent to the upper lip of the collar, and the z-axis is centered over the hole-collar center. Figure 11.9 shows the design-in-context arrangement of the shell panel's axis system (red) and the new axis system (yellow) is used to model the cover. Once the axis system is defined, the cover design then proceeds from there.

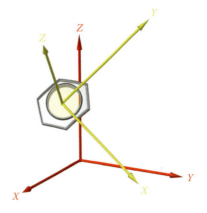

Figure 11.9. Design-in-context local axis.

The final results of the two methods appear to be the same. So what are the implications of choosing one method over the other? In the first approach in which constraints are used, the model's axis and coordinate systems are all relative to the first anchored part. A hole-collar cover model needs constraints to transform its xyz-axis system to the anchored system of the shell panel. However, the hole-collar model alone bears no relation to the context in which it is being used. Without the constraints, you have no sense of where this element belongs in the final design. This could lead to major design management problems if there were trivial differences between many similar looking covers—each belonging to only one particular panel's hole-collar.

In fact, the simple shell sphere design, shown in Figure 10.50, calls for two different covers, and their differences would be quite subtle. In the design-in-context approach, the cover is actually built in a local-axis system based on the panel with which it will be used. If the cover model is later retrieved by itself, it will be oriented and positioned in three dimensions, exactly where it will be used in the overall design.

In large projects, design-in-context methods tend to reduce placement errors and the need to rework the design. When a model is brought into the design, it takes its correct position without needing constraints. Models designed this way are unambiguous, and there is no confusion about which parts go where. Furthermore, it is common practice to include a logical link between the design-in-context model and the model on which its context is based—in this case, the hex panel with a hole-collar shell panel. The logical link keeps the context of the two models synchronized. If the hex shell panel is moved, the hole-collar moves with it.

Which approach is best—design-in-context or constraints? The answer depends on the uniqueness of each design and the results you need. If the design element is a standard catalog item such as a geometric detail, or a piece of hardware like a sensor or motorized unit, and the unit is used in many places throughout the design, then the best approach probably is to instantiate its geometry into the design. Use constraints to fix the location of the models in the overall assembly. However, if the item is one-of-a-kind, or is only used in a particular context, such as the hole-collar cover and its corresponding hex-panel, then it is probably better to design it in the context of where it will be used.

Design-in-context is just one application of creating a local axis system. We will revisit this topic again when we discuss how to instantly change a reference model's xyz-axis system.

11.10 Mirrored Enantiomorphs

In the last chapter, two spherical designs have 3D elements within their design units that are symmetrical, but not tangent; that is, they were enantiomorphs and had left- and right-handed versions of the same element. On the panel and hyperbolic spheres, two of the reference model triangles are symmetrical around the triacon diamond's long axis (which is also one of the spherical icosahedron's reference great circles). The shell design also has several enantiomorphic cases. Of the ten shell panels that make up a full icosahedral face, two triangular vertex shells, two triangular mid-edge shells, and two of the hexagonal shells with holes and collars were mirror images (enantiomorphs) of each other.

There is a simple solution to these geometric cases—mirroring. Basically a mirror turns geometry "inside out." Every feature—points, lines, surfaces, solids, or dress-up—is reflected around a plane, which you define. Left-handed models become right-handed and vice versa. Every point on one side of the mirror has a mirrored point on the other side of the mirror plane and both are the same perpendicular distance from the plane of the mirror.

The most common use of mirroring is to reuse a part that is perfectly symmetrical to a plane—design one side and use a mirror transformation to mirror the geometry to the other side. Figure 11.10 shows the three enantiomorphic shells (two triangular, one hexagonal) from the Ford shell sphere example being mirrored around the xy-plane. The original shells are on the right, the +y side of the zx-plane, and their mirrors are on the left, the −y side.

In CATIA, the mirrored versions are a special cut-paste application that creates a child model of the original parent model—the mirror. Both models are separate, but a logical link maintains the parent-child association.

Mirroring is powerful. First, it is extremely easy to create a mirrored model (child) from another model (parent). Once made, any change made to the parent model—like adding fillets, or enlarging the hole-collar detail—is instantly reflected in the child model. This is where the *logical association link* comes into play. The link keeps the mirrored model in synchronization with its parent model; this eliminates many sources of error when updating models. For changes, however, the link works only one way: from parent to child. A change to the child model does *not* change the parent.

Figure 11.10. Mirroring enantiomorphs.

Perhaps the best payoff of mirrored models can be found in assemblies. Since the original and mirrored versions are two independent models, they can be positioned and constrained in any way needed in the assembly. Also, if only one version is needed in the design (either the parent or the mirrored child), there is no obligation to use the other version. Since the parent and mirrored child are two independent models, they can each be assigned unique part numbers, and bills of material (BOMs) will correctly count and include them in their calculations.

11.11   Power Copy

As its name implies, *power copy* comes into play when you need to repeat a part or a detail over and over in a design. The "power" in the name refers to the fact that each copy need not be an exact replica; each instantiation can adapt to the local conditions.

Power copy is a perfect solution when you have a series of 3D features, such as a connector, which are used repeatedly, but each instance needs a slightly different layout each time it is used. Power copy models can be as simple as a small fastener detail or as complex as a large part with many features such as holes, fillets, or shells. The time savings and consistency in each instantiation make power copies worth using. However, it is their adaptive capability that makes them so powerful. In a sense, a power copy encapsulates a 3D design for reuse; this qualifies it to be called knowledgeware. A simple example will demonstrate how power copy works and show some of the problems it can solve.

Geodesic applications often involve a great number of repetitive details, where each detail is almost identical, but where there are small differences in angles, chord length, orientation, or some other design aspect. Consider the problem of detailing a connector that joins struts on a geodesic dome. Figure 11.11 shows an arrangement of slotted pipes and hub-spoke connectors for a single 8^v triacon spherical icosahedral diamond. The exaggerated scale of the connectors and pipes relative to the grid is intentional; it makes it easier to see how the connectors relate to each other.

Let's assume the connector is a metal hub-spoke design that is cast as a single unit.[1] One is shown in Figure 11.12. Spokes radiate out from a central hub and join pipes that are used for the dome's struts. Pipe ends are slotted to fit over the spokes, and they are welded to the

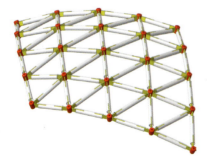

Figure 11.11. Hub and pipe spherical truss.   **Figure 11.12.** Hub-spoke connector layout.

---

[1] The choice of a casting is for illustration purposes only. Dome builders might favor a design in which one flexible connector design can solve all joinery conditions, rather than using a unique rigid casting for every new combination of joinery angles.

spokes during construction. (Something similar to this hub-spoke design was used in the Expo'67 geodesic dome at the Montreal World's Fair, although those connectors were far more complex than the simple one shown here.)

In the triacon layout, shown in Figure 11.11, there are both big and small differences from connector to connector. For instance, there are two-, three-, four-, and six-way hub connectors and, for the most part, they have a mirror image companion on the opposite side of the diamond. Even within connectors with the same number of spokes, there are different planar and *axial angles* between the spokes from one connector to another. In the design, shown in Figure 11.12, the hub (red) is the central body of the connector; it has an alignment hole to help position it during construction. Two to six spokes (yellow) radiate out from the hub. Each spoke has chamfered edges to provide a recess for welding its pipe strut. Fillets (gray) are added to every spoke-hub joint to help relieve stresses. Because the connector is a casting, all parts require draft angles (tapered) to facilitate removal from a mold. The draft angles or tapered spoke cross sections are noticeable in Figure 11.12.

Although the concept for the hub-spoke connector is simple, the wide variation in the number of spokes and the slight variations from connector to connector means that a different design (and casting mold) is required for each unique case. Designing dozens of these connectors is tedious and error-prone work, even with a computer. However, this makes it an ideal application for power copy.

11.12   Power Copy Prototype

Power copy can easily copy a prototype connector design and instantiate it where needed. Each instantiation will automatically adapt to the geometric conditions—the number of spokes, angles, and so on—where it is placed.

The first step in power copy is to build a prototype connector based on the most complex case expected—in this example, a six-way connector. The design parameters for the prototype are straightforward. They are the vertex position (*xyz*-coordinates) where the connector is located, a vertex-origin reference line, and the six triacon grid lines, or chords, that converge on the vertex.

Figure 11.13 shows a stylized version of the power copy hub-spoke connector prototype in transparency. Geometric elements such as reference axes or planes are represented

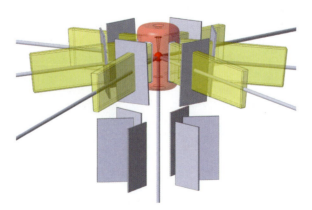

Figure 11.13. Hub-spoke connector power copy model.

with slender rods and blue-gray pads. These make it easier to see how the design from Figure 11.12 relates to its power copy input parameters. The hub is centered on a vertex, which is represented by a small red sphere. The hub's centerline axis is tangent to the vertex-origin line, which is shown by a long gray vertical cylinder. The centerline of each spoke (yellow) is tangent to the corresponding triacon grid line converging on the hub's vertex. They are shown as six radiating gray cylinders.

Each spoke is oriented vertically on a reference plane defined by its corresponding grid line and the hub's vertex-origin line. These reference planes are shown as gray pads below each spoke. One additional plane, called a cutback plane, is added to each spoke. It is perpendicular to the spoke's longitudinal axis and defines the limit of the pipe end when it is placed over the spoke. The pipe cannot extend beyond this plane towards the hub.

The distance between the cutback plane and the end of the spoke is determined by structural analysis, which defines the minimum welding line needed to join a pipe to the spoke and the minimum distance allowed between pipes on adjacent spokes to avoid interferences. The spoke reference plane and pipe cutback plane normally are not visible, but they are shown in the figure because their position is derived from the power copy input parameters and the planes need to be included each time the power copy is instantiated.

Dress-up features are added when the general connector layout is complete. For illustration purposes, they include the hub's alignment hole, the hub and spoke draft angles, spoke welding chamfers, and stress relief fillets between the spokes and hub. All of these features are shown in Figure 11.12. Other features, such as material strength, density, color, and texture, could be added as well. When the power copy prototype connector is completed, it is saved in a catalog where it is available for general use.

To use power copy, the designer simply selects the connector prototype from the catalog and instances it in the design. The power copy master maintains a list of the parameters it needs the user to input or to select from geometry already created in the model. Each instantiation of the hub-spoke connector calls for seven inputs: six define the axes of the pipes to be joined and one is the hub's centerline axis. When all seven selections have been made, power copy dynamically creates a hub-spoke arrangement, which meets this particular condition. The hub-spoke connector layout is tailored for this location, orientation, and so forth because its design is totally based on the input parameters.

Unlike the mirroring example shown earlier where a logical link is kept between the original object(s) and the mirrored ones, power copy instantiations have no logical link back to the original power copy prototype. Once the power copy has been instantiated, it is its own unique and complete 3D model. It has its own unique part number and will be properly counted in a bill of material. An instantiated power copy is an independent model (or group of features). Once instantiated, it can be further refined by adding or removing dress-up features; and those changes do not affect any other instantiations. This is a big advantage when some, but not all, instantiations need additional features.

For example, dome connectors near a foundation or adjacent to an opening may require drains, sensors, or extra fastener details. These can be easily applied. However, once the power copy has been instantiated, the opposite is also true. A change made to the power copy instantiation does not change the power copy prototype, so the catalog copy of the power copy prototype is protected. After the hub connectors are instantiated, the pipe-struts are added. No power copy is needed here, although a simple one could be made to define the lengths of each pipe and to slot them to fit over their connecting hub spokes. A pipe is selected from a structural components catalog and positioned by selecting the triacon chord joining two vertices. When the pipe is placed, its two ends are trimmed to the cutback planes defined on each spoke, which is why cutback planes were included in each power copy model.

Slots are created in the ends of the pipes by removing material where they intersect the spokes. The slots' width is determined by the width of the spoke; the spoke's draft angle and additional clearances are added to make the pipe easier to handle during construction, yet fit tightly enough against the spoke to ensure a good welded joint. If there were enough variation in the diameter of the pipe struts, overall length cuts, or unique slot cuts for different spokes, a power copy pipe placement model would be an advantage.

The truss layout in Figure 11.11 shows connectors that join two, three, and four pipes. How does the six-way connector design adapt to these other cases? The answer is power copy used with *knowledge rules*. Power copy can be used in combination with knowledge rules, and the result is an even more powerful power copy. Knowledge rules adds logic and decision-making ability, and the power copy can automatically choose the best design based on the input parameters it receives. What's more, knowledge rules can direct the power copy to automatically instantiate the prototype for a collection of vertices—an entire diamond or sphere—eliminating the need for the user to instantiate each vertex individually. Knowledge rules can cause the power copy to modify connector features, such as the hub's radius or the spoke's dimensions when it is instantiated. A smaller hub or spoke might serve when there are just a couple of pipes to be joined.

The adaptive characteristic of power copy makes it a very powerful and productive tool. The example in Figure 11.11 uses a triacon subdivision grid, but *any* grid, within reason, is acceptable as long as the basic input parameters are provided. This means that standard connector designs could be created and used on a wide variety of subdivision layouts—triacon, Ford, or even random ones. It also means that the best practices from one project can be preserved and used in the next one. Power copies can be packaged and redistributed in collaborative projects and are tremendous aids in promoting consistent design practices and standards within the team. They are stored in electronic catalogs and applied by anyone needing them. They can also be archived with project records so that the design criteria of the connector is maintained with the resulting truss layout. Power copy, particularly when used with knowledge rules, is one of the most powerful CAD methods available today. With power copy, the spherist stays focused on design rather than on the mechanics of solving repetitive details.

11.13   Macros

Few CAD systems offer spherical subdivision applications. However, most systems do offer a way to make your own geometric applications by combining geometric tools already available on the system—these are called macros. Marcos provide one solution to customizing a general CAD system to perform specific geodesic work.

Macros are user-written programs that automate a series of geometric tasks as simple as creating a point or as complex as defining the intersecting curve of two warped surfaces. Most CAD programs include macro-writing capabilities or at least a way to record common user interactions and reuse them later. The syntax used to write macros might be similar to commands the user might input to accomplish a task, or the syntax might be similar to a computer programming language, but keep in mind that macro syntax does vary from CAD vendor to CAD vendor.

CATIA allows users to record work sequences for replay later. Users can edit these simple recordings and fashion them into very powerful programs anyone can use. Macro programming ability is part of the standard CATIA installation and users have access to most of the object classes and methods for creating, manipulating, and querying geometry created by the application. When a CATIA add-on application is installed, it simply adds to the object classes already available. These object classes and methods are made available to the programmer in a series of class libraries that are directly accessible from Microsoft's Visual Basic programming language or C++. The object classes are declared as data types, and all the regular programming features of both languages—like Windows application programming interfaces (APIs), function libraries, display forms, dialog boxes, menus, and file input/output—are also available.

There are many benefits to using macros. Macros result in an enormous time savings (less time needed to learn and program) and they produce efficient, consistent, and repeatable results. Most importantly, macros provide tools that match the task at hand and are geodesic-specific.

A case in point is the 31 great circles sphere example, which described a technique for defining new spherical points. It involved intersecting the planes of two great circles. The sequence was simple and started with a user selection of two points, each on two great circles. The third point is the origin and is implied. Two planes are created, each passing through the origin and the pair of points on their corresponding great circle. An infinite line results where the two great circle planes intersect and this line, in turn, intersects the surface of the sphere at two opposite points. The selection points, planes, and computed great circle points are all named and grouped together. This procedure is straightforward, but not when dozens of new points must be computed. This and other routines are ideal applications for macros. Figure 11.14 illustrates some sample geodesic macro outputs. Every one of these examples can be done without macros. However, when there are hundreds and even thousands of cases to deal with, macros are clearly the most productive and efficient way to go. The following macros have been useful in illustrating this book:

- Import 3D points with their names from files outside the CAD package. Expose all lines and points (a).

- Expose all planes in a subdivision schema (a).

- Generate spherical surface, vertices, and arc edges (a) and (b).

- Compute the incenter or centroid for a set of closed spherical polygons; optionally, create holes at face centers (c).

- Build arc elements between selected points on the surface of a sphere (d).

- Define great circles or portions of them, their planes, and surfaces from two-point input.

- Define antipodal points on a sphere's surface where two great circles intersect (31 great circles sphere design example).

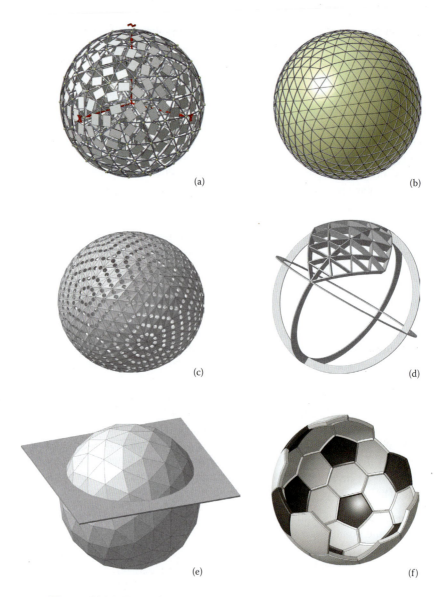

Figure 11.14. Examples of spherical models created with macros.

- *Truncate* a subdivided sphere, profile all cut elements, and generate a profile of the outer limits (e).

- Generate spherical panels based on a closed polygon of great circle arcs (f).

Macros and power copy are both powerful customization tools for the spherist. However, one may well wonder which tool is best for a given application. The author's criteria for deciding are quite simple. If the modeling requirement involves large numbers of repetitive sequences, such as importing vast number of points, generating hundreds of lines or great circles between points, or adding small parts to visualize a reference model, a macro is the most efficient and productive way to work. However, if the problem involves complex geometric shapes, topology, surface intersections, model dress-up features (shells, fillets, chamfers, slots, etc.) or geometry that must adapt to a variety of geometric situations, a power copy approach is the best choice.

Another criterion that helps when deciding between macros and power copy is the conceptual level at which you want to work. While macros give you the most detailed access to object classes and their methods, power copy gives you a higher-level, more intuitive way to work. And when power copy is used with knowledge rules, it is hard to match this power without significant macro programming that is both detailed and highly specialized.

11.14   Publication

Publication is the capability to assign meaningful names to geometric features or parameters in a 3D model and to make these names the only way users can select their geometry. Published names for points, lines, planes, surfaces, or entire configurations of solid bodies make it easier to understand 3D models by highlighting and distinguishing their key elements. Meaningful names make the elements easier to associate with design tasks. Users can understand a complex layout more easily because they do not have to decipher the meaning of automatically generated CAD names, such as Point.1, Plane.38, or Line.16. Instead, the user can select geometry with names, such as Ground_Plane, Apex_Point, or Sphere_Origin.

Published names help eliminate potential inputting errors by restricting a user's geometry selections and forcing him or her to use only published elements. Publishing also helps in collaborative projects. A published list of geometric elements can be created for a specific design task. A design team will see the entire design but will only be able to add new elements that are based on the team's published elements.

Publishing is a tremendous help when geometric features are close together, or when several geometric elements occupy the same location. It is common to have several points at the same location. When 3D models are being created, it is essential to select the proper feature (point, line, surface, plane, etc.). Other geometry may be linked to the element you select. If that geometry moves, so does the element you selected.

Publishing also helps clarify which geometry belongs to which part. For instance, in orthographic views of some spherical designs, geometric elements on the back side of the sphere are perfectly superimposed on those displayed on the front side. When you are selecting an element, which one are you selecting, the front one or the back one?

In another case, a point is displayed. Is it the apex of a triangular face you need to detail, or is it the zenith reference point that is coincident to it? You need the former for your design, and you may need to move it. However, the zenith reference point is a permanent reference and should never be changed.

In yet another case, a constraint is needed between two points, which are near small edge pieces. Was an edge selected by mistake? You can't always tell because points and small edges near them may look alike until you zoom into the model for a really close look. Selecting the wrong element might not show up until you try to create an assembly. A small geometric discrepancy may be enough of an error that attempts to apply constraints may fail.

Of course, in this last case, the user can zoom into the model and closely inspect the geometry. And most CAD systems offer a way to visualize geometry in tight spaces. However, the real solution to all of these problems, and many more, is to use published geometry.

Publishing can be used without a reference model, but maximum benefits are achieved when reference models and publishing techniques are used together. Publishing is particularly helpful when there is a collaborative project and users need to guarantee that they are creating complementary geometry. Publishing also eliminates the need for users to have direct access to geometry to control and manage links between parts. Users only need to reference the published name of an element. Project administrators can also restrict user access to only the published elements, thus ensuring that no one has inadvertently changed any of the values.

Publishing promotes design consistency by allowing parts to be positioned in an assembly without direct access to positioning geometry. Designs that use published geometry tend to maintain better synchronization with the original source. This is a genuine advantage when layouts become complex.

11.15   Data Structures

The choice of data structures is highly dependent on the geodesic application you have in mind. How you structure your data has as much to do with early design uses as with downstream uses such as manufacturing, publishing, or database interactions. Naming and grouping data into meaningful sets must be addressed early in any geodesic project. A structure suitable for building a geodesic dome might be inappropriate for performing structural finite element analysis of a spherical pressure vessel or mapping atmospheric pollution.

In this work, we hope to demonstrate the value of a geodesic reference model and provide criteria for developing your own. We will do this through a number of simple application cases where the pros and cons of various data structures become clear.

11.16   CAD Alternatives: Stella and Antiprism

Not everyone has access to CAD. There are alternatives to CAD for designers needing to develop spherical geometry, Stella and Antiprism are good choices.

Despite its modest price, Stella is one of the most powerful and comprehensive 3D polyhedral modelers available today.[2] Developed by Robert Webb, this personal computer program provides an extensive library of polyhedra, including Platonic, Archimedean, and *Kepler-Poinsot solids*, convex prisms and *antiprisms, Johnson solids, Stewart toroids* (regular-faced polyhedra with holes), and compounds. Stella's library also includes one-, two-, three-, four-, and six-frequency hemispherical subdivision models for tetrahedra, cubes, octahedra, and icosahedra, with similar models and frequencies for full spheres.

Users can display polyhedra, analyze their symmetries, and compare them to their duals. Stella can morph (undergo a dynamic transformation from one form to another) between dual polyhedra and show the results in real time. One particularly useful feature enables users to create 2D flat patterns (nets) of the polyhedron's faces. Nets can be viewed folding and unfolding in 3D. The nets are a valuable aid for making templates and fabricating polyhedral models from paper, wood, plastic, or other materials.

---

[2] Stella is available at http://www.software3d.com/Stella.php.

Figure 11.15. Stella subdivides an octahedron with 2^v, 4^v, 8^v, and 16^v grids.

For spherical work, Stella can tessellate any polyhedron whose faces are not on planes that pass through the polyhedron's center.[3] Stella tessellates a polyhedron by first subdividing its faces with a triangular grid. Any even or odd frequency is possible. The face grids are projected onto the polyhedron's circumsphere using gnomonic projection. If the polyhedron is a deltahedron, Stella's tessellation is equivalent to the Class I Equal-chords schema. Figure 11.15 shows an octahedron (a deltahedron) in progressive subdivision of 2^v, 4^v, 8^v, and 16^v frequency grids. At each stage, the resulting tessellation is Equal-chords. All figures are shown from the same viewpoint and scale.

With Stella, the user has tremendous flexibility to edit faces and create a wide variety of patterns and shapes. Figure 11.16 shows some of the many combinations that are possible based on different polyhedra and tessellations and on differences in how the faces are edited. The program's default color scheme is based on assigning the same color to faces that have the same symmetry relationship. This feature makes it easy to identify rotational and reflection symmetry in the subdivided sphere. Users are free to assign any color to any face. In addition to traditional pan, zoom, rotate, and multiple views, the user can select

Figure 11.16. Stella spherical models.

---

[3] When a face passes through the center of the polyhedron, there is no symmetry-maintaining way to project the face and maintain paired faces at each edge.

(a)   (b)   (c)

Figure 11.17. 4^v spherical icosahedron with face dual.

any face, edge, or vertex to display, accent, or hide. Stella will display any polyhedron and symmetry axes, reflection planes, reflection plane normals, reciprocation spheres, and intersections with spheres. Optional display annotations include vertex angles and face/vertex text. Color codes accent faces based on symmetry and, by default, all faces that are equivalent within the symmetry group are colored the same way, as shown in the top row of spheres in Figure 11.16. Color options allow the user to paint colors manually across faces and still maintain symmetry, if so desired. The bottom row of spheres in the figure shows two examples. Some spherical subdivisions are visually complex, and one display projection technique may be better than another at clarifying the geometry. All the examples in Figure 11.16 are perspective views; features nearer the viewer are somewhat larger than those farther away.[4]

Stella can display a polyhedron's dual, and this capability applies to the dual of geodesic spheres as well. Figure 11.17(a) shows a 4^v spherical icosahedron with the default color scheme, which gives symmetrical faces the same color. The vertices of the polyhedron all lie on a common sphere, known as the circumsphere. Figure 11.17(b) shows the dual of (a). Its faces are good approximations of a sphere, except that instead of all vertices in the dual lying on a single sphere, all face planes in the dual are tangent to a single insphere. The dual's colors indicate symmetrical faces. Figure 11.17(c) shows the original icosahedral subdivision from (a) superimposed on its dual from (b).

With Stella, you can resubdivide a subdivided polyhedron as many times as you like. A unique grid results each time a polyhedron is resubdivided and the grid at any point in the sequence is not the same grid that would result from single subdivision of the polyhedron at the product of all the resubdivision frequencies. For example, a 2^v subdivision of a polyhedron already subdivided 3^v results in a 6^v grid. This final 6^v grid is different than creating a 6^v grid in a single subdivision of the polyhedron. And neither of these two 6^v grids are the same as a grid that results when the polyhedron is subdivided in the opposite order; that is, starting with a 2^v subdivision and then resubdividing it 3^v.

The important use of polyhedron resubdivision is that in certain resubdivision sequences, Stella's basic Equal-chords schema can create grids that approximate another technique, Mid-arcs. An example demonstrates how. Keeping in mind that Stella uses only the Equal-chords subdivision method, a polyhedron first subdivided as 2^v and then resubdivided 2^v again results in a 4^v subdivided polyhedron. If this 4^v polyhedron is again subdivided 2^v, it creates an 8^v subdivided polyhedron. The repetitive application of Equal-chords subdivision combined with doubling the subdivision frequency each time (2^v) results in progres-

---

[4] Models courtesy of Robert Webb and Magnus Wenninger.

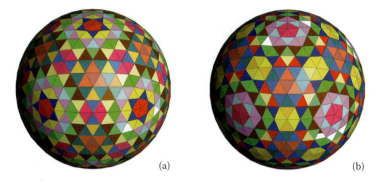

(a)   (b)

Figure 11.18. Single subdivision versus progressive resubdivision of a polyhedron.

sive polyhedral subdivisions 2^v, 4^v, 8^v, 16^v, 32^v, and so on. Each frequency, in succession, is a power of two. Figure 11.18 shows two 8^v icosahedra side by side. Both are from the same viewpoint. In (a), the icosahedron was subdivided in one single step, whereas in (b), a progression of resubdivisions was used where each stage's frequency was a power of two. Although the differences between the two spheres are subtle, the five triangles that form red pentagons around each of the icosahedron's vertices are noticeably smaller in (a) than in (b). There are also many other differences, but these are easy to spot. When subdivisions of a subdivided polyhedron double in frequency each time, such as in (b), the resulting grid at any step is approximate to what the Mid-arcs technique would produce even though the Equal-chords technique is used (see Section 8.5.8).

The subdivision schemas presented earlier in the book all produce triangular grids. When a dual is made from a triangular grid, most of the resulting faces are hexagonal because most vertices in geodesic spheres have six triangles meeting at them. Pentagons or squares appear over the vertices of the original polyhedron, depending on the valence of those original vertices. Geodesic duals are a unique feature of Stella; no CAD package has a built-in function to create such duals, though macros can be written to create them.

Visualization is very important when working with complex 3D polyhedra or spherical subdivisions. In addition to perspective projection, Stella offers orthographic and stereo projections (standard, cross, or anaglyph). Stereo projection is particularly useful in visualizing stellated polyhedra, polyhedra whose faces penetrate one another, or hollow spheres in which faces on the far side could be confused with those in the foreground. Figure 11.19 shows four examples of *cross-stereo*.[5] The first two examples, (a) and (b), are subdivided spheres within a sphere. Selected faces are removed, leaving chains of faces that interlock in a basket weave pattern. Figure 11.19(a) shows an intricate geometry with a single color (red) scheme. Its geometry would be difficult to understand without the use of stereo. Although perspective helps separate features nearer the viewer from those farther away,

---

[5] Cross-stereo displays do not require a stereopticon or any other viewing device to see 3D effects. To see a cross-stereo image in stereo, simply position the image pair at a comfortable viewing distance in front of you. Place your thumb on the display between the two images and focus on it. Slowly move your thumb towards your nose while maintaining your focus on your thumb. A virtual 3D image will form under your thumb as your thumb gets closer to your nose. As soon as you see a distinct third image under your thumb between the original two, quickly remove your thumb, without changing your eye focus. Your mind merges the two views into a virtual 3D image. The effect is quite pronounced and very clear. Your eyes will be somewhat cross-eyed at this point; hence, the technique is named "cross-stereo." Your right eye views the left image, while your left eye views the right image.

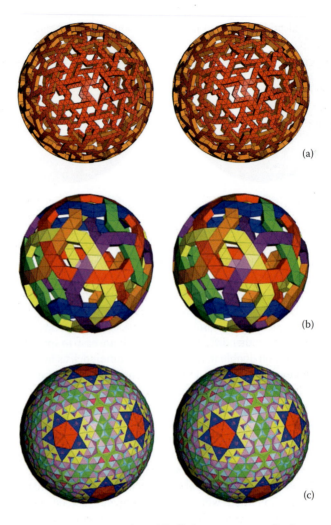

Figure 11.19. Examples of Stella's cross-stereo displays.

the monochromatic red faces do not help highlight symmetry groups. The intricate basket weave pattern in Figure 11.19(b) is based on a 6^v icosahedron and is much easier to understand than (a) because of its excellent use of color and stereo. The last stereo example shows a 5^v spherical icosahedron and its dual superimposed. The subdivided icosahedron uses identical colors for symmetrical faces and purple for the faces of the dual. Here, too, stereo projection helps separate one polyhedron from the other and which faces are above or below another.

In addition to excellent visualization options, *Stella* provides useful analysis tools. Spherists using Stella can make face, stellation face and stellation/cell diagrams, vertex figures, faceting diagrams, and cross sections of polyhedra. In addition, users can generate the convex hull of any polyhedron and display any face angle or output vertex coordinates and a list of the vertices around every face.

Users of spherical programs often need images and data for presentations, reports, or input to other programs. Stella display images can be output in all major formats in a wide range of resolutions. Three-dimensional data export files can be output for use in other display or analysis programs using standard exchange formats such as Autodesk's DXF or Alias/Wavefront formats, *Persistence of Vision Raytracer (POV-Ray)* or Virtual Reality Modeling Language, better known as VRML. Metric data output includes face/edge/vertices, circumsphere and insphere radii, edge lengths, facet types, area, edge types, adjacent faces and dihedrals, lengths, miter, and face angles.

11.17   Antiprism

In addition to Stella, Antiprism is another alternative to CAD. Antiprism is an amazing PC general purpose polyhedral modeler by Adrian Rossiter that incorporates many of his powerful algorithms and some by others.[6] Formerly called Packinon, Antiprism is a treasure trove of polyhedra and programs that can transform their geometry, subdivide spherical versions of them, and graphically show their beautiful forms. Antiprism can model and transform any polyhedron from its own library or from user input.

Antiprism's extensive polyhedral library includes the Platonic, Archimedean, uniform, Catalan, uniform dual, Johnson, prisms, antiprisms, pyramids, dipyramids, snub-antiprisms, cupolas, bicupolas, deltohedra, trapezohedra, Waterman, *zonohedra*, and *sphericons*.[7] Polyhedral geometry can be used as is, or can be transformed into spherical versions. You can combine distinct polyhedra in the same file or bind a face from one polyhedron to the same face type in another polyhedron so the common face is removed. The result is a new single polyhedron. Any polyhedron or its transformation can be displayed, rotated, pan-zoomed, labeled, or colored with Antiview, a built-in viewer, or passed to other popular 3D viewers such as Geomview, JavaView, and Springie, or photo-realistic display programs such as POV-Ray, VRML, or LiveGraphics3D. Many display examples can be seen at the Antiprism website.[8]

What makes Antiprism such an amazing resource is that the source programs can be freely downloaded and installed on your own computer. You can also redistribute or incorporate all or parts of Antiprism into other projects subject to the generous terms of the open source licenses used.[9] All programs are well documented and include help options. The package also includes HTML documentation with further program details and examples. Antiprism is maintained with regular enhancements and user forums discuss program features, new ways to use Antiprism, and suggestions for future enhancements.

Antiprism offers outstanding spherical subdivision functions. All three Class subdivision schemas are supported. Figure 11.20 shows POV-Ray displays of Antiprism Class I, II, and III subdivisions of a spherical icosahedron. Figure 11.20(a) and (f) are

---

[6] An antiprism is a polyhedron composed of two parallel copies of a polygon in which the polygons are rotated with respect to one another so their vertices are not lined up. An octahedron is an antiprism with triangular bases because any one of its faces is parallel to another opposing face, which is rotated so the faces between them are triangles.

[7] Discovered by Colin Roberts of Hertfordshire, England, the sphericon is a 3D solid with one side and two edges.

[8] www.antiprism.com

[9] The programs are supplied under a variety of permissive licenses granting free of charge, unrestricted rights to use, copy, modify, merge, publish, distribute, sublicense, and/or sell copies of the software, so long as the full copyright notice and permission notice is included in all copies or substantial portions of the software. See Antiprism's copying notice for details.

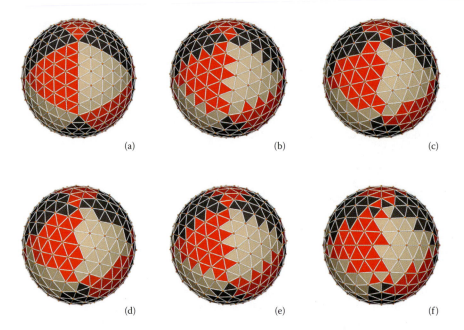

Figure 11.20. Examples of Class I, II, and III subdivisions.

Class I and II 6^v, respectively. Illustrations (b) through (e) show Class III subdivisions progressively evolving from (a) to (f).

In addition to the classic techniques based on subdividing spherical polyhedra with equilateral triangles, such as the tetrahedron, octahedron, and icosahedron, Antiprism can subdivide *any* polyhedra regardless of face type. If the polyhedron has nontriangular faces, they are initially converted to triangles by joining the edges to the face's center. Regardless of the base polyhedra and the shape of its faces, all faces are subdivided by equal angle division of the edges of the original polyhedron. Figure 11.21 shows how Antiprism accomplishes this. The nontriangular faces of a rhombicosidodecahedron in (a) are converted to triangles in (b) prior to further subdivision with a Class $III_{2,1}$ skew grid in (c). The ability to subdivide any polyhedra, regardless of face type, is a very powerful feature not offered by other geometric modelers.

Antiprism offers a rich modeling environment with many geometric options not found in other modelers. Some key spherical features include the following:

- *Spherical duals*. Voronoi cells of points in space.

- *Convex hull*. Volume of a polyhedron.

- *Minmax*. Minimize a spherical tessellation's maximum edge length, turn face data into a spherical tessellation.

- *Repulsion*. Distribution of *n* points on a sphere, equilibrium of points repelling on a sphere uses various repelling formulae.

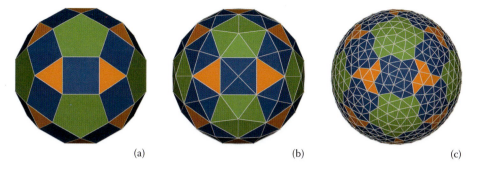

(a)   (b)   (c)

Figure 11.21. Subdivision of polyhedra with mixed face types.

- *Polar reciprocals*. A technique for creating polyhedron duals.

- *Rings and spiral points on spheres*.

The first four features can provide unique solutions to difficult spherical problems.

We first saw subdivision duals when we looked at grids for climate and weather fore-casting models (see Section 3.5). Recall that a polyhedron's dual (planar face versions of subdivided spheres are polyhedra) is another polyhedron that has a symmetry relation-ship—the vertices in one correspond to face positions in the other and vice versa. The edges in one correspond to the edges in the other; thus, both polyhedra have the same number of edges while the number of faces and vertices are equal but reversed in the other (see Section 6.2.1). These relationships are easily seen in Figure 11.22 where the Class I icosahedron 8^v grid shown in (a) appears next to its dual (Voronoi) in (b). The grid points in (b) have the same color as their dual faces in (a). Notice that every vertex in one is a face position in the other and that the edges in one correspond to the edges in the other. Less obvious is the fact that the area of each hexagon-pentagon polygon in (b) represents all points closest to the vertex in (a) that is within the polygon.

Convex hulls can quickly define a spherical layout. For example, you can input a sim-ple list of 3D points and let Antiprism's convex hull feature define the "wrapper" surface

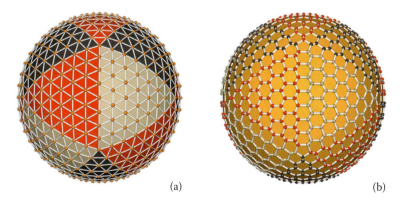

(a)   (b)

Figure 11.22. Class I Icosahedron 8^v and its spherical dual (Voronoi).

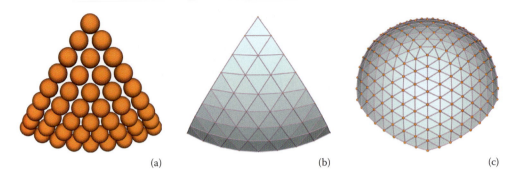

(a) (b) (c)

Figure 11.23. Spherical subdivision developed from convex hull of vertex points.

around them. If the input points are already on the surface of the sphere, the convex hull surface is, in effect, the planar triangle faces of a geodesic grid. Figure 11.23(a) is a display of 45 points.[10] The spheres around each point are as large as possible without interfering; thus, they give a visual indication of how even the points are distributed on the spherical triangle. Figure 11.23(b) shows the convex hull surface of the same set of points. Each triangular face is planar and they look like triangular tiles. Antiprism can replicate this tiling with akaleidoscope transformation and create the symmetrical pent-cap shown in (c) where the points and chords are emphasized and resemble a geodesic dome. What is impressive is that all three stages in Figure 11.23 are automatically handled by Antiprism and begin with just a simple list of points.

Users can combine Antiprism's convex hull and dual features to define still other spherical subdivisions. We have just seen how Antiprism can convert any set of points on a sphere to a convex hull surface. The result was an initial tiling of triangles. It is possible to define another subdivision by finding the tiling's dual. The dual relationship was illustrated in Figure 11.22. The dual becomes a new spherical subdivision, one derived from the original convex hull.

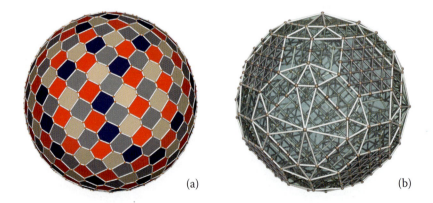

(a) (b)

Figure 11.24. Spherical subdivision based on convex hull of points.

---

[10] Antiprism models produced from Equal-arcs (three great circles) 8^v PPT vertex coordinates; see Table D.2.

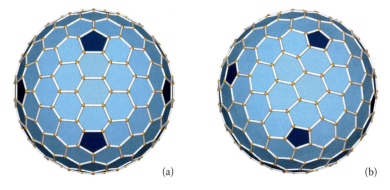

(a)   (b)

Figure 11.25. Minmax before and after optimal spherical tessellation.

Figure 11.24 shows two additional ways spherical subdivisions can be derived from convex hulls. In (a), the convex hull of a set of 3D points is created. The dual of the tiling creates another set of polygons and their vertices are projected onto the sphere's surface. In Figure 11.24(b), an octet truss sphere based on a Waterman polyhedra (see Section 6.4.8) is created from a convex hull. The other frame is the convex hull of a root 160, and the inner frame is the convex hull of a root 135.

Minmax is another sophisticated feature of Antiprism. It can optimize a tessellation by deriving another one from it where the maximum edge length of the new tessellation is constrained to be within a specified range. The length-changing algorithm can constrain the shortest and longest of all edges, or the shortest and longest edges attached to a vertex, to be within a range or a percentage of the original length. The resulting points in the new tessellation are on the surface of the sphere. Figure 11.25(a) and (b) show before and after optimization; the differences are subtle. In this example, you can see how the pentagons in (a) have wandered to new positions in (b) and that their edges have shrunk as they comply with constraints placed on edge length. Minmax can help reduce the number or size range of parts or the number of different subassemblies needed to manufacture a product. The cost savings are obvious.

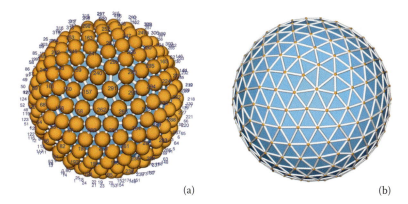

(a)   (b)

Figure 11.26. Repulsion distribution of 320 points and resulting spherical subdivision.

We have said that it is easy to distribute an arbitrary number of points on the circumference of a circle but evenly distributing such points on the surface of a sphere is quite difficult. Antiprism can find the equilibrium position of an arbitrary number of points based on several user-selected algorithms: inverse of distance, square of distance, cube of distance, or square root of distance. Antiprism's repulsion techniques are similar to the particle repulsion method described in the introductory chapter (see Section 1.3). Repulsion is an important technique because it can distribute *any* number of points, not just fixed sets of points based on subdivided polyhedra. There may be applications where it is desirable not to have symmetrical relationships such as vertex rotation, mirror planes, or antipodal points. The number of points distributed, the initial random point distribution, the number of balancing cycles, and the repulsion algorithm employed all influence the outcome. Figure 11.26(a) and (b) show 320 points distributed by repulsion using Antiprism's inverse square of distance algorithm. In (a), the largest equal-sized and numbered spheres possible are located at every point; none overlap but some just touch (kiss). Figure 11.26(b) is the same model and view as (a), but with small spheres at each vertex to make the convex hull more apparent.

Almost every spherical application requires global measures such as area, volume, centroid, maxima (for example, longest and shortest edges), and counts of faces, edges and vertices, as well as specific element data, such as polygon vertex order, central angle of edges, and face angles. Antiprism's standard reports provide them. Additional program options control formatting, significant digits for output, and file output for subsequent processing by other programs or entry into spreadsheets.

In addition to standard reports, Antiprism provides many types of graphical and numeric analysis. Symmetry detection is particularly useful when spherical designs are used as the basis for industrial product design or where symmetry (or lack of it) is required, as we have seen in golf ball dimple layouts.

Figure 11.27 shows two examples of *symmetry detection*. In (a), the full icosahedral symmetry is displayed. Note the special symbols displayed at each face vertex and at every midface and mid-edge point. These symbols indicate where reflection and rotation symmetry occur. You can see at a glance where the model has *mirror symmetry*. Now compare this illustration with the symmetry detection of the Skew icosahedral subdivi-

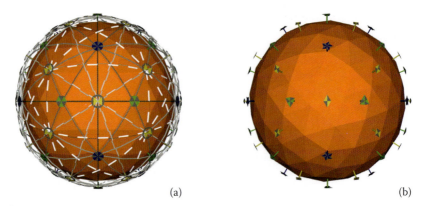

(a)   (b)

Figure 11.27. Examples of symmetry detection.

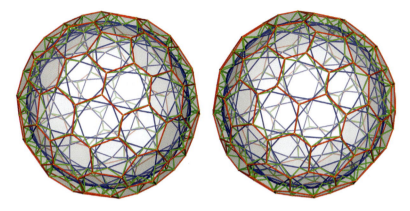

Figure 11.28. POV-Ray cross-stereo image pair of Antiprism spherical subdivision.

sion shown in (b). As expected, Skews have less symmetry. In this example, the $\{3,+5\}_{3,1}$ tessellation has only icosahedral rotational symmetry.

In addition to symmetry detection, Antiprism provides two other kinds of analysis for any polyhedra or spherical subdivision—*planarity* and *isoperimetric quotient*. Planarity is how close to flat the faces are. This is useful when subdivision grids are derived from triangular grids. For example, a spherical Voronoi grid from a triangular grid will be mostly spherical hexagons and a few spherical pentagons, none of which are planar. Planarity indicates how much each polygon deviates from being a perfect plane.[11] The isoperimetric quotient is a measure of how spherical the surface of a polyhedron is.

Visualization and good graphics control are very important when working with spheres. Antiprism's own viewer, Antiview, is fast, easy to use, and can pan-zoom, rotate, color control, add backgrounds, show/hide vertices, edges, faces, vertex and face numbers, and display in perspective or orthogonal projection. In addition, Antiprism can output OFF format files, a standard format used by complementary 3D viewers such as Geomview, JavaView, and Springie. The user can also export Antiprism geometry direct to the POV-Ray, Virtual Reality Modeling Language (VRML), or LiveGraphics3D programs for photo-realistic rendering, or simply use the OFF file as a coordinate geometry file. Figure 11.28 shows Antiprism output rendered by POV-Ray as a cross-eyed stereo pair of images.

11.18   Summary

CAD is a key resource for the spherist. It can be used in many ways. Some applications create pure geometry, enabling the spherist to derive complex geometry. Other CAD applications result in a physical design, such as the ones developed in the previous chapter.

In this chapter, we did not focus on a particular spherical design. Instead, we addressed methodology and best practices. A general approach to reference models and the use of local coordinate systems showed how a reference and assembly model framework can be created and oriented in any way that an application requires. Design-in-context and mirroring techniques showed ways to base one design element on another, and demonstrated

---

[11] See (Aravind 2011) for related spherical analysis of Archimedean solids.

that adding logical links between them can keep them consistent should changes on one affect the other.

Most geodesic applications generate a lot of geometry, and when a spatial area is filled with many planes, points, lines, or surfaces, it is a challenge to know which one is the one you need when building parts. Publishing makes this geometry easier to understand by giving geometric entities meaningful names. Options within CATIA can restrict user selections to only published elements. In congested areas, this can significantly reduce errors.

As rich as CATIA's geometric functions are, there are few geodesic-specific tools. However, power copy and macros can provide the needed functions. Power copy packages existing geometry for reuse elsewhere, and it has the power to adapt geometry to the local conditions where they are instanced. This is extremely powerful. Macro programming, on the other hand, adds high-performance functions to the CAD system, and macros can provide specific geodesic functions most CAD systems do not include. CATIA macros can be written in Visual Basic or C++ and have all their native capabilities, plus direct access to CATIA's geometric object classes and methods. A skilled programmer can produce highly efficient and powerful functions that can customize any CATIA workbench. There are alternatives to using CAD. Several well-written, inexpensive (even free) graphics and polyhedral modelers for personal computers are available. They provide many modeling functions that more expensive CAD packages offer and include specific spherical capabilities. Stella and Antiprism programs are among the best today.

Additional Resources

Huybers, Pieter. "Computer-Aided Design of Polyhedral Building Structures." *Design Studies* (Civil Engineering Department, Delft University of Technology) 14.1, 1993.

Nooshin, Hoshyare, P. L. Disney, and O. C. Champion. "Computer-Aided Processing of Polyhedric Configurations." *Beyond The Cube: The Architecture of Space Frames and Polyhedra.* ed. J. Francois Gabriel, New York: John Wiley & Sons, Inc., 1997: 343–384.

Rossiter, Adrian. Antiprism. http://www.antiprism.com, 2011.

Shea, Nicholas. TesselSphere—Geodesica, a PC computer program for designing geodesic domes, http://sourceforge.net/projects/geodesica/, 2010.

Webb, Robert. Stella, a polyhedral and spherical modeler, http://www.software3d.com/, 2011.

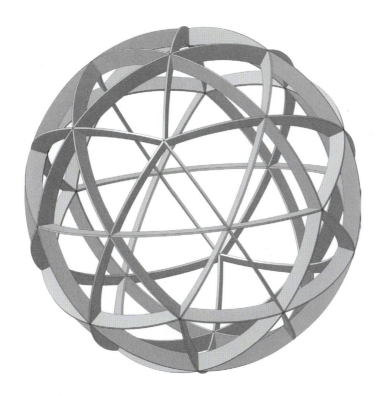

Ⓐ Spherical Trigonometry

T his appendix summarizes the main trigonometric relationships for computing the angles and sides of spherical triangles. The basic trigonometric functions are stated first. These functions are the building blocks for more complex trigonometric relationships that follow, such as the law of cosines, law of sines, solving right triangles, Napier's rule, and polar triangles. This appendix does not include proofs; however, several references with extensive proofs are listed in the additional resources.

A.1   Basic Trigonometric Functions

The six basic trigonometric functions are heavily used in subdividing spheres. Figure A.1 shows how the sine, cosine, tangent, cotangent, secant, and cosecant functions relate to the unit circle of radius r. "Co" is from complement, meaning the function is for the *complementary angle* $90°-a$. If you can find the value of either sine or cosine, you can easily derive the others:

$$\text{sine: } \sin\theta = \frac{y}{r} = \frac{1}{\csc\theta},$$

$$\text{cosine: } \cos\theta = \frac{x}{r} = \frac{1}{\sec\theta},$$

$$\text{tangent: } \tan\theta = \frac{y}{x} = \frac{1}{\cot\theta} = \frac{\sin\theta}{\cos\theta},$$

$$\text{cotangent: } \cot\theta = \frac{1}{\tan\theta} = \frac{\cos\theta}{\sin\theta},$$

$$\text{secant: } \sec\theta = \frac{1}{\cos\theta},$$

$$\text{cosecant: } \csc\theta = \frac{1}{\sin\theta}.$$

Trigonometric functions sine, cosine, secant, and cosecant repeat every 360°, while tangent and cotangent repeat every 180°. As a result, many useful relationships exist between the angles in the four quadrants (0°–90°, 90°–180°, 180°–270°, and 270°–360°). The formulas that follow show how the often-used sine and cosine functions relate to each other. They are useful when you know one angle, but need to use it in another way.

Trigonometric identities are equations involving trigonometric functions that are true for all values of the occurring variables. Identities are useful when you need to simplify more complex trigonometric functions. Only the ones most useful in geodesics are shown here. Remember that notationally, $\sin^2\theta$ is the conventional way of writing $(\sin\theta)^2$.

$$\sin^2\theta + \cos^2\theta = 1,$$
$$\sin(-\theta) = -\sin\theta,$$
$$\cos(-\theta) = \cos\theta,$$
$$\tan(-\theta) = -\tan\theta;$$

$$\sin(A+B) = \sin A \times \cos B + \cos A \times \sin B,$$
$$\sin(A-B) = \sin A \times \cos B - \cos A \times \sin B,$$
$$\cos(A+B) = \cos A \times \cos B - \sin A \times \sin B,$$
$$\cos(A-B) = \cos A \times \cos B + \sin A \times \sin B;$$

$$\tan(A+B) = \frac{\tan A + \tan B}{1 - \tan A \times \tan B},$$
$$\tan(A-B) = \frac{\tan A - \tan B}{1 + \tan A \times \tan B}.$$

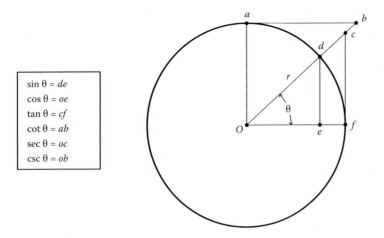

$$\sin \theta = de$$
$$\cos \theta = oe$$
$$\tan \theta = cf$$
$$\cot \theta = ab$$
$$\sec \theta = oc$$
$$\csc \theta = ob$$

Figure A.1. Six trigonometric functions.

A.2   The Core Theorems

For any *oblique spherical triangle*, if you know any three of the six quantities (three sur-face and three central angles), the triangle can be solved. Two laws—the law of cosines and the law of sines—provide almost all the relationships you will ever need. Of the two, the law of cosines proves its utility repeatedly and, if applied often enough, will solve almost any spherical triangle. The law of sines, as we will see, is also powerful, but its results may be ambiguous and generate not one, but two possible solutions. Figure A.2 shows a typi-cal spherical triangle with the angle and edge labels we will use for the remainder of this appendix.

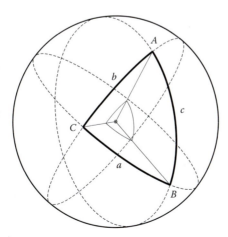

Figure A.2. Spherical triangle *ABC*.

A.3 Law of Cosines

The law of cosines, sometimes called the cosine law, is one of the most useful sets of side-angle-side relationships ever devised. You use the law of cosines to solve triangles that are not right-angled. You use it when you know any two sides and their included angle and you want to solve for the side opposite the angle. A variation of this law allows you to find the angles if you know only the sides. The law of cosines is also historically important. It is the fundamental equation for marine and aviation celestial navigation. It is used to determine directions and distances from port to port or to convert sights taken with a sextant to find a latitude and longitude position. Using the notation presented in Figure A.2, the cosine law for the sides of any spherical triangles (central angles) is expressed as

$$\cos a = \cos b \times \cos c + \sin b \times \sin c \times \cos A,$$
$$\cos b = \cos a \times \cos c + \sin a \times \sin c \times \cos B,$$
$$\cos c = \cos a \times \cos b + \sin a \times \sin b \times \cos C.$$

The law of cosines also applies to surface angles. It is useful when you know two surface angles and the side between them and you want to find the third surface angle:

$$\cos A = -\cos B \times \cos C + \sin B \times \sin C \times \cos a,$$
$$\cos B = -\cos C \times \cos A + \sin C \times \sin A \times \cos b,$$
$$\cos C = -\cos A \times \cos B + \sin A \times \sin B \times \cos c.$$

A variation of the law of cosines proves quite useful when you know the sides of a spherical triangle but not the surface angles:

$$\cos A = \frac{\cos a - \cos b \times \cos c}{\sin b \times \sin c},$$
$$\cos B = \frac{\cos b - \cos a \times \cos c}{\sin a \times \sin c},$$
$$\cos C = \frac{\cos c - \cos a \times \cos b}{\sin a \times \sin b}.$$

A.4 Law of Sines

The companion to the law of cosines is the law of sines. It defines another very useful set of relationships. The sides of a triangle are to one another in the same ratio as the sines of their opposite angles:

$$\frac{\sin a}{\sin A} = \frac{\sin b}{\sin B} = \frac{\sin c}{\sin C}.$$

About the only situation the law of sines cannot solve is when we are given two sides and the nonincluded angle (*SSA* case). This corresponds to the ambiguous case in plane geometry because there can be two solutions, one, or none at all. The reason for the ambiguity is that the sine of an *acute angle* is the same as the sine of the *obtuse angle*.

A.5   Right Triangles

We have said that you can solve any oblique triangle with the law of cosines or the law of sines. But spherical triangles with only one right angle are even easier, and are very common in geodesics. The Triacon Class II subdivision method is largely based on subdividing spherical right triangles. The most frequently used trigonometric relationships for right triangles are listed here. Although the list looks long, it includes only a few unique relationships between the sides and surface angles. Here are all the useful combinations where angle C is 90°:

$$\sin a = \sin A \times \sin c,$$
$$\sin b = \sin B \times \sin c;$$

$$\tan a = \cos B \times \tan c,$$
$$\tan a = \tan A \times \sin b,$$
$$\tan b = \cos A \times \tan c,$$
$$\tan b = \tan B \times \sin a;$$

$$\cos A = \sin B \times \cos a,$$
$$\cos B = \sin A \times \cos b,$$
$$\cos c = \cos a \times \cos b,$$
$$\cos c = \cot B \times \cot A.$$

Another short set of relationships is quite helpful in situations where you know all three surface angles of a right triangle and need to find the sides or central angles:

$$\cos a = \cos A / \sin B,$$
$$\cos b = \cos B / \sin A,$$
$$\cos c = \cos a \times \cos b.$$

A.6   Napier's Rule

John Napier (1550–1617), the renowned mathematician, invented logarithms as a way of reducing multiplication and division to addition and subtraction and also developed a new set of formulas to solve right spherical triangles. When one of the surface or interior angles is 90° (this triangle is said to be *quadrantal*), a simple set of relationships and formulas can be used. Figure A.3 shows the six parts of a right triangle: three surface angles (A, B, C) and three interior angles, which make the sides (a, b, c). If we eliminate right angle C, the

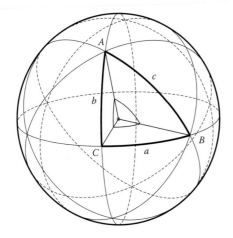

Figure A.3. Parts of a spherical right triangle.

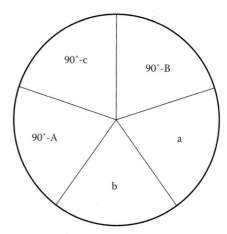

Figure A.4. Napier's circular parts diagram.

five remaining parts are used in Napier's rule, which states:

- The sine of any angle or side is equal to the product of the cosines of the opposite two angles.

- The sine of any angle or side is equal to the product of the tangents of the two adjacent angles.

To help remember these rules, use Napier's circular parts diagram, shown in Figure A.4. It graphically shows how the adjacent (near) and opposite (far) parts of the triangle relate to each other. The parts are shown in their clockwise sequence, as they appear in the triangles illustrated above. Again, right surface angle C is not used.

Notice that Napier's diagram expresses three angles as complementary angles—that is, their sum is 90°—instead of giving their direct angular value. Referring to Napier's two rules and Figure A.4, it is also possible to write out each of the relationships. Ten formulas result:

$$\sin a = \tan b \times \cot B = \sin c \times \sin A,$$
$$\sin b = \tan a \times \cot A = \sin c \times \sin B,$$
$$\cos c = \cot A \times \cot B = \cos a \times \cos b,$$
$$\cos A = \tan b \times \cot c = \cos a \times \sin B,$$
$$\cos B = \tan a \times \cot c = \cos b \times \sin A.$$

A.7 Using Napier's Rule on Oblique Triangles

Napier's rule can also be used to solve oblique (nonright) triangles, if you divide the oblique triangle into two right triangles. You do this by adding another arc from one of the vertices perpendicular to the opposite side. Figure A.5 shows how oblique triangle ABC becomes

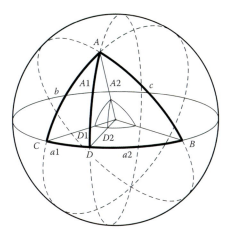

Figure A.5. Breaking a spherical oblique triangle into two right triangles.

two back-to-back right triangles, *ACD* and *ACB*, when side *AD* is added. Initially, you will not know the arc length of the new side *AD*, but you do know that by including its side, you now have two 90° angles (*D1* and *D2*). You can now apply Napier's rule to solve sides *a1* and *a2* based on the other knowns such as angles *B*, *b*, *C*, or *c*. When you use Napier's rule on oblique triangles, you will have two triangles to solve, thus more equations, but the calculations themselves will be easier.

A.8   Polar Triangles

A *polar triangle* is a special triangle derived from another spherical triangle. Once created, each spherical triangle is the other's polar triangle.

Every great circle has two poles and all points on the circle are equidistant from these poles. The poles lay on the surface of the sphere and on a line that is perpendicular to the

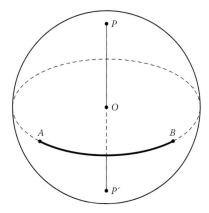

Figure A.6. Pole of a great circle.

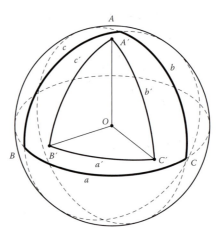

Figure A.7. Spherical triangle $A'B'C'$ and its polar triangle $A'B'C'$.

plane of the great circle at its origin (see Section 4.3). This relationship is analogous to the earth's equator (a great circle), its axis, and the north and south poles. Figure A.6 shows the origin (O) of a great circle, its two poles (P and P'), and an arc (AB). Only one of the poles of the great circle is needed, so point P' will be ignored.

The sides of a spherical triangle are each arcs of great circles and the poles of these three arcs—one for each side—define the three vertices of another triangle, its polar triangle. Figure A.7 shows this relationship between spherical triangle ABC and its polar triangle $A'B'C'$. The poles A', B', and C' and the sphere's origin O are highlighted in the figure. Vertex A' in triangle $A'B'C'$ is the pole of great circle arc BC in triangle ABC. Vertex A' lays on the surface of the sphere and on the normal $A'O$ to the plane of BC at the origin O. Likewise, B' and C' are poles of arcs AC and AB and are on normals $B'O$ and $C'O$.

Every spherical triangle is the polar triangle of another. Thus, polar triangles have a reciprocal relationship to the triangle from which they are derived. In Figure A.7, spherical triangle ABC is the polar triangle of triangle $A'B'C'$ because its vertices are the poles of the sides of $A'B'C'$. Every side or angle in one triangle is related to the angles and sides of the other.

The sides or angles of one triangle also are supplemental to the angles and sides of the other. For example, in Figure A.7, $A + a' = 180°$ and $A' + a = 180°$. This reciprocal relationship also means that, if one of the spherical triangles is *trirectangular* (has three right angles), its polar triangle is also trirectangular because a 90° angle in one corresponds to a 90° side in the other and their sum is 180°. The two trirectangular triangles are also congruent.

The reciprocal relationship between a spherical triangle and its polar triangle can be useful in solving some difficult cases. For example, if you know two angles and the included side, it is far easier to solve this triangle by creating its polar triangle from the supplements. Apply the law of cosines to get the missing parts you need and then take the supplement of their values to get the value of those angles on the original triangle you were trying to solve.

There is a spherical triangle case, however, that may have two solutions, one solution, or none at all, and its polar triangle does not help much. This is a spherical triangle where two sides and the angle opposite one of them are known. Its polar triangle is just as ambiguous because its opposite parts are two angles and a side opposite one of them. The law of sines and Napier's rules may help.

Additional Resources

Leighton, Henry Leland Chapman. *Solid Geometry and Spherical Trigonometry.* New York: D. Van Nostrand, 1943.

Maor, Eli. *Trigonometric Delights.* Princeton, NJ: Princeton University Press, 1998.

Smart, William Marshall. *Textbook on Spherical Astronomy, Second Edition.* Ed. R. M. Green. New York: Cambridge University Press, 1990.

Sperry, Pauline. *Short Course in Spherical Trigonometry.* Johnson's Mathematics Series. Atlanta, GA: Johnson Publishing Co., 1928.

Todhunter, I. *Spherical Trigonometry, for the Use of Colleges and Schools.* London: Macmillan and Co., 1871. (Project Gutenberg free e-book.)

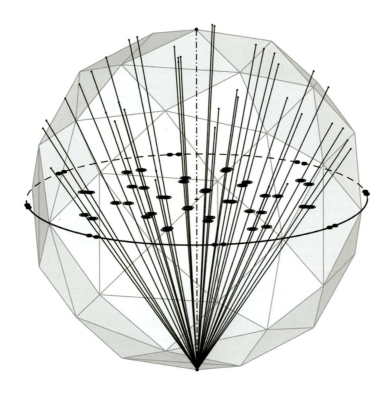

B Stereographic Projection

In many spherical designs, it is helpful to visualize points and great and lesser circles in two dimensions or as a flat drawing. As the illustrations in this book demonstrate, there are many ways to project spherical geometry onto a 2D surface. However, there is one projection method in particular that offers both visualization and direct analysis capabilities. *Stereographic projection* is a way to show the 3D relationship of points and arcs on a sphere in a 2D diagram called a stereogram. Geodesic subdivisions distribute many points and arcs on the surface of a sphere, and stereographic projection is the best way to represent them in a 2D way that preserves angular relationships.

Stereograms are like maps or graphs of the points and arcs on a sphere. We use stereographics to visualize how well a particular subdivision schema performs and to make comparisons among schemas. With experience, stereograms are easy to read. They immediately show symmetry, if it exists; they also show whether points are well distributed or if the subdivision faces are oriented uniformly. Often, you can identify the subdivision schema and the base spherical polyhedra being used simply by inspecting the stereogram.

Stereographics were studied by ancient Greek mathematicians and used to solve spherical trigonometry problems. Later, they were used to make astronomical instruments, such as the astrolabe. Today, the technique is used by cartographers, who make maps of the earth or maps from satellite images of moons and planets. Geologists and civil engineers use stereographics to analyze large-scale rock formations, while crystallographers use the technique to study small crystals and their chemical composition.

B.1   Points on a Sphere

Stereographic projections depict points and arcs on the surface of the sphere. Points on the surface of the sphere are positions in 3D space, and we identify each of these points with the vector that points from the center of the sphere to this point. A vector has direction and magnitude. A line established by the point and the origin of the sphere define the direction. The magnitude is the length or distance between the point and the origin. For almost every case example in this book, the magnitude is one unit because we choose to work with a generalized sphere of one radius unit. That unit can be one millimeter or one mile; it does not matter.

In the geodesic applications developed here, points on the sphere can be the vertices of triangular grids or vectors that are normal to each face. Face normals are very useful vectors because their direction passes through the origin and is perpendicular to the face, or more precisely, to the plane of the face. The face normal has a magnitude as well. This magnitude is directly proportional to the distance of the face from the origin. However, we can project (or scale) the normal point and define a new point that *does* lie on the surface of the unit sphere. The new point is called a unit normal and maintains the original direction of the normal; thus, it remains on a line from the origin and remains perpendicular to the plane of the face. The stereographic projection of either geodesic tessellation vertices, or the unit normals of their faces, is the best way of representing the inter-relationship of their directions and orientations on a plane.

B.2   Stereographic Properties

Many techniques exist for projecting shapes on a sphere onto 2D surfaces. Cartographers have invented scores. But unlike other projection methods, stereographic projection is particularly suitable when analyzing and comparing spherical subdivisions.

First, stereographic projections are conformal. That is, the stereogram preserves the angular relationship between points or intersecting arcs on the surface of the sphere. If lines of latitude and longitude, which meet at right angles on a globe, are projected, they will meet at right angles on the stereoprojection. This also means that the angles between points or between crossing arcs can be directly measured on the stereogram. *Conformality* is an important property because it preserves the general shape of projected objects.

Second, circles on the sphere project as circles on the projection plane. If a great circle passes through the center of a projection, it projects as a straight line (a segment of a circle with an infinite radius). Other spherical projections can display angular relationships between poles on the sphere, but the stereographic projection is the only one that projects any circle on the surface of the sphere into a circle on the projection.[1] Geodesics distribute many great circle arcs and spherical caps, lesser circles around points on the sphere, and are

---

[1] Some circles project as straight lines or circles with an infinite radius.

often used to analyze how uniformly points are distributed. All of them project as circles on the stereogram.

Projections can be made from either the north or south pole of the sphere, and the projection plane, called the primitive, can be the equator (actually any great circle) or tangent to the sphere at either pole (or any plane passing through the sphere). Each combination changes the appearance of the stereogram, but the information it contains and how it is used remain the same. These choices make stereographic projection flexible, although some combinations are more useful to geodesic work than others. We discuss some of these choices in this section, and we adopt some standard uses that we use to analyze subdivisions.

B.3   A History of Diverse Uses

The stereographic projection has its origins no later than the third century BC. Appollonius of Perga included theorems with direct application to proving circle preservation in his *Conics*, which dates from the late third century BC. Applications of the stereographic projection are certain to have been made by the Greeks no later than about 225 BC in the form of the anaphoric clock, which has a disk with stereographically projected stars rotating behind a grid including a stereographically projected horizon and tropics. Hipparchus certainly used the stereographic projection to make star maps and is incorrectly (in my opinion) credited by Ptolemy as being the originator of the projection method. Ptolemy's *Planisphaerium* demonstrates that a solid theoretical foundation for the projection was in place before 150 AD.

The Greeks called the process of representing a sphere on a plane "unfolding the sphere." The stereographic projection was named by François d'Aguilon (Franciscus Aguilonius) in *Opticorum libri sex* (Antwerp, 1613) from στερεοζ (solid). Hence, stereographic means "drawing solids (on a plane)." It is clear that the Greeks did not prove the conformal properties of the projection, but may have assumed it. It is equally clear that the Greeks did know and use the preservation of circles property. Preservation of angles (conformality) under the stereographic projection was first proved by Thomas Harriot (1560–1621).

Islamic astronomers used the projection and proved its basic properties no later than the ninth century. A notable entirely European contribution was *De plana spera*, an early thirteenth-century work by Jordanus de Nemore that presented the theoretical foundation for the stereographic projection. The stereographic projection was applied to astrolabes, quadrants, and other astronomical instruments in both Islamic countries and in Christian Europe in medieval times. Claudius Ptolemy (ca. 150 AD) wrote extensively on stereographics in his work entitled *Planisphaerium*, and in 1613, François d'Aguilon gave the technique its name, "stereographic."

B.4   The Astrolabe

Stereoprojection is a natural choice for star charts and reference grids for familiar objects or patterns in the night sky. Stereographics was the fundamental projection in the most important astronomy instrument ever devised: the astrolabe. An astrolabe is used to solve problems related to time and the positions of the sun and stars. The first astrolabes date to before 400 AD, and they remained very popular in Europe and the Islamic world into the seventeenth century. No other instrument enjoyed such a long life. The earliest astrolabes

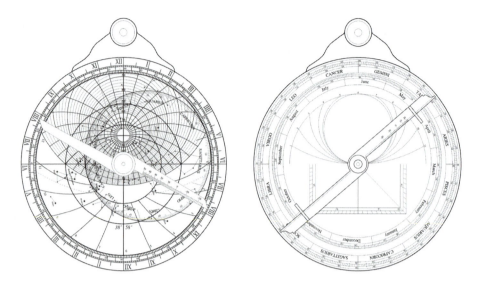

Figure B.1. Astrolabe front and back.

that evolved in Greek antiquity were refined in the Islamic world and were virtually perfected by the tenth century AD. They were introduced into Europe in the eleventh century from Moslem Spain (al-Andalus). Figure B.1 shows the front and back sides of a classic astrolabe.[2]

The earliest known astrolabes, brass disks 6 to 12 inches in diameter, were made in Alexandria. Muslims first encountered the astrolabe in the eighth century in Harran, currently in Turkey near the Syrian border. Even the earliest designs were highly sophisticated instruments. Every part was carefully designed and engraved with scales for specific functions. The main body, the *mater* (Latin for "mother"), was a round plate with a large, hollow circular pocket that accommodated a set of thin, circular disks called *climates*, *plates*, or *tympans* (Latin for "plate"). On the climate disk was engraved a stereoprojection celestial grid of the sky above the observer's local horizon, at the latitude from which the user would make his observations. The center of the climate disk corresponds to the north celestial pole. Well equipped astrolabes would also include a set of climate plates, stored in the mater, each engraved with a sky reference grid for a latitude where the instrument might be used.

Centered over the climate disk is a pierced disk called the *rete* (Latin for "net") which is free to rotate over the climate disk to simulate the daily motion of the stars. It is a frame with openings so the climate disk underneath can be seen. Pointers mark the positions of bright stars. The location of the sun is shown by the projection of the ecliptic as an offset circle on the rete. Scales on the back side of the astrolabe are used to determine the position of the sun on the ecliptic (i.e., the sun's longitude).

Early astronomers believed the earth to be fixed in the heavens and that the stars revolved around it. Although we now know this is not true, the concept is still effectively used today by ocean voyagers who use sextants to navigate. The star's position over the

---

[2] Astrolabe illustrations courtesy of James Morrison.

stereographic reference lines indicates where in the sky (azimuth and altitude) the sun or a star can be found.

Many European astrolabes included a rotating rule over the rete. It served many purposes, acting like the hand of a clock, providing an altitude scale, and aiding in the interpretation of the rete's position over the climate. It also was an index to scales engraved on the outer rim, or *limb*, of the astrolabe.

Progressive design refinements and advanced craftsmanship, plus the widespread dissemination of astrolabes in Europe in the fifteenth and sixteenth centuries, led to instrument refinements and contributed to the astrolabe being applied to surveying, orienteering, and timekeeping along with astronomical uses. Many of these additional features required engraved scales to be added to the back of the instrument. The astrolabe's popularity for casting horoscopes guaranteed its continued use even to the present day, and astrolabes were even available in paper and other affordable forms. The astrolabe was truly the most elegant and widely used scientific instrument ever made. Despite its many design variations, the fundamental use of stereoprojection remained constant.

B.5   Crystallography and Geology

New uses for stereographics have been found in the fields of crystallography, civil engineering, and structural geology. Crystallographers study the atomic structure of solids. Before the use of X-ray diffraction techniques, stereographic techniques were used to graph and analyze the geometry of crystals. Stereograms are made of the orientation of a crystal's planes and the angles between them. In order to make the stereogram, crystal is located inside an imaginary sphere. Points on the sphere are located directly above each face (on an axis normal to the face). These points are then stereoprojected from a pole in the hemisphere opposite the point. Figure B.2 shows the basic arrangement. The result is the dot-like pattern on the primitive, the equatorial plane where points are projected. Each dot is the projection of a point on the sphere directly above a crystal's face. The angles between dots indicate the angles between crystal faces. Other dot patterns indicate symmetrical relationships between vertices, edges, and crystal faces.

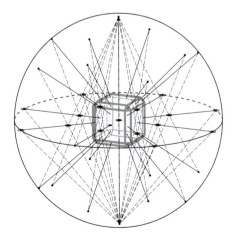

Figure B.2. Stereoprojection in crystallography.

Stereographic projection is particularly useful in analyzing crystals. An ideal crystal is a perfect geometric shape. Many are microscopic, and crystal forms are found in four of five platonic shapes. For example, some forms of calcite galena resemble small cubic shapes, while pyrite can take cubic, octahedral, or pyritohedral (a dodecahedron with pentagonal faces) forms.

In nature, few crystals are perfectly formed. Neighboring bodies can interfere, and inconsistent flow of the crystalline liquid or multiple formations occurring in close proximity all distort a crystal's growth. The number of faces in a crystal and their orientations, however, are a function of the crystal's chemistry, and interference during their formation does not affect the angles between faces.[3] These angles are also independent of the face's size. Thus, samples from imperfect crystals, where only a small portion of a face has formed, can oftentimes be used to make accurate stereograms, as if the crystal were perfectly formed.

On a much larger scale than crystallography, structural geologists and civil engineers use stereographics to analyze the orientation of linear and planar geological features. Reference lines over geological features—such as the horizontal reference line of a sloping rock shelf or the dip slope angle of the shelf—are 3D references to the orientation of the geological feature. Stereographic projections of these features show how plates and strata are positioned with respect to one another. A stereogram can identify faults, folds, and other geological discontinuities where minerals form or oil and gas might be trapped. In structural geology, stereograms provide common visualization and allow comparisons. They also summarize large amounts of 3D data and serve as a predictive tool important in exploration.

B.6   Cartography

Azimuthal stereographic projection is a classic projection for mapmakers. The earliest known map of the world using stereographics was made by Walther Ludd (Galtier Lud) of St. Dié, Lorraine, in the early sixteenth century. The projection was made famous when Gerardus Mercator, a sixteenth-century Flemish cartographer, used the equatorial stereographic projection for the world maps of the atlas of 1595.

Today, stereographic maps of the earth are still made and are sometimes used to display satellite images of celestial bodies, such as planets and moons. The projection plane and the point of projection can be placed at any location on a sphere, but two positions are commonly used. When these two positions are at the equator and projections are made from the north or south pole, the map is called a *polar stereographic* or *polar aspect* projection. It is the most common projection of the polar areas of the earth, moon, and planets since it is conformal. Directions measured from the center are true; the meridians are shown as straight lines radiating from the pole, while latitude circles appear as concentric circles around the pole.

Figure B.3(a) shows a polar stereographic map viewed from the north pole. When the projection plane is through the middle of the earth from the north to the south pole, instead of at the equator, and projections are made from a point on the equator, it is called an *equatorial aspect* map. One is shown in Figure B.3(b), while (c) shows an oblique view. Later we will see a universal stereoprojection called the *Wulff net*, which is an equatorial stereographic projection.

---

[3] (Whittaker 1984, 1)

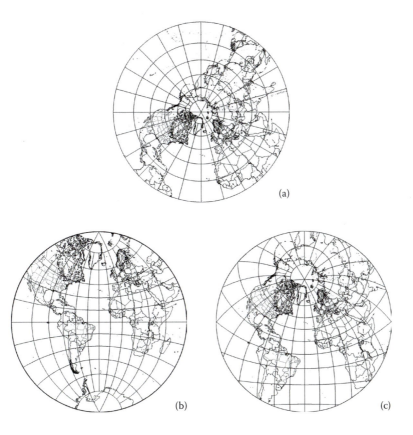

Figure B.3. (a) Polar, (b) equatorial, and (c) oblique stereographic maps.

Azimuthal stereoprojections are not distortion-free maps—no map projection is—but they preserve angular relationships, and circular features projected as circles. The scale of azimuthal stereographic maps is distorted, however, and becomes more pronounced as one moves away from the center of the map. These drawbacks are minimized, however, when the geographic area is smaller than a hemisphere.

From ancient astronomy and classic mapmaking to crystallography and structural geology, the stereographic projection continues to be relevant and a preferred projection in spherical work. In the next section, we look closer at this technique and examine the resulting stereograms in detail. In a later chapter, we apply stereographic projections in a novel way to evaluate spherical subdivisions. Stereograms are excellent ways to analyze how well a spherical subdivision distributes points on the sphere or how uniformly the resulting faces are oriented in three dimensions. We begin with stereoprojection basics.

B.7   Projection Methods

Stereographic projections are made by passing infinite rays (lines) from a projection point, usually the north or south pole of a sphere, to other points on the sphere. Rays will pass through a plane that is either tangent to one of the projection poles or at the equator of the

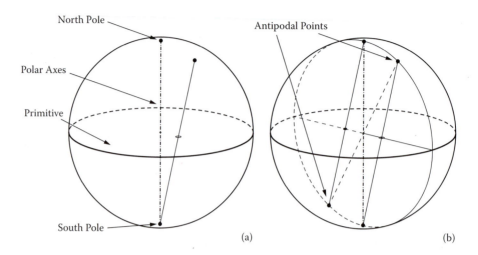

Figure B.4. Stereographic poles, primitive, and projection of points.

sphere. The projection plane is called the primitive. Stereoprojection points are defined where rays intersect the primitive.

For geodesic applications, it is convenient to locate north and south projection poles at the positive and negative z-axis of the 3D coordinate space and to use the xy-plane as the primitive. This is also the arrangement used by many crystallographers.

Figure B.4 shows the basic stereoprojection arrangement and a single point (on the sphere) being projected from the south pole. Depending on the type of stereogram desired, either pole can be used for projection. If one pole is used for all projections, points near the pole will project near infinity and the pole point itself cannot be projected at all. To avoid these problems, we adopt the practice of projecting from the pole opposite the point's hemisphere.

Figure B.4(a) represents the projected point with a small circle on the primitive. A dot in the middle of the circle indicates the exact projection point. Points projected from the pole of their hemisphere will project outside the circumference of the primitive. Cases such as this are common when celestial stereograms are made for astrolabes. Points, which are tangent to a pole, must be projected from the opposite pole. We will see some examples shortly.

Figure B.4(b) shows another fundamental principle of stereoprojection: antipodal points always project to points that are equidistant and opposite from the center of the primitive. Antipodal points are the circumference of the same great circle, and this is also shown in the figure.

We adopted the graphic convention to represent projected points from the north pole as small circles with a dot in the middle. This choice is deliberate. If points from opposite hemispheres project to the same point on the primitive, the larger circle graphic for points projected from the south pole will surround the smaller circle graphic from the points projected from the north pole. In both cases, a dot will mark the exact projection point in the center. The resulting bull's-eye graphic is easy to spot on the stereogram, and the exact point of projection (the dot) is maintained.

B.8   Great Circles

The stereoprojection of a great circle is a circle. If the great circle is tangent to a projection pole (it will be tangent to both), it will project as a line on the primitive; or more correctly, it will project as an arc of a circle with infinite radius. Figure B.5 shows this case where a great circle is tangent to both projection poles. Its stereoprojection is a line through the center of the primitive across the entire diameter of the primitive. Two points on the great circle are projected separately, just to show the relationship between a great circle arc segment and the resulting line (infinite radius arc) segment on the primitive.

The equator of a sphere is also a great circle. Because its plane is coincident with the plane of the primitive, the equator is projected as a circle with its center at the projection of the pole.

In geodesics, most great circles or great circle arcs do not pass through the poles or are coincident to the equator (primitive). All of them, however, will project as arcs or segments of circles on the primitive. Depending on the choice of projection poles, the projection circle may be totally or only partially within the limits of the primitive.

Figures B.6 and B.7 show the difference the choice of projection pole can make. In Figure B.6 the entire great circle is projected from one pole—the south pole. All the projected points intersect the plane of the primitive, but some projected points fall outside its circumference. The dihedral angle between the great circle and primitive in this figure is 30°. As the dihedral angle approaches 90°, the radius of the projected circle becomes infinite. Great circles projected from a single projection pole are only practical if the projected circle is graphically clipped or the dihedral angles are relatively small.

In Figure B.7, the projection pole is always opposite the hemisphere of the point being projected. The resulting stereogram is compact and lies entirely on or within the circumference of the primitive. Rays from the north pole are shown dashed, while rays from the south pole are solid.

Both figures illustrate another fact about stereoprojection of great circles: the line of intersection of two great circles in three dimensions—the primitive great circle and example

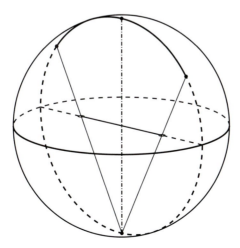

Figure B.5. Great circle tangent to projection axis.

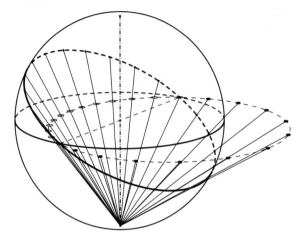

Figure B.6. Great circle projection from the same pole.

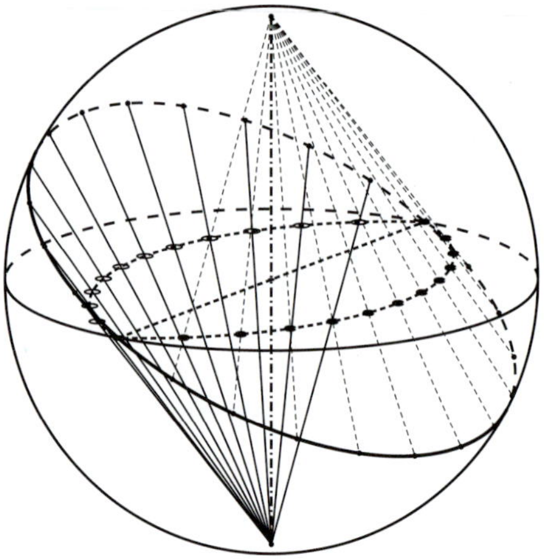

Figure B.7. Great circle projection from poles opposite points.

great circle projected—always passes through the origin of the sphere and intersects the surface of a sphere at two antipodal points. If one of these great circles is the equator, these two antipodal points are tangent to the primitive. A line joining them passes through the origin of the sphere and, likewise, through the center of the primitive.

B.9   Lesser Circles

Like great circles, lesser circles also project as circles on the primitive. Figure B.8 shows several orientations of lesser circles and their projections. A lesser circle, entirely in one hemisphere, is projected from the pole in the opposite hemisphere, as shown in Figure B.8(a). Figure B.8(b) shows cases where the centers of lesser circles are tangent to the projection axis. These "latitude" circles project as concentric circles centered in the primitive. Some sample rays are shown as points on their circumference.

Figure B.8(c) and (d) illustrate projections where the lesser circle cuts through the primitive. If a single projection pole is used (c) for the entire lesser circle, the resulting projection points are on the plane of the primitive, but some points fall outside its circumference and the distance is dependent on how close the center of the lesser circle is to the origin of the sphere. It approaches infinity, as the lesser circle gets larger and approaches a great circle. If the lesser circle is projected from the pole opposite the point on its circumference (d), a compact projection results and all projection points are on or inside the primitive.

There are three additional subtleties about lesser circles. The first is how their position on the sphere affects their projected radius; the second is the projection of their centers; and the third is the projection of circles that cross but are not perpendicular to the primitive.

Figure B.9 shows the stereoprojection of four lesser circles with the same radius. The projected lesser circles, however, do not have the same radius. The smallest projected radius belongs to the lesser circle at the zenith centered on the projection axis. The projected

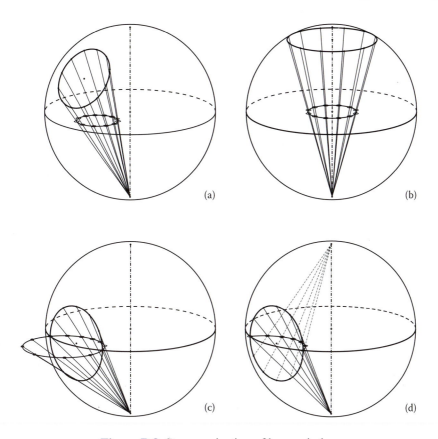

Figure B.8. Stereoprojection of lesser circles.

circle radius increases as the lesser circle on the sphere approaches the primitive. The largest projected lesser circle radius will be the lesser circle whose center is tangent to the primitive.

Figure B.10 shows the projection ray (dashed) of a lesser circle's center, point a, from the south pole. The ray's intersection with the primitive is marked by a small dot at a'. This point, however, does not qualify as a stereoprojection of point a because a is not on the surface of the sphere; it is on the plane of the lesser circle and inside the sphere. Point b is the projection of a from the origin to the sphere's surface. Mathematically, point (vector) a has been normalized to have a magnitude of one; thus, point b is on the surface of our unit radius sphere. Point b qualifies for stereoprojection and its projection ray (solid) to the south pole intersects the primitive at b'. Both b and b' are indicated with larger dots.

When the points a' and b' are compared, neither one is in the center of the lesser circle on the primitive. It turns out that the true center of the projected lesser circle is point c'; it is the projection of point c, which is outside the sphere altogether and at the apex of a cone, that is tangent to the sphere at the lesser circle. This was discovered by Michel Chasles, a nineteenth-century French mathematician, who developed theories of projective geometry.

Unlike the lesser circle examples, shown in Figure B.8(c) and (d), stereoprojection of lesser circles, which intersect the primitive but are not perpendicular to it, have somewhat

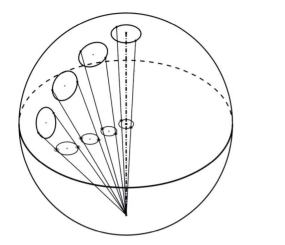

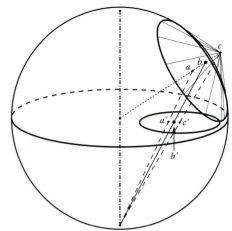

Figure B.9. Stereoprojection of equal lesser circles.   **Figure B.10.** Stereoprojection of lesser circle centers.

unexpected representations. Figure B.11 shows this case. In (a), the whole lesser circle is projected from the south pole, and part of its projection extends beyond the primitive. But when points on the lesser circle are projected from the pole opposite their hemisphere, as shown in Figure B.11(b), the projected lesser circle is totally within the primitive, and a part of the projection is "folded back" at the point where the lesser circle is tangent to the circumference of the primitive. This characteristic is not apparent in Figure B.8(d) because the plane of the lesser circle is perpendicular to the perfect. The "folded" halves of its projected circle (one from each pole) are perfectly superimposed and appear as just one arc on the primitive.

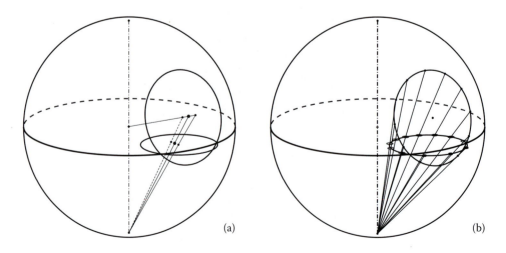

(a)   (b)

Figure B.11. Stereoprojection of lesser circles crossing the primitive.

B.10   Wulff Net

A Wulff net is a single, simultaneous stereoprojection of a large number of great and lesser circles. A classic Wulff net is shown in Figure B.12. In a Wulff net, the orientation of the great and lesser circles to be projected is very specific—they all share the same axis, and their axis lies on the primitive. For an analogy, imagine that a sampling of the earth's latitude and longitude circles are stereoprojected. Also assume the perfect bisects the earth not at its equator,[4] but from the north to the south pole. Now imagine that you stereoproject the set of longitude (great) circles and latitude (lesser) circles from one of the primitive's poles. Keep in mind that the primitive is cutting the earth in half from north to south. Thus, the primitive's stereoprojection poles are on the earth's equator, and these poles are on opposite sides of the earth. Because of the symmetry in both sets of longitude and latitude circles, a stereoprojection from either the primitive's north or south pole will produce the same stereogram.

The Wulff net, shown in Figure B.12, projects sample longitude and latitude circles every 2°, with emphasis every 10°. The number of longitude and latitude circles projected is arbitrary, limited only by application needs and the graphic standards employed. In this example, more than 1,700 combinations of circles are shown in one stereogram. For this reason, the Wulff net is also called the universal stereogram.

In the Wulff net illustration, latitude circles are labeled in degrees from each of the poles, contrary to the convention of referencing the earth's latitude and measuring north and south from the equator.[5] The labeling on the Wulff net figure is for zenith distance and is used for setting the latitude of a place on a saphea astrolabe. When the rule on a saphea instrument is set to a number on the scale, the rule represents the horizon for that latitude.

So why is the Wulff net useful? It's because it is a totally generalized stereographic framework. In a sense, it is a general-purpose stereographic protractor. It can overlay other

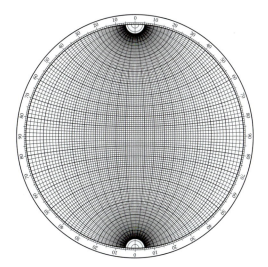

Figure B.12. Stereographic Wulff net.

---

[4] A polar net is a companion to the Wulff net and does bisect the earth at the equator.
[5] Illustration courtesy of James Morrison.

stereograms to measure angles or be used like graph paper when plotting points and arcs. It can measure surface and central angles, and within the accuracy of its graphics, it can be used to solve many spherical trigonometry problems. Angles between great circles (dihedral angles) correspond to interior spherical angles. Lesser circle projections measure angles along great circles and can be used for rotating points or arcs.

B.11 Polyhedra Stereographics

We introduced stereographic projection in the context of astronomy, cartography, and crystallography. Now we will place it in a polyhedral context and use stereographics to map the distribution of subdivision grid points, geodesic arcs between them, and the orientation of polyhedra faces. We treat polyhedra as if they were crystals, and our approach is similar to the way crystallographers use stereographics.

The vertices of regular polyhedra and the grid points resulting from spherical subdivision are on the surface of the polyhedron's circumsphere and can be stereoprojected directly.

Less obvious is the fact that a vector describing the orientation of a polyhedron's face can be used to define a point on the surface of the sphere that can also be stereoprojected. This vector is called a *normal*, and its direction is perpendicular to the plane of the face. Depending on how the normal is calculated, the direction can point out from the origin or in toward it. We can use normals to polyhedral faces, or to the plane of triangles created by subdivision grids, to tell us something about the shape and orientation of the face. What's more, we can make stereograms of them.

A normal vector is perpendicular to all points on the face and the plane the face defines. The normal is independent of the size of the face (or triangle on a subdivided sphere) no matter how large or small the face is. The normal vector to the plane of a face, a triangular face for example, does not necessarily point to the face's center. When the face is a regular polygon (equilateral triangles, squares, and pentagons) and its vertices are on the surface of the sphere, its normal will point to the polygon's centroid. This can be seen in the normals to the faces of the five platonic solids. But regardless of the shape of the polyhedron's face, a normal vector to that face will always point perpendicular to the plane of the face.

Stereographic projection adds little value for analyzing the regular polyhedra because the number of faces and vertices is small and their orientations are well known. However, when spherical subdivisions are developed from them, there is a dramatic increase in the number of vertices, faces, and diversity in the shape of faces. It is here, then, that stereographics become a powerful summary and analytical tool. Stereograms provide a visual, nonstatistical metric for comparing subdivisions, revealing symmetry, and identifying the fundamental polyhedra and subdivision techniques in use. As such, stereographics is a preferred method for determining which subdivision schemas are best when the application requires uniform or near-uniform face orientations. Given the simplicity of the technique and its useful results, it is difficult to understand why this technique is not exploited more often in geodesics. Building on the notion of polyhedra vertex and face vectors, we now make stereographic projections of them.

B.12 Polyhedra as Crystals

Figure B.13 shows the stereographic projection arrangement used by crystallographers. The primitive is at the equator, and its polar axis defines north and south polar projection

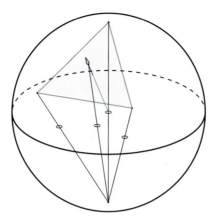

Figure B.13. Vertex and face stereographic projection.

points on the sphere. A single icosahedron face (equilateral triangle) is shown, and all three of its face vertices are tangent to its circumsphere. It is helpful to place the polyhedra vertex-zenith whenever possible. We will demonstrate later why this orientation makes it easier to measure the angles between face normals on the stereograms. In this figure, one vertex of the triangle is at the zenith and is coincident to the north pole.

In the figure, the triangle's normal direction is shown as an arrow from the origin through the middle of the triangle. The arrow indicates the normal's direction, and it points out and away from the origin of the sphere. A normal vector cannot be stereoprojected directly. Instead, it is used to define a Cartesian point that is on the surface of the sphere. This is accomplished by normalizing the normal vector; that is, scaling the vector's magnitude to one, the radius of the unit sphere. The resulting derived point is now on the surface of the sphere and on a ray from the origin that is perpendicular to the plane of the triangle. The triangle's vertices were already on the sphere; we can now stereoproject both the vertices and the derived point since they are all on the surface of the sphere.

To find the stereographic projection or a set of points, lines are drawn from each point to the pole in the opposite hemisphere. The stereographic projection of each point falls where its projection line intersects the primitive. In Figure B.13, all four projected points are shown as open circles on the primitive. Any point on the sphere coincident to a pole will project to the center of the primitive. All other points on the sphere project somewhere on the primitive, but not on its circumference or its center.

B.13 Metrics and Interpretation

Stereographic projections are *conformal projections*. This means the angle between any arcs or great circles on the sphere remains the same when projected onto the stereogram. Great circles passing through the projection poles project as straight lines and can be measured directly from the stereogram.

Earlier, we adopted a convention of orienting polyhedra vertex-zenith. This always places one vertex at the north projection pole. For all platonic polyhedra, except the tetrahedron, one vertex will always be at the south pole if another is at the north pole. By

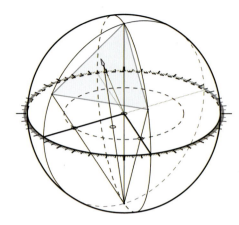

Figure B.14. Angle measurements from stereographic projections.

orienting vertex-zenith, we take advantage of the fact that any great circle passing through both the north and south projection poles project as straight lines onto the primitive and this makes it very easy to measure angles on the stereogram.

Figure B.14 adds great circles for two of the triangle's edges. Their projection lines are shown on the primitive as solid and dashed lines. A compass rose is added as a visual indication of the angle between lines; it measures 72°, the surface angle of the equilateral triangle face angle. Later, we will compare angles between projected face unit normals. This will alert us to the presence of gaps or unevenly distributed face orientations. Figure B.14 adds a concentric reference circle to the perfect. This illustrates the fact that projected points equidistant to the center of the primitive have the same north or south latitude.

B.14 Projecting Polyhedra

Let's examine the stereographic process for complete polyhedra. The icosahedron, tetrahedron, and octahedron form the basis of almost every subdivision developed in this book. Since their projected face normals will be found in a great many of the tessellations based on them, it is worth examining their stereograms.

Figure B.15(a) shows the stereographic arrangement for an icosahedron in a unit circumsphere. The icosahedron is oriented vertex-zenith, and two of its vertices are coincident to the north and south poles of the primitive. Figure B.15(b) shows the normal directions to the planes of all 20 faces and their derived points on the surface of the circumsphere. Since the icosahedron's faces are equilateral and their vertices are tangent to the sphere's surface, the direction of face normals passes through the centroid (and incenter) of each face.

Figure B.15(c) shows the stereographic projection of the icosahedron's 20 points derived from each of its face normals. Points in the northern hemisphere are projected from the south pole and indicated on the primitive with small open circles. Normals to each of the polyhedron's faces in (b) and the projection rays from derived points in (c) are shown to better explain how the resulting stereoprojection in (d) is done. The derived points from face unit normals in the southern hemisphere are projected from the north pole and these points are shown as solid-fill circles. Their projection rays are omitted to make the figure

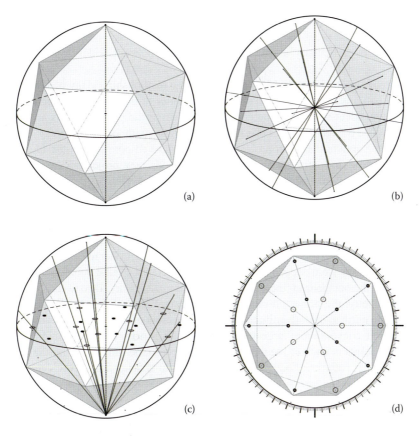

Figure B.15. Polyhedron stereographic projection sequence.

easier to read. The choice of projection symbols is deliberate and follows conventions used by crystallographers. If derived from opposite hemispheres projected to the same primitive point, they can still be distinguished. We will see cases such as this shortly, when we make a stereogram of an octahedron.

Interpreting the stereogram in Figure B.15(d) is easy. The shaded icosahedron super-imposed over the stereogram associates projected points with their polyhedra faces. The icosahedron's rotational symmetry is immediately evident in the stereogram. As expected, there are 20 projected points (10 hollow circles from derived points in the northern hemi-sphere and 10 solid-fill ones from points in the southern hemisphere).

In this stereogram, every vertex of the icosahedron has a symmetrical one directly op-posite the origin. This means that for the icosahedron's 12 vertices, there are six pairs of antipodal points, each creating a rotation axis. One pair of vertices is on the polar axis of the stereogram and they project to the center of the stereogram. In Figure B.15(d), a top view of the stereogram, we see that five equilateral triangles share one vertex coincident to the north pole projection point, and their projections show the five-fold rotation symmetry. Rotation symmetry means that the polyhedron can be rotated about the stereographic axis any multiple of 72°, and the appearance of its stereogram is not changed. We see that the symmetry in the stereogram and the compass rose shows the projected unit normals for adjacent icosahedron faces are 72° apart. Rotation symmetry is a powerful concept in 3D geometry and is particularly useful in geodesics. The concept is developed in detail in Chapter 6.

The stereogram in Figure B.15(d) shows another relationship. Many of the points are arranged in patterns and are equidistant from the center of the primitive. When projection points are coincident with the same concentric circle on the primitive, it is an indication that derived face points have the same latitude, north or south, on the sphere.

Now, let's examine the stereoprojections of the derived points from the face normals of the tetrahedron and octahedron. Each example adds new projection cases and helps develop the skill required to read more complex ones quickly.

B.15   Octahedron

Figure B.16 develops a stereogram for an octahedron. The stereographic rays are shown for all eight derived points from the octahedron's face normals to demonstrate what happens when derived points from opposite hemispheres project to the same point on the primitive. In the case of the octahedron, there are four superimposed pairs. Due to the octahedron's rotation symmetry, the projected points are equidistant from the center because they all have the same north and south latitude on the sphere. Sets of them are 90° apart, as expected. The graphic convention of large and small symbols is particularly helpful here because four derived points in each of the hemispheres project to the same point on the primitive. The now-familiar patterns of equal latitude and angular spacing are also evident.

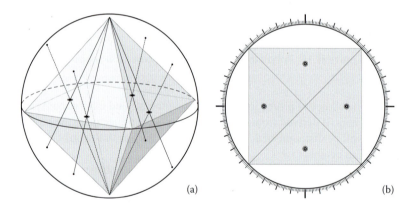

(a)   (b)

Figure B.16. Octahedron stereogram.

B.16   Tetrahedron

Figure B.17 shows a stereoprojection and stereogram of each derived point from the tetrahedron's face normals. The tetrahedron is arranged vertex-zenith. As in the previous examples, its vertices are on the unit circumsphere. Three of the four face derived points are in the northern hemisphere and are projected from the south pole. Their projection points are indicated on the primitive with open circles. The bottom face's derived points projects from

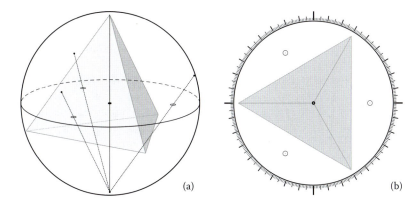

(a) (b)

Figure B.17. Tetrahedron stereogram.

the north pole and is shown as a solid fill circle in the middle of the primitive. Since the polyhedron is simple, all projection rays are shown as dotted lines. Figure B.17(b) shows the resulting stereogram with the image of the tetrahedron superimposed. As expected, the projections of the derived points in the northern hemisphere are 120° apart. The fact that these three projected points are equidistant from the center of the primitive also indicates that the unit normals all have the same or complementary latitude.

B.17 Geodesic Stereographics

The previous polyhedra examples demonstrate the concepts of stereographics. Now let's take a quick look at what we can expect when we apply stereographics to geodesic subdivisions.

Figure B.18 shows the face orientations of a subdivided spherical octahedron. Arrows indicate the normal vector's direction, which is perpendicular to the plane of each triangular face. Although all subdivision vertices are tangent to the unit radius sphere, none

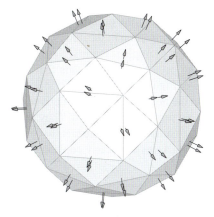

Figure B.18. Face orientations of a subdivided spherical octahedron.

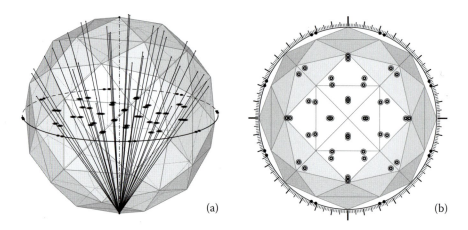

Figure B.19. Geodesic stereogram.

of the subdivision triangles themselves are equilateral. Therefore, the perpendicular point (from the origin) is not the face's centroid or incenter, but somewhere to the side of it. It is possible, though not in this example, for the intersection of the normal's ray and the face's plane to be completely outside the triangle, even though the normal vector is perpendicular to the plane of the triangle.

Figure B.19(a) shows the projection arrangement for the spherical octahedron. The primitive, its polar axis, and its projection points are included. Derived points from face normals in both hemispheres are projected onto the primitive using the technique already established. The projection rays from the south pole to derived points in the northern hemisphere are shown, but the rays from the north pole are not depicted to make the figure easier to read. The graphic standards for projected points on the primitive are the same as before; points in the northern hemisphere are represented by the largest circles, while projections of southern hemisphere points use the smaller ones.

The stereogram's plan view is shown in Figure B.19(b). The high degree of symmetry in face orientation between the two hemispheres is apparent. With the exception of the derived points from normals tangent to the primitive's circumference, every point in both hemispheres has a symmetrical companion in the other hemisphere, and both project to the same point on the primitive. This example is the first to show derived points from normals coincident to the plane of the primitive. This means that their corresponding faces are perpendicular to the primitive. Figure B.19(b), a top view of the stereogram, shows that 16 faces are perpendicular to the normal. Since their derived points are already on the circumference of the primitive, no projection is needed. They are shown as small dark circles around the circumsphere of the compass rose. The fact that pairs of points are close to each other indicates that two adjacent faces are nearly coplanar; we see by inspection that this is the case.

B.18   Spherical Icosahedron

All of the subdivision schemas developed in earlier chapters use a spherical polyhedral framework. Although the stereograms shown in Chapter 9 were used to analyze the grid

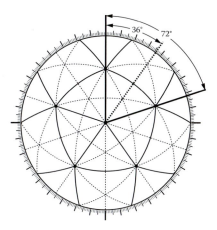

Figure B.20. Icosahedron great circle stereogram.

point distributions and subdivision face orientations of triangles, it is possible to analyze the icosahedron's symmetry as well.

Figure B.20 shows the stereoprojection of the 15 great circles of a vertex-zenith icosahedron. The zenith vertex is tangent to the north stereographic projection pole and our eye-point is above the north projection pole looking towards the primitive. A compass rose surrounds the circumference of the primitive and allows us to make direct measurements of the angles between the great circles.

Great circles, or arcs from great circles, project as arcs of circles on the stereogram primitive. In Figure B.20 icosahedron edges are shown as solid lines, while face medians are shown as dashed lines. It is easy to see the 15 great circles when they are combined.

Because the icosahedron's faces are equilateral, their median arcs intersect in the middle of each face. Great circle arcs (edge or median) that pass through the projection pole appear as straight lines, though they are actually arcs of circles with infinite radii. Due to the high degree of symmetry in an icosahedron, edge and median arcs intersect exactly at the equator and extend to the circumference of the primitive.

The compass rose has sufficient resolution to measure surface angles, the angle between two intersecting great circles. The surface angle of every icosahedral face is 72°, and this is confirmed by extending adjacent pairs of edges from the center of the primitive (the icosahedron's zenith point) to the primitive. Face medians are midway (dashed line) at 36°, as expected, (see Figure B.20). The stereogram of the icosahedron's great circles will become a familiar sight whenever the face orientations are analyzed.

B.19 Summary

Stereographic projection, one of the oldest techniques in descriptive geometry, is highly relevant to advanced geodesics and spherical subdivisions. The technique is simple to apply and results in a graphic that summarizes the distribution of points on a sphere. When those points are the vertices of polyhedra or points on the sphere derived from normal vectors to their faces, the resulting stereogram provides a complete map of how uniformly the vertices are distributed or the faces oriented.

A stereogram is the projection of one or more points, circles, or other features, such as arcs or derived points from face normals onto the primitive. A key characteristic of stereographic projections is that they are conformal; that is, they preserve the angular relationships between points and intersecting arcs. Angles can be measured directly from the stereogram; thus, stereograms can be used to graphically solve spherical trigonometry problems (within the precision of their graphics). All points, projection poles, and points to be projected must be on the surface of the sphere. Either the north or south pole can be used to project a point. The stereogram application determines which is most suitable. A standard technique is to project points from the pole in the opposite hemisphere. This will ensure that projected points will lie inside the primitive or on its circumference. Points projected from the pole of their own hemisphere will project outside the circumference of the primitive. Points that are tangent to a pole cannot be projected from that pole because their projection is undefined. However, they can be projected from the pole in the opposite hemisphere. They project to the center of the primitive.

Antipodal points lie on the same great circle and project to points equidistant from and opposite to the center of the primitive. Different graphic symbols can be used to distinguish points projected from the north and south poles. Crystallographers employ conventions, such as concentric circles, to identify which projection pole was used for a point and to distinguish points projecting to the same location on the primitive. All circles (greater or lesser) or arcs between points on the surface project as arcs of circles on the primitive. Any points coincident with the poles project to the center of the primitive. Antipodal points are on the same great circle and project to points opposite and equidistant from the center of the primitive. Great circles will intersect the circumference of the primitive at two opposite points. Great circles, passing through the projection poles, project as straight lines through the center of the primitive. These lines are arcs of infinite radii circles. The equator of a sphere is a great circle and coincident to the primitive. It is not projected. The line of intersection of two great circles passes through the origin of the sphere and intersects the surface of a sphere at two antipodal points. These two points are always on the primitive. A line joining them passes through the origin of the sphere and through the center of the primitive. Lesser circles project as circles on the primitive. The center of the lesser circle is not the center of the projected circle because it not a point on the sphere. The lesser circle's center is on a plane of the lesser circle and is actually inside the sphere. Lesser circles with centers tangent to the polar axis (latitude circles) project as circles centered in the primitive. A Wulff net is the stereographic projection of a series of greater and lesser circles.

Additional Resources

Lisle, Richard J. and Peter R. Leyson. *Stereographic Projection Techniques for Geologists and Civil Engineers.* Cambridge [Eng.], New York: Cambridge University Press, 2004.

Morrison, James E. *The Astrolabe.* Rehoboth Beach, DE: Janus, 2007.

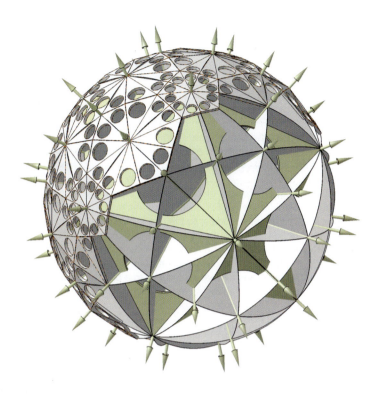

C Geodesic Math

This appendix presents the geodesic math necessary to subdivide the principle poly-hedral triangle (PPT) or the least common denominator triangle (LCD) based on Class I, II, and III schemas presented earlier in this book. With the algorithms shown here, you can derive the coordinates of grid points, arc, and chord distances between them and all edge and face angles. Math techniques are described for each of the classes. In addition, this appendix includes some handy algorithms to solve common 3D geo-metric problems, such as finding the perpendicular to a plane, or the shortest distance between two nonintersecting lines in three dimensions.

The angles that result from subdividing a PPT or LCD do not change if the PPT or LCD set of points are moved together to a different location on the surface of the sphere. How-ever, moving them does change the coordinates of each point. The procedures developed here create *one* spherical PPT or LCD grid oriented either vertex-, edge-, or face-zenith. These points are sufficient for most analysis and simple displays. If you need to extend this grid to cover the rest of the sphere, there are at least two ways to achieve this. First, you can

Geodesic Algorithms	
Subroutine	**Operation**
gcsect	calculate the two points and dihedral angle where two great circles intersect
gdihdrl	return the dihedral angle between two planes
gtricent	calculate the centroid of a triangle
stABC	calculate the surface angles of a spherical triangle
vabs	calculate the absolute value of a point (also called vector length)
vadd	add two vectors
varcv	locate a point at an arbitrary place, a geodesic arc between two other points
vcos	return the cosine of the angle between two vectors
vcrs	calculate a vector perpendicular to the plane of two other vectors
vdir	locate a point between two others at a perscribed distance
vdis	calculate the distance between two points
vdot	return the dot product (also called scalar or inner product) of two vectors
vnor	calculate the normal vector (perpendicular) to a plane
vrevs	reverse vector's sense, makes an antipode point if vector is point on sphere
vscl	scale a vector or point (moves it along infinite line through origin)
vsub	subtract one point from another
vuni	project a point to the surface of the unit sphere
vzero	initialize a range of point coordinates to (0,0,0)

Table C.1. Summary of geodesic algorithms.

input the points into a CAD system and use its assembly features to replicate the PPT or LCD points elsewhere on the sphere. The CAD system computes the necessary coordinate transformations for you. All of the examples in Chapter 10 were created this way. Another way is to transform copies of the original PPT or LCD point set, the copy is rotated to a new position elsewhere on the sphere; Appendix E tells you how to achieve this.

In the appendix, we take a cookbook approach to avoid using specialized math notation. We will explain the required subdivision steps at the same time we make references to simple formulas or algorithms that implement the step. The formulas are presented as a subdivision step is explained; the algorithms are listed in their entirety later in the chapter. In addition to explaining algorithms with each step, a separate table lists them all and their function in case you need to look one up quickly. Anyone doing geodesic calculations will use a computer or calculator, thus all algorithms are written in C. Its language structure and syntax is easy to understand and convert to other computer languages or notations used by programmable calculators. Some algorithms are used by other algorithms and we include the full set. Each algorithm is fully documented and designed for clarity, not necessarily for maximum efficiency. Computer programmers can improve the data structures and efficiency of these algorithms by using computer language features, such as object classes or specific data structures (vectors, maps, lists, etc.) offered in the C++ and Java languages.[1]

The objective of this appendix is to list the algorithms and their sequence of use. Most will access a single master array of Cartesian coordinates (x, y, z). Some algorithms are performing vector operations, others are trigonometric solutions for spherical triangles,

---

[1] Vector containers and predefined algorithms are standard template library features in ANSI Standard C++.

and still others are housekeeping routines. The algorithms are listed in alphabetical order in Table C.1. The actual code of each vector routine is listed at the end of this appendix.

C.1   Class I: Alternates and Fords

Class I subdivisions are the result of points created by the intersection of two or three great circles. The intersecting great circles are defined by reference points along the PPT's edges (and the origin of the sphere). Class I schemas differ only in the way their PPT edge reference points are located and the combination of great circles intersected to define grid points. The following section tells you how to define edge reference points and intersect great circles for each of the schemas presented in this book. Points for the Equal-arcs schema can be created by intersecting two or three great circles. Three great circles result in a superior grid, but an additional step is required and the procedure is described here.

C.1.1   Step 1: Define the PPT Apex Coordinates

Orient the deltahedron's equilateral PPT face in a vertex-zenith position symmetrical to the *xz*-plane, as shown in Figure C.1(a). The example here is a single icosahedron PPT; PPTs for the tetrahedron and octahedron will look similar. The PPT's vertex-zenith Cartesian coordinate for unit sphere will always be (0, 0, 1). The other two PPT apex coordinates are found by converting their spherical coordinates to Cartesian coordinates. See Table 8.2 for their coordinates.

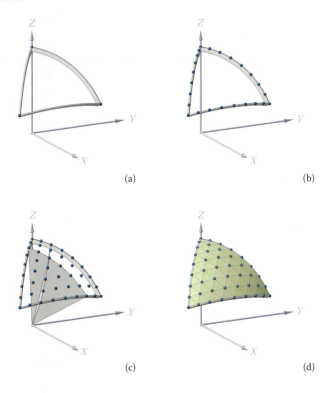

(a)　　　　　　　　　　(b)

(c)　　　　　　　　　　(d)

Figure C.1. Class I subdivision sequence.

C.1.2 Step 2: Define PPT Edge Reference Points

Locate $v - 1$ points between each of the PPT apex points. The total number of required reference points for all edges of the PPT, excluding the PPT's apices, will be $3(v - 1)$. Figure C.1(b) shows the PPT apex and reference points necessary for an Equal-chords 8^v icosahedron subdivision. Reference points for the other schemas will look similar. Define PPT edge reference points as follows:

- *Equal-chords.* Find the PPT's chord length with `vdix()`. Use `vdir()` to compute the coordinates of $v - 1$ evenly-spaced reference points between the vertices of each PPT edge chord. These reference points are along the polyhedron's chords, thus they are inside the sphere. Project (normalize) each reference point to the sphere's surface with `vuni()`.

- *Equal-arcs (two orthree great circles).* Evenly space $v - 1$ reference points along the PPT's arc edges. Algorithm `varcv()` can create them on the surface of the sphere. They do not need to be projected.

- *Mid-arcs.* Mid-arcs is a recursive subdivision technique that uses a triangle's medial triangle to break it into four subtriangles. The triangle's three mid-arc points define the vertices of its medial triangle. A triangle's edge length is found with `vdis()` and its midpoint is found with `vdir()`. Project each chord mid-point to the surface of the sphere with `vuni()`.

- *PPT center point.* It is useful to have a reference point in the center of the PPT's face. Some schemas for certain frequencies define a grid point at the PPT's center but not every frequency-schema combination does so. Use `gtricent()` to define the PPT's center point, given the triangle's vertex coordinates, and then use `vuni()` to project the centroid point from the plane of the PPT face to the surface of the sphere.

C.1.3 Step 3: Subdivide the PPT

Subdivision grid points within the PPT are the result of the intersection of two or three great circles, as defined in Chapter 8. A great circle is defined by three points; two are reference points along the PPT's edge, and the third is the sphere's origin. Figure C.1(c) shows how a typical grid point is defined at the intersection of two great circles, themselves defined by a pair of edge reference points (and the center of the sphere). The choice of reference points, which define great circles, depends on the schema you are creating. Chapter 8 explains how edge reference points are selected, when defining the intersection of great circles. Here, we describe how to define grid points where great circles intersect.

- *Equal-chords.* In Equal-chords, great circles defined, as described in Chapter 8, produce a highly regular set of gridpoints over the PPT. Use `gcsect()` to locate the two points where these great circles intersect. One of the two points will have a negative z component. Discard this point, since it is not on the surface of the PPT. Continue intersecting pairs of great circles until all grid points have been defined.

- *Equal-arcs.* In Equal-arcs, you have your choice of defining grid points by intersecting two or three great circles. Intersecting two great circles always produces a single grid point within the PPT and two intersections are simpler to process than three; however, the resulting point set in the PPT is not rotationally symmetric about the PPT's centroid. Using three great circles produces superior results and the points are rotationally symmetric. Intersecting three great circles produces three points within the PPT. Additional processing is required to yield the single point needed. Use `gtricent()` to find the centroid of this window triangle and then project the centroid point (from the plane of the window triangle, which is inside the sphere) to the surface of the sphere with `vuni()`. Use `gcsect()` to find the two points where a pair of great circles intersect. One of the two points will have a negative *z* component. Discard this point, since it is not within the surface of the PPT. Continue intersecting pairs of great circles until all grid points have been defined.

- *Mid-arcs.* Mid-arcs is the easiest subdivision technique because you are simply subdividing triangles with their medial triangle. The vertices of a triangle's medial are simply the mid-arc points of their three edges. Mid-arcs is a recursive subdivision technique. In the first cycle, the PPT's equilateral triangle is subdivided by its medial triangle and results in a 2^v grid. If a higher frequency subdivision is required, the four triangles from the first cycle are each subdivided by their *medials*, resulting in a 4^v grid. Each subdivision cycle doubles the overall subdivision frequency, thus Mid-arcs schema always produce grids with frequencies that are powers of two. To find the three vertices of a triangle's medial, compute the midpoint of each PPT edge chord, using `vdir()` and a scale of 0.5. This midchord point is inside the sphere, so project it to the sphere's surface, using `vuni()`. The three mid-arc points of a triangle define its medial's vertices. Repeat the process of defining medial triangles until the desired frequency of subdivision is obtained.

C.1.4 Class I Summary

As the result of the steps followed, the PPT will be fully subdivided and resemble the grid in the figure to the right. The spherical arcs of the icosahedra are shown for reference only. Appendix D lists the grid points for each Class I schema developed in this book. You can use these coordinates as is for analysis or display, or you can cover the entire surface of the sphere by rotating copies of them to predefined locations elsewhere on the sphere. See Appendix E for details.

C.2 Class II: Triacon

Unlike Classes I and III, the Class II triacon method can subdivide all five spherical Platonics. The following steps show you how this is done. The example developed here is an 8^v subdivision of a spherical icosahedron but the process for subdividing any of the other four spherical Platonics are the same, only their LCD angles differ in the initial step.

In Class I, we began by locating reference points around the edges of the PPT. Based on these points, two or three great circles are defined and points on the surface of the PPT are derived from their intersections. All other spherical information, such as arcs and chords or triangle angles, are computed from the coordinates of these PPT grid points. The algorithms used in Class I schemas are mostly based on coordinate geometry.

In Class II, the sequence or steps are just the opposite of Class I. Subdivision starts by solving the parts of the spherical polyhedron's LCD right triangle. Instead of using coordinate geometry, the geodesic math is exclusively trigonometric solutions to right triangle and these types of triangles are the easiest to solve. The coordinates of grid points are solved by using spherical trigonometry (Appendix A provides a quick refresher). Once the LCD's grid points are defined, it is trivial to define the coordinates of the other three LCDs that make up the edge-zenith triacon diamond. You can use these grid points as is, input them into a CAD system, or cover the rest of the sphere. Appendix E describes how this is done.

C.2.1   Step 1: Position and Define the Triacon LCD

The first step in triacon subdivision is to position the spherical polyhedron edge-zenith. The triacon diamond is symmetrical to this edge; the LCD we subdivide is one fourth of a diamond and located in its positive (x, y, z) quadrant. Figure C.2 shows our example spherical icosahedron and triacon diamond in this position; the LCD portion we subdivide is accented.

- *LCD principal parts.* Label the LCD spherical parts, a right triangle, as shown in Figure C.2, and find sides a, b, and c, given its three surface angles A, B, and C. The basic relationships are shown in Figure C.2.

It's easy to establish the icosahedron's LCD surface angles A, B, and C; we can do this by inspection. They are also listed in Table 8.4. Through symmetry and knowing that the sum of the surface angles around any point on the sphere is always $360°$, the LCD angles for a spherical icosahedron are $A = 60°$ and $B = 36°$ and, by definition $C = 90°$. We now use the formulas in Figure C.2 and surface angles A, B, and C to find sides (central angles) a, b, and c. They are $31.7175°$, $20.9052°$, and $37.3774°$, respectively.

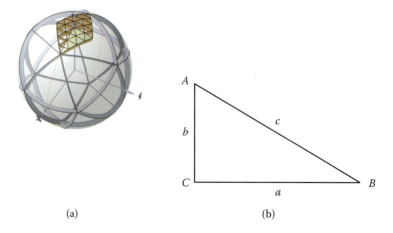

(a)   (b)

Figure C.2. Orientation of triacon LCD triangle.

We also need the arc and chord factors for the LCD's sides as well as the full icosa-hedron's full polyhedral edge and face meridian arc length. The arc factors for any of the LDC's sides are simply the radian value of their central angle, or $2\pi\theta/360$, where θ is the arc's central angle expressed in degrees. The chord factor between two points on the unit sphere, is the straight-line distance between them and is simply $2r\sin(\theta/2)$, where r is the radius of the sphere and θ is the central angle between the two points. Given LCD sides, the polyhedron's edge and face medians are simply

$$\text{PPT polyhedron edge} = 2a, \text{ and}$$
$$\text{PPT face median} = b + c.$$

C.2.2   Step 2: Subdivide the Triacon LCD

- *Define nested right triangles.* The triacon schema subdivides the LCD by nesting a series of right spherical triangles. Figure C.3 shows how an eight frequency grid is developed. The first subdivision step is to divide half the polyhedron's edge (arc) into $v/2$ equal arc segments. Each segment forms all or part of the base of a series of right spherical triangles. The result is $(v/2) - 1$ nested spherical right triangles within the LCD right triangle. All of these right triangles share angle B. In this example, there are three nested right triangles within the overall LCD right triangle. We will label their angles and sides the same way we labeled the overall LCD triangle. Side a of each nested right triangle is an integral number of polyhedron edge segments. In Figure C.3, it appears that each of the nested right triangles has the same angle A, but this is not the case. There are no similar triangles on a sphere; each nested right triangle has a slightly different value for A, but angles B and C remain constant and are $36°$ and $90°$, respectively. Given a, B, and C for each nested triangle, we find parts A, b, and c using these three relationships:

$$\cos A = \cos a \times \sin B,$$
$$\cos b = \cos B / \sin A, \text{ and}$$
$$\cos c = \cos a \times \cos b.$$

- *Subdivide LCD sides.* With all the nested triangle parts solved, divide LCD side b into arc segments equal to the b sides of each of the nested right triangles, as shown in Figure C.3(b). The b sides for each nested right triangle were found in the previous step. Each side b is slightly different and the overall LCD side b will be subdivided into $v/2$ arc segments.

- *Complete the triangulation.* Triangulate the grid with diagonals by alternating nested triangle c arcs across the LCD. Each row of c arcs is the same length. The result is a series of triangles, each with their own a, b, and c arcs, as shown in Figure C.3(c).

Diagonals are constructed with nested triangle c arcs to alternate points, as shown in Figure C.3(d). When completed, all interior tessellation vertices have three-way intersections of

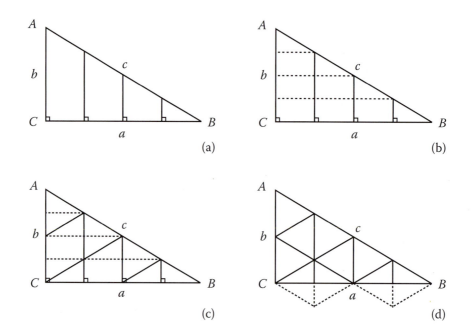

Figure C.3. LCD subdivision sequence.

two diagonals and a perpendicular to side *a*. Sides *a* and *c* of the overall LCD triangle will be subdivided *v*/2 times, while *b* is subdivided *v*/4 times.

Solving the LCD edge perpendiculars above solves all six parts (three angles, three sides) of each nested right triangle within the LCD. Subdividing the LCD sides, striking diagonals, and completing the triangulation replicate these various arcs across the LCD to create the three-way grid. The elegance of the triacon schema lies in the fact that parts *A*, *b*, and *c* of the nested triangles are replicated across the LCD to define the angles and sides of the other triangles in their respective rows. Figure C.4 shows how the replication works.

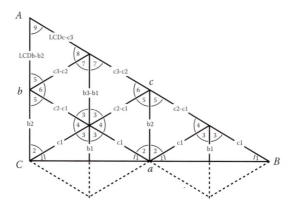

Figure C.4. LCD 8^v subdivision summary.

Starting at the smallest nested right triangle, the one immediately next to LCD vertex B, notice how its angles 1 and 3 and side $b1$ carry across the LCD to all other triangles in its row. Notice that angle 2 is the complement of angle 1. Other angles are found this way; they are either complements or supplements of one or more nested triangle parts. For example, angle 4 is the supplement of two angle 3s. Some diagonal c arcs are found by factoring out c arcs from another nested triangle. Figure C.4 shows how this is done.

C.2.3   Step 3: Define Grid Points

Vertex references greatly facilitate geodesic work, and you have complete freedom to name chords, triangles, surfaces, and central angles any way that is convenient to your application. For illustration purposes, we number triacon diamond quadrants 1 through 4 and name grid points with numbered a-columns and b-rows. This is the convention used in Figure C.5. Other conventions might be more convenient for calculator or PC programming. The polyhedral edge, not part of the final grid, is shown dashed in the figure. Points 1,1 in all quadrants are the vertices of the subdivided polyhedron. Points 5,5 are points in the middle of the icosahedron face.

   Class I and III schemas start by computing the coordinates of the PPT's vertices and end by finding the central and surface angles of all the grid triangles. Class II is just the opposite. We start with the surface angles of the LCD and end by finding the coordinates of the grid points. It is easy to find the Cartesian coordinates of grid points in the triacon diamond. Only two arc angles, a and b, are needed for each point and we already have their values when we solved the parts of each nested right triangle in the previous steps. Note that angles a and b of each point correspond to the φ and λ in the coordinate (ρ, φ, λ) system described in Chapter 4:

$$x = \sin a \times \cos b,$$
$$y = \sin b,$$
$$z = \cos a \times \cos b.$$

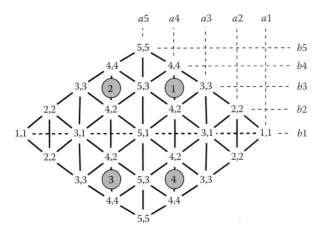

Figure C.5. Triacon grid vertex references.

LCD Vertex Angles	
Vertex	Angle°
a1	31.71748
a2	23.78811
a3	15.85874
a4	7.92937
a5	0
b1	0
b2	5.72354
b3	11.22946
b4	16.33348
b5	20.90516

Icosahedron 8v LCD Vertices			
a,b	X	Y	Z
1,1	0.52573	0.00000	0.85065
3,1	0.27327	0.00000	0.96194
5,1	**0.00000**	**0.00000**	**1.00000**
2,2	0.40134	0.09973	0.91048
4,2	0.13726	0.09973	0.98550
3,3	0.26804	0.19474	0.94352
5,3	0.00000	0.19474	0.98085
4,4	0.13238	0.28123	0.95046
5,5	0.00000	0.35682	0.93417

Table C.2. Icosahedron 8^v LCD vertices, angles, and coordinates.

Angle a is measured along the xy-plane and is positive if it is on the $+x$-axis side of the plane; otherwise, it is negative. Angle b, measured on the yz-plane, is positive, when it is on the $+y$-axis side and negative if on the $-y$ side. We simply find the cumulative a and b values for each vertex in the LCD grid and apply the above formula. Using the vertex reference system suggested in Figure C.5, the a and b vertex angles listed in Table C.2 and the formulas above, we find the coordinates of each LCD grid point. Figure C.2 also lists the coordinates of every vertex in the LCD. Notice that vertex 5,1 (highlighted in the table) is at the zenith of the sphere with coordinates (0, 0, 1), as expected.

To find vertex coordinates for grid points in quadrants 2, 3, and 4, simply change the signs of the coordinates in quadrant 1. The z-coordinate for every point in the edge-zenith diamond will always be positive. The sign changes for each quadrant are shown in Table C.3. Appendix D lists the coordinates for the triacon diamond to higher precision and for all four quadrants of the diamond.

Triacon LCD Coordinate Signs			
Quadrant	X	Y	Z
1	+	+	+
2	+	+	+
3	+	-	+
4	-	-	+

Table C.3. Triacon quadrant coordinate signs.

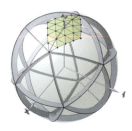

C.2.4   Class II Summary

After following the previous steps, the triacon LCD is fully subdivided and its coordinates reflected to define grid points in the other three diamond quadrants. The fully subdivided diamond will resemble the grid in the figure to the left. You can leave the diamond's grid as is for analysis or simple displays, or you can cover the entire sphere by rotating copies of the coordinates to predefined locations elsewhere on the sphere. Appendix E explains how this is done. You can also use these grid points with any of the coordinate geometry algorithms listed in this appendix to find centers of triangles, face normals, distances between arbitrary pairs of points, and so on.

C.3   Class III: Skew

This section describes how to create a Class III skew subdivision. Unlike Class I and III schemas where 3D grid points are created by intersecting great circles or by solving spherical right triangles, Class III defines points in two dimensions and then projects them onto the surface of the sphere. Skew grids are not parallel or perpendicular to the edges of their base polyhedron; thus, we do not have the convenience of locating reference points or defining right triangles along the edges of the foundation polyhedron. In projection, we essentially develop a flat triangular grid that meets the subdivision specification and then we project it onto the surface of the sphere, much like a movie screen, except that our screen is spherical.

The steps that follow describe how to presubdivide a flat, 2D, equilateral PPT with a skewed grid. A simple coordinate system based on a triangular grid (also called an *isometric grid*) creates the points needed for any (b,c) subdivision specification. Once the PPT's 2D grid points are defined, the entire set or points is centered over the Cartesian origin, scaled to fit the face of the spherical polyhedron we are subdividing, and then projected onto the surface of the unit sphere. The projected PPT and its grid are oriented face-zenith. The edges of the equilateral PPT are not retained as subdivision grid members. Projection subdivides a single PPT. Its grid can be used as is, input to a CAD system for further development, or copied and rotated to cover the remainder of the sphere's surface.

The projection method described here can also create Class I and II grids as well, depending on the subdivision grid's (b,c) specification. If b and c are zero, these steps create a Class I grid. Likewise, if $b = c$ and neither is zero, a Class II grid results. Class I subdivisions made by projection are the same as Class I Equal-chords discussed earlier in the book. Class II subdivisions made by projection resemble the triacon, but do not minimize the number of parts (chords, arcs, face shapes) as triacon does.

C.3.1   Step 1: Define the PPT Grid

The PPT grid points are initially defined on the xy-plane, thus the z-coordinate of each grid point is zero. The triangular grid axis system was detailed in Section 8.7.6, illustrated in Figure 8.43. It is an example of an 8^v ($b = 2$, $c = 6$) grid. The trigrid coordinates of all points within the PPT are found with these formulas:

$$x = \cos 60°\,(c + b),$$
$$y = \sin 60°\,c, \text{ and}$$
$$z = 0.$$

The PPT's apex points will be $(0,0)$, (b,c) and (c,b). Define the remainder of the points within the PPT with the same triangular grid axis system.

C.3.2   Step 2: Position PPT for Projection

The orientation of the PPT and its point set relative to the Cartesian x- and y-axis depends on the (b,c) specification used to make the grid in step one. In this step, the PPT and its point set is reoriented, centered, and scaled to a standard position suitable for projection.

- *PPT edge perpendicular to the x-axis.* One vertex of the PPT is already at the origin of the coordinate system. In this step, the entire PPT and its point set are rotated around this vertex to position the PPT side opposite it perpendicular to the x-axis.[2] The rotation angle will never exceed 30°. To find the exact rotation angle needed to place the PPT edge perpendicular to the x-axis, use `vdir()` to find its midpoint. Then use `vcos()` to find the angle between its mid-edge point and the x-axis. Use `zrot()` to rotate the PPT's point set counterclockwise (right hand rule) around the origin (z-axis). The command `zrot()` is listed in Appendix E.

- *Center the PPT on the z-axis.* The PPT must be centered over the origin (z-axis). Use `gtricent()` to find the PPT's centroid. Since the PPT edges are already symmetrical to the x-axis, the result of the rotation in the previous step, the PPT's centroid x-coordinate is the distance the PPT's center is from the origin (the centroid's y- and z-coordinates are zero). To center the PPT over the origin, subtract the centroid's x-coordinate from the x-coordinate of all points in the PPT point set. The PPT point set is now centered on the Cartesian origin with one of the PPT's edges perpendicular to the x-axis.

- *Scale the trigrid PPT.* The PPT edge points were defined by a unit triangular grid, thus it is not the same size as the edge chord of the deltahedron we are subdividing. To correct this, the PPT and its points set must be rescaled. The scale needed is the ratio of the PPT's edge length to the polyhedral edge length (inscribed in a unit sphere). Use `vdis()` and any two vertices of the PPT to find the PPT's edge length. Table C.4 lists the chord lengths of the deltahedra inscribed in a unit sphere. Use the scale and algorithm `vscl()` to scale the PPT. After scaling the edges of the PPT equilateral triangle are now equal to the chord lengths of the deltahedron we are subdividing.

To review, the PPT equilateral triangle and its point set is positioned with one apex coincident to the origin, and the PPT edge opposite that vertex is perpendicular to the x-axis. The PPT's centroid is found and the entire PPT point set is translated in the $-x$ direction to

Deltahedron Unit Sphere Characteristics		
Deltahedron	**Chord Length**	**Inradius**
Tetrahedron	1.63299316	0.33333333
Octahedron	1.41421356	0.57735027
Icosahedron	1.05146222	0.79465447

Table C.4. Deltahedron chord lengths and inradius.

---

[2] If you are using the projection method to make a Class II grid ($b = c$), the edge opposite the PPT apex at the origin is already perpendicular to the x-axis, and you do not need to rotate the trigrid point set.

center it over the origin. The PPT and its points set are scaled, so the PPT's edges are the same length as the chord length of the deltahedron we are subdividing.

C.3.3   Step 3: Project Trigrid Points

The PPT point set is on the xy-plane and centered at the origin (z-axis). The z-coordinate of every point is zero. They must be moved off the xy-plane before they can be projected. Translate the PPT point set along the $+z$-axis, a distance equal to the unit insphere radius of the deltahedron being subdividing. This is accomplished by simply adding the inradius distance to the z-coordinate of every point in the PPT set. Table C.4 lists the inradius for each deltahedron. Once translated, the apex points of the trigrid PPT will be on the surface of the unit sphere and tangent to the vertices of the face-zenith deltahedron being subdividing. The trigrid points are on the plane of the PPT grid and, thus, are inside the sphere. Use vuni() to project the grid points onto the surface of the sphere. A single face of the base deltahedron is now subdivided.

C.3.4   Class III Summary

At the conclusion of step three, the icosahedron's Class III $\{3, \}_{2,6}$ PPT will be fully subdivided and look like the figure to the right. Notice that a number of triangles straddle the polyhedral edge. Thus, to define all of their vertices, you must copy the grid points within the PPT and rotate them to define the points in the adjacent polyhedral faces. Appendix E explains how this is done. The next section describes how to find the chord and arc factors as well as the central and surface angles of any triangle.

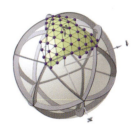

C.4   Characteristics of Triangles

Almost every Class I, II, or III spherical application requires supporting information regarding the breakdown. It is easy to extract chord and arc factors or the central and surface angles of grid members and triangles from the coordinates of the grid vertices. The procedures that follow are based on grid points located on the surface of the unit sphere.

- *Chord factors.* The chord factor, or chord length on a unit sphere, is simply the straight line distance between two grid points; use vdis(). A chord factor is typically found for chords between neighboring points, but there are no computational reasons why a chord factor between any arbitrary pair of points cannot be computed. This might be needed in some applications, for example, the edge of an opening spanning two nonneighboring points.

- *Arc factors and central angles.* Arc factors are simply the radian measure of the central angle between two points (measured from the origin). Like chord factors, arc factors are typically the geodesic arc distance between neighboring points but an arc factor can be computed the same way between any two points on the sphere. Use vcos() to find the cosine of the arc factor. The arccosine is the central angle.

- *Surface and dihedral angles.* The spherical surface angle between any two great circles is the same as the dihedral angle between their planes. Use gdihdrl() to find the angle (in decimal degrees). Three points define each plane. For great circles, one of these points will be the origin of the sphere. Use `gdihdrl()` to find the dihedral angle between two adjacent planar triangles (with vertices on the unit sphere). If the six point input to `gdihdrl()` are the vertices of two adjacent triangles, the dihedral angle between them is returned.

C.5   Storing Grid Points

The computer programs that follow assume all PPT or LCD grid points are stored in a single double precision array of triplets called *xyz*[SIZE][3] where the minimum size is determined by the subdivision frequency "*v*" plus working area. The formula for SIZE is

$$ \text{SIZE} = \frac{(v + 1)(v + 2)}{2}. $$

This 2D array is easy to visualize. The Cartesian *xyz*-coordinates of the fifth point in a grid, for example, would be *xyz*[4][0], *xyz*[4][1], and *xyz*[4][2], respectively. Note that in C, as in many other computer languages, array elements are indexed by their offset from the beginning of the array. Thus, the first array element index starts at zero [0].

Some algorithms need temporary locations to hold coordinates. Those algorithms use a simple integer array, *r*[], that contain indices to available unused elements in the master *xyz*-coordinate array. The actual index contained in them is not important to understanding the algorithm. How the temporary location is used is documented.

C.5.1   gcsect(): Intersection Points of Two Great Circles

The command `gsect()` returns two points, *p*0 and *p*1, and the dihedral angle between two intersecting great circles. Two great circles are defined by points v1,v2 and v3,v4 and the origin of the sphere. The command `gsect()` is based on normals to the planes of the great circles. See Figure 4.41 for an explanation of how normals of two planes define another plane perpendicular to their line of intersection. Note, also, that the dihedral angle returned is also the surface angle at their intersection.

```
int gcsect (int v1,int v2, int v3,int v4, int p0,int p1, double *dihedral)
{

    // find normals to both great circles and then normal to the normals
    vzero(r[0],r[0]);                       // r[0] is origin of sphere
    vnor(v1,r[0],v2,r[1]);                  // r[1] normal to first gc's plane
    vnor(v3,r[0],v4,r[2]);                  // r[2] normal to second gc's plane
    // find normal to plane created by above normals, the resulting normal
    // is
    // parallel to the line of intersection of the two gc planes
    vnor(r[1],r[0],r[2],r[3]);              // r[3] normal to normals,
    // normal 1 / origin / normal 2
    // check that the normal to normals is non-zero. A zero normal means
    // the input points are from coincident great circles
```

```
     if (vabs(r[3])==0.0) {      // test for coincident great circles
               vzero(p0,p0);     // initialize the return n to
               vzero(p1,p1);     // zero and
      *dihedral = 0.0;           // initialize dihedral angle to zero
               rput(count,r);    // return working registers to free pool
               return 9;         // signal error with return code
     }
     vuni (r[3],p0);             // unitize the normal, this is intersect 1
     vrevs(p0,p1);               // intersection 2 is antipodal of 1
     // find dihedral, return smallest angle
     // Note that cos function for vector dot products can return an
     // angle 0>angle<180. To insure angle is always the smallest dihedral
     // take absolute value of cos. Result is always the supplement.
     *dihedral = acos( fabs(vdot(r[1],r[2]) / (vabs(r[1]) * vabs(r[2]))) );
     return 0;                   // two valid points found, return
}
```

C.5.2   gdihdrl(): Dihedral Angle between Two Planes

The command gdihdrl() is a general function for finding the dihedral angle (also surface angle if points are on intersecting great circles) between any two planes. Plane 1 is defined by points p1, p2, and p3. Plane 2 is defined by points p4, p5, and p6.

```
double gdihdrl (int p1, int p2, int p3, int p4, int p5, int p6)
{
     extern double xyz[6500][3];      // xyz coordinate space
     double dihedral = 0.0;           // dihedral angle initialized
     vnor(p1,p2,p3,r[0]);             // find normal to first plane
     vnor(p4,p5,p6,r[1]);             // find normal to second plane

// compute angle between normals
     dihedral = degree(acos(vdot(r[0],r[1]) / (vabs(r[0])*vabs(r[1]))));
     if (dihedral <= 90.0) dihedral = 180.0 - dihedral;
     // check for smallest
     return dihedral;
```

C.5.3   gtricent(): Centroid of a Triangle

The command gtricent() returns the centroid point v4 of a triangle defined by points v1, v2, and v3. Point v4 is on the plane of the triangle.

```
void gtricent (int v1, int v2, int v3, int v4)
{
     extern double xyz[SIZE][3];       // xyz coordinate space
     xyz[v4][0] = (xyz[v1][0] + xyz[v2][0] + xyz[v3][0]) / 3.0;
     xyz[v4][1] = (xyz[v1][1] + xyz[v2][1] + xyz[v3][1]) / 3.0;
     xyz[v4][2] = (xyz[v1][2] + xyz[v2][2] + xyz[v3][2]) / 3.0;
     return;
}
```

C.5.4   stABC(): Surface Angles of a Spherical Triangle

The command `stABC()` returns the three surface angles of spherical triangle and the triangle's spherical excess (its area in steradians) given its three vertices on the surface of the unit sphere. The command `gdihdrl()` computes the dihedral angle.

```
void stABC(int v1, int v2, int v3, double *A, double *B, double *C,
        double *Excess)
{
    *A = radian(gdihdrl(v2,v1,v3));
    *B = radian(gdihdrl(v1,v2,v3));
    *C = radian(gdihdrl(v2,v3,v1));
    *Excess = (*A + *B + *C) - PI;
    return;
}
```

C.5.5   vabs(): Length of a Vector

The command `vabs()` returns the absolute value (also called vector length) of vector v1.

```
double vabs (int v1)                      // find absolute value (lengths) of v1
{
    extern double xyz[SIZE][3];
    double sum = 0.0;                     // init sum or squares
    for (int i=0;i<3;i++) sum += xyz[v1][i] * xyz[v1][i];// sum the squares
    sum = sqrt(sum);                      // keep summing the squares
    return sum;                           // return the final total
}
```

C.5.6   vadd(): Add two Vectors

The command `vadd()` adds vector v1 and v2 placing sum in v3.

```
void vadd (int v1, int v2, int v3)
{
    extern double xyz[6500][3]; // xyz coordinate space
    int i; // general loop index
    double temp[3];
    for (i=0;i<3;i++) temp[i] = xyz[v1][i] + xyz[v2][i];
    for (i=0;i<3;i++) xyz[v3][i] = temp[i];
    return;
}
```

C.5.7   varcv(): Locate a Point on a Geodesic Arc between Two Other Points

The command `varcv()` is a very useful algorithm for placing a point on a geodesic arc between two other points on the surface of the sphere. The point is placed a percentage of the distance from v1 towards v2. If the percent is 0.5, v3 is placed on the arc midway between v1 and v2. If the percent is 1.0, v3 will have the same value as v2. The command `varcv()` returns the central angle between v1 and v3 in radians. No error checking is performed.

```
double varcv(int v1, int v2, double scaleDistance, int v3)
{
    extern double xyz[SIZE][3];         // xyz coordinate space
    double alpha, beta, angle;
    double arcangle;
    vzero(r[2],r[2]);                    // create an origin point
    angle = acos(vcos(v1,r[2],v2));      // find great circle v1-v2 arc
    alpha = sin((1.-scaleDistance)*angle)/sin(angle);
    beta = sin(scaleDistance*angle)/sin(angle);
    vscl(v1,alpha,r[0]);                 // scale v1
    vscl(v2,beta, r[1]);                 // scale v2
    vadd(r[0],r[1],v3);                  // addition is desired point,
                                         // returned in v3
    vzero(r[2],r[2]);                    // create an origin point
    arcangle=acos(vcos(v1,r[2],v3));     // return v1 to v3 arc angle (in
                                         // radians)

    return arcangle;                     // return central angle
}
```

C.5.8 vcos(): Cosine of Angle between two Vectors

The command `vcos()` returns the cosine of the angle between two vectors, v1 and v3, measured from v2. If v1 and v3 are points on the surface of a unit radius sphere and v2 is the sphere's origin, the cosine is the central radian angle. This subroutine can be used in many situations.

```
double vcos (int v1, int v2, int v3)
{
    extern double xyz[SIZE][3];         // xyz coordinate space
    double cosine;                       // cosine of angle between two
                                         // vectors
    vsub(v2,v1,r[0]);                    // r[0] free v2-v1 edge vector
    vsub(v2,v3,r[1]);                    // r[1] free v2-v3 edge vector
    cosine = vdot(r[0],r[1]) / (vabs(r[0]) * vabs(r[1]));
// divide dot product by
// product of absolutes
    return cosine;
}
```

C.5.9 vcrs(): Cross Product of Two Vectors

The command `vcrs()` computes the cross product of vectors v1 and v2 and returns a vector, v3, that is perpendicular to their plane. v3's magnitude is also returned. The cross product of two vectors returns a normal vector (one that is 90°) to the plane defined by the two vectors.

```
double vcrs (int v1, int v2, int v3)
{
    extern double xyz[SIZE][3];         // xyz coordinate space
// column diagonals multiplied
    xyz[v3][0] = xyz[v1][1] * xyz[v2][2] - xyz[v1][2] * xyz[v2][1];// v3x
    xyz[v3][1] = xyz[v1][2] * xyz[v2][0] - xyz[v1][0] * xyz[v2][2];// v3y
    xyz[v3][2] = xyz[v1][0] * xyz[v2][1] - xyz[v1][1] * xyz[v2][0];// v3z
    return (vabs(v3));
}
```

C.5.10   vdir(): Point on a Parametric Line

The command `vdir()` defines a point, v3, on an infinite line defined by points v1 and v2 at a scalar distance from v1, which is sometimes called the anchor point. If the scalar value is 0 or 1, v3 will equal v1 or v3, respectively. If scalar value is 0.5, v3 will be a point midway between v1 and v2. Negative scales place v3 away from v1 opposite the direction from v2. This algorithm is useful in Equal-chords schemas that space reference points along a chord. Reference points are inside the sphere and need to be projected to its surface. See `vuni()` to project them.

```
void vdir (int v1, int v2, double scalar, int v3)
{
    extern double xyz[SIZE][3];// xyz coordinate space
    vsub(v2,v1,r[0]);              // translate v2 by v1 relative to origin
    vscl(r[0],scalar,r[1]);       // scale v2, put in temp registers at 3
    vadd(v1,r[1],v3);             // translate scaled v2 back relative to v1
    return;
}
```

C.5.11   vdis(): Distance between two Points

The command `vdis()` returns the distance between points v1 and v2.

```
double vdis (int v1, int v2)
{
    extern double xyz[SIZE][3];         // xyz coordinate space
    double dist;
    vsub(v2,v1,r[0]);                   // find difference between n
    dist = vabs(r[0]);                  // return the absolute value
    return dist;
}
```

C.5.12   vdot(): Dot Product

The command `vdot()` returns the dot product, also called the scalar or inner product, of two vectors.

```
double vdot (int v1, int v2)
{
    extern double xyz[SIZE][3];         // xyz coordinate space
    double sum = 0.0;                   // temp summation variable
    for (int i=0;i<3;i++) sum += xyz[v1][i] * xyz[v2][i];
    return sum;                         // return sum of squares
}
```

C.5.13   vnor(): Normal Vector to a Plane

The command vnor finds the normal vector, v4, to a plane defined by points v1, v2, and v3. v4's magnitude is returned.

```
double vnor (int v1, int v2, int v3, int v4)
{
    extern double xyz[SIZE][3];         // xyz coordinate space
    vsub(v1,v2,r[0]);                   // define t1 vector on the plane
    vsub(v3,v2,r[1]);                   // define t2 vector on the plane
    vcrs(r[1],r[0],v4);                 // find cross product t2 X t1
    return vabs(v4)/2.0;                // area is half the abs value of
                                        // normal

    // vector
}
```

C.5.14   vrevs(): Reverse a Vector's Sense

With the command vrevs(), the reverse sense of v1 is returned in v2 placing. If v1 is a vector, it now points the opposite direction and its magnitude remains the same. If v1 is a point, it is relocated to a new position on the opposite side of the origin along an infinite line between its old position and the origin of the sphere. If v1 is on the surface of the unit sphere, v2 will be its antipode.

```
void vrevs (int v1, int v2)
{
    extern double xyz[SIZE][3];         // xyz coordinate space
    for (int i=0;i<3;i++) xyz[v2][i] = xyz[v1][i] * -1.0; // reverse sense
    return;
}
```

C.5.15   vscl(): Scale a Vector or Point

With the command vscl(), the vector or point v1 is scaled along an infinite line through the origin and returned in v3. The scale factor can be any real (positive value) number. The command vscl() is very useful in scaling up or down all points on a sphere or simply to locate a new point in the same direction.

```
void vscl (int v1, double scalar, int v3)
{
    extern double xyz[SIZE][3];         // xyz coordinate space
```

```
    int i;                              // general loop index
    for (i=0;i<3;i++) xyz[v3][i] = xyz[v1][i] * fabs(scalar);
    if (scalar<0) for (i=0;i<3;i++) xyz[v3][i] = xyz[v3][i] * -1;

    return;
}
```

C.5.16 vsub(): Subtract a Vector or Point

The command `vsub()` subtracts v2 from v1. The difference is returned in v3.

```
void vsub (int v1, int v2, int v3)
{
    extern double xyz[SIZE][3];        // xyz coordinate space
    for (int i=0;i<3;i++) xyz[v3][i] = xyz[v1][i] - xyz[v2][i];
    return;
}
```

C.5.17 vuni(): Normalize a Vector

The command `vuni()` normalizes the vector or point v1 and returns its position on the surface of the unit sphere in v2. This is a highly used algorithm.

```
void vuni (int v1, int v2)
{
    extern double xyz[SIZE][3];        // xyz coordinate space
    int i;                             // general loop index
    double len = 0.0;                  // init sum or squares
    for (i=0;i<3;i++) len += sq(xyz[v1][i]);// sum the squares
    len = sqrt(len);                   // find length of vector
    for (i=0;i<3;i++) xyz[v2][i] = xyz[v1][i] / len;
    return;
}
```

C.5.18 vzero(): Initialize a Vector

The command `vzero()` initializes the vector or point v1 to zero and returns the value in v2.

```
void vzero (int v1, int v2)
{
    extern double xyz[SIZE][3];        // xyz coordinate space
    for (int i=v1;i<=v2;i++) {         // for every vector row element
    for (j=0;j<3;j++) xyz[i][j] = 0.0;// set its coordinates to zero
    }
    return;
}
```

Additional Resources

Clinton, Joseph D. *Advanced Structural Geometry Studies Part I—Polyhedral Subdivision Concepts for Structural Applications.* National Aeronautics and Space Administration contractor report NASA CR-1734, Carbondale, IL: Southern Illinois University, 1971.

——. "Geodesic Math." *Domebook 2,* ed. Lloyd Kahn et al., Bolinas, CA: Pacific Domes (1971): 106–113.

Kenner, Hugh. *Geodesic Math and How to Use It.* Berkeley: University of California Press, 1976.

Kitrick, Christopher J. "A Unified Approach to Class I, II and III Geodesic Domes." *International Journal of Space Structures* 5.3–4 (1990): 223–246.

Messer, Peter W. "Mathematical Formulas for Geodesic Domes." Appendix to *Spherical Models* by Magnus J. Wenninger (New York: Dover 1999): 145–149.

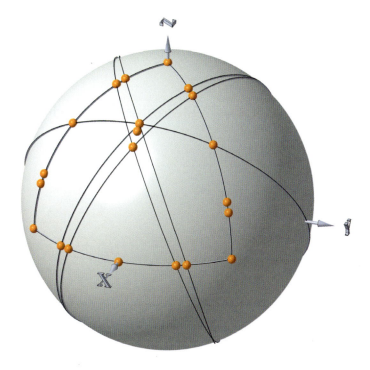

D Schema Coordinates

Chapter 8 developed six methods for distributing points on a sphere. The resulting vertex coordinates for a single principal triangle or diamond for a subdivided 8^v spherical
icosahedron are listed here. Appendix E describes methods for transforming these coordinates to cover the entire sphere without gaps or overlaps. A vertex number diagram and a
3D image indicate the position and orientation of individual vertices within the PPT or diamond. Vertex listings appear in the same order as the schemas presented in earlier chapters:

- Class I—Alternates and Ford

    - Equal-chords

    - Equal-arcs (two great circles)

    - Equal-arcs (three great circles)

    - Mid-arcs

- Class II—Triacon

- Class III—Skew

D.1   Coordinates for Class I: Alternates and Ford

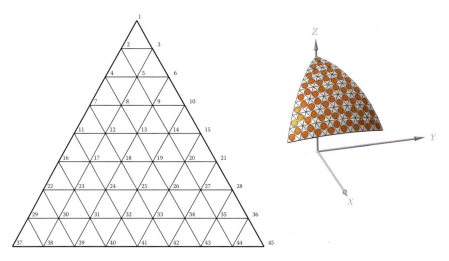

Figure D.1. Alternates and Ford PPT vertex numbers and *x,y,z* orientation.

Vertex	X	Y	Z	Vertex	X	Y	Z
1	0.00000000	0.00000000	1.00000000	24	0.67082039	-0.16245985	0.72360680
2	0.09647148	-0.07009063	0.99286482	25	0.67985212	0.00000000	0.73334923
3	0.09647148	0.07009063	0.99286482	26	0.67082039	0.16245985	0.72360680
4	0.20318274	-0.14762090	0.96794878	27	0.64574253	0.31277294	0.69655558
5	0.20543347	0.00000000	0.97867108	28	0.60954823	0.44286271	0.65751317
6	0.20318274	0.14762090	0.96794878	29	0.67530038	-0.49063444	0.55067889
7	0.31525325	-0.22904489	0.92095267	30	0.71901714	-0.37314038	0.58632807
8	0.32287127	-0.07819324	0.94320728	31	0.75336629	-0.23457971	0.61433834
9	0.32287127	0.07819324	0.94320728	32	0.77249586	-0.08017873	0.62993770
10	0.31525325	0.22904489	0.92095267	33	0.77249586	0.08017873	0.62993770
11	0.42532540	-0.30901699	0.85065081	34	0.75336629	0.23457971	0.61433834
12	0.44142621	-0.16035746	0.88285242	35	0.71901714	0.37314038	0.58632807
13	0.44721360	0.00000000	0.89442719	36	0.67530038	0.49063444	0.55067889
14	0.44142621	0.16035746	0.88285242	37	0.72360680	-0.52573111	0.44721360
15	0.42532540	0.30901699	0.85065081	38	0.77177186	-0.42054381	0.47698124
16	0.52542208	-0.38174148	0.76039797	39	0.81273098	-0.29524181	0.50229537
17	0.55178276	-0.24053618	0.79854751	40	0.84067532	-0.15269659	0.51956592
18	0.56654344	-0.08232358	0.81990936	41	0.85065081	0.00000000	0.52573111
19	0.56654344	0.08232358	0.81990936	42	0.84067532	0.15269659	0.51956592
20	0.55178276	0.24053618	0.79854751	43	0.81273098	0.29524181	0.50229537
21	0.52542208	0.38174148	0.76039797	44	0.77177186	0.42054381	0.47698124
22	0.60954823	-0.44286271	0.65751317	45	0.72360680	0.52573111	0.44721360
23	0.64574253	-0.31277294	0.69655558				

Table D.1. Equal-chords 8^v PPT vertex coordinates.

Vertex	X	Y	Z	Vertex	X	Y	Z
1	0.00000000	0.00000000	1.00000000	24	0.68862897	-0.14929788	0.70957754
2	0.11160571	-0.08108629	0.99043888	25	0.70003892	0.00000000	0.71410469
3	0.11160571	0.08108629	0.99043888	26	0.68862897	0.14929788	0.70957754
4	0.22107727	-0.16062204	0.96193836	27	0.65438044	0.29511029	0.69620123
5	0.23942149	0.00000000	0.97091573	28	0.59719638	0.43388857	0.67460893
6	0.22107727	0.16062204	0.96193836	29	0.66677672	-0.48444164	0.56632597
7	0.32632133	-0.23708633	0.91504342	30	0.72388502	-0.35321544	0.59264604
8	0.36050851	-0.08030431	0.92929265	31	0.76211876	-0.21481432	0.61076166
9	0.36050851	0.08030431	0.92929265	32	0.78129547	-0.07208689	0.61998457
10	0.32632133	0.23708633	0.91504342	33	0.78129547	0.07208689	0.61998457
11	0.42532540	-0.30901699	0.85065081	34	0.76211876	0.21481432	0.61076166
12	0.47153216	-0.15747050	0.86767532	35	0.72388502	0.35321544	0.59264604
13	0.48696452	0.00000000	0.87342175	36	0.66677672	0.48444164	0.56632597
14	0.47153216	0.15747050	0.86767532	37	0.72360680	-0.52573111	0.44721360
15	0.42532540	0.30901699	0.85065081	38	0.77838243	-0.40335535	0.48106680
16	0.51619630	-0.37503857	0.76999185	39	0.81827364	-0.27326653	0.50572092
17	0.57001145	-0.22955130	0.78891897	40	0.84251764	-0.13795224	0.52070454
18	0.59688506	-0.07726320	0.79859791	41	0.85065081	0.00000000	0.52573111
19	0.59688506	0.07726320	0.79859791	42	0.84251764	0.13795224	0.52070454
20	0.57001145	0.22955130	0.78891897	43	0.81827364	0.27326653	0.50572092
21	0.51619630	0.37503857	0.76999185	44	0.77838243	0.40335535	0.48106680
22	0.59719638	-0.43388857	0.67460893	45	0.72360680	0.52573111	0.44721360
23	0.65438044	-0.29511029	0.69620123				

Table D.2. Equal-arcs (two great circles) 8^v PPT vertex coordinates.

Vertex	X	Y	Z	Vertex	X	Y	Z
1	0.00000000	0.00000000	1.00000000	24	0.66660202	-0.14969073	0.73022903
2	0.11160571	-0.08108629	0.99043888	25	0.67543709	0.00000000	0.73741762
3	0.11160571	0.08108629	0.99043888	26	0.66660202	0.14969073	0.73022903
4	0.22107727	-0.16062204	0.96193836	27	0.64031186	0.29555655	0.70897606
5	0.22913741	0.00000000	0.97339409	28	0.59719638	0.43388856	0.67460893
6	0.22107727	0.16062204	0.96193836	29	0.66677672	-0.48444164	0.56632597
7	0.32632133	-0.23708633	0.91504342	30	0.71644992	-0.35405326	0.60112046
8	0.34342031	-0.07985221	0.93578102	31	0.75021900	-0.21570434	0.62501447
9	0.34342031	0.07985221	0.93578102	32	0.76730807	-0.07245314	0.63717256
10	0.32632133	0.23708633	0.91504342	33	0.76730807	0.07245314	0.63717256
11	0.42532540	-0.30901699	0.85065081	34	0.75021900	0.21570434	0.62501447
12	0.45170423	-0.15684647	0.87827244	35	0.71644992	0.35405326	0.60112046
13	0.46057041	0.00000000	0.88762318	36	0.66677672	0.48444164	0.56632597
14	0.45170423	0.15684647	0.87827244	37	0.72360680	-0.52573111	0.44721360
15	0.42532540	0.30901699	0.85065081	38	0.77838243	-0.40335535	0.48106680
16	0.51619630	-0.37503857	0.76999185	39	0.81827364	-0.27326653	0.50572092
17	0.55142743	-0.22929961	0.80209069	40	0.84251764	-0.13795224	0.52070453
18	0.56925430	-0.07714631	0.81853405	41	0.85065081	0.00000000	0.52573111
19	0.56925430	0.07714631	0.81853405	42	0.84251764	0.13795224	0.52070453
20	0.55142743	0.22929961	0.80209069	43	0.81827364	0.27326653	0.50572092
21	0.51619630	0.37503857	0.76999185	44	0.77838243	0.40335535	0.48106680
22	0.59719638	-0.43388856	0.67460893	45	0.72360680	0.52573111	0.44721360
23	0.64031186	-0.29555655	0.70897606				

Table D.3. Equal-arcs (three great circles) 8^v PPT vertex coordinates.

Vertex	X	Y	Z	Vertex	X	Y	Z
1	0.00000000	0.00000000	1.00000000	24	0.67082039	-0.16245985	0.72360680
2	0.11160571	-0.08108629	0.99043888	25	0.67985212	0.00000000	0.73334923
3	0.11160571	0.08108629	0.99043888	26	0.67082039	0.16245985	0.72360680
4	0.22107727	-0.16062204	0.96193836	27	0.64056733	0.30125888	0.70634027
5	0.22398550	0.00000000	0.97459248	28	0.59719638	0.43388856	0.67460893
6	0.22107727	0.16062204	0.96193836	29	0.66677672	-0.48444164	0.56632597
7	0.32632133	-0.23708633	0.91504342	30	0.71704513	-0.35822879	0.59792843
8	0.33760223	-0.08114185	0.93778501	31	0.75224951	-0.22011703	0.62102268
9	0.33760223	0.08114185	0.93778501	32	0.77021827	-0.08224247	0.63245553
10	0.32632133	0.23708633	0.91504342	33	0.77021827	0.08224247	0.63245553
11	0.42532540	-0.30901699	0.85065081	34	0.75224951	0.22011703	0.62102268
12	0.44170765	-0.15643447	0.88341531	35	0.71704513	0.35822879	0.59792843
13	0.44721360	0.00000000	0.89442719	36	0.66677672	0.48444164	0.56632597
14	0.44170765	0.15643447	0.88341531	37	0.72360680	-0.52573111	0.44721360
15	0.42532540	0.30901699	0.85065081	38	0.77838243	-0.40335535	0.48106680
16	0.51619630	-0.37503857	0.76999185	39	0.81827364	-0.27326653	0.50572092
17	0.55490470	-0.23867693	0.79694046	40	0.84251764	-0.13795224	0.52070453
18	0.56654344	-0.08232358	0.81990936	41	0.85065081	0.00000000	0.52573111
19	0.56654344	0.08232358	0.81990936	42	0.84251764	0.13795224	0.52070453
20	0.55490470	0.23867693	0.79694046	43	0.81827364	0.27326653	0.50572092
21	0.51619630	0.37503857	0.76999185	44	0.77838243	0.40335535	0.48106680
22	0.59719638	-0.43388856	0.67460893	45	0.72360680	0.52573111	0.44721360
23	0.64056733	-0.30125888	0.70634027				

Table D.4. Mid-arcs 8^v PPT vertex coordinates.

D.2   Coordinates for Class II: Triacon

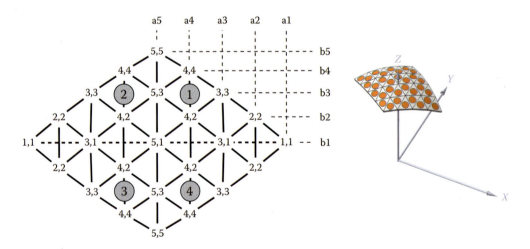

Figure D.2. Triacon 8^v diamond vertex numbers and x,y,z orienation.

Quadrant	Vertex a, b	Coordinate X	 Y	 Z
1	1, 1	**0.52573149**	**0.00000000**	**0.85065057**
1	3, 1	**0.27326675**	**0.00000000**	**0.96193830**
1	5, 1	**0.00000000**	**0.00000000**	**1.00000000**
1	2, 2	0.40134480	0.09972860	0.91048150
1	4, 2	0.13726462	0.09972860	0.98550121
1	3, 3	0.26803508	0.19473891	0.94352210
1	5, 3	**0.00000000**	**0.19473891**	**0.98085511**
1	4, 4	0.13238474	0.28122769	0.95046582
1	5, 5	**0.00000000**	**0.35682236**	**0.93417226**
2	1, 1	**-0.52573149**	**0.00000000**	**0.85065057**
2	3, 1	**-0.27326675**	**0.00000000**	**0.96193830**
2	5, 1	**0.00000000**	**0.00000000**	**1.00000000**
2	2, 2	-0.40134480	0.09972860	0.91048150
2	4, 2	-0.13726462	0.09972860	0.98550121
2	3, 3	-0.26803508	0.19473891	0.94352210
2	5, 3	**0.00000000**	**0.19473891**	**0.98085511**
2	4, 4	-0.13238474	0.28122769	0.95046582
2	5, 5	**0.00000000**	**0.35682236**	**0.93417226**
3	1, 1	**-0.52573149**	**0.00000000**	**0.85065057**
3	3, 1	**-0.27326675**	**0.00000000**	**0.96193830**
3	5, 1	**0.00000000**	**0.00000000**	**1.00000000**
3	2, 2	-0.40134480	-0.09972860	0.91048150
3	4, 2	-0.13726462	-0.09972860	0.98550121
3	3, 3	-0.26803508	-0.19473891	0.94352210
3	5, 3	**0.00000000**	**-0.19473891**	**0.98085511**
3	4, 4	-0.13238474	-0.28122769	0.95046582
3	5, 5	**0.00000000**	**-0.35682236**	**0.93417226**
4	1, 1	**0.52573149**	**0.00000000**	**0.85065057**
4	3, 1	**0.27326675**	**0.00000000**	**0.96193830**
4	5, 1	**0.00000000**	**0.00000000**	**1.00000000**
4	2, 2	0.40134480	-0.09972860	0.91048150
4	4, 2	0.13726462	-0.09972860	0.98550121
4	3, 3	0.26803508	-0.19473891	0.94352210
4	5, 3	**0.00000000**	**-0.19473891**	**0.98085511**
4	4, 4	0.13238474	-0.28122769	0.95046582
4	5, 5	**0.00000000**	**-0.35682236**	**0.93417226**

Table D.5. Triacon 8^v diamond vertex coordinates.

Note: A complete triacon diamond contains four back-to-back LCDs. Vertex reference numbers have three parts: quadrant number, a range, and b range (see the table above). Triacon coordinates are initially computed for the LCD in quadrant 1. The signs of quadrant 1 coordinates are changed to define symmetrical points in LCD quadrants 2, 3, and 4. As a result, some vertices are shared by LCDs in two or three quadrants. Shared vertex coordinates are indicated in bold in the table above. Depending on your application, you may need to eliminate duplicate coordinates.

D.3   Coordinates for Class III: Skew

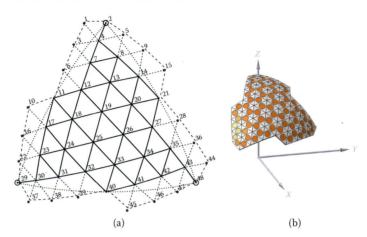

(a)   (b)

Figure D.3. Skew PPT vertex numbers and x,y,z orienation.

Vertex	X	Y	Z	Vertex	X	Y	Z
1	-0.00440119	-0.12830445	0.99172506	25	0.66173043	-0.07591189	0.74588888
2	0.00000000	0.00000000	1.00000000	26	0.62690251	0.10121585	0.77249504
3	**0.13469185**	**-0.19026520**	**0.97244911**	27	0.57329431	0.26951468	0.77375414
4	0.12066474	-0.04383403	0.99172506	**28**	**0.51098051**	**0.41879015**	**0.75067552**
5	0.07897610	0.10121353	0.99172506	29	0.72360680	-0.52573111	0.44721360
6	**0.28195914**	**-0.25239638**	**0.92563228**	30	0.75415464	-0.41642332	0.50778184
7	0.25797622	-0.09371535	0.96159540	31	0.77392866	-0.28114604	0.56744281
8	0.22257502	0.06930438	0.97244911	32	0.77563347	-0.12250667	0.61918077
9	**0.16884763**	**0.21639033**	**0.96159540**	33	0.75376853	0.04978590	0.65525138
10	**0.43348138**	**-0.46195067**	**0.77375414**	34	0.70818708	0.22051201	0.67070523
11	0.42532540	-0.30901699	0.85065081	35	0.64494055	0.37486139	0.66598095
12	0.40467833	-0.14700801	0.90256528	**36**	**0.57984189**	**0.51368564**	**0.63238473**
13	0.37095514	0.02450133	0.92832751	**37**	**0.80568464**	**-0.46025735**	**0.37287456**
14	0.32717339	0.19016430	0.92563228	**38**	**0.84171947**	**-0.32342044**	**0.43232806**
15	**0.26486541**	**0.33944399**	**0.90256528**	39	0.85837432	-0.16639377	0.48529026
16	**0.55581196**	**-0.49753637**	**0.66598095**	40	0.85065081	0.00000000	0.52573111
17	0.55619476	-0.35655807	0.75067552	41	0.81793347	0.16639377	0.55072493
18	0.54819529	-0.19914360	0.81229536	42	0.76311437	0.32342044	0.55951379
19	0.52241876	-0.02530396	0.85231353	43	0.69382227	0.46025735	0.55387167
20	0.47967088	0.14935770	0.86464335	**44**	**0.62908871**	**0.58856180**	**0.50778184**
21	0.42532540	0.30901699	0.85065081	**45**	**0.90068595**	**0.12250667**	**0.41684161**
22	**0.65213363**	**-0.51763685**	**0.55387167**	**46**	**0.85364769**	**0.28114604**	**0.43845470**
23	0.66772507	-0.39272482	0.63238473	**47**	**0.79144210**	**0.41642332**	**0.44744947**
24	0.67446389	-0.24501335	0.69646746	48	0.72360680	0.52573111	0.44721360

Table D.6. Skew 8^v PPT vertex coordinates.

Note: Vertices in bold are outside the PPT; they are listed to complete the definition of PPT triangles that straddle the PPT's edge.

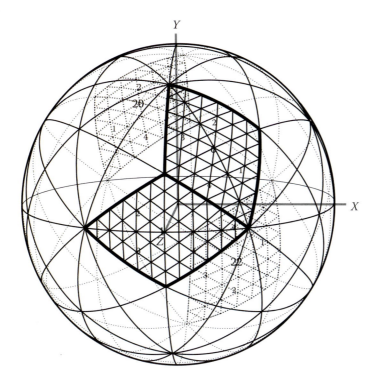

E Coordinate Rotations

R otations are mathematical transformations that reposition one or more points on the surface of the sphere. Rotations were key transformations in the symmetry groups discussed in Chapter 6. Rotations are very useful when covering the entire surface of a sphere with a grid because we can subdivide a small section of the sphere, make a copy of the geometry, and then rotate the copy to cover another part of the sphere. By repeating this copy-rotate sequence, the entire sphere is covered without gaps or overlaps. Rotation does not add any new geometry to the original subdivided section; it only replicates what's there to another place on the sphere. Rotations are one of several isometry transformations that preserve distance between points; thus, the angles between points and the chords, and the arcs between them, remain the same as before.[1]

This appendix outlines general methods for rotating points on a sphere. We show simple and compound rotations around the x-, y-, and z-axes, as well as a more advanced technique to rotate points around an arbitrary axis. In this appendix, we show how the single PPT or

---

[1] Reflections, translations, and glide reflections, like rotations, are isometry transformations; they do not change the distance or angles between points.

diamond coordinates, such as those developed earlier in the book, are rotated to cover the entire sphere. All three class subdivision schemas for icosahedra, octahedra, and tetrahedra are illustrated, as well as the additional Class II dodecahedra and cube schemas. For each polyhedra and schema, a table lists the required angle-axis rotation sequences, and a companion figure shows step-by-step results. All information needed to rotate a single vertex- or face-zenith PPT and edge-zenith diamond to all other positions on the sphere is described here.

Mathematically, there are several ways to implement rotations. We use matrix techniques because they are easy to visualize and to program in computers or calculators. Complex rotation sequences involving more than one rotation can be performed in a series of single operations or by combining the matrices into a single matrix via matrix multiplication.

Rotations are not the only way to reposition geometry. Most subdivisions have highly symmetrical geometry; thus, PPTs and diamonds tend to have symmetrical counterparts elsewhere on the sphere. In these cases, geometry can be reflected to another side of a Cartesian plane, or vertex antipodes can place geometry on the opposite side of the sphere, simply by manipulating the signs of coordinates. Reflections and antipodal points are easy to figure out and can substitute for rotations in some cases. But both techniques have side effects, however, that you need to be aware of. We begin our discussion of coordinate rotations with general concepts.

E.1   Rotation Concepts

Rotation is a simple concept.One or more points are repositioned on the sphere by rotating them together around one of the Cartesian axes by some angle. Rotation does not affect the component of the points corresponding to the axes of rotation. That is, the x-component of points rotated around the x-axis do not change; only the y- or z-values change, depending on the amount of rotation. Rotating points, plus or minus any multiple of 360° around an axis, does not change the point's position.

Rotations can be simple or compound. A simple rotation involves only one axis and a positive or negative rotation angle around it. A compound one is several simple ones in sequence. When compound rotations are performed, the result depends entirely on the amount of each simple rotation and the sequence they are applied. Compound rotations are not *commutative.* A rotation by some amount around x followed by another around y usually does not result in the same position if the sequence of rotations is reversed, even if the angles specified are the same. This key point is discussed in detail in the following section.

E.2   Direction and Sequences

Mathematicians, physicists, and engineers use standard conventions to designate which direction is positive when rotating. We use the *right-hand rule (RHR)* convention because it is commonly found in mathematics texts and computer graphics literature. Use the following steps: Grasp the axis to be rotated around by your right hand. Point your thumb in the positive axis direction (away from the coordinate system origin). Your fingers curl around the axis, indicating the positive direction (angle) the object will be rotated. Figure E.1 shows the RHR positive angle direction around the z-axis, and Figure E.2 shows the before (a) and after (b)

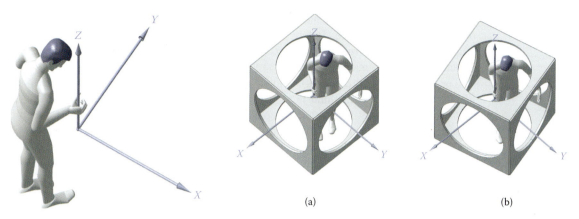

Figure E.1. The *z*-axis right-hand rule.

Figure E.2. Cube rotated +15° around the *z*-axis.

effect of rotating a cube +15° around the *z*-axis. Note that the cube rotates; the *xyz*-reference axis does not.[2]

E.3   Simple Rotations

A simple rotation is a single axis-angle rotation. The choice of axis is important because the axis-angle combination usually affects the final result. Figure E.3 shows an example of

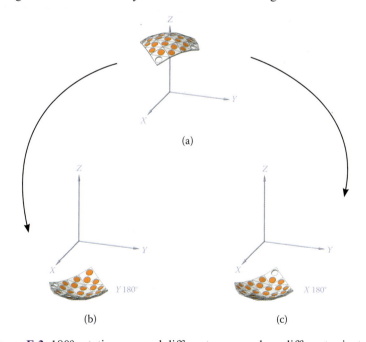

Figure E.3. 180° rotations around different axes produce different orientations.

[2] It is also possible to rotate the axis system. This is often used in CAD systems (see Section 11.9).

a simple rotation of an edge-zenith Class II triacon diamond (octahedron 8^v) rotated to the opposite side of the sphere. Two red circular surfaces have been removed—one from the long-axis and another from the short-axis of the diamond—to make it easier to compare the effects of the two rotation sequences.

The initial diamond position is shown in Figure E.3(a). Two different sequences are illustrated. In (b), the diamond is rotated 180° around the *y*-axis. In (c), the rotation is 180° around the *x*-axis. Both sequences placed the original edge-zenith diamond (a) on the opposite side of the sphere, as expected. Notice that the axes of the two diamonds are opposite one other in (b) and (c), as indicated by the missing red surfaces. For some geodesic applications, this difference might not be important, but at other times it may be. Next, let's look at how reflections can accomplish similar results without any rotation at all.

E.4   Reflections

Like rotations, reflections can reposition PPT and diamond coordinates elsewhere on the sphere. Reflections are isometry transformations and will not change the distance or angle between points within the set being reflected. In a reflection, all points are exactly the same distance from the plane they are reflected through; they are just on the opposite side of the plane. Polyhedra have multiple face-, edge-, and vertex-symmetries, as do their subdivisions (see Chapter 6). We can take advantage of symmetry where PPTs and diamonds have counterparts opposite one of the Cartesian planes. Points can be reflected to the other side of a Cartesian plane simply by reversing the sign of the component perpendicular to that plane.

Figure E.4 shows a number of these cases for an edge-zenith Class II icosahedron. In the figure, only the outline of the triacon diamond and their quadrant numbers are shown. The view is from the +*z*-axis looking towards the origin; the *x*- and *y*-axes are labeled. Notice that diamonds 6 to 11 on the positive *x*-side (right) of the *yz*-plane are reflected to the negative *x*-side, making diamonds 12 to 17 simply by reversing the sign of their *x*-components. We could continue this process and reflect all the diamonds we see in Figure E.4 to the opposite side of the sphere (*xy*-plane) by reversing the *z*-component of all the coordinates.

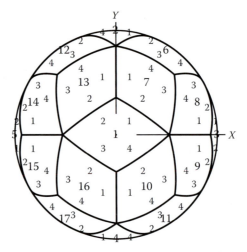

Figure E.4. Reflecting diamonds.

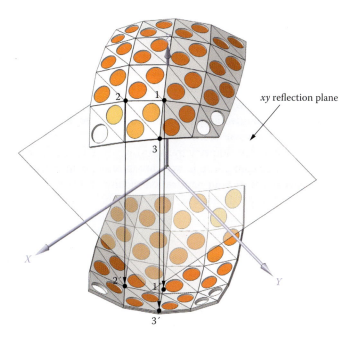

Figure E.5. Reflecting coordinates.

To be sure, reflection is the easiest way to cover the sphere where there are symmetric relationships around any of the three Cartesian planes.

Reflecting a PPT or diamond's coordinates has the effect of turning it inside out when it is placed on the opposite side of the reflection plane. This can have negative effects on polygons if you are using left- or right-hand point sequences to know which face of the polygon faces away from the origin of the sphere.

Many 3D applications, such as hidden line removal or face-shading, depend on standard sequences of vertices when defining face boundaries. A sequence, in effect, "walks around" the perimeter of the polygon and references each vertex in the proper order necessary to define the polygon's perimeter.

Figure E.5 shows an example of a single edge-zenith diamond reflected through the xy-plane. The z-component of each vertex in the zenith diamond is reflected to the opposite side of the xy-plane by reversing the sign of the z-component of every vertex.

Two problems surface with the reflected diamond: its orientation is unique, and the handedness has changed. Notice that the three holes in the diamond remain on the same side of the xz-plane after the reflection. This is unique to reflections as no combination of rotations of the edge-zenith diamond around the surface of the sphere can produce this orientation. This can be proved by comparing this diamond to the orientations of the two in Figure E.3.

Another problem with the reflection concerns polygon vertices. We list polygon vertices using the right-hand rule, just as we do when specifying rotation angles. Imagine an axis perpendicular to the face and pointing away from the origin. You are outside the sphere. As before, you grasp the axis with your thumb pointing away from the origin of the sphere.

Your fingers curl counterclockwise, defining the polygon. In Figure E.5, the three vertices of a single triangle in the upper diamond are numbered 1, 2, and 3 in RHR sequence. Their reflected points in the lower diamond are numbered 1′, 2′, and 3′. Notice that the reflected points are now left-hand ruled! If you were outside the sphere, their RHR order would be 1′, 3′, and 2′. These two triangles are symmetric but not congruent.

To see more examples of how handedness changes with reflection, take another look at Figure E.4. Notice how the quadrant numbers 1 to 4 change hand when diamonds are reflected across any of the Cartesian planes.

To summarize, reflections are isometry transformations: a kind of transformation that preserve point distances. They are a simple way to replicate geometry that is symmetrical to a Cartesian plane. However, if your application requires consistent handedness where point sequences define polygons, do not use reflection techniques; use rotations instead.

E.5   Antipodal Points

An antipode is a point diametrically opposite a point on the other side of the sphere; it is the point furthest away, yet still on the surface of the sphere. A line between any point and its antipode passes through the sphere's origin. Figure E.6 shows an example of two diamonds where every tessellation vertex in one is the antipode of the other.

Antipodal points have the same *xyz*-component magnitudes, but each has opposite signs. Unlike rotations or reflections, antipodes can only replicate geometry to the opposite side of the sphere. In this sense, antipodes are less flexible than reflections and rotations, but the technique can replicate geometry within certain limits and is easy to use.

Antipodal points have the same two drawbacks as reflections: the geometry is unique and polygon handedness is reversed. In Figure E.6, notice the position of the holes where

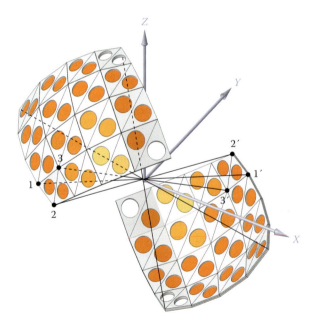

Figure E.6. Replicating geometry with antipodal points.

selected red surfaces have been omitted. In the lower antipodal diamond the holes are not in the same position as the reflected example, shown in Figure E.5, nor the simple rotation examples, shown in Figure E.3. The orientation of the antipode diamond is unique, demonstrates opposite reflection orientations, and cannot be replicated by any combination of rotations about the x-, y-, or z-axis. Notice, also, that the handedness of the polygons changes, as demonstrated by the vertex number sequences in the two triangles. The upper diamond is right-hand rule, while the lower one is *left-hand rule*, when viewed from outside the sphere.

Antipodal points can be useful shortcuts in certain situations. But in general, we do not recommend using them to replicate geometry on the sphere if orientation or handedness is important to your application.

E.6   Compound Rotations

Most of the time, a PPT or a diamond's coordinates must be rotated two or three times to properly position it elsewhere on the surface of the sphere. The sequence of rotations is critical because compound rotations are *not* commutative. A rotation around one axis and then another does not always result in the same positioning, if the sequence is reversed. We will use an edge-zenith diamond in Figure E.7(a) with some missing red surfaces to illustrate this concept. Again, we use the right-hand rule to specify rotation angle directions.

In column one, the edge-zenith diamond is initially rotated 90° around the y-axis and then again 45° around the z-axis. In column two, the same axis-angles rotations are used, but the sequence is reversed (commuted). The z-axis rotation is performed first, then the y-axis rotation. It is clear that the final rotated diamond positions (b) and (c) are not the same.

To undo a compound sequence of rotations, you apply the rotation sequence that created them in reverse order and use opposite rotation angles in each step. For example, to return the diamond in Figure E.7(c) to its original position at (a), rotate the diamond −90° around the y-axis, and then −45° around the z-axis.

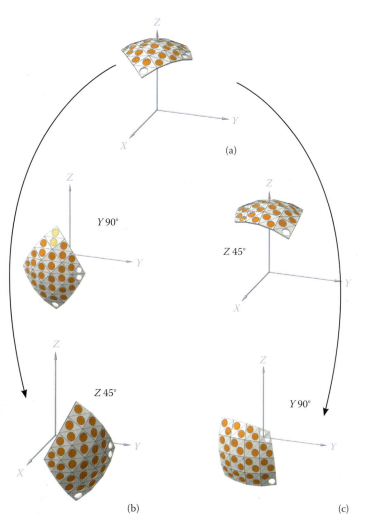

Figure E.7. Compound rotations are not commutative.

E.7   Rotation around an Arbitrary Axis

A PPT or LCD set of points can be repositioned anywhere on the surface of the sphere and maintain the handedness of any polygons within them with simple and compound rotations. A more intuitive and less error-prone approach for compound rotation cases might be to specify rotations around an arbitrary axis, not just the standard ones. This means you could define a new axis from the origin to any point on the sphere and then rotate any amount of geometry (PPT, diamond, etc.) around it. The procedure for doing this still relies on compound rotations, but excepting the arbitrary axis-angle specification, you do not have to know ahead of time what the other axis-angles are. This is because they are derived from the axis point you select.

Arbitrary axis rotation involves four steps. First, define the new axis by selecting a point on the surface of the sphere. The new axis is a line between the point and the sphere's origin. Second, the geometry you are rotating must be rotated twice in a way that makes the new axis coincident with the $+z$-axis. These rotation angles and axes will be described shortly. Third, the geometry is rotated around the z-axis, the specified positive or negative angle for the arbitrary axis. And last, the geometry is rotated twice more, opposite the order of the second step, to return the axis point back to its original position. This last sequence is the reverse of the rotations originally used to make the axis point coincident with the positive z-axis.

To repeat, a rotation around an arbitrary axis rotates the geometry to place the new axis coincident to the $+z$-axis. The geometry is rotated around the $+z$-axis the specified amount, and then rotated back so that the arbitrary axis is returned to its original position. This entire procedure sounds like a variant of compound rotations; it is. What is different is the simplicity of your original rotation specification. You define a convenient axis and rotation angle and let the computer/calculator handle the transformations.

Figure E.8 shows an example of arbitrary axis rotation. We want to copy-rotate the edge-zenith diamond to a new position immediately adjacent to one of its edges. The intended position is indicated by a half-tone diamond in (a). A convenient new axis is to use the edge-zenith diamond, point P, and to rotate the copy a negative angle around it. To do this, the new axis and a copy of the edge-zenith diamond is rotated the amount needed to bring the new axis coincident with the $+z$-axis, as in Figure E.8(b). The copied diamond is now rotated the specified negative angle around the z-axis (c). And finally the diamond is rotated back the amount needed to restore the arbitrary axis to its original position. The new diamond is now where we want it (d).

The idea behind arbitrary axis rotation is to use the axis point itself to provide the compound rotation specification needed to make the arbitrary axis coincident with the z-axis and then to return it to its original position.

But what angles are needed to make the arbitrary axis coincident to the positive z-axis and to reverse the sequence to return it to its original position? Only two angles are needed, and they are simply the spherical φ and θ angles of the selected axis point (see Section 4.15).

It couldn't be easier. The compound rotation sequence is always the same. Find the spherical φ and θ angles of the axis point. Rotate the geometry $-\theta$ around the z-axis and then $-\varphi$ around the y-axis. The arbitrary axis is now coincident with the $+z$-axis, and the geometry moves with it. Yet the points maintain their positions relative to one another and to the axis point. Now rotate the geometry around the $+z$-axis by the angle specified for the

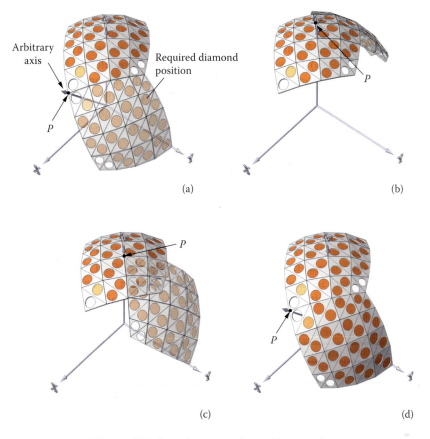

Figure E.8. Rotation around an arbitrary axis.

arbitrary axis. Next, apply the reverse rotation sequences used to make the arbitrary axis coincident. This moves the axis point back to its original location and brings the rotated geometry along with it. The reverse rotation sequence is $+\varphi$ around the y-axis, followed by $+\theta$ around the z-axis.

The convenience of arbitrary axis rotations is obvious, but what about those five rotation steps? They can all be combined into one single matrix and applied only once to the geometry. Five transformation matrices are combined by multiplying each one together: two will move the arbitrary axis coincident with the $+z$-axis, one rotates geometry around the z-axis, and two more return the arbitrary axis and rotated geometry to where the arbitrary axis was originally defined. The algorithm for arbitrary axis rotation is at the end of this appendix. Its simplicity might surprise you.

E.8   Polyhedra and Class Rotation Sequences

Simple and compound rotation sequences give us all the tools necessary to rotate a single PPT or diamond's vertices to all the positions needed to cover the sphere without overlap or gaps. These sequences preserve point distance and handedness of polygon definitions.

We will not use reflection techniques here for the reasons discussed earlier. The following section details the sequences for all the Classes and polyhedra covered in Chapter 8. We use 4^v subdivisions to make it easier to visualize rotation specifications (axis and angles) and their effects. Most sequences involve two rotations; a few require three. The matrix math techniques used are outlined at the end of this appendix. We begin with rotations for icosahedron Classes I and III schemas, the most frequently used in spherical subdivision.

E.9   Icosahedron Classes I and III

Our strategy for covering the entire sphere with a grid is simple. We make copies of the single PPT's (or LCD's) point set we developed in Chapter 8 and then rotate each copy to a unique place on the sphere's surface. We repeat the copy-rotate process until the entire sphere is covered and there are no gaps or overlaps. In effect, we derive all new PPT point sets by copying a previous point set and rotating it to cover a new area of the sphere's surface. For Class I and III subdivisions of the spherical icosahedron, this means we need 19 copy-rotation sequences of the original PPT. Recall that in every Class I schema presented, we oriented the PPT vertex-zenith and symmetric to the *yz*-plane. The orientation of the original subdivided PPT is important because it determines the angles of rotation required for copies to cover the rest of the sphere. Although our schema examples in Chapter 8 were 8^v grids with 64 grid points each, the copy-rotate sequence described here is the same for any number of points in the PPT.

Table E.1 summarizes the copy-rotations necessary. We start with one PPT point set, the one we computed in Chapter 8, and refer to it as PPT 1. Each table row defines how a

PPT	Ref PPT	Rotation 1			Rotation 2		
		X°	Y°	Z°	X°	Y°	Z°
6	1	0	0	180	0	116.5651	0
7	1	0	63.43495	0	0	0	36
16	1	0	-180	0	0	0	-144
2	1	0	0	72	0	0	0
3	6	0	0	72	0	0	0
4	7	0	0	72	0	0	0
5	16	0	0	72	0	0	0
8	1	0	0	144	0	0	0
9	6	0	0	144	0	0	0
10	7	0	0	144	0	0	0
11	16	0	0	144	0	0	0
12	1	0	0	216	0	0	0
13	6	0	0	216	0	0	0
14	7	0	0	216	0	0	0
15	16	0	0	216	0	0	0
17	1	0	0	288	0	0	0
18	6	0	0	288	0	0	0
19	7	0	0	288	0	0	0
20	16	0	0	288	0	0	0

Table E.1. Icosahedron Class I vertex-zenith PPT rotation constants.

new PPT is derived from a previously computed PPT. The new PPT's number is listed in column one and shown in Figure E.9. The new PPT is derived from a previously computed PPT whose reference number is listed in column two. The reference PPT must be copied and then rotated one or two times to establish the position of the new PPT. The angles (listed in decimal degrees) and direction of rotation (right-hand rule) around the x-, y-, or

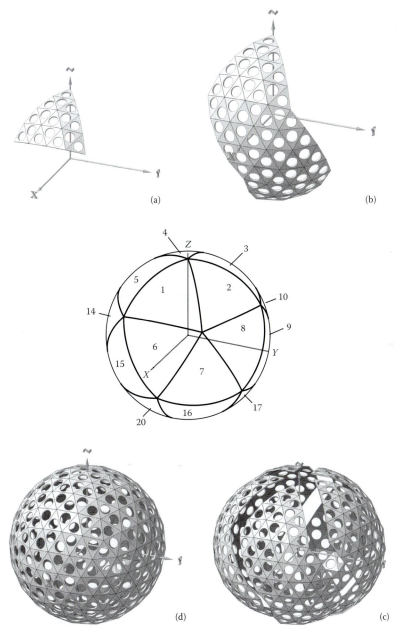

Figure E.9. Icosahedron Class I vertex-zenith PPT rotation sequences.

z-axis for one or two rotations are listed in the same table row. Keep in mind that in most computer applications, rotation angles must be specified in radians, thus the rotation angles listed in the table must be converted.

Let's take an example. The first row in Table E.1 specifies that PPT 6's point set (column-none) refers to a copy of PPT 1's point set (column two) that is rotated once around the *z*-axis 180° in a positive direction and then around the *y*-axis 116.5651° in a positive direction. Figure E.9 shows where these PPT references are located on the sphere and the images show rotation progress. When the copy-rotate sequence for PPT 6 is completed, two PPT point sets, our original PPT 1 and a new PPT 6 derived from it, cover the sphere.

The process for the remaining PPTs follows the same sequence. Each new PPT is a copy-rotate of a previous PPT, thus each new PPT must be found in the order specified in the table. Notice, too, that several PPTs—for example, PPTs 6, 7, 16, and 2—are all derived by copy-rotations of our original PPT 1's point set. Notice, too, that some PPTs further down the table only require a single rotation.

E.10   Icosahedron Class II

Icosahedron Class II rotations are a little more complex than their Class I and II counterparts, simply because more diamonds than PPTs are needed to cover the entire sphere—30

Dia	Ref Dia	Rotation 1 X°	Rotation 1 Y°	Rotation 1 Z°	Rotation 2 X°	Rotation 2 Y°	Rotation 2 Z°	Rotation 3 X°	Rotation 3 Y°	Rotation 3 Z°
2	1	0	0	-90	0	90	0	0	0	0
3	1	0	0	-90	-90	0	0	0	0	0
4	1	0	0	-90	0	-90	0	0	0	0
5	1	0	0	90	90	0	0	0	0	0
6	1	0	180	0	0	0	0	0	0	0
7	1	0	31.7175	0	0	0	108	0	31.7175	0
8	1	0	-31.71749	0	0	0	-108	0	-31.71749	0
9	1	0	-31.71749	0	0	0	108	0	-31.71749	0
10	1	0	31.7175	0	0	0	-108	0	31.7175	0
11	1	0	-31.71749	0	0	0	198	-58.28255	0	0
12	1	0	-31.71749	0	0	0	-144	0	31.7175	0
13	1	0	-31.71749	0	0	0	144	0	31.7175	0
14	1	0	-31.71749	0	0	0	-198	58.2826	0	0
15	1	0	31.7175	0	0	0	198	58.2826	0	0
16	1	0	31.7175	0	0	0	-144	0	-31.71749	0
17	1	0	31.7175	0	0	0	144	0	-31.71749	0
18	1	0	31.7175	0	0	0	-198	-58.28255	0	0
19	7	0	180	0	0	0	0	0	0	0
20	8	0	180	0	0	0	0	0	0	0
21	9	0	180	0	0	0	0	0	0	0
22	10	0	180	0	0	0	0	0	0	0
23	11	0	180	0	0	0	0	0	0	0
24	12	0	180	0	0	0	0	0	0	0
25	13	0	180	0	0	0	0	0	0	0
26	14	0	180	0	0	0	0	0	0	0
27	15	0	180	0	0	0	0	0	0	0
28	16	0	180	0	0	0	0	0	0	0
29	17	0	180	0	0	0	0	0	0	0
30	18	0	180	0	0	0	0	0	0	0

Table E.2. Icosahedron Class II edge-zenith diamond rotation constants.

in all. The strategy we will use is to define five diamonds at each of the orthogonal positions. Table E.2 shows the angle-axis sequences for 2–5. Diamond 6 copy-rotates the original edge-zenith diamond to the opposite side of the sphere. Figure E.10(b) shows the positions of the first six diamonds.

With the orthogonal diamonds in position, an upper semihemisphere of 12 diamonds (7–18) is formed (see Figure E.10(c)). These diamonds require three rotation sequences,

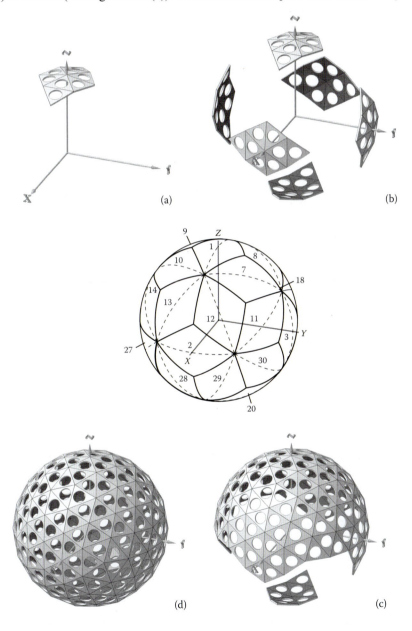

Figure E.10. Icosahedron Class II edge-zenith diamond rotation sequences.

each based on the original edge-zenith diamond 1. The icosahedral symmetry helps us significantly, which is why the specifications in Table E.2 have a symmetrical angle-axis pattern.

With the upper hemisphere diamonds in position, the entire set is simply copy-rotated 180° around the y-axis to complete the lower hemisphere diamonds 19–30.

When all rotations are completed, diamond 1 will have been copy-rotated a total of 29 times to cover the entire sphere.

E.11 Octahedron Classes I and III

Class I and II octahedra are the easiest rotation sequences of any polyhedra because each PPT is one-eighth of a sphere and all rotations are some multiple of 90°. Once again, we will use the gore strategy and define two PPTs (1 and 5) that span from the north to the south pole of the sphere. We will then copy-rotate the gore around the z-axis 90°, 180°, and 270° times to define 1 and 6, 3 and 7, and finally 4 and 8. Table E.3 and Figure E.11 show the sequence.

The gore strategy is quite useful and is used often in geodesics, especially in CAD assemblies (see Chapter 10 for additional examples).

PPT	Ref PPT	Rotation 1			Rotation 2		
		X°	Y°	Z°	X°	Y°	Z°
5	1	0	0	180	0	180	0
2	1	0	0	90	0	0	0
6	5	0	0	90	0	0	0
3	1	0	0	180	0	0	0
7	5	0	0	180	0	0	0
4	1	0	0	270	0	0	0
8	5	0	0	270	0	0	0

Table E.3. Octahedron Class I vertex-zenith PPT rotation constants.

E.12 Octahedron Class II

An octahedron has 12 edges; thus, we will need 11 rotation specifications to copy-rotate the original edge-zenith diamond to cover the entire Class II sphere. The strategy we use

Dia	Ref Dia	Rotation 1			Rotation 2			Rotation 3		
		X°	Y°	Z°	X°	Y°	Z°	X°	Y°	Z°
2	1	0	90	0	0	0	0	0	0	0
3	1	0	180	0	0	0	0	0	0	0
4	1	0	270	0	0	0	0	0	0	0
5	1	0	45	0	0	0	90	0	-45	0
6	1	0	-45	0	0	0	-90	0	45	0
7	5	0	0	180	0	0	0	0	0	0
8	6	0	0	180	0	0	0	0	0	0
9	5	180	0	0	0	0	0	0	0	0
10	6	180	0	0	0	0	0	0	0	0
11	7	180	0	0	0	0	0	0	0	0
12	8	180	0	0	0	0	0	0	0	0

Table E.4. Octahedron Class II diamond-zenith rotation constants.

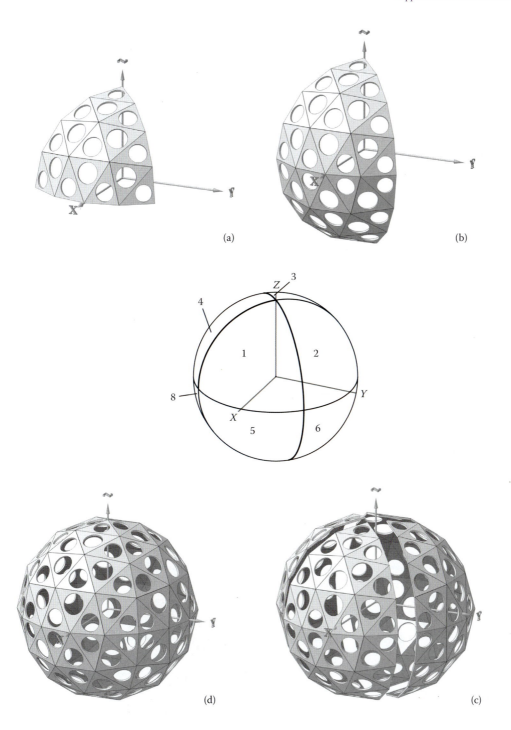

Figure E.11. Octahedron Class I vertex-zenith PPT rotation sequences.

here is to form a "ring" of diamonds around the sphere and then fill in the areas on either side. Table E.4 shows that the angles are increments of 90° and the rotations for the first three instances are around the *y*-axis. Figure E.12(a) and (b) show the ring being formed.

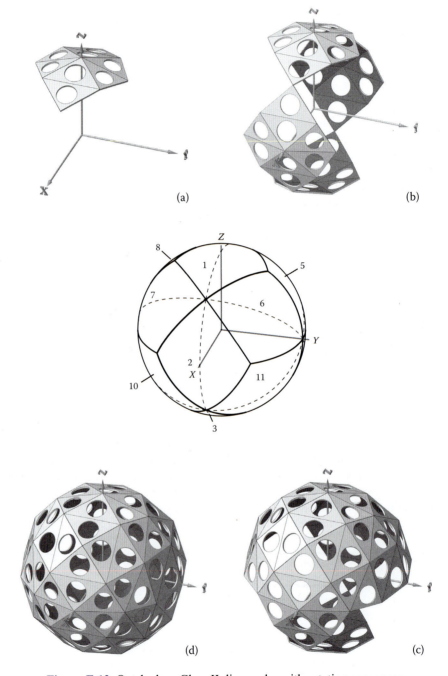

Figure E.12. Octahedron Class II diamond-zenith rotation sequences.

To fill in the areas on either side of the ring, diamonds 5 and 6 are rotated three times to complete two quadrants of the sphere. Diamonds 7 and 8 reference diamonds 5 and 6 and copy-rotate them 180° around the x-axis. This places diamonds 7 and 8 on the opposite side of the sphere and completes the upper hemisphere, as shown in (c).

The remaining quadrants in the lower hemisphere, diamonds 9 through 12, are copy-rotations of diamonds 5 through 8 from the upper hemisphere. This fills in the remaining gaps in the lower hemisphere. All rotations in this set are 180° degrees around the x-axis. When the entire sphere is covered in Figure E.12(d), diamond 1 will have been copy-rotated 11 times.

E.13   Tetrahedron Classes I and III

Class I and III tetrahedron rotations are the easiest and require the fewest rotation sequences of any subdivided polyhedra. Table E.5 lists the rotation sequences for PPTs 2, 3, and 4; each is a copy-rotation of the original PPT 1.

The vertex-zenith PPT 1, shown in Figure E.13(a), is copy-rotated around the z-axis 120° and 240° (b) to define PPTs 2 and 3. PPT 4 on the −z-axis is created by copying PPT 1 and rotating it 180° around the z-axis and then 250.5°. This angle is the spherical height of the PPT around the y-axis. Figure E.13(c) shows the relationship of the PPTs in the completed sphere (d).

	Ref	Rotation 1			Rotation 2		
PPT	PPT	$X°$	$Y°$	$Z°$	$X°$	$Y°$	$Z°$
2	1	0	0	120	0	0	0
3	1	0	0	240	0	0	0
4	1	0	0	180	0	250.5288	0

Table E.5. Tetrahedron Class I and III vertex-zenith PPT rotation constants.

E.14   Tetrahedron Class II

Class II tetrahedron rotations are quite simple. Only five rotations of the edge-zenith diamond are needed to cover the sphere. Except for the first one, they are all compound sequences with a similar pattern. Table E.6 shows the sequence and this repetitive pattern of angle-axis specifications. Notice that all diamonds are based on copy-rotations of diamond 1.

In this example, the initial edge-zenith diamond 1, shown in Figure E.14(a), is rotated 90° around the z-axis and then 180° around the x-axis, defining the bottom diamond

	Ref	Rotation 1			Rotation 2			Rotation 3		
Dia	Dia	$X°$	$Y°$	$Z°$	$X°$	$Y°$	$Z°$	$X°$	$Y°$	$Z°$
2	1	0	0	90	180	0	0	0	0	0
3	1	-90	0	0	0	-45	0	0	0	-45
4	1	-90	0	0	0	45	0	0	0	45
5	1	90	0	0	0	45	0	0	0	-45
6	1	90	0	0	0	-45	0	0	0	45

Table E.6. Tetrahedron Class II diamond-zenith rotation constants.

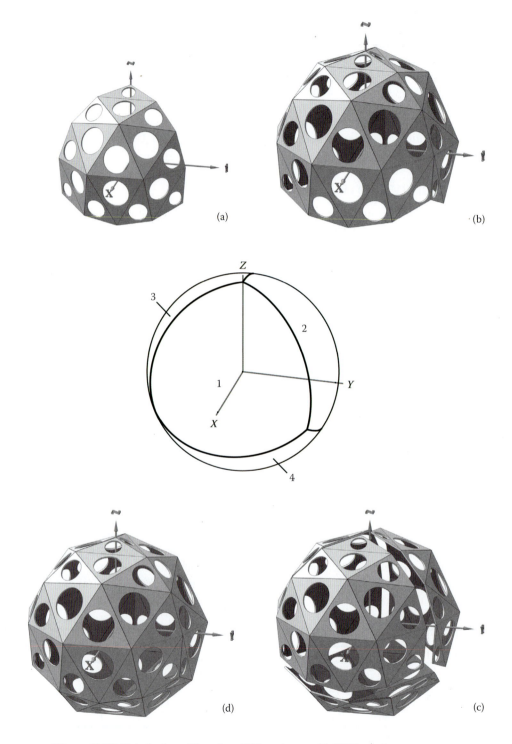

Figure E.13. Tetrahedron Class I and III vertex-zenith PPT rotation sequences.

PPT 2. It is shown in Figure E.14(b). The four diamonds that make up the middle of the sphere, diamonds 3–6, all have the same general rotation sequences, where diamond 1 is rotated 90° plus or minus around the x-axis, followed by plus or minus 45° around the y-axis, and then 45° around the z-axis.

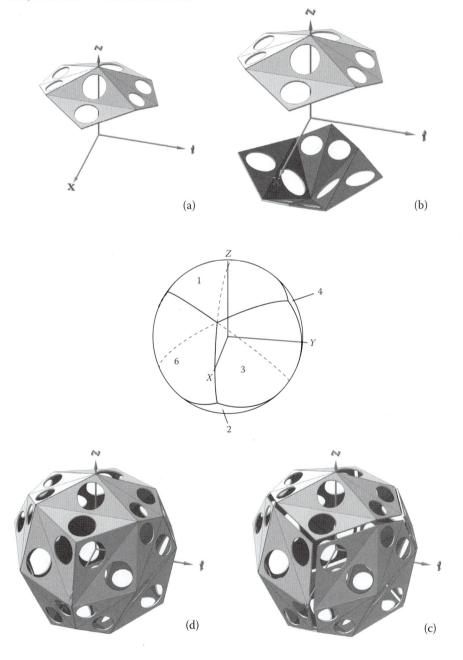

Figure E.14. Tetrahedron Class II diamond-zenith rotation sequences.

E.15   Dodecahedron Class II

Class II dodecahedra require more rotation sequences than any other subdivided spherical polyhedra. There are 30 edges in a dodecahedron; thus, it will take 29 copy-rotation specifications of the initial edge-zenith diamond to cover the rest of the sphere. The strategy we will use here is to define an upper hemisphere of diamonds and then copy-rotate the entire set to complete the lower hemisphere. A few diamonds are needed to span the equator, and we treat these separately.

Table E.7 shows the angle-axis sequences we need. Notice the repetitive pattern of rotation angles and axes for diamonds 2 through 13. Diamonds 14 to 17 span the equator and have 90° rotation angles. Notice, too, that all rotated diamonds thus far reference the original edge-zenith diamond.

Figure E.15 shows this sequence graphically. Notice that we build the upper hemisphere first and then copy it to the lower hemisphere. The lower hemisphere's 13 diamonds is easily specified by simply copy-rotating every upper hemisphere diamond that does not span the equator. The entire set is rotated 180° around the x-axis to complete the sphere.

Dia	Ref Dia	Rotation 1 $X°$	$Y°$	$Z°$	Rotation 2 $X°$	$Y°$	$Z°$	Rotation 3 $X°$	$Y°$	$Z°$
2	1	31.7170	0	0	0	0	72	-31.7170	0	0
3	1	-31.7170	0	0	0	0	-36	-31.7170	0	0
4	1	-31.7170	0	0	0	0	36	-31.7170	0	0
5	1	31.7170	0	0	0	0	-72	-31.7170	0	0
6	1	-31.7170	0	0	0	0	72	31.7170	0	0
7	1	31.7170	0	0	0	0	-36	31.7170	0	0
8	1	31.7170	0	0	0	0	36	31.7170	0	0
9	1	-31.7170	0	0	0	0	-72	31.7170	0	0
10	1	-31.7170	0	0	0	0	-18	0	58.2350	0
11	1	-31.7170	0	0	0	0	18	0	-58.2350	0
12	1	31.7170	0	0	0	0	-18	0	-58.2350	0
13	1	31.7170	0	0	0	0	18	0	58.2350	0
14	1	0	0	90	0	90	0	0	0	0
15	1	0	0	90	-90	0	0	0	0	0
16	1	0	0	90	0	-90	0	0	0	0
17	1	0	0	90	90	0	0	0	0	0
18	1	180	0	0	0	0	0	0	0	0
19	2	180	0	0	0	0	0	0	0	0
20	3	180	0	0	0	0	0	0	0	0
21	4	180	0	0	0	0	0	0	0	0
22	5	180	0	0	0	0	0	0	0	0
23	6	180	0	0	0	0	0	0	0	0
24	7	180	0	0	0	0	0	0	0	0
25	8	180	0	0	0	0	0	0	0	0
26	9	180	0	0	0	0	0	0	0	0
27	10	180	0	0	0	0	0	0	0	0
28	11	180	0	0	0	0	0	0	0	0
29	12	180	0	0	0	0	0	0	0	0
30	13	180	0	0	0	0	0	0	0	0

Table E.7. Dodecahedron Class II diamond-zenith rotation constants.

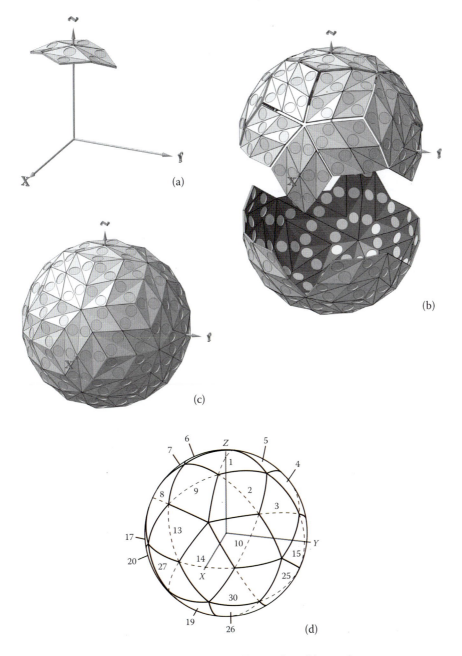

Figure E.15. Dodecahedron Class II diamond-zenith rotation sequences.

E.16   Cube Class II

The approach used to cover a Class II cube is similar to that used with the dodecahedron. The rotation sequences are shown in Table E.8 and illustrated in Figure E.16. The sequence

Dia	Ref Dia	Rotation 1			Rotation 2			Rotation 3		
		X°	Y°	Z°	X°	Y°	Z°	X°	Y°	Z°
2	1	0	0	35	0	60	0	0	0	35
3	1	-90	0	0	0	0	0	0	0	0
4	1	0	0	-35	0	-60	0	0	0	-35
5	1	0	0	35	0	-60	0	0	0	35
6	1	90	0	0	0	0	0	0	0	0
7	1	0	0	-35	0	60	0	0	0	-35
8	7	180	0	0	0	0	0	0	0	0
9	5	180	0	0	0	0	0	0	0	0
10	4	180	0	0	0	0	0	0	0	0
11	2	180	0	0	0	0	0	0	0	0
12	1	180	0	0	0	0	0	0	0	0

Table E.8. Cube Class II diamond-zenith rotation constants.

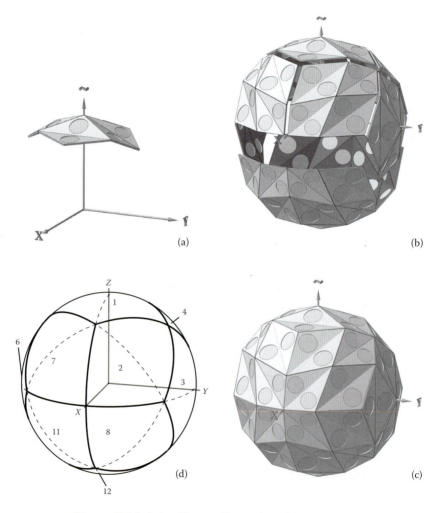

Figure E.16. Cube Class II diamond-zenith rotation sequences.

begins by copy-rotating the basic edge-zenith diamond 1 six times to create diamonds 2 to 7. This forms an upper hemisphere. Two diamonds, 3 and 6, span the equator. Once this is completed, the five diamonds that do not span the equator (1, 2, 4, 5, and 7) are copy-rotated 180° around the *x*-axis to complete the lower hemisphere's diamonds 8 through 12. All 12 diamonds are in place and cover the sphere perfectly.

E.17   Implementing Rotations

This book develops six spherical subdivision schemas. Appendix C presented the algorithms and steps necessary to subdivide a single principal polyhedral triangle (PPT) or least common denominator triangle (LCD). The result was a set of points stored in an array *xyz*[ ] [ ]. In order to cover the entire sphere, it is necessary to rotate copies of these grid points to specified locations elsewhere on the sphere. The first part of this appendix discussed the theory of rotations and listed the required rotation angles for each subdivision class and polyhedral foundation.

   In this section, we present the algorithms necessary to rotate gridpoints to the other parts of the sphere's surface. To rotate one or more grid points, we will multiply their *xyz*-coordinates by a 4 × 4 transformation matrix. It is assumed that the initial position of the PPT or LCD points is the recommended position described in the text. For illustration purposes, we also assume that the PPT or LCD grid points are stored in the same *xyz*[ ] [ ] of triplets we presented in Appendix C.

- Define the required rotation angles (in radians). Table E.1 through Table E.9 list the required angles and the sequence they are applied to a master rotation matrix to rotate an initial set of PPT or LCD points to all other positions for each of the schemas and polyhedral frameworks presented.

- Initialize a master 4 × 4 transformation matrix. Initialization requires setting it to the identity matrix. Function `midn()` performs the initialization.

- Define the required rotation angle sine and cosine values in the appropriate transformation matrix cells. Functions `xrot()`, `yrot()`, and `zrot()` perform this operation and multiply the set of points by the transformation matrix.

- Define any desired translations. A translation is simply moving a set of grid points along the *x*-, *y*-, or *z*-axis, no rotation is performed. None of the schemas developed in this book require translation; we show the algorithm to demonstrate the flexibility of transformation matrices. See function `mtrans()`.

- If a set of PPT or LCD points need to be rotated around more than one axis to achieve the required positioning, repeat the prior step for each additional axis rotation. One way to do this is to create a temporary 4 × 4 transformation matrix, initialize it (set it to the identity matrix), and then use `xrot()`, `yrot()`, or `zrot()` to define and perform each rotation one by one. A better approach, however, is to combine the separate rotations into a single 4 × 4 transformation matrix that accomplishes the same thing with just a single move. Matrix multiplication combines two transformation matrices into a single one capable of performing the combined operation. There is no limit to the number of matrices multiplied together to accomplish complex transformations;

matrix multiplication can be more efficient than performing a series of separate ones. Function `mcat()` multiplies two 4 × 4 matrices (sometimes matrix multiplication is called *concatenation*), creating a third 4 × 4 matrix with the combined specifications. As explained below, if you choose to multiply matrices, you must multiply them in the reverse order of use.

- Apply the final transformation matrix to the set of PPT or LCD grid points. In most applications, a copy of the original PPT or LCD points is made and only this set of points is transformed. Transformation is performed by function `xform()`.

E.18   Using Matrices

E.18.1   Identity

Before any transformation matrix can be used, the matrix must be initialized to the identity matrix. This sets matrix diagonal values to 1.0 and all others to 0.0. Always initialize transformation matrices before using them.

E.18.2   Specifying Angles

This book adopts the right-hand rule convention for the direction of angles. All angles are expressed in radians and rotations around the x-, y-, and z-axis. Use the following convention:

Axis of rotation	Direction of positive rotation
x	from y to x
y	from z to x
z	from x to y

E.18.3   Matrix Multiplication

Most PPT or LCD points need to be rotated around more than one axis to position them elsewhere on the sphere. Instead of performing each transformation separately, you can compute the necessary transformation matrix for each one and then combine them into a single matrix that has the combined rotational effect on the PPT or LCD point set. If you multiply matrices, you must do so in the *reverse* order in which they are needed. For example, if you need to rotate points around the x-axis and then the y-axis, you must define the y-axis rotation matrix first and then multiply it by the x-axis rotation matrix before you transform the PPT or LCD points.

Matrix multiplication is sometimes referred to as concatenation, a term that means joining together. We use the terms matrix multiplication and concatenation interchangeably in algorithm comments.

The sections that follow define the basic algorithms needed to define, multiply, and apply the rotational transformation to a set of PPT or LCD points.

E.19   Rotation Algorithms

The following C algorithms show how to initialize a transformation matrix, define transformations to rotating points around the x-, y-, and z-axis, translate points along the x-, y-, and *z*-axis, and then apply them to a set of points. You can define and apply these transformation matrices one at time or you can combine them into a single matrix and apply them to a set of PPT or LCD points once. The last algorithm is an example of how to use them.

E.19.1   Identity

```
//---------------------------------------------------------------------
// midn
// initialize the master transformation matrix mtx[][]. Diagonal cells set
// to 1.0,
// all other cells set to 0.0.
//
// Identity matrix
// +-         -+
// | 1 0 0 0 |
// | 0 1 0 0 |
// | 0 0 1 0 |
// | 0 0 0 1 |
// +-         -+
//
//---------------------------------------------------------------------
void midn (double mtx[4][4])
{
    int i, j;                    // general loop indices
    for (i=0;i<4;i++)            // for each row and
    for (j=0;j<4;j++) mtx[i][j] = 0.0;// column, set element to 0.0
    for (j=0;j<4;j++) mtx[j][j] = 1.0;// set matrix diagonal to 1.0
    return;
}
```

E.19.2   Rotation around the *X*-Axis

```
//---------------------------------------------------------------------
// xrot
// define a rotation around x-axis and concatenate
// it to the master transformation matrix, mtx[][]. Angle is expressed in
// radians.
//
// X Rotation Matrix
// +-              -+
// | 1   0    0  0 |
// | 0  COS>  SIN> 0 |
// | 0 -SIN>  COS> 0 |
// | 0   0    0  1 |
// +-              -+
//
//---------------------------------------------------------------------
```

```
void xrot (double angle, double mtx[4][4])
{
    double xm[4][4],temp[4][4];// working matrices
    midn(xm);                   // initialize working matrix to identity
    xm[1][1] = cos(angle); xm[1][2] = sin(angle);
    xm[2][1] = -sin(angle); xm[2][2] = cos(angle);
    mcat(mtx,xm,temp);          // concatenate master and x rotation
    mcpy(temp,mtx);             // copy temporary 4×4 into master matrix
    return;
}
```

E.19.3   Rotation around the *Y*-Axis

```
//------------------------------------------------------------------------
// yrot
// define a rotation around y-axis, concatenate it to the master
// transformation matrix, mtx[][].
//
// Y Rotation Matrix
// +-              -+
// | COS> 0 -SIN> 0 |
// | 0    1   0   0 |
// | SIN> 0  COS> 0 |
// | 0    0   0   1 |
// +-              -+
//
//------------------------------------------------------------------------
void yrot (double angle,double mtx[4][4])
{
    double ym[4][4],temp[4][4];// working matrices
    midn(ym);                   // initialize working matrix to identity
    ym[0][0] = cos(angle); ym[0][2] = -sin(angle);
    ym[2][0] = sin(angle); ym[2][2] = cos(angle);
    mcat(mtx,ym,temp);          // concatenate master and y rotation
    mcpy(temp,mtx);             // copy temporary 4×4 into master matrix
    return;
}
```

E.19.4   Rotation around the *Z*-Axis

```
//------------------------------------------------------------------------
// zrot
// define a rotation around z-axis, concatenate it to the master
// transformation matrix, mtx[][].
//
// Z Rotation Matrix
```

```
// +-             -+
// |  COS> SIN> 0 0 |
// | -SIN> COS> 0 0 |
// |  0    0   1 0 |
// |  0    0   0 1 |
// +-             -+
//
//----------------------------------------------------------------------
void zrot (double angle, double mtx[4][4])
{
    double zm[4][4],temp[4][4];// working matrices
    midn(zm);                     // initialize working matrix to identity
    zm[0][0] = cos(angle); zm[0][1] = sin(angle);
    zm[1][0] = -sin(angle); zm[1][1] = cos(angle);
    mcat(mtx,zm,temp);            // concatenate master and z rotation
    mcpy(temp,mtx);               // copy temporary 4×4 into master matrix
    return;
}
```

E.19.5 Translation

```
//----------------------------------------------------------------------
// mtrans
// translate (move) a point along the x-, y- and/or z-axis and concatenate
// transformation to the master transformation matrix
//
// Translation Matrix
// +-             -+
// |  1    0    0 0 |
// |  0    1    0 0 |
// |  0 .  0    1 0 |
// | T(X) T(Y) T(Z) 1 |
// +-             -+
//
//----------------------------------------------------------------------
void mtrans (double x_tr, double y_tr, double z_tr, double mtx[4][4])
{
    double xm[4][4],temp[4][4];// working matrices
    midn(xm);                     // initialize working matrix to identity
    xm[3][0] = x_tr;              // insert the translation x,
    xm[3][1] = y_tr;              // y and
    xm[3][2] = z_tr;              // z values into transform matrix
    mcat(mtx,xm,temp);            // concatenate master and x rotation
    mcpy(temp,mtx);               // copy temporary 4×4 into master matrix
    return;
}
```

E.19.6   Matrix Multiplication (Concatenation)

```
//-----------------------------------------------------------------------
// mcat
// concatenate two 4×4 transformation matrices, m1[][] and m2[][], into
// m3[][]
//
// m1 * m2 -> m3
//-----------------------------------------------------------------------
void mcat (double m1[4][4],double m2[4][4],double m3[4][4])
{
    int i, j, k;                    // local loop indices
    for (i=0;i<4;i++) {             // for every row
     for (j=0;j<4;j++) {            // and every column
     m3[i][j] = 0.0;                // use all in that row vs all
                                    // in that column
     for (k=0;k<4;k++) m3[i][j] += ( m1[i][k] * m2[k][j] );
     }
    }
    return;
}
```

E.19.7   Transform Points

```
//-----------------------------------------------------------------------
// xform
//
// transform a set of points defined in an array of triplets, xyz[*][3].
// the array of triplets is defined externally, for example
// double xyz[][3];
//
//-----------------------------------------------------------------------
void xform (int from, int to, double mtx[4][4]) // from to xyz range to
                                                // xform
{
    extern double xyz[6500][3];// reference external PPT or LCD points
    double a[4], b[4];         // homogenous matrix working areas
    int i, j, k;               // local loop indices
    for (i=from;i<=to;i++) {   // convert every PPT or LCD point
     a[0] = xyz[i][0];         // from a triplet to a 1x4 set
     a[1] = xyz[i][1];
     a[2] = xyz[i][2];
     a[3] = 1.0;               // add homogenous scale factor
     for (j=0;j<4;j++) {       // multiply point by 4×4 transform matrix
     b[j] = 0.0;               // all elements in that row vs all
     for (k=0;k<4;k++) b[j] += a[k] * mtx[k][j];// elements in that column
     }
     xyz[i][0] = b[0];         // update point coordinates with
```

```
    xyz[i][1] = b[1];              // resulting transformed version
    xyz[i][2] = b[2];
    }
    return;
}
```

E.20   An Example

This short code segment assumes that a set of PPT or LCD grid points have already been computed and now need to rotate the set to another position on the sphere. An actual implementation would copy them and perform the transformation on the copy, leaving the initial set of points unchanged.

```
int main()
{
    double xyz[2500][3];
    double vmtx[4][4];
    .
    .
    .
    // code to compute PPT or LCD grid points goes here
    // assume xyz[0][*] to xyz[125][*] have them

    midn(vmtx);                    // initialize master matrix to identity

// rotate points 0 to 125 around the y axis -15 degrees and then
// 20 degrees around the x. Note that the rotations are concatenated
// in reverse order needed.

    xrot(radian(20.),vmtx);       // concatenate x rotation to identity
    yrot(radian(-15.),vmtx);      // concatenate y rotation to master
    xlate(xTrans,yTrans,zTrans,vmtx);  // x,y and z translations
    // rotate points 1 to 125 using transformation matrix vmtx[][]
    xform(0, 125,vmtx);


    // PPT or LCD points 0 to 125 are now positioned

    return 0;
}
```

E.21   Summary

Rotations are angular specifications around the x-, y-, and/or z-axes. Angles can be specified in either the left-hand rule (LHR) or right-hand rule (RHR). This book uses the right-hand rule throughout. If more than one is to be performed, they can be accumulated by multiplying individual rotation matrices together and performed as a single transformation. This can be simple (a single angle-axis rotation) or complex where a sequence of angle-axis rotations is required. Compound rotations, however, are not commutative. A rotation around one axis followed by a rotation around another generally does not result in the same position if the sequence is reversed. Rotating a point any multiple of 360° around an axis does not change its position either. Rotated geometry can be restored to its original position by applying the reverse sequence of axis rotation and the opposite angle specification. Arbitrary axis rotation allows a single point to define a new axis, which can greatly simplify angle-axis specifications. Arbitrary axis rotation still relies on rotations around the orthogonal axis, but intermediate rotations are computed automatically.

Additional Resources

Penna, Michael A. and Richard R. Patterson. *Projective Geometry and Its Applications to Computer Graphics.* Englewood Cliffs, NJ: Prentice-Hall, 1986.

Van Verth, James M. and Lars M. Bishop. *Essential Mathematics for Games and Interactive Applications: A Programmer's Guide.* Amsterdam: Morgan Kaufmann, 2008.

Glossary

A

acute angle: An angle that is less than 90° but more than 0°; in other words, smaller than a right angle. *See also* obtuse angle.

acute triangle: A plane triangle in which all three angles are less than right angles (less than 90°).

alternate subdivision: A spherical subdivision method developed and improved on by Don Richter, Jeffrey Lindsay, and Duncan Stuart in the 1950s in which the triangular subdivision grid runs "parallel" to the sides of the Platonic face edges. The method is one of many similar techniques grouped together and called Class I subdivisions. *See also* Class I; Ford subdivision.

anaglyph stereo: A 2D display technique that provides a 3D visual effect. Two views of the sphere are presented. One is colored red and is the view displayed for the left eye; the other view is cyan and for the right eye. Wearing red-cyan glasses to separate the left-right views from each other, the viewer experiences a 3D visual effect.

anchor constraint: In computer-aided design, a type of mathematical transformation that fixes a 2D or 3D geometric entity so that it has no degrees of freedom; its position cannot be changed. *See also* angle constraint; coincident constraint; constraint; contact constraint; degrees of freedom (DOF); offset constraint.

angle constraint: In computer-aided design, a type of mathematical transformation between two geometric entities that forces them to always maintain a certain angular relationship with respect to each other. *See also* anchor constraint; coincident constraint; constraint; contact constraint; degrees of freedom (DOF); offset constraint.

angular radius: The angle to any point on the circumference of a lesser or greater circle from the circle's nearest pole measured from the origin of the sphere. The angular radius of a great circle is 90°.

angular separation: The arc distance between two points on intersecting great circles.

antipodal point(s): A pair of points that lie on opposite sides of the sphere. A line between them passes through the center of the sphere. They are perfectly opposite each other. For a 2D circle, they are the two ends of a diameter. *See also* antipodes.

antipodes: Points opposite each other on a sphere. A line between two antipodal points passes through the center of the sphere. *See also* antipodal point(s).

antiprism: A polyhedron composed of two parallel copies of a polygon in which the polygons are rotated with respect to one another so that their vertices are not lined up. An octahedron is an antiprism with triangular bases because any one of its faces is parallel to another opposing one, which is rotated so that the faces between them are triangles.

apices: Plural of apex.

apothem: For a circle, the apothem is the perpendicular distance from the midpoint of a chord to the circle's center. *See also* chord; inradius; sagitta.

arc angle: The angle of an arc measured from the origin of the circle or sphere's great circle.

arc factor: Radian measure of a great circle arc measured from the origin of a unit sphere. It can be converted into an arc length by multiplying it by the radius of the sphere. *See also* arc length.

arc length: Length of a portion of the circumference of a circle or of a great circle arc between two points on the surface of the sphere. *See also* arc factor.

Archimedean solids: The family of polyhedra composed of two or more kinds of regular polygons. There are 12 Archimedean solids.

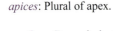

associative geometry: In computer-aided design, constructing 3D objects based on an axis system defined by some other 3D object. When geometric objects are associated, a transformation of one—such as rotations, translations, or scaling—affects the other.

astrolabe: An astronomical instrument with roots in antiquity used to solve problems related to time and the position of the sun and stars. Astrolabes were by far the most popular astronomical instruments in both Europe and the world of Islam until they were replaced by more accurate devices in the seventeenth century.[1] Important stereographic projection techniques developed for astrolabes are useful today for visualizing subdivided spheres.

axial angle: Angle formed between the radius of the sphere and the end of a chord between two points on a sphere.

azimuth: In the horizontal coordinate system, azimuth is one of the two coordinates. The other is altitude, sometimes called elevation above the horizon. Measured from the north, increasing clockwise (USA), or measured +/− from the south (Europe).

---

[1] Illustration courtesy of James Morrison.

B

barycenter (of a triangle): The center of gravity of a triangle. The point where the three medians of a triangle intersect, assuming that the triangle defines an area of uniform density. This same point is also called the triangle's centroid. *See also* centroid; circumcenter; incenter; orthocenter.

bigon: A two-sided polygon that can only be formed on a sphere. An alternative name for a digon or lune. *See also* biangle; digon; dihedron; hosohedron; lune.

bilateral symmetry: *See* reflection symmetry.

birectangular: A spherical triangle with two right angles.

Buckyball or *Buckminster Fullerene*: A large carbon-cage molecule discovered by Robert Curl, Harold Kroto, and Richard Smalley[2] in 1985 and named in honor of Buckminster Fuller because of its spherical geodesic-like form.

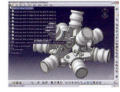

C

CAD: Acronym for computer-aided design.

Cartesian coordinates or *Cartesian rectangular coordinates*: Named after René Descartes (1596–1650), the Cartesian coordinate system describes any 2D point on a plane using two numbers, and any point in 3D space using three numbers, by using a set of two or three orthogonal axes, respectively, at right angles to one another and measuring along these axes.

Catalan solids or *Catalan polyhedra*: Polyhedra duals of Archimedean solids. They are convex solids with uniform faces. The faces are not regular polygons (squares, equilateral triangles, etc.). They are named after the French mathematician, Eugene Catalan, who first described them in 1865.

CATIA: Acronym for Computer-Aided 3D Interactive Application, a computer-aided design and analysis system developed by Dassault Systèmes, Vélizy Villacoublay, France. CATIA is a registered trademark of Dassault Systèmes or its subsidiaries in the United States and/or other countries.

central angle: An angle formed by two radii or an angle 0–180° between the endpoints of an arc of a circle. The angle is measured at the center of the circle. For great circles, this is the center of the sphere. *See also* arc factor; arc length.

centroid: In planar triangles, the point where the three medians of a triangle meet (line from the apex to the midpoint of the opposite edge). The centroid is always inside the triangle; it is also called the triangle's center of gravity or barycenter if the area defined by the triangle has uniform density. *See also* barycenter; circumcenter; incenter; orthocenter.

chiral: An asymmetry property of a polyhedron that cannot be made into its mirror image by rotations and translations alone. Your hands are chiral objects. No combination of rotations and

---

[2] Carbon, C_{60}, was discovered in 1985 by scientists Robert Curl, Harold Kroto, and Richard Smalley. They were awarded the Nobel Prize in Chemistry in 1996 for discovering this new class of compounds.

translations of one hand will produce the other. *See also* chiral polyhedra; enantiomorph; mirror images; mirror symmetry.

chiral polyhedra: Polyhedra that have two distinct forms, mirror images, or enantiomorphs of each other. The snub cube and snub dodecahedron are chiral polyhedra. *See also* chiral; enantiomorph; mirror images; mirror symmetry.

chord: Straight-line segment joining two points on the circumference of a circle or on the surface of the sphere. *See also* inradius.

chord factor: The chord length between two points on the surface of a unit sphere. The straight-line distance between any two points on a sphere is equal to $2r \sin(\theta/2)$, where r is the radius of the sphere and θ is the angle between the two points measured from the sphere's center. *See also* chord length.

chord distance or *chord length*: The straight-line distance between any two points on a sphere. It is always less than or equal to the sphere's diameter. *See also* chord factor.

circle packing: A convention where circles of the same diameter are centered over points in a 2D plane or 3D spherical surface. The radius of all circles is the same and the largest possible without causing any circles to overlap.

circlespheres: A characterization by artist Kenneth Snelson as "an organization of identical, non-overlapping small circles on a sphere."[3] Snelson employs circlespheres with polyhedral forms from circlespheres made from magnetic rings and wheels in his sculptures.

circumcenter: The point where the three perpendicular bisectors of a triangle meet. The point falls inside acute triangles, outside obtuse triangles, and at the midpoint of the hypotenuse of right triangles. A circle centered at the circumcenter can be tangent to all vertices. *See also* barycenter; centroid; circumcircle; incenter; orthocenter.

circumcircle: The circle that passes through all three of the triangle's vertices. *See also* circumcenter.

circumsphere: If all the vertices of a polyhedron lie on a sphere and the sphere's origin is also the origin of the polyhedron, then the sphere is said to circumscribe the polyhedron, and it is called its circumsphere.[4] The five Platonic solids (tetrahedron, octahedron, cube, dodecahedron, and icosahedron) can be perfectly circumscribed by a sphere; its radius is called the circumradius. *See also* incenter; incircle; inradius; insphere; midsphere.

Class I: A term coined by Joseph Clinton[5] to describe the broad family of subdivision grids that run approximately "parallel" to the edges of the principal triangle they subdivide. Class I subdivisions can be any positive frequency. *See also* Class II; Class III; Equal-arcs; Equal-chords; Ford subdivision; frequency; Mid-arcs; principle polyhedral triangle (PPT); skew subdivision; triacon subdivision.

Class II: A term coined by Joseph Clinton to describe the broad family of spherical subdivisions in which the subdivision grid runs approximately perpendicular to the edges of the principal tri-

[3] (Heartney 2009, 110)
[4] (Cromwell 1997, 52)
[5] (Clinton 1971, 106)

angle they subdivide. The triacon subdivision is a Class II grid. Class II subdivisions are limited to even frequencies. *See also* Class I; Class III; Equal-arcs; Equal-chords; Ford subdivision; frequency; Mid-arcs; principle polyhedral triangle (PPT); skew subdivision; triacon subdivision.

Class III: A broad family of spherical subdivisions related to the snub dodecahedron. Class II spherical subdivision grids are not "parallel" or perpendicular to the edges of the principal triangles they subdivide. The Class III term was first used by Magnus Wenninger. They can be any positive frequency. *See also* Class I; Class II; Equal-arcs; Equal-chords; Ford subdivision; skew subdivision; triacon subdivision.

climate modeling: Computer intensive mathematical simulation of weather and climate considering heating and cooling of the atmosphere, temperature, moisture, air, land and water masses, and the dynamics, physics, and chemistry of their interactions.

climates: Alternative term for an astrolabe's plate or tympan. Also, the division of the earth by Claudius Ptolemy (ca. 150 AD) into bands defined by the length of the longest day. *See also* astrolabe; limb; mater; plate; tympan.

closure deficit or *closure defect*: René Descartes' (1596–1650) law of closure defect states that for any convex polyhedron, the sum of the face angles about any vertex is always less than 360°, and that the sum of the defects for all vertices of the polyhedron equals 720°.

coincident constraint: In computer-aided design, a type of mathematical transformation between two geometric entities in two or three dimensions that forces them to share a common place. *See also* anchor constraint; angle constraint; constraint; contact constraint; degrees of freedom (DOF); offset constraint.

colunar: When the two sides of a lune are cut by an arc of a great circle, two spherical triangles are formed and they are said to be colunar to one another. *See also* congruent triangles; lune.

commutative: Mathematical property where two variables (scalars, matrices, etc.) can be used in arithmetic operations in either order (i.e., $AB = BA$).

complementary angle or *complement*: Two angles are complementary angles if the sum of their degree measurements equals 90°. Each of the complementary angles is said to be the complement of the other. *See also* supplementary angles.

complete truncation: A polyhedral truncation that cuts through the mid-edge of every edge that meets at that vertex. *See also* truncate; uniform truncation.

concatenate: To join together or combine. Separate matrix operations like rotation, translation, and scaling can be combined (concatenated) into a single matrix by multiplying individual matrices together. The single matrix performs all transformations simultaneously. The order in which matrix multiplication is performed is important.

conformal mapping or *conformal projection*: A type of projection that preserves angles and the fidelity of shape. The angle between any two lines on the sphere must be the same between their projected counterparts on a 2D map. For example, a globe's reference lines of latitude and longitude would cross at 90° on a conformal map of that globe. *See also* stereographic projection.

conformality: Cartographic measure of how accurately a map projection maintains the shapes of places like land masses, rivers, lakes, and so on. An ideal projection would introduce no distortion at all. *See also* conformal mapping; stereographic projection.

congruent: Identical in size and shape.

congruent spherical triangles: Two or more spherical triangles with identical surface angles in the same order when viewed from outside the sphere; they can be overlayed perfectly. There are no similar triangles on a sphere—they are either congruent (identical in size and shape) or they are different. *See also* congruent; spherical triangles; symmetric spherical triangles.

conjecture: A mathematical statement that appears likely to be true, but has not been formally proven or rigorously tested and proved to be true by the rules of logic.

constraint: As used in computer-aided design, a mathematical transformation applied to limit one or more geometric relationships between two 3D objects. For example, the relative positions of two objects may be constrained to ensure that they are always a certain distance apart or angle between them. Common constraints are contact, distance, angle, or axis-aligned. Constraints can be applied to points, edges, faces, or surfaces. *See also* degrees of freedom (DOF).

contact constraint: In computer-aided design, a type of mathematical transformation between two geometric entities that forces them to always remain in geometric contact. *See also* anchor constraint; angle constraint; coincident constraint; constraint; degrees of freedom (DOF); offset constraint.

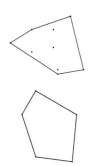

convex hull: In two dimensions, it is the outermost boundary surrounding a set of points and analogous to a rubber band stretched tight around the set of points. All the points within the boundary are contained in the convex hull. In three dimensions, the convex hull is the smallest convex polyhedron that entirely contains a set of points in three dimensions.

convex polygon: A polygon in which all internal angles are less than 180°. A line between any two vertices defining the polygon is either tangent to one of its edges or totally within the polygon.

convex polyhedron: A polyhedron is convex if every dihedral angle is less than 180° and a line connecting any two (noncoplanar) points on the surface always passes through the interior of the polyhedron. Convex polyhedra include the Platonic and Archimedean solids.

coplanar: Lying in the same plane.

cross-stereo: In the context of stereo graphics, your eyes' lines of sight are crossed and are focused on a point in front of the image.

cubic close packing (CCP): An arrangement (or lattice) of identical spheres occupying the greatest amount of volume in three dimensions; that is, they are packed as densely as possible. There are several ways to pack or stack spheres so as to fill up the most space.

Cundy-Rollett symbols: A system for summarizing a polyhedron's regular face types that surround any vertex order. The Cundy-Rollett symbol {3.5.3.5} means that every vertex of this solid is surrounded by alternating triangles and pentagons. *See also* Schläfli symbols.

curvature: In describing circles, curvature is the reciprocal of the circle's radius ($1/r$). A circle's curvature is large when r is small and small when r is large.

D

decagon: A polygon with ten sides.

degree: 1/360th of a circle. *See also* radian; solid angle; spherical degree.

degrees of freedom (DOF): The number of independent forces necessary to completely restrain a body in space. In computer-aided design, there are six degrees of freedom between any two rigid bodies: three translational and three rotational. The two bodies are perfectly constrained if there are exactly six degrees of constraint. Less constraint means the bodies can move relative to one another. *See also* constraint.

deltahedron (pl. *deltahedra*): Named by H. Martyn Cundy and A. P. Rollett,[6] a deltahedron is a polyhedron with faces made only of congruent equilateral triangles. There are an infinite number of deltahedra, but only eight are convex. The tetrahedron, octahedron, and icosahedron are the most familiar deltahedra.

density factor: The ratio of the total area of all the spherical caps that surround subdivision grid points to the surface area of the whole sphere. All caps must have the same and largest possible radius without causing the caps to overlap.

digon: A two-sided polygon that can only be formed on a sphere. An alternative name for a lune or bigon. *See also* bigon; hosohedron; lune.

dihedral angle: Angle formed by two planes meeting in a common line. The angle is measured in a plane perpendicular to the line. *See also* lunar angle; surface angle.

dodecagon: A polygon with 12 sides.

dual polyhedron: The dual of a polyhedron has a vertex for each face and a face for each vertex of the original polyhedron. Duals can be planar or spherical. A dual polyhedron has as many vertices as the other polyhedron has faces and vice versa. The symmetry of a dual polyhedron is the same as that of the original polyhedron.

duals or *duality*: A mapping between the vertices, faces, and edges of two polyhedra. For example, the face centers of a cube define the vertices of an octahedron. *See also* dual polyhedron.

---

[6] (Cundy and Rollett 1961)

Dymaxion: Buckminster Fuller's trademark word meaning "doing more with less" was coined by marketer Waldo Warren in the late 1920s to promote an exhibit featuring Fuller's revolutionary house design. *See also* synergy.

Dymaxion projection: A type of cartographic projection onto a Platonic solid such as the cuboctahedron or icosahedron developed by Buckminster Fuller. *See also* gnomonic projection; Mercator projection; orthographic projection; perspective projection; polar stereographic projection; polar aspect projection; stereographic projection.

E

edge: Any boundary line or side of a polygon or polyhedron's face.

edge-zenith: Orientation of a polyhedron where one of its edges is the uppermost part. Thus, a top view of the polyhedron is looking directly at an edge. The zenith edge is normal to the *z*-axis in the Cartesian coordinate system. *See also* face-zenith.

enantiomer: Two geometric objects that are mirror images of each other but are not identical in that they cannot be perfectly superimposed on one another; that is, no rotation or translation of one will create the other. Your hands are enantiomers: they are mirror images of each other with a left- and right-handed version.

Equal-arcs: A Class I subdivision method that tessellates a PPT based on equally spaced reference points along the PPT's arc sides. *See also* Class I; Class II; Class III; Equal-chords; Ford subdivision; frequency; principle polyhedral triangle (PPT); skew subdivision; triacon subdivision.

equal-area: In cartography, a characteristic of a projection of spherical areas in two dimensions that preserves area relationships. Some projections, such as the Mercator map projection, greatly enlarge the relative size of landmasses near the poles (Greenland and Antarctica), making their projected size far larger relative to other places on the globe with similar or equal area.

Equal-chords: A Class I spherical subdivision method that tessellates a PPT based on equally spaced reference points along the PPT's chord sides. *See also* Class I; Class II; Class III; Equal-arcs; Ford subdivision; frequency; Mid-arcs; principle polyhedral triangle (PPT); skew subdivision; triacon subdivision.

equator: A great circle where every point on it is equidistant from the great circle's two poles.

Euler line: A line that passes through the orthocenter, centroid, and circumcenter of a triangle. *See also* centroid; circumcenter; orthocenter.

Euler's law: A law that states that in 3D space divided into polyhedral cells, the number of vertices minus the number of edges plus the number of faces minus the number of polyhedral cells is equal to one. Euler's theorem for any polyhedron states that the number of polygons (P), plus the number of vertices (V), equals the number of edges (E) plus two. Stated algebraically: $P + V = E + 2$. René Descartes also discovered this same relationship between a polyhedron's polygons, vertices, and edges. Some scholars call the law the Descartes-Euler formula.

F

face: A polygon planar surface.

face angle: *See* interior angle.

face normal: A vector (direction) that is perpendicular to the plane of a polyhedron's face. *See also* normal vector.

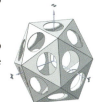

face-zenith: Orientation of a polyhedron where one of its faces is the uppermost part. Thus, a top view of the polyhedron is looking directly at a face. The zenith face is normal to the z-axis in the Cartesian coordinate system. *See also* edge-zenith.

fill surface: In CAD, a minimum surface in three dimensions spanning a bounded perimeter.

five-fold symmetry: A symmetry commonly found in nature (flowers, shells, fruit). The icosahedron has five-fold symmetry; from an initial position, four more rotations are possible about each pair of its six vertex axes. After rotation, the object is invariant—its appearance remains the same. *See also* icosahedral symmetry.

Ford subdivision: A spherical subdivision similar to the 1950s Ford Rotunda Dome subdivision and one of several Class I subdivisions in which the resulting tessellation grid runs approximately parallel to the edges of the faces of the polyhedra it is based on. This method is limited to deltahedra, spherical polyhedra composed only of equilateral faces such as the tetrahedron, octahedron, or icosahedron. *See also* Class I; Class II; Class III; Equal-arcs; Equal-chords, frequency; Mid-arcs; principle polyhedral triangle (PPT); skew subdivision; triacon subdivision.

frequency: The number of times the edges of a spherical polyhedra are segmented by a tessellation grid or the number of subdivisions per edge. Some notations use the Greek lower case nu (v) to represent frequency.

Fullerenes: *See* Buckyball.

fundamental region: *See* Schwarz triangle.

G

geodesic: The shortest distance on the surface of a sphere between two points. A geodesic is an arc segment of a great circle.

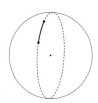

geodesic notation: A shorthand notation introduced by H. M. S. Coxeter to describe geodesic dome grids.[7]

geodesy: The study of the shape of the earth and the definition of exact positions of geographical points.

glide reflection: A plane symmetry combination that results from a reflection and a translation in either sequence. *See also* center of rotation; mirror; plane; rotation; reflection.

---

[7] (Coxeter 1971)

gnomonic projection: Also called central projection, the gnomonic projection is a nonconformal map projection obtained by projecting points onto the surface of the sphere from a sphere's center. *See also* Dymaxion projection; Mercator projection; orthographic projection; perspective projection; polar stereographic projection; polar aspect projection; stereographic projection.

golden ratio: A very important irrational number that reoccurs in nature and many number series. Often symbolized as phi or the Greek letter φ, it is the ratio of $(a + b)/a = a/b = \varphi$. Its approximate value is 1.6180339887... and can be computed from $\left(1+\sqrt{5}\right)/2$. *See also* golden section.

golden section or *golden rectangle*: A rectangle with sides whose lengths have the ratio 1:φ where φ is the golden ratio, approximately equal to 1.6180339887.... *See also* golden ratio.

gore: A series of 2D polygons, each defined on their own plane, that form a strip that approximates a 3D surface. Subdivision triangles approximate the surface of the sphere. A contiguous series of triangles can wrap around the sphere. One can "peel off" gores of triangles, much like peeling the skin off an orange. *See also* lune.

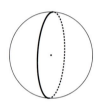

great circle: A circular line formed on the surface of a sphere by a plane that passes through the center of a sphere. Examples include the earth's equator and lines of longitude. A great circle through any point also passes through its antipodal point. *See also* geodesic; great circle arc; great circle distance; great circle route; lesser circle.

great circle arc: A segment of a great circle. *See also* geodesic; great circle; great circle distance; great circle route.

great circle distance: The shortest distance between any two points on a sphere. *See also* geodesic; great circle; great circle arc; great circle route.

great circle route: A route that follows the shortest distance between any two points on a sphere. *See also* geodesic; great circle; great circle arc; great circle distance.

greatest common factor (GCF): Also called the greatest common divisor (GCD) or highest common factor (HCF) of two nonzero integers, it is the largest positive integer that divides two other unequal integers without a remainder.

H

hemisphere: Half of a sphere formed when a plane passes through its center.

hemispherical assemblies: In computer-aided design, a geometric model of exactly half a full sphere and may itself be an assembly of smaller geometric subassemblies. The symmetry of its design features may determine if two of them can be joined to make a symmetrical spherical model.

hexagon: A polygon with six sides.

hexahedron: A polyhedron with six faces. The cube with its six regular square faces is one type of hexahedron. A pent-cap, more precisely the pentagonal pyramid, one of the Johnson solids, is also a hexahedron with five equilateral triangles and one pentagonal face.

hexavalent: *See* valence.

hosohedron: A degenerate spherical polyhedron made of two-sided spherical polygons. A hosohedron is only possible on a sphere and resembles a beach ball. *See also* bigon; digon; lune.

I

icosahedral symmetry: Designated (I), objects with icosahedral symmetry remain invariant (appear the same) within three sets of rotations: four about six pair of opposite vertices, two about ten pairs of opposite faces, and one about fifteen pairs of opposing edges. When identity and reflection symmetry are considered, the objects will have 120 symmetries. *See also* five-fold symmetry; identity symmetry; mirror symmetry; octahedral symmetry; reflection symmetry; rotation symmetry; symmetry; symmetry detection; symmetry group; tetrahedral symmetry.

icosahedron: A Platonic solid composed of 20 equilateral triangular faces, 12 vertices, and 30 edges. *See also* Platonic solids.

icosahedron's 31 great circles: A system of fundamental great circles based on the symmetry of the icosahedron. Fifteen pass through opposing edges, ten pass through the midpoints of each edge (and none pass through any of the icosahedron's vertices), and six pass through the mid-edges of adjacent faces. A key set of great circle relationships at the core of Buckminster Fuller's synergetic geometry.

identity symmetry: Sometimes called the "do nothing" transformation, a symmetry where a body is rotated 360° about any axis leaving it invariant (its appearance is unchanged). All objects have identity symmetry, even asymmetric objects; it has the same effect in symmetry transformations as 0 when adding numbers. *See also* five-fold symmetry; icosahedral symmetry; mirror symmetry; octahedral symmetry; reflection symmetry; rotation symmetry; symmetry; symmetry detection; symmetry group; tetrahedral symmetry.

incenter: The intersection point of two or all three angle bisectors of a triangle. The incenter is always inside a triangle and is also the center of an inscribed circle. *See also* incircle; inradius; insphere.

incircle or *inscribed circle* (of a triangle): The largest circle that can be constructed inside a triangle. It is tangent to all three sides. *See also* incenter; inradius; insphere.

inradius: The radius of a triangle's incircle or a polyhedron's insphere. *See also* incenter; incircle; insphere.

insphere or *inscribed sphere*: A sphere inscribed in a polyhedron such that the surface of the sphere is tangent to the center or centroids of faces on the polyhedron. *See also* circumsphere.

instance: In computer-aided design, a logical link in a 3D model to geometry stored in a catalog. Instanced geometry in the model references the catalog geometry with a logical link and provides a transformation (rotation, translation, etc.) to position the catalog geometry at the desired location within the model. Instance geometry is not a copy of catalog geometry, thus minimizing model size in memory. If the catalog geometry is updated later, all model instances (references to it) automatically update by way of their logical links.

interior angle: For a polygon, the angle between these two sides measured from their common vertex.

interstices: In the context of geometry, the spaces (areas or volume) between things. In circle or sphere packing, it refers to the areas or volumes between packed, nonintersecting circles or spheres. *See also* circle packing; sphere packing.

inverse-square law: A physical quantity like the intensity of light on a surface or the repulsion strength of two like-charged particles is inversely proportional to the square of the distance from the source of that quantity. Double the distance to a source of light, and only one quarter the amount of light will reach the subject.

isomeric: Polyhedra that have the same number and types of faces, but arranged in different ways. For example, an icosidodecahedron's edges define six great circles. The two halves of the polyhedra on either side of any great circle can be rotated 36° relative to each other, thus defining new polyhedra.

isometric grid: *See* triangular grid.

isometry: A transformation that preserves distance between points. Reflections, rotations, translations, and glide reflections are isometry transformations. *See also* reflection; rotation; translation; glide reflection.

isoperimetric quotient: In three dimensions, a mathematical measure of how spherical a polyhedron's surface is. In two dimensions, it is the ratio of the area enclosed by a curve to the area of a circle with the same perimeter. *See also* planarity; symmetry detection.

isosceles triangles: Triangles with two sides of equal length and a third side—sometimes called the base—of a different length. The two interior angles common to the base are equal.

isotropic vector matrix: Term used by Buckminster Fuller to refer to cubic close packing. *See also* cubic close packing (CCP).

J

Johnson solids or *Johnson polyhedra*: Named after Norman W. Johnson who enumerated them, their faces are regular polygons, but not uniform overall. There is no requirement that each face or a Johnson solid must be the same polygon, or that the same polygons join around each vertex. An example is the square pyramid, a single square face, and four equilateral triangles.

K

Kepler solids or *Kepler-Poinsot solids*: Four regular concave polyhedra that have intersecting face planes made of regular concave polygons.

kissing: A term used in billiards to describe two balls that just touch each other but do not interfere. *See also* kissing number.

kissing number: Refers to the largest possible number of equal-size spheres that can touch a given sphere in *n*-space (dimensions) without any intersections. In the most restrictive use, the kissing number refers to the maximum number of unit spheres that can touch the surface of another unit sphere. *See also* kissing; sphere packing; spherical code.

kissing distance: In the context of spherical subdivision, the maximum radius of spheres of the same radius centered at each grid vertex that just touch and do not interfere.

knowledge rules: In computer-aided design, a method of defining or detecting geometric relationships between objects. Knowledge rules can reflect design experience, regulations such as safety codes, or how objects are manufactured. CAD programs can use knowledge rules to suggest design choices, to automate geometry creation, or to insure that geometry comply with specifications.

knowledgeware: In the context of computer-aided design, they are computer program functions that automate certain design tasks or check a design for compliance to some predefined standard or accepted practice.

L

least common denominator (LCD): A term popularized by Buckminster Fuller in the context of spherical subdivision, LCD refers to the smallest Schwarz triangle tile that can cover the entire sphere a finite number of times without overlap. *See also* Schwarz triangle.

left-hand rule (LHR): In reference to geometric rotations of points or objects in space, the analogy is the left hand grasping the axis of rotation with the thumb pointing in the positive axis direction. The fingers of the left hand curl around the axis, indicating the positive direction of rotation. *See also* right-hand rule (RHR).

lesser circle or *small circle*: A circular line on the surface of a sphere formed by the intersection of a plane through the sphere that does not pass through the center of the sphere. The latitude reference lines on a globe are lesser circles except for the equator, which is a great circle. *See also* great circle.

limb: The outside ring of hour numbers and degree scale on an astrolabe. *See also* astrolabe; climates; mater; plate; tympan.

logical association link: See instance.

loxodrome or *loxodrome curve*: A line on the surface of a sphere that crosses all meridians at the same angle. It is the line on which a ship sails when her course is always in the direction of the same point of the compass. A loxodrome is a great circle only if it is along the equator or a meridian; any other rhumb line between two points is longer than the great circle route between the same two points. *See also* rhumb line.

lunar angle: Also called the dihedral angle, it is the angle between two intersecting great circles. *See also* dihedral angle.

lune: A spherical figure formed when two intersecting planes pass through its center. These planes subdivide the sphere into four lunes or biangles. All four are equal if the planes of the great circles are perpendicular; otherwise, two pairs of equal lunes are created. *See also* bigon; decagon; digon; gore.

M

macro(s): In computer-aided design, a macro is a user interface facility that allows users to record common sequences of key strokes, mouse movements, menu selections, or other CAD commands and save them in a library for future use. Once created, they can be invoked to perform whatever task they were designed to do. They can also be shared by other users. Macros greatly improve productivity, consistency of results, and reduce the chances of design errors. *See also* power copy.

mater: Latin for "mother," it's the main body of an astrolabe that provides storage for plates and features the limb that contains degree scales on all astrolabes and time scales on European instruments. *See also* astrolabe; climates; limb; plate; tympan.

medial triangle: Also called an auxiliary triangle, it is the triangle formed by joining the midpoints of the sides of a triangle. In two dimensions, the sides of the medial triangle are parallel to and half the length of the sides of the original triangle.

medials: A polyhedron generated from another polyhedron in which its vertices are defined by the mid-edge points of the other polyhedron. The cuboctahedron is a medial of the cube.

median (of a triangle): A line segment drawn from one vertex to the midpoint of the opposite side. The three medians of a triangle intersect at the triangle's centroid. *See also* barycenter; centroid.

Mercator projection: A type of cylindrical map projection of the earth's features. It is one of the most common map projections, especially for nautical purposes, because lines on it represent constant course directions. The course line is called a rhumb line or loxodrome. *See also* Dymaxion projection; gnomonic projection; loxodrome; orthographic projection; perspective projection; polar stereographic projection; polar aspect projection; rhumb line; stereographic projection.

meridian: Also called longitude, meridians are great circles through the two poles of a sphere.

meridian plane: A plane that passes through two poles and another point on the sphere. The meridian plane defines a great circle. In astronomy or celestial navigation, the observer's position and the earth's two poles define his meridian plane.

Mid-arcs: A Class I subdivision that tessellates a PPT by recursively defining medial triangles within other medial triangles. Only even-numbered frequencies are possible. *See also* Class I; Class II; Class III; Equal-arcs; Equal-chords; Ford subdivision; frequency; medial triangle; medials; principle polyhedral triangle (PPT); skew subdivision; triacon subdivision.

midsphere: Also called an intersphere, it is a sphere whose origin is the center of the polyhedron and whose surface intersects its edges at their midpoints or some other symmetrical point on every edge. The midsphere radius is called the midradius. *See also* circumsphere; insphere; inscribed sphere.

mirror images: Two images, with one appearing to be the reflection of the other in a mirror. For 3D shapes, the mirror can result in left- and right-hand versions of a form. *See also* chiral; chiral polyhedra; enantiomorph; mirror symmetry.

mirror symmetry: Also called mirror-image symmetry, reflection symmetry, or bilateral symmetry. In two dimensions, there is an axis of symmetry. In three dimensions, there is a plane of symmetry. The object or figure is indistinguishable from its reflected image. Two-dimensional examples include the reflection of a circle, a line, or any other regular polygon. The Platonic solids are 3D examples. *See also* five-fold symmetry; icosahedral symmetry; identity symmetry; octahedral symmetry; reflection symmetry; rotation symmetry; symmetry; symmetry detection; symmetry group; tetrahedral symmetry.

N

n-gone: A polygon with *n* number of sides.

nadir: The point on a sphere directly opposite the zenith. *See also* pole; zenith.

Napier's rule: A set of formulas developed by John Napier (1550–1617) for solving right spherical triangles.

net: The 2D pattern of the faces of a polyhedron. The net of some polyhedra is a single pattern that can be folded to form the polyhedron. In more complex cases, several patterns may have to be combined.

network: The pattern that results from lines connecting discrete points on a plane or in space.

nonuniform rational basis splines (NURBS): A type of mathematical representation for complex curves and surfaces, particularly free-form curves and surfaces. Product designers who use computer-aided design systems often use NURBS mathematics to represent sculptured surfaces on products the same as the contours of automobile bodies or consumer electronics.

normal: Perpendicular to. A line normal to a plane is perpendicular to the plane and everything lying on the plane.

normal vector: A mathematical representation that has direction and a length of one. If a vector is "normalized," its length becomes one, but its direction remains the same. *See also* face normal.

nu (ν): Greek symbol commonly used to designate the frequency of spherical subdivision, the number of segments the side of the principal spherical triangle is divided.

O

oblique triangle: A triangle that does not contain a right angle.

obtuse angle: An angle that is greater than 90° but less than 180°. *See also* acute angle.

obtuse triangle: A triangle with one angle greater than a right angle.

octahedral symmetry: Designated (O), objects with octahedral symmetry remain invariant (appear the same) within three sets of rotations: three about three pairs of opposite vertices, two about four pairs of opposite faces, and one about six pairs of opposing edges. When identity and reflection symmetry are considered, the objects will have 48 symmetries. *See also* five-fold symmetry; icosahedral symmetry; identity symmetry; mirror symmetry; reflection symmetry; rotation symmetry; symmetry; symmetry detection; symmetry group; tetrahedral symmetry.

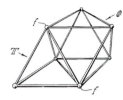

octet truss: A 3D structural system made of octahedron-tetrahedron units. An assemblage of octahedrons and tetrahedrons in a face-to-face relationship. Thus, when four tetrahedrons are grouped to define a larger tetrahedron, the resulting central space is an octahedron; together, these figures are comprised in a single or "common" octahedron-tetrahedron system.

offset constraints: In computer-aided design, a type of mathematical transformation between two geometric entities that forces them to always remain at a given distance from each other. *See also* anchor constraint; angle constraint; coincident constraint; constraint; contact constraint; degrees of freedom (DOF).

orthocenter: The point where the three altitudes of a triangle intersect. The point falls inside acute triangles, outside obtuse triangles, and on right triangles. *See also* barycenter; centroid; circumcenter; incenter.

orthogonal: Any two lines, edges, faces, planes, and so forth that are oriented 90° to each other.

orthographic projection: In computer-aided design, a projection convention for viewing 3D objects in two dimensions. For example, imagine a 3D object place in a box and each side of the box is a viewing screen of what is inside. The sides of the box are orthogonal (at right angles) to one another and all viewing lines are perpendicular to the screens. Standard orthographic views are the top, front, and sides of an object. *See also* Dymaxion projection; gnomonic projection; Mercator projection; perspective projection; polar stereographic projection; polar aspect projection; stereographic projection.

P

packing: An assemblage of forms, usually assembled with the objective of eliminating spaces between forms.

parallax: The difference in apparent position or direction of an object as seen from different viewpoints. *See also* perspective projection; stereographic projection.

parallelogram: A quadrilateral with two pairs of parallel sides.

pentagon: A polygon with five sides.

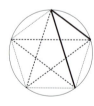

pentagonal cap: A reference to the five equilateral triangles that cluster around any of the vertices of an icosahedron.

pentagram: Pentagonal star where the ratio of the lengths between any short and long line is the golden ratio.

pentavalent: *See* valence.

Persistence of Vision Raytracer (POV-Ray): An open-source free computer program for producing very high-quality 3D graphic displays. POV-Ray is available for most computer operating systems; see http://www.povray.org.

perspective projection: A 2D representation (drawing, geometry, etc.) of objects in three dimensions that give the visual illusion of three dimensions. *See also* Dymaxion projection; gnomonic projection; Mercator projection; orthographic projection; parallax; polar stereographic projection; polar aspect projection; stereographic projection.

phi (φ): Greek symbol often used to represent the golden ratio 1.618… in mathematics or the angle to the *z*-axis in spherical polar coordinates.

planarity: A mathematical measure of how much a face deviates from a perfect plane. Planarity is used to detect symmetry in polyhedra or subdivisions of them. *See also* isoperimetric quotient.

plate: The part of the astrolabe with stereographic projection of the local horizon and circles of altitude and azimuth. *See also* astrolabe; climates; limb; mater; stereographic projection.

Platonic solids: Also called the regular solids or regular polyhedra, they are convex polyhedra with equivalent faces composed of congruent convex regular polygons. There are exactly five such solids[8] —the cube, dodecahedron, icosahedron, octahedron, and tetrahedron, as was proved by Euclid in the last proposition of the *Elements*.

polar aspect projection: Also called polar normal aspect, a form of planar map projections in which the center of the map projection is a pole and the plane of projection is tangent to the sphere at that pole. Parallels (lines of latitude) are represented as concentric circles and meridians (lines of longitude) are straight lines with true angles radiating from the pole point in the map projection center. Different types of polar projections result from the choice of projection point that may be within the sphere or at the opposite pole. *See also* Dymaxion projection; gnomonic projection; Mercator projection; orthographic projection; perspective projection; polar stereographic projection; stereographic projection.

polar distance: Arc distance on the sphere between the point and either the north or south pole.

polar stereographic projection: A mapping projection where the North or South Pole is the central point. It is the only polar 2D projection that is conformal, that is, accurately represents local shapes. *See also* Dymaxion projection; gnomonic projection; Mercator projection; orthographic projection; perspective projection; polar aspect projection; stereographic projection.

polar triangle: A spherical triangle whose three angular points are the poles of the sides of another spherical triangle. The two spherical triangles are each other's polar triangle. *See also* pole; polar triangle; spherical triangle; symmetric triangle.

pole: A point on the surface of the sphere and on a line that is normal to the plane of the great circle at the origin. *See also* nadir; polar triangle; spherical triangle; zenith.

polygon: From the Greek *poly* (many) and *gwnos* (angle), a finite, closed, 2D area bounded by edges and vertices. Polygons always have as many angles as sides. The sum of the interior angles of a simple polygon is $(n-2)\pi$ radians, which is also $(n-2)180°$, where *n* is the number of sides.

polyhedral packing: A packing of one kind of polyhedron in 3D space where the packing arrangement fills all space and leaves no holes or gaps. Identical sized cubes are space-filling polyhedra.

[8] (Steinhaus 1999, 252–256)

polyhedron (pl. *polyhedra*): A finite, connected set of plane polygons, in which every side of each polygon belongs to just one other polygon.[9] *See also* Platonic solids; regular polyhedra.[10]

POV-Ray: *See* Persistence of Vision Raytracer (POV-Ray).

power copy: In computer-aided design, a facility to record the characteristics of geometry and its arrangement by using variable geometry that is sensitive to where the geometry is placed and how it relates to geometry around it. In effect, power copy's geometry "adapts" its final form according to where it is placed and it proximity to other geometry. Power copies are one of the most powerful features of CAD systems and greatly improve productivity, consistency of results, and error reduction. *See also* macro(s).

primitive or *primitive circle*: In stereographic projection, the primitive is the equatorial projection plane where the stereogram is created. The projection pole axis is perpendicular to the primitive and tangent to its center. Points on the surface of the sphere are projected to the primitive from one or the other of the primitive's poles. *See also* stereographic projection.

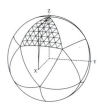

principle polyhedral triangle (PPT): A selected equilateral face of a spherical polyhedron that serves as the computational area for developing the subdivision grid. Once subdivided, copies of the PPT's grid are positioned, through rotation transformations, to cover the rest of the sphere without causing gaps or overlaps. The initial orientation of the PPT (vertex-, edge- or face-zenith) can greatly simplify subdivision mathematics and depends on the desired subdivision class. *See also* Class I; Class II; Class III; Equal-arcs; Equal-chords; Ford subdivision; frequency; Mid-arcs; skew subdivision; triacon subdivision.

product data management (PDM): In computer-aided design, a reference to managing and publishing product data resulting from design activities as data evolves from requirements and specifications, to early conceptual design and matures in refined functional and detail design. Often, it includes manufacturing and production data as well. Managing data includes access and change control, translation to different formats, publication, communication to different user communities, backup, and long-term archiving.

publication: In computer-aided design, a restrictive meaning is the naming and access to collaborative design team members to common geometry in a 3D design. For example, in complex geometry, many points, lines, or planes from different geometric objects may overlap or have the same position. Publication is a way to name specific geometry so that designers can correctly reference the one they intend. Publication can give access to certain parts of design geometry without disclosing other details, which may be proprietary. Publication can prevent users from changing geometry and make it read- or reference-only. *See also* reference model.

Q

quadrantal triangle: A spherical triangle that has at least one side equal to a "quadrant" or an arc equal to 90°.

---

[9] (Coxeter 1973)
[10] Robert Webb has developed a comprehensive glossary of polyhedral terms at www.software3d.com/Glossary.php#p.

quadrilateral: A polygon with four sides.

quasi-regular polyhedra: A polyhedron with faces from two sets of regular polygons arranged so that one type of face is surrounded by the other type. Quasi-regulars have equal dihedral angles and equal sides. One Platonic solid, the octahedron (two pair of equilateral triangles), and two Archimedean solids, the cuboctahedron (square and triangle sets) and the icosidodecahedron (triangle and pentagon sets), are quasi-regular.

quaternary subdivision: A subdivision of a triangle created by joining the midpoints of each side to create four equilateral subtriangles on a 2D plane.

R

radian or *rad*: A unit-free angular measurement based only on the radius of a circle, the angle made by taking the radius and wrapping it along the edge of the circle. One radian is equal to $360°/2\pi$. One radian equals approximately $57° 17' 44.8062''$ or $57.2957795°$. *See also* degree.

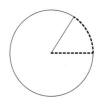

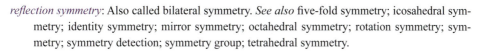

radome: A geodesic dome enclosing a radar antenna.[11]

random subdivision: A subdivision method sometimes used in radar domes to reduce or eliminate parallel or near-parallel structural members. The regular surface pattern of geodesics causes radar signal loss and deflects the signal direction when leaving the antenna. Despite their appearance, random subdivisions are typically based on selective use of vertices and chords from highly uniform subdivisions of spheres.

reference model: In computer-aided design, a 3D model that defines geometric entities such as points, lines, and planes or surfaces that designers will use to build their own geometry from. Often the reference model can only be changed by systems administrators. Geometric features may be published. *See also* publication.

reflection: A geometric transformation of coordinates that produces the mirror set (the reflection) on the opposite side of a line in two dimensions or a plane in three dimensions. All points in the reflection maintain their same perpendicular distance from the line or plane as the original point. Reflection is mathematically analogous to looking at your image in a mirror. The reflected image is chiral or opposite-handed. *See also* chiral; mirror symmetry; reflection symmetry.

reflection symmetry: Also called bilateral symmetry. *See also* five-fold symmetry; icosahedral symmetry; identity symmetry; mirror symmetry; octahedral symmetry; rotation symmetry; symmetry; symmetry detection; symmetry group; tetrahedral symmetry.

regular polygon: A closed 2D area having all edges straight and equal in length and all vertex angles identical.

regular polyhedra: Polyhedra composed of only one type of regular polygon (polygons with equal edge length and face angles). The five Platonic solids—tetrahedron, octahedron, cube (hexahedron), dodecahedron, or icosahedron—are regular polyhedra.

---

[11] FOX-Main Radome 1962; photo courtesy of Brian Jeffrey, www.VE3UU.com.

rete: Movable part of the astrolabe that contains pointers to the stars.[12] *See also* azimuth; astrolabe; stereographic projection.

rho (ρ): Greek symbol often used in mathematics to represent the radius in the system of spherical polar coordinates.

rhombus: A quadrilateral parallelogram with equal sides. *See also* parallelogram; quadrilateral; trapezoid.

rhumb line or *rhumb*: A line of constant bearing on a sphere or on a map of a sphere. It cuts every meridian at the same angle. *See also* loxodrome.

right-hand rule (RHR): In reference to rotations, the analogy is the right hand grasping the axis of rotation with the thumb pointing in the positive axis direction. The fingers of the right hand curl around the axis, indicating the positive direction of rotation. *See also* left-hand rule (LHR).

rotation: A geometric transformation in which one or more geometric objects (lines, points, faces, planes, surfaces, etc.) are rotated around an axis to a new position in 3D space. The relation of the objects to one another is not affected by the rotation. *See also* reflection; scaling; translation.

rotational symmetry: A figure has rotation symmetry if it can be rotated around a center point by less than 360° and the result appears unchanged. For example, a square can be rotated three times about its center and remain identical to its original. *See also* glide reflection; mirror symmetry; reflection.

S

sagitta: The perpendicular distance from an arc's midpoint to the chord across it. The sagitta is equal to the radius minus the apothem. *See also* apothem; chord; inradius.

scaling: Uniform scaling is a geometric transformation in two or three dimensions in which all objects are uniformly increased or decreased in size. The distances between points changes accordingly, but angular relationships between points are preserved. *See also* reflection; rotation; translation.

Schläfli symbols: A symbolic notation of the form $\{p,q,p,...\}$ named after the Swiss mathematician Ludwig Schläfli, who in 1850 described regular polygons, polyhedra, and their higher-dimensional counterparts. A cube, for example, has a Schläfli symbol $\{4,3\}$, which means the solid is bounded by four-sided polygons where three meet at each vertex. *See also* Cundy-Rollett symbols.

Schwarz triangle: A spherical triangle that can be used to tile or cover the entire sphere through reflections and rotations a minimum number of times, leaving no gaps or causing overlaps. *See also* least common denominator (LCD).

semiregular polyhedra: Polyhedra in which all faces are regular polygons and their sequence around each vertex is the same. Examples include the snub cube and snub dodecahedron. Both of these polyhedra are chiral; that is, they each have a left-handed and right-handed version.

---

[12] (Morrison 2007, 411)

separation angle: The angle between two meridians at a given latitude measured from the origin of the sphere.

similar triangles: Plane triangles that have the same angles and sides that are proportional to one another. There are no similar triangles on a sphere.

skew subdivisions: *See* Class I; Class II; Class III; Equal-arcs; Equal-chords; Ford subdivision; frequency; Mid-arcs; principle polyhedral triangle (PPT); triacon subdivision.

small circle: *See* lesser circle.

snub: Refers to a chiral process where some polyhedra have two versions, a left- and right-handed form. Polyhedra examples include the left- and right-handed forms of the snub cube and snub dodecahedron. *See also* chiral.

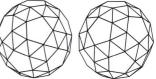

solid angle: The area or region defined by the projection of a surface onto a unit sphere. For a spherical triangle on the unit sphere, the solid angle is equal to the spherical excess of the triangle. *See also* spherical angle; spherical excess; steradians.

sphere: The locus of points in 3D space that are a fixed distance from a given point (called the center or origin). A sphere is an undevelopable surface that cannot be unfolded or flattened onto a plane without distortion.

sphere packing: An arrangement of nonoverlapping identical spheres that fill a given space. Generally applied to 3D Euclidean space, but the concept is generalized to 2D space with circles, or to *n*-dimensional space with hyperspheres. *See also* kissing number; spherical code.

spherical angle: *See* surface angle.

spherical cap: Portion of a sphere cut off by a plane resembling a contact lens. If the plane passes through the center of the sphere, the spherical cap is a hemisphere.

spherical code: Sometimes called sphere packing, the code refers to distributing *n* number of points on a sphere so as to maximize the minimal distance (or equivalently, the minimal angle) between them. *See also* kissing number; sphere packing.

spherical coordinates: A 3D coordinate system in which the position of a point is specified by three numbers: a radial distance from a fixed origin and two angles based on an axis perpendicular to a reference plane through the origin. The polar angle (also called the zenith angle) is measured from a fixed zenith direction on the perpendicular axis to the point. The azimuth angle is measured from a reference point on the plane, clockwise or counterclockwise, around the perpendicular axis to the point. *See also* azimuth; Cartesian coordinates; zenith.

spherical defect: Denoted (D), the difference between the sum of the radian side angles of a spherical triangle and 2π radians (360°). $D = 2\pi - (a + b + c)$. *See also* spherical excess.

spherical degree: 1/720th of the surface of any sphere. *See also* solid angle; spherical angle; steradians.

spherical excess: The difference between the sum of the surface angles of a spherical triangle and π radians or 180°. Excess is the solid angle of the spherical triangle measured in steradians. *See also* solid angle; steradians.

spherical polygon: Closed geometric figure on the surface of a sphere formed only by arcs of great circles.

spherical spiral: An alternative name for a loxodrome. *See also* loxodrome.

spherical triangle: A triangle on the surface of a sphere created by three great circle arcs intersecting pairwise in three vertices. Unlike plane geometry, three angles always define a unique spherical triangle. There are no similar spherical triangles on a sphere. They are either congruent or they are different. *See also* polar triangle; congruent triangles; symmetric triangles.

sphericon: A 3D solid with one side and two edges discovered by Colin Roberts of Hertfordshire, England.

spherist: One who is interested in or works with spheres.

square degree: An area unit based on a small spherical square one degree of arc on a side. *See also* steradians.

standard deviation: A statistical measure of how spread out data values are, a measure of dispersion from the average value or expected value.

Stella: A polyhedral modeling computer program by Robert Webb.[13]

stellation: Creating a new polyhedron from an existing one by extending its facial planes past the polyhedron's edges until they intersect and create another set of faces. Some polyhedra, such as the cube or tetrahedron, cannot be stellated because their extended face-planes never intersect to form closed volumes. Other polyhedra have more than one stellated form; the icosahedron, for instance, has 59. The new polyhedron is a stellation of the original one and often named after the original. The great stellated dodecahedron (shown at left) is one stellation of the dodecahedron.

steradians: A dimensionless unit of spherical area. The area of a unit sphere is 4π steradians. A steradian is the solid angle measured at the center of a sphere of radius r by a portion of the surface of the sphere having area r^2. The International System of Units (ISU) abbreviation for steradians is "sr." *See also* solid angle.

stereogram: A 2D projection (stereoprojection) of points on the surface of the sphere. The projection plane can be located at the equator or tangent to one of the sphere's poles. Spherical points are projected from the pole opposite the hemisphere that the spherical point is in. *See also* stereographic projection.

stereographic projection: A method for projecting points on a sphere onto a 2D plane. The plane of projection, called the primitive, can be located at several places and includes the equator or

---

[13] Image courtesy of Robert Webb.

tangent to one of the sphere's poles. Although any stereographic projection misses one point on the sphere (the projection point), the entire sphere can be mapped using two projections from opposite poles. A projected point in the primitive is the point of intersection of a ray from the spherical point to the pole in its opposite hemisphere. This projection is used extensively in the design of astrolabes. *See also* astrolabe; Dymaxion projection; gnomonic projection; Mercator projection; orthographic projection; perspective projection; plate; polar stereographic projection; polar aspect projection; stereogram; Wulff net.

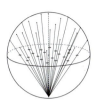

Stewart toroids: A polyhedron named after mathematician, B. M. Stewart, which has nonintersecting regular polygon faces that share an edge and are not coplanar but are "quasi-convex" (its convex envelope has no new edges).

supplementary angle or *supplement*: Two angles that add up to 180° or form a straight line are called supplementary angles. For example, 120° and 60° are supplements. *See also* complementary angles.

surface angle: Also called a spherical angle, the angle formed by the intersection of two arcs of great circles. A surface angle has the same measure as the dihedral angle formed by the planes of the two great circles. *See also* dihedral angle; lunar angle.

symmetric triangles: Two spherical triangles with identical angles but in opposite order when viewed from outside of the sphere. Two triangles whose vertices are antipodal are symmetric triangles. *See also* spherical triangle.

symmetry: An aesthetic and mathematical perspective. In mathematics, the study of symmetry identifies transformations of an object (polyhedra, spherical subdivisions, etc.) such as rotations, translations, and reflection, which leaves the object invariant, that is, looking exactly the same in shape, position, and orientation as it did before. For example, rotations, a symmetry operation, rotates an object about an axis defined by its faces, vertices, or edges. After the rotation, the object looks the same as it did before although its faces, vertices, and edges have been moved by the rotation. *See also* five-fold symmetry; icosahedral symmetry; identity symmetry; mirror symmetry; octahedral symmetry; reflection symmetry; rotation symmetry; symmetry detection; symmetry group; tetrahedral symmetry.

symmetry detection: A sophisticated capability in 3D geometry computer programs to detect symmetry within a geometric object by analyzing the relationships of edges, planarity of faces, and vertices to one another. *See also* five-fold symmetry; icosahedral symmetry; identity symmetry; isoperimetric quotient; mirror symmetry; octahedral symmetry; planarity; reflection symmetry; rotation symmetry; symmetry; symmetry group; tetrahedral symmetry.

symmetry group: The collection of all symmetry-preserving transformations that leave an object (polyhedra, subdivided sphere, etc.) invariant, that is, looking exactly the same in shape, position, and orientation as it did before. Rotation is a common symmetry-preserving transformation. *See also* five-fold symmetry; icosahedral symmetry; identity symmetry; mirror symmetry; octahedral symmetry; reflection symmetry; rotation symmetry; symmetry; symmetry detection; tetrahedral symmetry.

synergetic geometry: An invention of Buckminster Fuller, a design framework based on geometric relationships derived from nature and science. Synergetic geometry is sometimes used as a metaphor, a conceptual vehicle, for moving ideas from one context to another.

synergetics: A term coined by Buckminster Fuller to express systems of transformation and behavior based on the interaction of components not predicted by the behavior of individual components; the whole is greater than the sum of its parts. Today, the term is used in many fields such as design and humanistic thinking where creative and nontraditional problem framing, analysis, and solutions are emphasized. *See also* synergetic geometry; synergy.

synergy: A term defined by Buckminster Fuller: "The behavior of a system as a whole unpredicted by its parts."[14]

T

Tammes problem: The problem of finding a point set to maximize the minimum angle between the points. Named for the famous Dutch 1930s biologist, P. M. L. Tammes, to commemorate his work on the distribution of pollen grains in 1930.

tensegrity: A structural system composed of continuous-tension and discontinuous-compression members. The tensegrity concept was originally conceived by sculptor Kenneth Snelson[15] and later applied to entirely new spherical and linear truss construction by Buckminster Fuller.[16]

tessellation: Filling in mosaics with tiles or the repetition of a shape or pattern covering a plane or sphere without any gaps or overlaps. *See also* tiling.

tetrahedral symmetry: Designated (T), objects with tetrahedral symmetry remain invariant (appear the same) within two sets of rotations: two about four axes (vertex to opposite-face), and one rotation about the center of three pairs of opposite edges. When identity and reflection symmetry are considered, the objects will have 24 symmetries. *See also* five-fold symmetry; icosahedral symmetry; identity symmetry; mirror symmetry; octahedral symmetry; reflection symmetry; rotation symmetry; symmetry; symmetry detection; symmetry group.

tetrahedron: A regular or uniform Platonic solid composed of four equilateral triangles. *See also* Platonic solids.

tetravalent: *See* valence.

tiling: The repetition of a shape or pattern covering a plane without any gaps or overlaps. Three types of regular polygons—squares, equilateral triangles, or hexagons—can cover a 2D surface without overlaps or gaps. *See also* tessellation.

translation: A mathematical transformation along an axis of one or more objects in three dimensions. Only the absolute position of the objects in space changes; the relative position of the objects to each other does not change. *See also* reflection; rotation; scaling.

trapezoid: A quadrilateral with two parallel sides.

triacon subdivision: A method developed by Duncan Stuart for minimizing the number of different chords when subdividing spherical Platonic solids. The method results in grids that run perpen-

[14] (Fuller 1961, section 2, ara. 20)
[15] Image courtesy of Kenneth Snelson.
[16] (Fuller 1962)

dicular to the base polyhedral edges. Triacon subdivisions are one of several Class II methods. *See also* Class I; Class II; Class III; Equal-arcs; Equal-chords; Ford subdivision; frequency; Mid-arcs; principle polyhedral triangle (PPT); skew subdivision.

triangular grid: Also called an isometric grid, it is formed by a regular grid of equilateral triangles.

triangulation number (T): Devised by Michael Goldberg in the 1930s,[17] this number is the intensity of triangulation that results from subdividing a triangle. The formula $T = b^2 + bc + c^2$ is based on the orientation and intensity of the subdivision triangular grid. The grid orientation and frequency is specified by two coordinate parameters b,c specification. The T number is a convenient way to find the total number of faces, edges, and vertices that result from the polyhedron and grid specification combination. *See also* valence.

trirectangular triangle: A triangle with three right angles.

trivalent: *See* valence.

trixel: A single spherical triangle within a hierarchical triangular mesh, a technique for subdividing spheres by recursively subdividing triangles into smaller ones.[18]

truncate: In polyhedral transformations, to cut off a vertex, edge, or face of a polyhedron. Many polyhedra are formed by truncating another. *See also* complete truncation; uniform truncation.

tympan: Alternative name for an astrolabe's plates. *See also* astrolabe; climates; limb; mater; plate.

U

undevelopable surface: A surface that cannot be "flattened" out. The surface of a sphere is undevelopable; it cannot be flattened out on a plane without distortion.

uniform truncation: A polyhedral truncation that cuts every edge that meets at a vertex at a distance *less than* the mid-edge point. In most cases, all resulting edges after the truncation are equal. *See also* truncate; complete truncation.

unit circle: A circle where the radius is equal to one. *See also* unit radius; unit sphere.

unit radius: A circle or sphere where the radius is equal to one. *See also* unit circle; unit sphere.

unit sphere: A sphere with a radius of one. *See also* unit circle; unit radius.

V

valence: In the context of spherical subdivision, valence is the number of grid members (chords or arcs) that meet at a single grid point. Also called degree, in polyhedra it is the number of edges that meet at a vertex. A point or vertex could be univalent (1), divalent (2), trivalent (3), tetravalent (4), pentavalent (5), hexavalent (6), octavalent (8), etc. Pentavalent and hexavalent grid points are highlighted in the figure to the right.

---

[17] (Goldberg 1937)
[18] (Szalay 2005)

variance: See standard deviation.

vector equilibrium: Alternative name given to the cuboctahedron by Buckminster Fuller, who recognized that the distances between adjacent vertices are the same as between each vertex and the center of the polyhedron.

vertex transitive: A polyhedron is vertex transitive if any vertex can be carried to any other vertex by a symmetry operation.[19]

vertical circle: *See* azimuth.

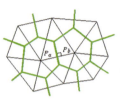

Voronoi cells: Named after Georgy Voronoi and sometimes called a Dirichlet cell, it is a distance cell (two shown in green to the right) in which all the points in the cell are closer to a given point (P_a and P_b in the diagram) than to any other given point. A drawing or schematic is called a Voronoi diagram.[20] Voronoi cells are the dual of triangular cells.

W

Waterman polyhedra: A type of polyhedron defined by the convex hull of a lattice point set where all points in the set are within the range of a given distance from the origin. The point set is defined by the centers of close packed spheres with a diameter of √2. One sphere in the set is centered at the origin of the Cartesian space. *See also* convex hull; cubic close packing (CCP).[21]

window: A term, first used by Duncan Stuart, to describe a small spherical triangle that results when great circle arcs almost, but do not quite, intersect at a single point.[22] In spherical subdivision schemas like Equal-arcs (three great circles), three intersecting great circles do not exactly intersect at a single point. Instead, they form a small spherical triangle called a window. The centroid of this small triangle is then adopted as the desired grid point.

Wulff net: The stereogram of latitude and longitude reference circles. The Wulff net is useful for making stereograms manually and for graphically solving spherical trigonometry problems. Other uses include celestial navigation and the design of astronomy instruments, like astrolabes. See Appendix B for details. *See also* stereogram; stereographic projection.[23]

Z

zenith: A point directly overhead or uppermost on the sphere. The point opposite the nadir. *See also* nadir; pole.

zonohedron: In 3D space (other dimensions are possible), a convex polyhedron whose faces are parallelograms, each face has an even number of edges and each edge of each face is parallel to its opposite edge, and each of its 2D faces have centers of symmetry.

---

[19] (Cromwell 1997, 369)
[20] Image courtesy of Ross Heikes.
[21] Illustration courtesy of Steven Waterman.
[22] (Stuart 1952)
[23] Illustration courtesy of James Morrison.

Useful Websites

American Masters. "Buckminster Fuller: Thinking Out Loud, Thirteen." *WNET New York.* http://www.thirteen.org/bucky/film.html, undated.

Blake, Trevor. synchronofile.com, an independent resource on R. Buckminster Fuller. http://synchronofile.com/, 2009.

Bourke, Paul. "Distributing Points on a Sphere." http://local.wasp.uwa.edu.au/~pbourke/geometry/spherepoints/, June 1996.

Buckminster Fuller Institute. http://www.bfi.org/, 2009.

Eppstein, David. "The Geometry Junkyard." *Theory Group, Information and Computer Sciences, University of California at Irvine.* http://www.ics.uci.edu/~eppstein/junkyard, 2009.

Fearnley, Christopher J. "CJ Fearnley's List of Buckminster Fuller Resources on the Internet." http://www.cjfearnley.com/buckyrefs.html, 2009.

Foundation for New Directions, a holistic living center focusing on the relationships between biology, physics, geometry, and genertics. http://www.fnd.org/.

Furuti, Carlos A. "Cartographical Map Projections." http://www.progonos.com/furuti/MapProj/Normal/TOC/cartTOC.html, 2008.

Gray, Robert W. "Great Circle and LCD Triangle Info." *The Projects of R. W. Gray.* http://www.rwgrayprojects.com/rbfnotes/greatc/greatc1.html, 2009.

Hart, George W. *Virtual Polyhedra: The Encyclopedia of Polyhedra.* http://www.georgehart.com/virtual-polyhedra/vp.html, 2000.

———. "The Pavilion of Polyhedreality." http://www.georgehart.com/pavilion.html, 2008.

Keys, Gene. "B. J. S. Cahill Butterfly Map Resource Page—Octahedral Map of the World." http://www.genekeyes.com/B.J.S._CAHILL_RESOURCE.html, 2009.

Moore, Joe S. "Buckminster Fuller Virtual Institute." http://www.buckminster.info/index.html, 2009.

Neubert, Karl-Dietrich. "The Entropy Reduction Laboratory." www.neubert.net, 2009.

Persistence of Vision Raytracer (POV-Ray). http://www.povray.org, 2009.

Roelofs, Rinus. "Mathematical Sculpture." http://www.rinusroelofs.nl, 2009.

Rossiter, Adrian. "Antiprism." http://www.antiprism.com, 2009.

Sahr, Kevin. "Discrete Global Grids (DGGs)," data structures for geospatial applications. http://webpages.sou.edu/~sahrk/dgg/, 2003.

Sloane, N. J. A. "Spherical Codes—Nice arrangements of points on a sphere in various dimensions." *Information Sciences Research, AT&T Shannon Lab.* http://www.research.att.com/~njas/packings, 2005.

Sherwood, Anton. "How Can I Arrange n Points Evenly on a Sphere." http://www.ogre.nu/sphere.htm, 2009.

Stanford University. *R. Buckminster Fuller Archive*, the most comprehensive collection of life history and media. This collection is a fully indexed chronological collection, searchable via the university's SULAIR system. http://www-sul.stanford. edu/depts/spc/fuller/index.html.

The Synergetics Collaborative (SNEC). http://www. synergeticscollaborative.org.

Syntropy: CJ Fearnley's Blog. An "Explorer in Universe" contributing thoughts, seeking feedback, building syntropy. http://blog.cjfearnley.com/, 2010.

Urner, Kirby. "Synergetics on the Web." http://www.grunch.net/synergetics.

Waterman, Steve. "Waterman's Polyhedron." http://www.watermanpolyhedron.com, 2009.

Webb, Robert. "Stella: Polyhedron Navigator." http://www.software3d.com/Stella.php, 2008.

Weisstein, Eric W. "Uniform Polyhedron." *Wolfram MathWorld.* http://mathworld.wolfram.com/UniformPolyhedron.html, July 28, 2009.

Wolff, John. "Web Museum of Calculators." http://home.vicnet.net.au/~wolff/calculators/CalcIndex.htm, 2009.

Bibliography

Aguilar, Rodolfo Jesus. "A Study of the Stability of Framed, Triangulated Geodesic Domes Under the Action of Concentrated Loads." Diss. NC State University at Raleigh, NC, 1964.

Ahern, William W. and Brigit T. Mathe. Spherical Structural Arrangement. Geometrics, Inc., assignees. USPTO Patent 4,026,078. 31 May 1977.

Ahern, William W. and William H. Wainwright. Spherical Structural Arrangement. Geometrics, Inc., assignees. USPTO Patent 3,392,495. 16 July 1968.

Alexander, James. "Loxodromes: A Rhumb Way to Go." *Mathematics Magazine* 77.5 (2004): 349–356.

Aoyama, Steven. Golf Ball. Wilson Sporting Goods Co., assignee. USPTO Patent 4,560,168. 24 Dec. 1985.

———. Golf Ball. Acushnet Co., assignee. USPTO Patent 4,948,143. 14 Aug. 1990.

———. Golf Ball Dimple. Acushnet Co., assignee. USPTO Patent 6,162,136. 19 Dec. 2000.

———. Golf Ball Dimple Pattern. Acushnet Co., assignee. USPTO Patent 5,957,786. 28 Sept. 1999.

———. Golf Ball Dimples. Acushnet Co., assignee. USPTO Patent 7,207,905. 24 Apr. 2007.

———. Three Parting Line Quadrilaterals Golf Ball Dimple Pattern. Acushnet Co., assignee. USPTO Patent 5,415,410. 16 May 1995.

Aoyama, Steven and Douglas E. Jones. Golf Ball with Improved Flight Performance. Acushnet Co., assignee. USPTO Patent 6,916,255. 12 July 2005.

Aoyama, Steven and William E. Morgan. Golf Ball Dimple Pattern with Overlapping Dimples. Acushnet Co., assignee. USPTO Patent 7,258,632. 21 Aug. 2007.

Aravind, P. K. "How Spherical Are the Archimedean Solids and Their Duals?" *The College Mathematics Journal* 42.2, March (2011): 98–107.

Aste, Tomaso and Swnia Weaire. *The Pursuit of Perfect Packing*. Second Edition, New York: Taylor & Francis, 2008.

Baer, Steve. *Zome Primer: Elements of Zonohedra Geometry; Two and Three Dimensional Growths of Stars with Five Fold Symmetry*. Albuquerque, NM: Zomeworks Corp., 1970.

Bagchi, Bhaskar. "How to Stay Away from Each Other in a Spherical Universe." *Resonance* September (1997): 18–26.

Ball, Walter William Rouse and H. S. M. Coxeter. *Mathematical Recreations and Essays*. New York: Dover Publications, 1986.

Bartholdi, John and Paul Goldsman. "Continuous Indexing of Hierarchical Subdivisions of the Globe." *International Journal of Geographical Information Science* 15.6 (2001): 489–522.

Berman, Joel D. and Kit Hanes. "Volumes of Polyhedra Inscribed in the Unit Sphere in *E3*." *Mathematische Annalen* 188.1 (1970): 78–84.

Blenkinsop, Tom. "Pedagogy of Stereographic Projection." *Journal of African Earth Sciences* 29.4 (1999): 897–902.

Bohlen, J. C. *Trigonometric Relationships for Geodesic Domes with Special Reference to the Dodecahedron*. Vancouver, BC: Western Forest Products Laboratory, Information Report VP-X-121, 1974.

Bourke, Paul. "Distributing Points on a Sphere." http://paulbourke.net/geometry/spherepoints/, 1996.

Bowditch, Nathaniel. *The American Practical Navigator: An Epitome of Navigation*. Bethesda, MD: National Imagery and Mapping Agency, 2002.

Brand, Stewart and James Baldwin. *Whole Earth Catalog*. Menlo Park, CA: Portola Institute, 1968.

Bromberg, Manuel. Personal interview. Woodstock, New York: 6 May 2004.

——. Telephone interview. 13 Dec. 2008.

Brook, David Louis Sterrett. "Henry Leveke Kamphoefner, the Modernist, Dean of the North Carolina State University School of Design 1948–1972." Masters. NC State University, Dept. of Adult and Community College Education, 2005.

Bullard, James Atkins and Arthur Kiernan. *Plane and Spherical Trigonometry with Stereographic Projections*. Boston: D. C. Heath, 1922.

Bunger, Donald J. and Joseph F. Stiefel. Golf Ball. Spalding & Evenflo Co., Inc. assignee. USPTO Patent 5,060,953. 19 Oct. 1991.

Cahill, Bernard J. S. "An Account of a New Land Map of the World." *Scottish Geographical Magazine* (1909): 449–469.

——. Geographical Globe. USPTO Patent 1,081,207. 9 Dec. 1913.

——. Map of the World. USPTO Patent 1,054,276. 25 Feb. 1913.

Calter, Paul A. *Squaring the Circle: Geometry in Art and Architecture*. Emeryville, CA: Key College Pub, 2006.

Caspar, Donald L. D. "Deltahedral Views of Fullerene Polymorphism." *Philosophical Transactions of the Royal Society of London. Series A: Mathematical, Physical & Engineering Sciences* 343, (1993): 133–144.

Caspar, Donald L. D. and A. Klug. "Physical Principles in the Construction of Regular Viruses." *Cold Spring Harbor Symposia on Quantitative Biology* 27 (1962): 1–24.

Cessna, Joseph B. and Thomas R. Bewley, "Honeycomb-Structured Computational Interconnects and Their Scalable Extension to Spherical Domains." *Proceedings of the 11th International Workshop on System Level Interconnect Prediction (SLIP).* American Computing Machinery, San Francisco (2009): 27–35.

Ch, Adam and Paul Tannery. "Progymnasmata de Solidorum Elementis." *Oeuvres des Descartes* 10 (1996): 265–276.

Chemical Rubber Company. *CRC Standard Mathematical Tables.* Boca Raton, FL: CRC Press, 1953.

Chu, Hsiao-yun and Roberto G. Trujillo. *New Views on R. Buckminster Fuller.* Stanford, CA: Stanford University Press, 2009.

Clinton, Joseph D. *Advanced Structural Geometry Studies Part I—Polyhedral Subdivision Concepts for Structural Applications.* National Aeronautics and Space Administration contractor report NASA CR-1734, Carbondale, IL: Southern Illinois University, 1971.

——. *Advanced Structural Geometry Studies Part II—A Geometric Transformation Concept for Expanding Rigid Structures.* National Aeronautics and Space Administration contractor report NASA CR-1735, Carbondale, IL: Southern Illinois University, 1971.

——. "Chord Factors and Angles." *Domebook 1*, ed. Lloyd Kahn. Los Gatos, CA: Pacific Domes, 1970.

——. "Clinton's Equal Central Angle Conjecture." PolyModular Ltd., 2006.

——. "Geodesic Math." *Domebook 2,* ed. Lloyd Kahn. Bolinas, CA: Pacific Domes, 1971.

——. "A Group of Spherical Tessellations Having Edges of Equal Length." *Space Structures 5* 2.105 (2002): 995–1004.

——. "A Limited and Biased View of Historical Insights for Tessellating a Sphere." *Space Structures 5,* ed. G. A. R. Parke and P. Disney. London: Thomas Telford, 2002.

——. "Lowest Common Frequency: b2 + b3 + c2." *International Journal of Space Structures: Special Issue on Geodesic Forms.* London: Multi-Science Publishing Co., 1990.

——. Method of Tessellating a Surface. USPTO Patent 2004/0257433. 23 Dec. 2004.

Clinton, Joseph D. and Jay Salsburg. "Geodesic Math." http://mr-fusion.hellblazer.com/pdfs/geodesicmath.pdf, undated.

Cohen, Albert, Phillip Davis, and Joseph F. Orabona. Radome Structural Devices. United States of America rep. Secretary of the Air Force, assignee. USPTO Patent 2,978,704. 4 Apr. 1961.

Collidge, Julian Lowell. *A Treatise on the Circle and the Sphere.* Bronx, NY: Chelsea Pub. Co., 1971.

Colorado State University. "Geodesic Climate Model Uses Different Mapping Technique, Coordinates and Supercomputing to Improve Predictions." *ScienceDaily* 2001.

Congleton, Carol A. *Navigational Applications of Plane and Spherical Trigonometry*. Centreville, MD: Cornell Maritime Press, 1980.

Conway, John H., Heidi Burgiel, and Chaim Goodman-Strauss. *The Symmetries of Things*. Wellesley, MA: A K Peters, 2008.

Coxeter, H. S. M. *Introduction to Geometry, Second Edition*. New York: John Wiley & Sons, 1969.

——. *Non-Euclidean Geometry, Spectrum Series, Sixth Edition.* Washington, DC: Mathematical Association of America, 1998.

——. "The Partition of a Sphere According to the Icosahedral Group." *Scripta Math* 4 (1936): 156–157.

——. "The Problem of Packing a Number of Equal Non-Overlapping Circles on a Sphere." *Transactions of The New York Academy of Sciences* (Department of Mathematics, University of Toronto) 24 (1962): 320–331.

——. *Projective Geometry, Second Edition*. New York: Springer-Verlag, 1987.

——. *Regular Complex Polytopes, Second Edition*. New York: Cambridge University Press, 1991.

——. *Regular Polytopes, Third Edition*. New York: Dover Publications, 1973.

——. *Twisted Honeycombs.* Regional Conference Series in Mathematics. Providence, RI: American Mathematical Society, 1970.

——. "Virus Macromolecules and Geodesic Domes." *A Spectrum of Mathematics,* ed. J. C. Butcher. Auckland: Auckland University Press, 1971.

Coxeter, H. S. M. and Patrick Du Val. *The Fifty-Nine Icosahedra.* New York: Springer-Verlag, 1982.

Coxeter, H. S. M., Michele Emmer, Roger Penrose, and Marianne L. Teuber. *M. C. Escher: Art and Science.* Proceedings of the International Congress on M. C. Escher. 26–28 March 1985. Elsevier Science Pub. Co., 1986.

Coxeter, H. S. M., M. S. Longuet-Higgins, and J. C. P. Miller. "Uniform Polyhedra." *Philosophical Transactions of the Royal Society of London. Series A: Mathematical, Physical & Engineering Sciences* 246.916 (1954): 401–450.

Coxeter, H. S. M. and L. Fejes Tóth. "The Total Length of the Edges of a Non-Euclidean Polyhedra with Triangular Faces." *Oxford Journals Mathematics & Physical Sciences Quarterly Journal of Mathematics* 14.1 (1963): 273–284.

Critchlow, Keith. *Order In Space: A Design Source Book*. New York: Viking Press, 1970.

Critchlow, Keith and Rod Bull. *Time Stands Still: New Light on Megalithic Science*. New York: St. Martin's Press, 1982.

Cromwell, Peter R. *Polyhedra*. Cambridge, UK: Cambridge University Press, 1997.

Cundy, H. M. and A. P. Rollett. *Mathematical Models, Second Edition*. Oxford: Clarendon Press, 1961.

Dawson, Robert J. MacG. "Some New Tilings of the Sphere with Congruent Triangles." Dept. of Mathematics and Computing Science, Saint Mary's University. http://www.mi.sanu.ac.yu/vismath/bridges2005/dawson/index.html, 2005.

Dorst, Leo, Daniel Fontijne, and Stephen Mann. *Geometric Algebra for Computer Science: An Object-Oriented Approach to Geometry*. The Morgan Kaufmann Series in Computer Graphics. Amsterdam: Elsevier, 2007.

Doskas, Gary. *Spherical Harmony—A Journey of Geometric Discovery.* LuLu Marketplace: Hedron Designs, 2011.

Dunham, William. *The Mathematical Universe—An Alphabetical Journey Through the Great Proofs, Problems, and Personalities*. New York: John Wiley & Sons, 1994.

Dutton, Geoffrey H. "Encoding and Handling Geospatial Data with Hierarchical Triangular Meshes." *7th Symposium on Spatial Data Handling*. Delft, Netherlands (1996): 8B.15–28.

——. "Geodesic Modeling of Planetary Relief." *Cartographica* 21.2–3 (1984): 188–207.

——. *A Hierarchical Coordinate System for Geoprocessing and Cartography*. Berlin, Germany: Springer-Verlag, 1985.

——. *A Hierarchical Coordinate System for Geoprocessing and Cartography*. Lecture Notes in Earth Sciences. New York: Springer, 1999.

——. "Polyhedral Hierarchical Tessellations: The Shape of GIS to Come." *Geographical Information Systems* 1.3 (1991): 49–55.

——. "Universal Geospatial Data Exchange via Global Hierarchical Coordinates." *International Conference on Discrete Global Grids*. US National Center for Geographic Information and Analysis, Santa Barbara, CA: 2000.

Easton, Robert and Lloyd Kahn. *Domebook One*. Los Gatos, CA: Pacific Domes, 1970.

Edmondson, Amy C. *A Fuller Explanation: The Synergetic Geometry of R. Buckminster Fuller*. Design Science Collection. Boston: Birkhäuser, 1987. Reprinted *Back-In-Action* book series (Pueblo, Colorado: Emergent World Press, 2007).

Emerson, Brent D. et al. Distance Golf Ball - DDH Steel Distance. Dunlop Slazenger Manufacturing, assignee. USPTO Patent 6,572,494. 3 June 2003.

Euclid and Thomas L. Heath. *The Thirteen Books of Euclid's Elements*. New York: Dover Publications, 1956.

Faux, I. D. and M. J. Pratt. *Computational Geometry for Design and Manufacturing*. New York: Halsted Press, 1979.

Fearnley, Christopher J. "CJ Fearnley's List of Buckminster Fuller Resources on the Internet." http://www.cjfearnley.com/buckyrefs.html, 2011.

Fekete, György. "Rendering and Managing Spherical Data with Sphere Quadtrees." *IEEE Visualization, Proceedings of the 1st conference on Visualization* 1 (1990): 176–186.

Fisher, N. I., Toby Lewis, and B. J. J. Embleton. *Statistical Analysis of Spherical Data*. New York: Cambridge University Press, 1987.

Fitzgibbon, James W. "The Design History of the Baton Rouge Dome," Raleigh, NC: Synergetics, Inc., Department of Special Collections and University Archives, Stanford University Libraries, 1957.

——. Letter to R. A. Lehr, Union Tank Car Company. Raleigh, NC: Synergetics, Inc., Department of Special Collections and University Archives, Stanford University Libraries, 2 Aug. 1957.

——. Letter to William H. Wainwright, Geometrics, Inc. Raleigh, NC: Synergetics, Inc., Department of Special Collections and University Archives, Stanford University Libraries, 6 Sept. 1957.

——. Memo to R. Buckminster Fuller Re: Current Work Assignments. Raleigh, NC: Synergetics, Inc., Department of Special Collections and University Archives, Stanford University Libraries, 29 Oct. 1957.

Food and Agriculture Organization (FAO) of the United Nations. *The State of World Fisheries and Aquaculture, 2006 (SOFIA)*. Ed. Fisheries and Aquaculture Department. New York: United Nations, 2007.

Fowler, P. W., T. Tarnai, and Zs. Gáspár. "From Circle Packing to Covering on a Sphere with Antipodal Constraints." *Philosophical Transactions of The Royal Society of London. Series A: Mathematical, Physical & Engineering Sciences* 458.2025 (2002): 2275–2287.

Fowler, P. W., T. Tarnai, and S. Kabai. "Packing Regular Triplets of Circles on a Sphere." *Philosophical Transactions of The Royal Society of London. Series A: Mathematical, Physical & Engineering Sciences* 461 (2005): 2355–2367.

Freeman, Ira and Mae Freeman. *Fun with Figures: Easy Experiments for Young People*. New York: Random House, 1946.

Fuller, R. Buckminster. "3 Way Geodesic Grid B Calculations on Present." *Noah's Ark II (Notebook)*. Department of Special Collections and University Archives, Stanford University Libraries, 1948.

——. Building Construction. USPTO Patent 2,682,235. 29 June 1954.

——. Cartography. USPTO Patent 2,393,676. 29 Jan. 1946.

——. "Fuller Projection Dymaxion Air-Ocean World." Los Angeles: Buckminster Fuller Institute, 1992.

——. Geodesic Structures. USPTO Patent 3,197,927. 3 Aug. 1965.

——. Geodesic Tent. USPTO Patent 2,914,074. 24 Nov. 1959.

——. Laminar Geodesic Dome. USPTO Patent 3,203,144. 31 Aug. 1965.

——. Motor Vehicle. The Dymaxion Corp., assignee. USPTO Patent 2,101,057. 7, Dec. 1937.

——. Non-Symmetrical Tension-Integrity Structures. USPTO Patent 3,866,366. 18 Feb. 1975.

——. Octahedral Building Truss. USPTO Patent 3,354,591. 28 Nov. 1967.

——. Prefabricated Bathroom. Phelps Dodge Corp., assignee. USPTO Patent 2,220,482. 5 Nov. 1940.

——. "Project Noah's Ark #2 Discovering New Man Advantage (Notebook)." Department of Special Collections and University Archives, Stanford University Libraries, 1950.

——. Self-Strutted Geodesic Plydome. USPTO patent 2,905,113, 22 Sept. 1959.

——. Synergetic Building Construction. USPTO Patent 2,986,241. 30 May 1961.

——. Tensile-Integrity Structures. USPTO Patent 3,063,521. 13 Nov. 1962.

——. *Utopia or Oblivion: The Prospects for Humanity.* Toronto: Bantam Books, 1971.

Fuller, R. Buckminster and E. J. Applewhite. *Synergetics Dictionary: The Mind of Buckminster Fuller with an Introduction and Appendices.* 4 vols. New York: Garland, 1986

——. *Synergetics: Explorations in The Geometry of Thinking.* New York: Macmillan, 1975.

Fuller, R. Buckminster and Carl Solway Gallery. *Buckminster Fuller: Inventions, Twelve Around One.* Cincinnati, OH: Carl Solway Gallery, 1981.

Fuller, R. Buckminster and James Ward. *The Artifacts of R. Buckminster Fuller: A Comprehensive Collection of His Designs and Drawings.* 4 vols. New York: Garland, 1985.

——. *Inventions.* New York: St. Martin's Press, 1983.

Fuller, R. Buckminster and Thomas Tse Kwai Zung. *Buckminster Fuller: Anthology for the New Millennium.* New York: St. Martin's Press, 2002.

Gabriel, J. François (ed.). *Beyond the Cube: The Architecture of Space Frames and Polyhedra.* New York: John Wiley & Sons, 1997.

Geodesics, Inc. "Largest Rigid Plastic Structure in History (press release)." Department of Special Collections and University Archives, Stanford University Libraries, 1955.

Gobush, William. Golf Ball. Acushnet Co., assignee. USPTO Patent 4,765,626. 23 Aug. 1988.

——. Multiple Dimple Golf Ball. Acushnet Co., assignee. USPTO Patent 4,804,189. 14 Feb. 1989.

——. Multiple Dimple Golf Ball. Acushnet Co., assignee. USPTO Patent 4,960,283. 2 Oct. 1990.

——. Multiple Dimple Golf Ball. Acushnet Co., assignee. USPTO Patent 5,060,954. 29 Oct. 1991.

Goldberg, Michael. "A Class of Multi-Symmetric Polyhedra." *Tohoku Mathematics Journal* 43 (1937): 104–108.

——. "Viruses and a Mathematical Problem." *Journal of Molecular Biology* 24.2 (1967): 337–338.

Graver, Jack E. "Encoding Fullerenes and Geodesic Domes." *Society for Industrial and Applied Mathematics* 17.4 (2004): 596–614.

Gray, Robert W. "Great Circle and LCD Triangle Info." *The Projects of R. W. Gray.* http://www.rwgrayprojects.com/rbfnotes/greatc/greatc1.html, 2009.

Hardin, R. H., N. J. A. Sloane, and W. D. Smith. "Tables of Spherical Codes with Icosahedral Symmetry." www.research.att.com/~njas/icosahedral.codes, 2008.

Hart, George W. "The Encyclopedia of Polyhedra." www.georgehart.com/virtualpolyhedra/vp.html, 2008.

——. "Icosahedral Constructions." *Bridges: Mathematical Connections in Art, Music, and Science.* Southwestern College, Winfield, KS (1998): 195–202.

——. "Pavilion of Polyhedreality." www.georgehart.com/pavilion.html, 2008.

Hart, George W. and Henri Picciotto. *Zome Geometry: Hands-on Learning with Zome Models.* Emeryville, CA: Key Curriculum Press, 2001.

Heartney, Eleanor and Kenneth D. Snelson. *Kenneth Snelson: Forces Made Visible*. Lenox, MA: Hard Press Editions, 2009.

Heikes, Ross et al. "Continuing Development of Models Based on Spherical Geodesic Grids." *PDE Sphere 2007*, Dept. of Atmospherical Science, Colorado State University (2007).

Henderson, David W. and Eduarda Moura. *Experiencing Geometry: On Plane and Sphere*. Englewood Cliffs, NJ: Prentice Hall, 1996.

Henderson, David W. and Daina Taimiņa. *Experiencing Geometry: Euclidean and Non-Euclidean with History, Third Edition*. Upper Saddle River, NJ: Pearson Prentice Hall, 2005.

Hoberman, Charles. Radial Expansion/Retraction Truss Structures. USPTO Patent 5,024,031. 18 Jan. 1990.

———. Reversibly Expandable Doubly-Curved Truss Structure. USPTO Patent 4,942,700. 24 July 1990.

Holden, Alan. *Shapes, Space, and Symmetry*. New York: Dover Publications, 1991.

Hoschek, Josef and Dieter Lasser. *Fundamentals of Computer-Aided Geometric Design*. Wellesley, MA: A K Peters, 1993.

Hotchkiss, John F. and John Stuart Martin. *500 Years of Golf Balls: History and Collector's Guide*. Dubuque, IA: Antique Trader Books, 1997.

Howard, T. C. *Possible Ways the Random Geometric Grid Developed by Lincoln Laboratories may be Covered by Patent 2,682,235 Owned by Inventor R. Buckminster Fuller*. Raleigh, NC: Geodesics, Inc., 1958.

Hume, Andrew. *Exact Descriptions of Regular and Semi-Regular Polyhedra and Their Duals*. Murry Hill, NJ: AT&T Bell Laboratories, Computing Science Technical Report No. 130, 1986.

Hunt, Jarvis. Golf Ball. USPTO Patent 1,517,514. 2 Apr. 1924.

Huybers, Pieter. "Computer-Aided Design of Polyhedral Building Structures." *Design Studies* 14.1 (1993).

———. "The Polyhedral World." *Beyond The Cube: The Architecture of Space Frames and Polyhedra*, ed. J. François Gabriel. New York: John Wiley & Sons, 1997.

Hwang, In H. Golf Ball. Hya Co., Ltd. (Seoul), assignee. USPTO Patent 5,564,708. 15 Oct. 1996.

———. Golf Balls. Hya Co., Ltd. (Seoul), assignee. USPTO Patent 5,575,477. 19 Nov. 1996.

Ihara, Keisuke. Golf Ball. Bridgestone Corp. (Tokyo), assignee. USPTO Patent 4,844,472. 4 July 1989.

Kabai, Sándor. "Rhombic Structures: Use of Mathematica in Combination with Solid Models for Studying Space Structures." *2006 Wolfram Technology Conference*, Champaign, IL (2006).

Kahn, Lloyd. *Domebook Two*. Bolinas, CA: Shelter Publications, 1974.

———, ed. *Refried Domes*. Bolinas, CA: Shelter Publications, 1989.

———. "The Wonder of Jeana." *Refried Domes*. Bolinas, CA: Shelter Publications, 1989.

Kaiser Aluminum & Chemical Sales, Inc. "Kaiser Aluminum Dome; Engineering Data for Mean Plan Diameter 125 Feet: A-80-11.5." Chicago: Kaiser Aluminum & Chemical Sales, Inc., 1958.

Kang, Hye Sook. Dimple Pattern on Golf Ball. Hanyoung Kangaroo Co., Ltd. (Seoul), assignee. USPTO Patent 7,278,933. 9 Oct. 2007.

Kasashima, Atsuki. Golf Ball. Bridgestone Sports Co., Ltd. (Tokyo), assignee. USPTO Patent Application 2003/0171167. 11 Sept. 2003.

———. Golf Ball. Bridgestone Sports Co., Ltd. (Tokyo), assignee. USPTO Patent Application 2004/0121858. 24 June 2004.

———. Golf Ball. Bridgestone Sports Co., Ltd. (Tokyo), assignee. USPTO Patent 6,971,962. 6 Dec. 2005.

———. Golf Ball. Bridgestone Sports Co., Ltd. (Tokyo), assignee. USPTO Patent 7,354,358. 8 Apr. 2008.

Kells, Lyman M., Willis F. Kern, and James R. Bland. *Plane and Spherical Trigonometry*. New York: McGraw-Hill Book Co., 1943.

Kennedy III, Thomas J. Golf Ball with Elevated Dimple Portions. The Top-Flight Golf Co., assignee. USPTO Patent 6,626,772. 30 Sep. 2003.

Kenner, Hugh. *Bucky; A Guided Tour of Buckminster Fuller*. New York: Morrow, 1973.

———. *Geodesic Math and How to Use It, Second Edition*. Berkeley: University of California Press, 2003.

Keyes, Gene. "B. J. S. Cahill Butterfly Map—Octahedral Map of the World." http://www.genekeyes.com/B.J.S._CAHILL_RESOURCE.html, 2009.

Kimerling, A. Jon et al. "Comparing Geometrical Properties of Global Grids." *Cartography and Geographic Information Science* 26.4 (1999): 271–288.

Kitrick, Christopher J. "Geodesic Domes." *Structural Topology* 11 (1985): 15–20.

———. "Nonlinear Analysis of Normal and Inverted Geodesic Domes Under the Action of Concentrated Loads." Cincinnati, OH: University of Cincinnati, 1984.

———. Tensegrity Module Structure and Method of Interconnecting the Modules. USPTO Patent 4,207,715. 17 June 1980.

———. "A Unified Approach to Class I, II and III Geodesic Domes." *International Journal of Space Structures* 5.3–4 (1990): 223–246.

Kovács, F. et al. "Double-Link Expandohedra: A Mechanical Model for Expansion of a Virus." *Philosophical Transactions of the Royal Society of London. Series A: Mathematical, Physical & Engineering Sciences* 460.2051 (2004): 3191–3202.

Kroto, Harold. "Symmetry, Space, Stars and C_{80}." *Reviews of Modern Physics* 69.3 (1997): 703–722.

Kunszt, Peter Z., Alexander S. Szalay, and Aniruddha R. Thakar. "The Hierarchical Triangular Mesh." *Proceedings of the MPA/ESO/MPE Workshop*. Garching, Germany: Springer-Verlag (2000).

Lalvani, Haresh. Building Systems with Non-Regular Polyhedra Based on Subdivisions of Zonohedra. USPTO Patent 5,623,790. 29 Apr. 1997.

——. "Continuous Transformations of Subdivided Periodic Surfaces." *Space Structures* 5.3–4 (1990): 255–279.

——. "Multidimensional Periodic Arrangements of Transforming Structures." University of Michigan, 1981.

——. "Origins of Tensegrity: Views of Emmerich, Fuller and Snelson." *International Journal of Space Structures* 11.1–2 (1996): 27–55.

——. Space Structures with Non-Periodic Subdivisions of Polygonal Faces. USPTO Patent 5,524,396. 11 June 1996.

——. Sports Ball. Milgo Industrial, Inc., assignee. USPTO Patent application US2008/0268989, 30 Oct. 2008.

——. "Structures on Hyperstructures: Multidimensional Periodic Arrangements of Transforming Space Structures." *Papers in Theoretical Morphology* 3 (1982): 112.

——. "Transpolyhedra: Dual Transformations by Explosion-Implosion." *Papers in Theoretical Morphology* 1 (1977).

——. "Transpolyhedra and Explosion-Implosion Principles for Dual Transformations." *IASS World Congress on Space Enclosures*. Montreal: University of Concordia, 1976.

Lane, Henry C. *Final Report Study of Shelter Logistics for Marine Corps Aviation*. Ed. US Marine Corps Aviation Logistics and Materiel Branch. Washington, DC: US Marine Corps, 1955.

Latypov, Nurakhmed Nurislamovich and Nurulla Nurislamovich Latypov. System for Placing a Subject into Virtual Reality. USPTO Patent 6,563,489. 13 May 2003.

Lavallee, Gerald A. and Edward F. Mendrala. Golf Ball. Lisco, Inc., assignee. USPTO Patent 5,356,150. 18 Oct. 1994.

Leighton, Henry L. C. *Solid Geometry and Spherical Trigonometry*. New York: D. Van Nostrand Company, Inc., 1943.

Lemons, Lane D. and Matthew B. Stanczak. Golf Ball with Secondary Depressions. Dunlop Maxfli Sports Corp., assignee. USPTO Patent 6,010,442. 4 Jan. 2000.

Leon, Bruno. Letter to the author discussing the first wooden lattice geodesic dome prototype. 3 July 2004.

——. Telephone interview. 6 May 2004.

Lettvin, Jonathan D. "Diffuse: A Computer Program for Distributing n Points on a Sphere." http://local.wasp.uwa.edu.au/~pbourke/geometry/spherepoints/, 2003.

Leytem, Charles. "Hidden Symmetries in the Snub Dodecahedron." *European Journal of Combinatorics* 17.5 (1996): 451–460.

Li, Yue et al. "Constructing Tensegrity Structures from One-Bar Elementary Cells." *Philosophical Transactions of the Royal Society of London. Series A: Mathematical, Physical & Engineering Sciences* 466.2113 (2010): 45–61.

LIFE. "Tape, Plastic, and Aluminum: Ford Builds a 'Geodesic Dome.'" *LIFE* 8 (June 1953): 67–70.

Lim, Dong-keun. Dimple Pattern and the Placement Structure on the Spherical Surface of the Golf Ball. Dong Sung Chemical Ind. Co., Ltd., assignee. USPTO Patent 5,441,276. 15 Aug. 1995.

Lisle, Richard J. and Peter R. Leyson. *Stereographic Projection Techniques for Geologists and Civil Engineers, Second Edition*. New York: Cambridge University Press, 2004.

Livio, Mario. *The Golden Ratio: The Story of Phi, the World's Most Astonishing Number*. New York: Broadway Books, 2002.

Loeb, Arthur L. "Buckminster Fuller and the Relevant Pattern." *Beyond the Cube: The Architecture of Space Frames and Polyhedra*, ed. J. François Gabriel. New York: John Wiley & Sons, 1997.

———. *Space Structures: Their Harmony and Counterpoint*. Reading, MA: Addison Wesley Pub. Co., Advanced Book Program, 1976.

Lorance, Loretta. *Becoming Bucky Fuller*. Cambridge: MIT Press, 2009.

Machin, Brian E. Golf Balls with Isodiametrical Dimples. Dunlop Ltd., assignee. USPTO Patent 5,377,989. 3 Jan. 1995.

Mackey, Gary T. Golf Ball. Gary T. Mackey, assignee. USPTO Patent 5,046,742. 10 Sep. 1991.

MacLean, Kenneth James Michael. *A Geometric Analysis of the Platonic Solids and Other Semi-Regular Polyhedra: With an Introduction to the Phi Ratio, for Teachers, Researchers and the Generally Curious*. Ann Arbor, MI: Loving Healing Press, 2007.

Maehara, Kazuto, Keisuke Ihara, and Atsuki Kasashima. Dimple Golf Ball and Dimple Distributing Method. Bridgestone Corp. (Tokyo), assignee. USPTO Patent 6,254,496. 3 July 2001.

Makai, Endre Jr. "On Some Geometrical Problems of Single-Layered Spherical Grids with Triangular Network." *Il International Conf. on Space Structures*. Guilford, England: University of Surrey, 1975.

Makai, Endre Jr. and T. Tarnai. "Morphology of Spherical Grids." *Acta Tecnica Acad. Sci. Hung.* 83 (1976): 247–283.

Maor, Eli. *e: The Story of a Number*. Princeton, NJ: Princeton University Press, 1994.

———. *Trigonometric Delights*. Princeton, NJ: Princeton University Press, 1998.

Marks, Robert W. and R. Buckminster Fuller. *The Dymaxion World of Buckminster Fuller*. Garden City, NY: Anchor Books, 1973.

Martin, Frank S. Golf Balls. United States Rubber Co., assignee. USPTO Patent 2,728,576. 27 Dec. 1955.

Martin, Frank S. and Thaddeus A. Pietraszek. Golf Ball. Uniroyal, Inc., assignee. USPTO Patent 4,090,716. 23 May 1978.

Martin, John Stuart. *The Curious History of the Golf Ball, Mankind's Most Fascinating Sphere*. New York: Horizon Press, 1968.

Matsko, Vincent J. "Polyhedra and Geodesic Structures." Aurora, IL: Illinois Mathematics and Science Academy, http:\\www.vincematsko.com, 2009.

McGuire, Kenneth Stephen. Golf Ball with Non-Circular Shaped Dimples. The Procter & Gamble Co., assignee. USPTO Patent 6,409,615. 25 June 2002.

McHale, John. *R. Buckminster Fuller*. Makers of Contemporary Architecture. New York: G. Braziller, 1962.

Melvin, Terence. Golf Ball. Spalding & Evenflo Companies, Inc., assignee. USPTO Patent 4,880,241. 14 Nov. 1989.

Melvin, Terence et al. Two-Piece Solid Golf Ball. Uniroyal, Inc., assignee. USPTO Patent 4,141,559. 27 Feb. 1979.

Messer, Peter W. "Mathematical Formulas for Geodesic Domes." *Spherical Models.* Ed. Magnus J. Wenninger. New York: Dover Publications, 1999.

Morell, Joseph. Golf Ball. Salomon S.A., assignee. USPTO Patent 4,973,057. 27 Nov. 1990.

Morgan, Gregory J. "Historical Review: Viruses, Crystals and Geodesic Domes." *Trends in Biochemical Sciences* 28.2 (2003): 86–90.

Morgan, William E. Dimple Pattern for Golf Balls. Acushnet Co., assignee. USPTO Patent 6,849,007. 1 Feb. 2005.

———. Golf Ball with Spherical Polygonal Dimples. Acushnet Co., assignee. USPTO Patent 7,309,298. 18 Dec. 2007.

Morrison, James E. *The Astrolabe*. Rehoboth Beach, DE: Janus, 2007.

Motro, René. *An Anthology of Structural Morphology.* Hackensack, NJ: World Scientific, 2009.

———. *Tensegrity: Structural Systems for the Future.* Sterling, VA: Kogan Page Science, 2003.

Museum of Modern Art. "Three Structures by Buckminster Fuller in the Garden of the Museum of Modern Art." New York: The Museum of Modern Art, Exhibit Guide, 1960.

Nardacci, Nicholas M. Dimpled Golf Ball and Dimple Distributing Method. Acushnet Co., assignee. USPTO Patent Application 2005/0176525. 11 Aug. 2005.

National Oceanic & Atmospheric Administration. "Offshore Aquaculture in the United States: Economic Considerations, Implications & Opportunities.*"* Silver Spring, MD: US Dept. of Commerce, 2008.

Nesbit, Dennis R. and Joseph F. Stiefel. Golf Ball. Spalding & Evenflo Companies, Inc., assignee. USPTO Patent 5,044,638. 3 Sep. 1991.

Nomura, June and Keisuke Ihara. Golf Ball. Bridgestone Corp. (Tokyo), assignee. USPTO Patent 4,869,512. 26 Sep. 1989.

Nooshin, Hoshyare, P. L. Disney, and O. C. Champion. "Computer-Aided Processing of Polyhedric Configurations." *Beyond the Cube: The Architecture of Space Frames and Polyhedra.* Ed. J. François Gabriel. New York: John Wiley & Sons, 1997.

NPD Group, Inc. "2008 Video Game Software Sales Across Top Global Markets Experience Double-Digit Growth (press release)." *NPD Group, Inc.*, 2009.

North Carolina State College. "Student Publications of the School of Design." Raleigh, NC: School of Design, 3.1 (1952).

Ogg, Steven S. Aerodynamic Surface Geometry for a Golf Ball. Callaway Golf Co., assignee. USP-
TO Patent 6,958,020. 25 Oct. 2005.

——. Aerodynamic Surface Geometry for a Golf Ball. Callaway Golf Co., assignee. USPTO Patent
7,198,577. 3 Apr. 2007.

——. Golf Ball with Multiple Sets of Dimples. Callaway Golf Co., assignee. USPTO Patent
6,482,119. 19 Nov. 2002.

——. Golf Ball with Pyramidal Protrusions. Callaway Golf Co., assignee. USPTO Patent 6,471,605.
29 Oct. 2002.

Oka, Kengo. Golf Ball. Sumitomo Rubber Industries, Ltd., assignee. USPTO Patent 5,145,180. 8
Sept. 1992.

Oka, Kengo and Shinji Ohshima. Golf Ball. Sumitomo Rubber Industries, Ltd., assignee. USPTO Patent 5,338,039. 16 Aug. 1994.

Otero, César and Reinaldo Togores. "Computational Geometry and Spatial Meshes." *Proceedings of
International Conference Computational Science ICCS 2002 Part II*. Ed. M. A. Sloot, Amsterdam, 2002.

Page, Stephen H. Containment Pens for Finfish Aquaculture. Ocean Farm Technologies, Inc., assignee. USPTO Patent 7,284,501. 23 Oct. 2007.

Pantke, Conrad. Space-Covering Structure. USPTO Patent 1,773,851. 26 Aug. 1930.

Pearce, Peter. *Structure in Nature Is a Strategy for Design*. Cambridge: MIT Press, 1978.

Pearce, Peter and Susan Pearce. *Polyhedra Primer*. New York: Van Nostrand Reinhold, 1978.

Pearson, Frederick. *Map Projection Methods*. Blacksburg, VA: Sigma Scientific, Inc., Computer Sciences Corporation, 1984.

Pocklington, Terence W. Golf Ball. Hansberger Precision Golf, Inc., assignee. USPTO Patent
5,536,013. 16 July 1996.

——. Golf Ball Dimple Construction. Hansberger Precision Golf, Inc., assignee. USPTO Patent
5,547,197. 20 Aug. 1996.

Popko, Edward S. *Geodesics*. Industrialization and Technology Course Supplement, vol. 1. Detroit:
University of Detroit Press, 1968.

——. *SPHERES: FORTRAN IV Programs for Spherical Subdivision, vers. 2.0*. Computer software,
University of Detroit, 1968.

Posamentier, Alfred S. and Ingmar Lehmann. *The (Fabulous) Fibonacci Numbers*. New York: Prometheus Books, 2007.

Puderbaugh, Homer L. "Projections for a Geodesic Sphere." *Architectural Science Review* March
(1964): 19–26.

Pugh, Anthony. *An Introduction to Tensegrity*. Berkeley: University of California Prses, 1976.

——. *Polyhedra: A Visual Approach*. Berkeley: University of California Press, 1976.

Radin, Charles. *Miles of Tiles.* Student Mathematical Library. Providence, RI: American Mathematical Society, 1999.

———. "Review of the Pursuit of Perfect Packing." *The Mathematical Association of America* 113 (2006): 88–90.

Randall, David A. et al. "Climate Modeling with Spherical Geodesic Grids." *Computing in Science and Engineering IEEE Computational Science and Engineering* 4.5 (2002): 32–41.

Riblet, Byron C. Game-Ball. USPTO Patent 906,932. 15 Dec. 1908.

Richeson, David S. *Euler's Gem: The Polyhedron Formula and the Birth of Topology.* Princeton, NJ: Princeton University Press, 2008.

Richter, Donald L. Building Construction. Kaiser Aluminum & Chemical Corp., assignee. USPTO Patent 3,026,651. 27 Mar. 1962.

———. Structural Unit. Kaiser Aluminum & Chemical Corp., assignee. USPTO Patent 3,058,550. 16 Oct. 1962.

Ringler, Todd D., Ross P. Heikes, and David A. Randall. "Climate Modeling with Spherical Geodesic Grids." *Computing in Science and Engineering IEEE Computational Science and Engineering* 4.5 (2002): 32–41.

———. "Modeling the Atmospheric General Circulation Using a Spherical Geodesic Grid: A New Class of Dynamical Cores." *American Meteorological Society* (2000): 2471–2489.

Roberts, Siobhan. *King of Infinite Space: Donald Coxeter, the Man Who Saved Geometry.* New York: Walker & Company, 2006.

Sadao, Shoji. *Buckminster Fuller and Isamu Noguchi: Best of Friends.* Milan: 5 Continents Editions, 2011.

———. Geodesic Pentagon and Hexagon Structure. Fuller & Sadao, Inc., assignee. USPTO Patent 3,810,336. 14 May, 1974.

Sadourny, Robert, Akio Arakawa, and Yale Mintz. "Integration of the Nondivergent Barotopic Vorticity Equation with an Icosahedral-Hexagonal Grid for the Sphere." *Monthly Weather Review* 96.6 (1968): 351–356.

Saff, E. B. and A. B. J. Kuijlaars. "Distributing Many Points on the Sphere." *Mathematical Intelligencer* 12.1 (1997): 5–11.

Sahr, Kevin, Denis White, and A. Jon Kimerling. "Geodesic Discrete Global Grid Systems." *Cartography and Geographic Information Science* 30.2 (2003): 121–134.

Sajima, Takahiro. Golf Ball. SRI Sports Ltd., assignee. USPTO Patent 7,223,183. 29 May 2007.

———. Golf Ball. SRI Sports Ltd., assignee. USPTO Patent 7,331,879. 19 Feb. 2008.

Sato, Katsunori. Golf Ball. Bridgestone Sports Co., Ltd. (Tokyo), assignee. USPTO Patent 7,252,601. 7 Aug. 2007.

Sato, Katsunori and Atsuki Kasahima. Golf Ball. Bridgestone Sports Co., Ltd. (Tokyo), assignee. USPTO Patent 7,160,212. 9 Jan. 2007.

Schläfli, Ludwig. *Gesammelte Mathematische Abhandlungen.* Vols I, II, and III. Verlag Birkhäuser, 1950.

Schneider, Michael S. *A Beginner's Guide to Constructing the Universe.* New York: Harper Perennial, 1995.

Shaw, Michael. Golf Balls. Dunlop Ltd., assignee. USPTO Patent 4,960,282. 2 Oct.1990.

Shaw, Michael and Robert C. Haines. Golf Balls. Dunlop Ltd., assignee. USPTO Patent 4,142,727. 6 Mar. 1979.

Shaw, Michael and Brian F. Machin. Golf Balls. Dunlop Ltd., assignee. USPTO Patent 4,722,529. 2 Feb. 1988.

Sherwood, Anton. "How Can I Arrange N Points Evenly on a Sphere?" http://www.ogre.nu/sphere.htm, 2007.

Shimosaka, Hirotaka et al. Golf Ball. Bridgestone Sports Co., Ltd. (Tokyo), assignee. USPTO Patent 5,902,193. 11 May 1999.

Sieden, Lloyd Steven. *Buckminster Fuller's Universe.* Cambridge: Perseus Pub., 2000.

Simonds, Vincent J. and Thomas J. Bergin. Aerodynamic Surface Geometry for a Golf Ball. Callaway Golf Co., assignee. USPTO Patent Application 2006/0122010. 8 June 2006.

———. Aerodynamic Surface Geometry for a Golf Ball. Callaway Golf Co., assignee. USPTO Patent 7,338,392. 4 Mar. 2008.

Skilling, J. "The Complete Set of Uniform Polyhedra." *Philosophical Transactions of the Royal Society of London. Series A: Mathematical, Physical & Engineering Sciences* 278 (1975): 111–135.

Smart, William Marshall. *Astronomical Navigation: A Handbook for Aviators.* New York: Longmans, 1942.

———. *Textbook on Spherical Astronomy, Sixth Edition.* Ed. R. M. Green. New York: Cambridge University Press, 1990.

Snelson, Kenneth D. "Circles, Spheres and Atoms." *Symmetry: Culture and Science* 13.1 (2002): 1–18.

———. Continuous Tension, Discontinuous Compression Structures. USPTO Patent 3,169,611. 16 Feb. 1965.

———. Magnetic Geometric Building System. USPTO Patent 6,017,220. 25 Jan. 2000.

———. Model for Atomic Forms. USPTO Patent 3,276,148. 4 Oct. 1966.

———. Model for Atomic Forms. USPTO Patent 4,099,339. 11 July 1978.

Snyder, John Parr. *Map Projections—A Working Manual.* Washington, DC: US Geological Survey, 1987.

Sobel, Dava. *Longitude: The True Story of a Lone Genius Who Solved the Greatest Scientific Problem of His Time.* New York: Walker, 1995.

Sobel, Dava and William J. H. Andrews. *The Illustrated Longitude: The True Story of the Lone Genius Who Solved the Greatest Scientific Problem of His Time.* New York: Walker, 1998.

Solheim, Karsten. Golf Ball. USPTO Patent 4,653,758, 31 Mar. 1987.

Song, Lian, Jon Kimerling, and Kevin Sahr. "Developing an Equal Area Global Grid by Small Circle Subdivision." *Discrete Global Grids.* Santa Barbara, CA: University of California, National Center for Geographic Information & Analysis, 2002.

Sperry, Pauline. *Short Course in Spherical Trigonometry.* Johnson's Mathematics Series. Atlanta, GA: Johnson Publishing Co., 1928.

Spunt, Leonard. "Modular Dome Structure." *IASS World Congress on Space Enclosures.* Build. Res. Centre Concordia, Univ. Montreal 1 (1976): 235–240.

——. Modular Dome Structures. USPTO Patent 3,959,937. 1 June 1976.

State University Colorado. "Geodesics Climate Model Uses Different Mapping Techniques, Coordinates and Supercomputing to Improve Predictions." *ScienceDaily.* www.sciencedaily.com/releases/2001/09/010926071704.htm, 2001.

Stewart, Ian. "Crystallography of a Golf Ball." *Scientific American* February (1997): 96–98.

Stiefel, Joseph F. Golf Ball and Method of Forming Dimples Thereon. Lisco, Inc., assignee. USPTO Patent 5,890,975. 6 Apr. 1999.

Stiefel, Joseph F. and Dennis Nesbitt. Tetrahedral Dimple Pattern Golf Ball. Lisco, Inc., assignee. USPTO Patent 5,890,974. 6 Apr. 1999.

Stuart, Duncan R. Letter to Buckminster Fuller. Raleigh, NC: Skybreak Carolina Corp., Department of Special Collections and University Archives, Stanford University Libraries, 11 June, 1952.

——. "The Orderly Subdivision of Spheres." *The Student Publications of the School of Design,* NC State University 5 (1963): 23–33.

——. "Polyhedra." *The Student Publications of the School of Design,* NC State University 3.1 (1962).

——. "Polyhedral and Mosaic Transformations." *The Student Publications of the School of Design,* NC State University 12.1 (1963): 2–28.

——. *A Report on the Triacon Gridding System for Spherical Surface.* Raleigh, NC: Skybreak Carolina Corp., Department of Special Collections and University Archives, Stanford University Libraries, 1953.

——. "Spherical Truss for Ford Motor Co. by R.B. Fuller." Raleigh, NC: Skybreak Carolina Corp., Department of Special Collections and University Archives, Stanford University Libraries, 1953.

——. "Topological Data." Raleigh, NC: Fuller Research Foundation, Department of Special Collections and University Archives, Stanford University Libraries, 1951.

——. "The Transformation of Polyhedra: An Experiment in Animated Film Making." *Journal of Architectural Education* 18.3 (1963): 41–43.

Stuart, Duncan R. and Fred Eichenberger. *The Mass Production of Unique Items.* Raleigh, NC: Design Research Laboratory of the School of Design, North Carolina State University, 1968.

Sullivan, Michael J. Golf Ball Dimples. Acushnet Co., assignee. USPTO Patent 6,569,038. 27 May 2003.

——. Golf Ball Dimples. USPTO Patent 6,709,349. 23 Mar. 2004.

——. Golf Ball with Elevated Dimple Portions. Spalding Sports Worldwide, Inc., assignee. USPTO Patent 6,139,448. 31 Oct. 2000.

Sullivan, Michael J., Steven Aoyama, and Edmund A. Hebert. Golf Ball Surface Patterns Comprising Multiple Channels. Acushnet Co., assignee. USPTO Patent Application 2009/0017941. 15 Jan. 2009.

Sultan, Cornel, Martin Corless, and Robert E. Skelton. "Symmetrical Reconfiguration of Tensegrity Structures." *International Journal of Solids and Structures* 39 (2002): 2215–2234.

Sutton, Daud. *Platonic & Archimedean Solids* (Wooden Books). New York: Walker & Co., 2002.

Szalay, Alexander S. et al. *Indexing the Sphere with the Hierarchical Triangular Mesh.* Microsoft Corp. 2005. Technical report MSR-TR-2005-123, 2005.

Szpiro, George. *Kepler's Conjecture: How Some of the Greatest Minds in History Helped Solve One of the Oldest Math Problems in the World.* Hoboken, NJ: John Wiley & Sons, 2003.

Tammes, P. M. L. "On The Origin of Number and Arrangement of the Places of Exit on the Surface of Pollen-Grains." *Recueil Des Travaux Botaniques Néerlandais* 27 (1930): 1–84.

Tarnai, Tibor. "Dense Sphere Packing on a Sphere." *Proceedings of the First International Seminar on Structural Morphology.* Ed. R. Motro and T. Wester. Montpellier, Le Grand Motte, France, 1992.

——. "Geodesic Dome with Three Different Bar Lengths." *Structural Topology* 8 (1983): 23–24.

——. "Geodesic Domes and Fullerenes." *Philosophical Transactions of the Royal Society of London. Series A: Mathematical, Physical & Engineering Sciences* 343.1667 (1993).

——. "Geodesic Domes and Golf Balls." *Space Structures* (1993).

——. *Geodesic Domes with Skew Networks.* Budapest: Hungarian Institute for Building Science, 1987.

——. "Ivory Shells and Polyhedra." *Proceedings of the International Association for Shell and Spatial Structures (IASS) Symposium.* Ed. A. Domingo and C. Laxaros. Universidad Politecnica de Valencia, Spain, 2009.

——. "Polymorphism in Multisymmetric Close Packings of Equal Spheres on a Spherical Surface." *Structural Chemistry* 13.3–4 (2002): 289–295.

——. "Problems Concerning Spherical Polyhedra and Structural Rigidity." *Structural Topology* 4 (1980): 61–66.

——. "Spherical Circle-Packing in Nature, Practice and Theory." *Structural Typology* 9 (1984): 39–58.

——. "Spherical Grid Structures: Geometric Essays on Geodesic Domes." Budapest: Hungarian Institute for Building Science, 1987.

——. "Spherical Grids of Triangular Network." *Acta Technica Acad. Sci. Hung.* 76 (1974), 307–336.

Tarnai, Tibor, P. W. Fowler, and S. Kabai. "Packing of Regular Tetrahedral Quartets of Circles on a Sphere." *Philosophical Transactions of the Royal Society of London. Series A: Mathematical, Physical & Engineering Sciences* 459 (2003): 2847–2859.

Tarnai, Tibor and Zsolt Gáspár. "Covering the Sphere with Equal Circles." *Colloquium on Intuitive Geometry.* Balatonszéplak, Hungary, 1985.

——. "Packing of Equal Regular Pentagons on a Sphere." *Philosophical Transactions of the Royal Society of London. Series A: Mathematical, Physical & Engineering Sciences* 457 (2001): 1043–1058.

Tarnai, Tibor, Zsolt Gáspár, and Lidia Szalait. "Pentagon Packing Models for 'All-Pentamer' Virus Structures." *Biophysical Journal* 69.2 (1995): 612–618.

Tarnai, Tibor and Koji Miyazaki. "Circle Packings and the Sacred Lotus." *Leonardo* 36.2 (2003): 145–150.

Tarnai, Tibor and Magnus J. Wenninger. "Spherical Circle-Coverings and Geodesic Domes." *IASS International Congress.* Moscow, 1985.

Taylor, William. Golf Ball. USPTO Patent 878,254. 4 Feb. 1908.

——. Golf Ball. USPTO Patent 1,286,834. 3 Dec. 1918.

Taylor, William W. Golf Ball. USPTO Patent 4,921,255. 1 May 1990.

Thompson, D'Arcy Wentworth. *On Growth and Form.* New York: Dover Publications, 1992.

Tickoo, Sham and Vivek Singh. *CATIA V5R18 for Designers.* Schererville, IN: CADCIM Technologies, 2008.

Todhunter, I. *Spherical Trigonometry for the Use of Colleges and Schools.* London: Macmillan and Co., 1871.

Tomlow, Jos, ed. *Polyhedra: From Pythagoras to Alexander Graham Bell.* Chapter 1. New York: John Wiley & Sons, 1997.

Tóth, L. Fejes. "On The Densest Packing of Spherical Caps." *American Mathematical Monthly* 56 (1949): 330–331.

Turner, Merle B. *Celestial for the Cruising Navigator.* Centreville, MD: Cornell Maritime Press, 1986.

Twarock, Reidun. "Mathematical Virology: A Novel Approach to the Structure and Assembly of Viruses." *Philosophical Transactions of the Royal Society of London. Series A: Mathematical, Physical & Engineering Sciences* 364 (2006): 3357–3373.

University of Minnesota Science and Technology Center. "The Geometry Center: A Project of the National Science Foundation." www.geom.uiuc.edu/, 2008.

University of New South Wales. "Distributing Points on the Sphere." School of Mathematics and Statistics, http://www.maths.unsw.edu.au/school/articles/me100.html, 2008.

Urner, Kirby. "The Invention Behind the Invention." *Synergetica Journal, Buckminster Fuller Institute* 1.1 (1991).

US Department of Commerce. "Table of Sines and Cosines to Fifteen Decimal Places at Hundredths of a Degree." Ed. Nat. Bureau of Standards, Applied Mathematics Laboratories. Washington, DC: US Govt. Printing Office, 1949.

US Department of Labor. "CPI Inflation Calculator." http://data.bls.gov/cgi-bin/cpicalc.pl, 2011.

Veilleux, Thomas A., Vincent J. Simonds, and Kevin J. Shannon. Golf Ball Dimple Pattern. Callaway Golf Co., assignee. USPTO Patent 7,179,178. 20 Feb. 2007.

Verheyen, H. F. "The Complete Set of Jitterbug Transformers and the Analysis of Their Motion." *Computers & Mathematics with Applications* 17.13 (1989): 203–225.

Vince, John. *Computer Graphics.* London: Design Council, 1992.

——. *Computer Graphics for Graphic Designers.* White Plains, NY: Knowledge Industry Publications, 1985.

——. *Geometric Algebra for Computer Graphics.* London: Springer, 2008.

——. *Geometry for Computer Graphics: Formula, Examples and Proofs.* London: Springer, 2005.

——. *Mathematics for Computer Graphics, Second Edition.* London: Springer, 2006.

Watkins, Christopher D. and Vincent P. Mallette. *Stereogram Programming Techniques.* Rockland, MA: Charles River Media, 1996.

Webb, Robert. "Great Stella Manual." http://www.software3d.com/StellaManual.php?prod=great, 2008.

——. "Stella: Polyhedron Navigator." *Symmetry: Culture and Science* 11.1–4 (2000): 231–268.

Wenninger, Magnus J. *Dual Models.* New York: Cambridge University Press, 1983.

——. *Polyhedron Models.* New York: Cambridge University Press, 1971.

——. *Polyhedron Models for the Classroom, Second Edition.* Reston, VA: National Council of Teachers of Mathematics, 1975.

——. *Spherical Models.* New York: Cambridge University Press, 1979.

——. *Spherical Models.* Mineola, NY: Dover Publications, 1999.

Wertz, James Richard and Wiley J. Larson. *Space Mission Analysis and Design, Third Edition.* Space Technology Library. El Segundo, CA: Microcosm, 1999.

Weyl, Hermann. *Symmetry.* Princeton, NJ: Princeton University Press, 1952.

White, Denis. "Global Grids from Recursive Diamond Subdivisions of the Surface of an Octahedron or Icosahedron." *Environmental Monitoring and Assessment* 64.1 (2000): 93–103.

White, Denis, A. Jon Kimerling, and W. Scott Overton. "Cartographic and Geometric Components of a Global Sampling Design for Environmental Monitoring." *Cartography and Geographic Information Systems* 10.1 (1992): 5–22.

White, Denis, A. Jon Kimerling, Kevin Sahr, and Lian Song. "Comparing Area and Shape Distortion of Polyhedral-Based Recursive Partitions of the Sphere." *International Journal of Geographical Information Science* 12.8 (1992): 805–827.

White, John. Golf Ball. USPTO Patent 1,418,220. 2 Oct. 1920.

White, Mark P. "Rafiki." Rafiki, Inc. http://www.codefun.com/, 2011.

——. Rafiki Model and Map to the Genetic Code. USPTO Patent Application US 2003/0190657 A1. 9 Oct. 2003.

Whittaker, E. J. W. *The Stereographic Projection.* Cardiff, Wales: University College Cardif Press, 1984.

Williams, Robert Edward. *The Geometrical Foundation of Natural Structure—A Source Book of Design.* New York: Dover Publications, 1979.

——. *Handbook of Structure.* Huntington Beach, CA: Douglas Advanced Laboratories, 1968.

Winfield, Douglas C. and Steven Aoyama. Pentagonal Hexecontahedron Dimple Pattern on Golf Balls. USPTO Patent 6,527,653. 4 Mar. 2003.

Wodehouse, Roger P. *Pollen Grains: Their Structure, Identification, and Significance in Science and Medicine.* New York: McGraw-Hill, 1935.

Woo, Young-Kyu. Golf Ball. Kumho & Co., Inc., assignee. USPTO Patent 5,192,078. 9 Mar. 1993.

Xiaochong, Tong et al. *The Expression of Spherical Entities and Generating of Voronoi Diagram Based on Truncated Icosahedron DGG.* Zhengzhou, China: Institute of Surveying and Mapping, 2002.

Yagley, Michael S. et al. Golf Ball. Callaway Golf Co., assignee. USPTO Patent 6,634,965. 21 Oct. 2003.

Yamada, Kaname. Golf Ball. Sumitomo Rubber Ind. Ltd., assignee. USPTO Patent 4,720,111. 19 Jan. 1988.

——. Golf Ball. Sumitomo Rubber Ind. Ltd., assignee. USPTO Patent 4,946,167. 7 Aug. 1990.

Index

About the Author

E d Popko is a graduate of the University of Detroit's School of Architecture and has both Masters and PhD degrees from Massachusetts Institute of Technology. He is a registered architect and former Fulbright Scholar. In the 1960s, he was an apprentice in Buckminster Fuller's affiliate office, Geometrics, Inc., in Cambridge, Massachusetts, and later authored *Geodesics,* a primer on geodesic domes, and *Transitions*, a documentary of urban settlements in developing countries.

For the past 25 years, he has held research, product development, and marketing positions for CAD technology within IBM. He has managed software development teams in computer mapping and CAD solutions for architects, industrial buildings, high-tech plants, and shipbuilding.

He is retired from IBM and lives with his wife, Geraldine, in Woodstock, New York.[1] They have two sons, Edward and Gerald.

---

[1] The author appears with a model of a Class III subdivision of a spherical icosahedron $\{3,5+\}_{5,2}$ made by Father Magnus Wenninger.